Editor-in-Chief

Prof. Janusz Kacprzyk
Systems Research Institute
Polish Academy of Sciences
ul. Newelska 6
01-447 Warsaw
Poland
E-mail: kacprzyk@ibspan.waw.pl

For further volumes:
http://www.springer.com/series/7092

Emilia Tantar, Alexandru-Adrian Tantar,
Pascal Bouvry, Pierre Del Moral, Pierrick Legrand,
Carlos A. Coello Coello, and Oliver Schütze (Eds.)

EVOLVE - A Bridge between Probability, Set Oriented Numerics and Evolutionary Computation

Springer

Editors
Emilia Tantar
Computer Science and Communications
Research Unit
Faculty of Sciences, Technology,
and Communication
University of Luxembourg
Luxembourg

Alexandru-Adrian Tantar
Computer Science and Communications
Research Unit
Faculty of Sciences, Technology,
and Communication
University of Luxembourg
Luxembourg

Pascal Bouvry
Computer Science and Communications
Research Unit
Faculty of Sciences, Technology,
and Communication
University of Luxembourg
Luxembourg

Pierre Del Moral
Bordeaux Mathematical Institute
INRIA Bordeaux-Sud Ouest
Université Bordeaux I
Talence cedex
France

Pierrick Legrand
Université Bordeaux II
Bëtiment Leyteire
UFR Sciences et Modélisation
Bordeaux
France

Carlos A. Coello Coello
CINVESTAV-IPN
Computer Science Department
Mexico City
Mexico

Oliver Schütze
CINVESTAV-IPN
Computer Science Department
Mexico City
Mexico

ISSN 1860-949X e-ISSN 1860-9503
ISBN 978-3-642-32725-4 e-ISBN 978-3-642-32726-1
DOI 10.1007/978-3-642-32726-1
Springer Heidelberg New York Dordrecht London

Library of Congress Control Number: 2012944264

Preface

This volume comprises a collection of research contributions that were presented at the international workshop EVOLVE 2011. The aim of the EVOLVE workshop, the originating point of this book, is to build a bridge between probability, statistics, set oriented numerics and evolutionary computing, as to identify new common and challenging research aspects. The event is also intended to foster a growing interest for robust and efficient new methods with a sound theoretical background and, last but not least, to unify theory-inspired methods and cutting-edge techniques that ensure performance guarantee factors. By gathering researchers with different backgrounds, e.g. ranging from computer science to mathematics, statistics and physics, to name just a few, a unified view and vocabulary can emerge where theoretical advancements may echo in different domains. What is more, the massive use and large applicability spectrum of evolutionary algorithms for real-life applications determined a need for establishing solid theoretical grounds. Only to offer one example, one may consider mathematical objects that are sometimes difficult and/or costly to calculate, in the light of acknowledged new results showing that evolutionary algorithms can act in some cases as good and fast estimators. Similarly, the handling of large quantities of data may require the use of distributed environments where the probability of failure and the stability of the algorithms may need to be addressed. What is more, common practice confirms in many cases that theory-based results have the advantage of ensuring performance guarantee factors for evolutionary algorithms in areas as diverse as optimization, bio-informatics or robotics. Summarizing, the EVOLVE focuses on challenging aspects arising at the passage from theory to new paradigms and practice, aiming to provide a unified view while, at the same time, raising questions related to reliability, performance guarantees and modeling.

EVOLVE 2011 was jointly organized by the University of Luxembourg, CINVESTAV, Mexico (Research and Advanced Studies Center of the National Polytechnic Institute of Mexico) and INRIA, France (National Institute for Research in Computer Science and Control), being a follow-up of the Workshop on Evolutionary Algorithms – New Challenges in Theory and Practice, organized by the ALEA Working Group in Bordeaux, France, March 2010, with the support of the EA association.

Focusing more on the direct scope and aims of this book, when considering the current advancements in the evolutionary computing domain, one has to conclude that evolutionary type algorithms are very often presented with limited regard to their rigorous mathematical foundations and without any strong analysis of the underlying random evolutionary processes. Nonetheless, under certain regularity conditions, most of the genetic type evolutionary computing algorithms converge towards some particular probability distributions. At the same time, although for the significant results shown in practice, most if not all of the existing approaches are often derived on intuition with nothing but numerical observations for support, making evolutionary computing research to recurrently fall into the exploration of obscure inspired paradigms. And while to some extent this makes sense and can be sustained for a real world setup where finite time and resource constraints are imposed, understanding the different paradigms and afferent convergence properties demands an accurate description of the underlying mathematical models. In this context, several questions arise: To what extent do we accept numerical evidence as a proof when dealing with complex systems of high risk as nuclear plants, health care management, security or defense systems? What is the sensitivity and the error those paradigms introduce as part of a solution's design? What is the impact and how uncertainties propagate w.r.t. time? Answering to all these questions amounts to rigorously analyzing the convergence and the performance of these stochastic gradients, or these genetic particle sampling models, as later detailed in this book. Where from the aims of this work in presenting not only current advancements in the evolutionary computing domain, or connected areas of interest, but also in providing a strong theoretical support for understanding and analyzing the behavior of these approaches, i.e. creating a bridge between probability, set-oriented numerics and evolutionary computation.

The book encloses several of the contributions of the EVOLVE 2011 invited speakers along with a selection of the best full-length papers submitted, peer-reviewed by an international program committee. In order to offer a structured and easy to follow view, the book was divided into four parts, standing as segments of the underlying idea of EVOLVE, i.e. of creating a bridge between probability, set oriented numerics and evolutionary computation.

The first part, Part I, consists of three contributions that make the passage from the probabilistic bases of evolutionary algorithms to probabilistic driven paradigms, applied as per se methods, namely estimation of distribution algorithms.

Throughout the first chapter, while adopting a pedagogical perspective, the aim is on setting the bases of evolutionary algorithms under a probabilistic formulation, hence connecting the foundations of those algorithms with the afferent results that echoed in practice. The covered areas range from theoretical aspects, with a brief outline of absorption models, Bayesian inference and interacting Kalman filters, to conceptual ideas and guidelines for a more practical framework, e.g. stochastic optimization, signal processing or analysis of convergence under uncertain behavior and rare events simulation and analysis.

For the next two chapters, Estimation of Distribution Algorithms (EDAs) are discussed along with novel extensions for using graph based models inside continuous

EDAs. As later described, the presented results allow a better understanding of the dependencies generated within this well-established branch of probabilistic based evolutionary algorithms, while introducing new powerful mechanisms for problem modeling.

Thus, Chapter 2 elaborates on the use of regular vines within EDAs, providing the means of learning the structure of a probabilistic model by taking into account dependencies among variables, and this regardless of the behavior marginal distributions have. Regular vines, originating from graph theory, are based on bivariate and conditional bivariate copula functions, and can be used to model the most important dependencies in a specific population. The chapter concentrates on the use of truncated regular vines (for C-vines and D-vines), which are adequate for modeling high dimensional distributions. Specifically, the approach relies on using copula entropy as a measure of dependency, the most important dependencies being represented as part of a graphical model through copula functions. Lower copula functions are also employed to factorize joint distributions. The results are applied in a numerical optimization context by making use of continuous EDAs, nevertheless being emphasized that they can also be used in other areas such as classification problems.

The third chapter represents a natural continuation of the second one, providing a Gaussian Poly-Tree Estimation that extends the idea of employing graph based models inside EDAs. In this approach, the graph based models encode the joint probability distribution of the population as a product of conditional distributions. Through this chapter the authors provide an answer to the question: to what extent does a graph model approximate the true distribution of an EDA's population? New extensions, including the use of Gaussian copula functions and that of local optimizers, are also provided, along with an extensive experimental validation on a set of 20 well-established convex (unimodal) and multimodal benchmark functions.

The second part of the book, Set Oriented Numerics, focuses on providing quality indicators, with an explicit aim for quantifying the quality of an experimental testing output result, section followed by set oriented methods for multi-objective optimization.

The fourth chapter and the first of this second part provides a thorough review of quality distance or biodiversity based indicators for the approximation of level sets. One of the straightforward applications of these indicators is in multi-objective optimization, given that Pareto front approximations, under certain conditions, can be recasted as level set problems. The authors provide an elaborated and motivational overview of the role quality indicators have, in a multi-objective decision theory inspired context. Furthermore, the focus is set on some selected quality indicators and their properties, providing the grounds to introduce novel quality indicators, e.g. Hausdorff distance (concept) based, that can be used in cases where knowledge of the level set is absent. Properties like continuity, spread, and monotonicity are studied, together with computational complexity considerations, followed by an analysis conducted on an indicator-based evolutionary algorithm for level set approximation.

The results on set oriented numerics are completed by Chapter 5, which presents the advancement on set oriented techniques, focusing on subdivision and continuation methods, applied to the computation of Pareto set approximations. The authors tackle the efficiency of using sequences of box collections (common characteristic for all the outlined methods) in a new context, that of multi-objective optimal control problems. Both synthetic and real-life problems are used for testing purposes, by means of continuation methods and by using a transformation into high-dimensional multi-objective optimization problems. In this chapter, the authors present subdivision and recover techniques for multi-objective optimization, oriented either toward the parameter or the image space. The continuation methods are furthermore modeled as to fit to the context of multi-objective optimal control, for both differential flat systems and Lagrangian systems. The efficiency of the approaches has been experimentally tested on several problems, for which promising results were obtained.

The third part of the book, "Landscape, coevolution and cooperation", focuses on providing a better understanding of what complexity means by modeling and analyzing fitness landscapes through complex networks that capture the neighborhood structure of a function as explored by cellular genetic algorithms. Through the coevolution paradigm, ways of using problem's information by bayesian network structure learning in a more decentralized manner are provided and the findings illustrated on a concrete example. The second and third chapters of this section discuss aspects related to co-evolutionary, respectively hybrid designs, as an example of how specific characteristics can be exploited, e.g. problems that allow a decomposition into subcomponents, or of how algorithms can be combined for a more flexible and exploration efficient design.

The sixth chapter deals with aspects from hard combinatorial search spaces using a complex-networks oriented perspective. As a main contribution, the chapter introduces a novel, graph-based approach for representing the fitness landscape of combinatorial search problems, where vertices stand for local optima and oriented edges are used to represent transition probabilities between those optima. A focus for exploiting local optima networks knowledge is considered with a final aim of advancing our current state of the art for local search heuristics. The chapter starts by introductory notions like neighborhood, fitness landscape, local optima and basin of attraction, followed by definitions and a brief outline of local optima networks and weighted complex networks. Next, after a short review of NK landscapes (specific family of multimodal fitness landscapes that can be adapted to include features from smooth to rugged and for which all feasible solutions can be enumerated), a detailed analysis and discussion of different topological attributes like shortest path to the global optimum, outgoing weight distribution, disparity or basins of attraction are presented. In addition, a similar discussion is carried for the Quadratic Assignment Problem fitness landscapes with a short comparative view of the NK and QAP families as well as insights for optima distribution and clustering. Analysis and results, as a final aspect, are shown to provide grounds for qualitatively relating topological features like node degrees, clustering coefficients or edge weight disparity, among others, to problem difficulty. Furthermore, it is shown through experimentation that the distribution of the basin of attraction size is not uniform but rather right skewed with a positive correlation of the optimum fitness and basin size.

In the seventh chapter, a cooperative co-evolution approach is presented, addressing a specific case where solutions are represented not by single individuals but as aggregations of several individuals. As a main advantage of this representation, a more economic simulation of the evolution principles becomes possible, although restricted to problems where a decomposition into subcomponents can be made and at the price of a more complex design phase. Furthermore, as a specific characteristic of the proposed algorithm, the interaction of the different subspecies evolved within a population is not constrained to act at evaluation only, e.g. independent, partial fitness vs an aggregation-based evaluation process, but can also be triggered via genetic operators. The approach is detailed and analyzed by considering an industrial agricultural-food process with a study on (1) a deterministic modeling problem and (2) a complex NP optimization problem in the form of a Bayesian network structure estimation.

The eight chapter, authored by Alba and Villagra, addresses a series of aspects from hybrid algorithms' design, having as case study a new approach where the active components of a population based evolutionary paradigm are inserted within a cellular genetic algorithm. After a brief outline of Particle Swarm Optimization (PSO) and Cellular Genetic Algorithms (cGA), i.e. the two metaheuristics considered within the chapter, a short discussion of the main ideas is given. A modified cGA is afterwards introduced, enclosing a specific operator that captures the main characteristics of PSO, i.e. by maintaining information about cognitive and social factors. Two different versions are proposed, a local neighborhood based one and, the second one, global oriented, based on exploiting the best solution found within the population. Results, comparable to or better than the counterpart results of a standard approach, are reported on a large set of benchmarks. Those benchmarks, among others, include Massively Multimodal Deceptive Problems, Frequency Modulation Sounds or the Error Correcting Code Design Problem. As a last side note, the proposed hybrid algorithm, as the authors mention, can be extended to include other metaheuristics like, for example, the Ant Colony Optimization, the Simulated Annealing or the Variable Neighborhood Search.

The fourth part, dedicated to multi-objective optimization theoretical aspects and real-life applications, encloses several ideas on what problems multiple objectives pose and how to address them by exploiting different approaches or hybrid techniques. A more abstract perspective is adopted for the first two chapters while the ending chapter, of this section and of the book, is more concerned with aspects one would expect more to find in a real-world context.

In Chapter 9, Lara, Schtze and Coello Coello detail on the possibility of employing the objective-function gradients in constructing local search operators for continuous multi-objective optimization problems. The chapter encloses practical approaches of using this type of information, through descent direction methods, hill climbers, but also directed movements. A sum of new ideas, including the use of gradient information in a side-step procedure, is proposed and their role and utility detailed. All these ideas make place for new means of hybridization that open the path for various possible uses of the gradient information in a variety of methods, such as population-based algorithms. Some interesting remarks are also provided regarding the already existing hybridization techniques and an overview regarding the early hybrids concludes the contribution.

Grimme, Kemmerling and Lepping make the passage in Chapter 10, from theoretical single-objective scheduling to multi-objective optimizers. The main foundation of the article stays on the construction of a framework capable of creating multi-objective optimizers through the modular composition of single-objective scheduling heuristics. The approach has as grounds the predator-prey model and it is based on cellular and agent-based paradigms. The properties of the agent-based system are carefully detailed on a spatial population, as preys are placed on a toroidal grid to represent a spatially distributed population, while the predators move across the spatial structure. Multiple types of operators (as mutation, recombination) can be used in the proposed model by simply defining different predator species. Different classes of coupling variation operators and predator agents are enclosed. The model was experimentally validated on a set of 100 synthetic test problem instances, by using the predator-prey model, but also an extended version englobing ε-constraint principles, against NSGA-II variants. The proposed method is modularly composed and escapes the monolithic structure of classical multi-objective evolutionary algorithms. Results are illustrated through two strongly NP-hard problems (variants of the identical parallel machine offline scheduling) with clear advantages acquired, e.g. the use of available knowledge from single-objective scheduling techniques mapped on a dedicated to multi-objective optimization structure.

Chapter 11, is dedicated to using multi-objectivisation in order to avoid stagnation, i.e. single-objective problems reformulated and solved in a multi-objective form. The method used for constructing bi-objective formulations relies on the aggregation of a new alternative function as the second objective, e.g. an Euclidean distance in the decision space. Multi-objectivisation, in this work, is applied and analyzed as part of a hybrid model constructed on top of a hyper-heuristic and DCN-THR. The approach has been tested on several benchmarks, the results obtained by single objective optimization being compared to the ones of the multi-objective counterparts. The chapter discusses two different versions, with and without parameter control, different adaptation levels being proposed for the mixture of parameters with hyper-heuristics and multi-objectivisation. A series of experiments that consider a large number of variables are also presented, reflecting not only the performance of the hybridization scheme but also revealing cases where multi-objectivisaiton can produce a negative effect.

The last chapter offers an insight into aspects related to applying evolutionary algorithms for real-world stock market problems. As a main area of study, the authors address the predictability of heuristic conversion algorithms in the German stock market, e.g. Moving Average Crossover and Trading Range Breakout. The nature of those heuristics, i.e. designed as a predefined set of buying and selling rules and based on data from technical analysis, leads to a static runtime behavior, where from the directory lines of the work. The study is constructed in the limits of a competitive analysis framework, with multiple online conversions of assets where time wise buying and selling decisions need to be taken and where the behavior of several algorithms like GP, VMA or TRB, is studied. Last but not least, several questions are raised for further study, among others pointing at non-trending market with statistical and worst-case analysis studies, positive excess returns or realistic entry and exit signals.

As an ending thought, our gratitude goes out to all the invited speakers for accepting to give an outstanding presentation and an overview of their latest work during the event as well as for contributing to this book, to all the authors, for sharing their knowledge and expertise, also summarized in this book, and last but not least, to all the participants for their extraordinary support and contribution. We would also like to express our foremost appreciation to all the referees and members of the program committee which, by their considerate work, contributed to the creation of a bridge between the different fields that stand at the basis of the event. The editors would furthermore like to thank Professor Janusz Kacprzyck (Editor-In-Chief, Springer Studies in Computational Intelligence Series) for the editorial assistance and excellent collaboration during the development of this volume. Finally, we gratefully thank the partner institutions and sponsors of the event for providing support for making from EVOLVE 2011 a success.

Luxembourg, Bordeaux, and Mexico City   Emilia Tantar
August 2012   Alexandru-Adrian Tantar
                                                                           Pascal Bouvry
                                                                        Pierre Del Moral
                                                                        Pierrick Legrand
                                                                 Carlos A. Coello Coello
                                                                        Oliver Schütze

Contents

Part II: Set Oriented Numerics

Part III: Landscape, Coevolution and Cooperation

Part IV: Multi-objective Optimization, Heuristic Conversion Algorithms

9   On Gradient-Based Local Search to Hybridize Multi-objective Evolutionary Algorithms 305

Adriana Lara, Oliver Schütze, Carlos A. Coello Coello

10   On the Integration of Theoretical Single-Objective Scheduling Results for Multi-objective Problems 333

Christian Grimme, Markus Kemmerling, Joachim Lepping

List of Contributors

Enrique Alba
Department of Computer Science, University of Málaga, Spain
e-mail: eat@lcc.uma.es

Olivier Barrière
INRIA Saclay - Ile-de-France, Parc Orsay Université, 4, rue Jacques Monod, 91893
Orsay Cedex, France
e-mail: olivier.barriere@inria.fr

Cédric Baudrit
UMR782 Génie et Microbiologie des Procédés Alimentaires. AgroParisTech,
INRA, F-78850 Thiverval-Grignon, France

Carlos A. Coello Coello
Computer Science Department, CINVESTAV-IPN, Av. IPN 2508, Col. San Pedro
Zacatenco, 07360 Mexico City, Mexico
e-mail: ccoello@cs.cinvestav.mx

Fabio Daolio
Faculty of Business and Economics, Information System Department, University
of Lausanne, Switzerland
e-mail: Fabio.Daolio@unil.ch

Pierre Del Moral
Centre INRIA Bordeaux Sud-Ouest, Institut de Mathématiques de Bordeaux ,
Université de Bordeaux I, 351 cours de la Libération 33405 Talence cedex, and
Applied Mathematics Department, CMAP École Polyetchnique Paris, France
e-mail: Pierre.Del-Moral@inria.fr

Michael Dellnitz
University of Paderborn, Chair of Applied Mathematics, Warburger Str. 100,
D-33098 Paderborn, Germany
e-mail: dellnitz@math.uni-paderborn.de

André Deutz
LIACS, Niels Bohrweg 1, 2333-CA Leiden, The Netherlands
e-mail: deutz@liacs.nl

Michael Emmerich
LIACS, Niels Bohrweg 1, 2333-CA Leiden, The Netherlands
e-mail: emmerich@liacs.nl

Christian Grimme
Robotics Research Institute, TU Dortmund, Germany
e-mail: christian.grimme@udo.edu

Arturo Hernández-Aguirre
Center for Research in Mathematics (CIMAT), Guanajuato, México
e-mail: artha@cimat.mx

Sebastian Jansen
University of Hohenheim, Banking and Financial Services, D-70593 Stuttgart,
Germany
e-mail: sebastian_yansen@yahoo.com

Markus Kemmerling
Robotics Research Institute, TU Dortmund, Germany
e-mail: markus.kemmerling@udo.edu

Johannes Kruisselbrink
LIACS, Niels Bohrweg 1, 2333-CA Leiden, The Netherlands
e-mail: jkruisse@liacs.nl

Adriana Lara
Mathematics Department ESFM-IPN, Edif. 9 UPALM, 07300 Mexico City, México
e-mail: adriana@esfm.ipn.mx

Joachim Lepping
INRIA Rhône-Alpes, Grenoble University, France
e-mail: joachim.lepping@inria.fr

Coromoto León
Dpto. de Estadística, I.O. y Computación. Universidad de La Laguna, La Laguna,
38271, Santa Cruz de Tenerife, Spain
e-mail: cleon@ull.es

Evelyne Lutton
INRIA Saclay - Ile-de-France, Parc Orsay Université, 4, rue Jacques Monod, 91893
Orsay Cedex, France
e-mail: evelyne.lutton@inria.fr

Esther Mohr
Saarland University, P.O. Box 151150, D-66041 Saarbrücken, Germany
e-mail: em@itm.uni-sb.de

Sina Ober-Blöbaum
University of Paderborn, Chair of Applied Mathematics, Warburger Str. 100,
D-33098 Paderborn, Germany
e-mail: sinaob@math.uni-paderborn.de

Nathalie Perrot
UMR782 Génie et Microbiologie des Procédés Alimentaires. AgroParisTech,
INRA, F-78850 Thiverval-Grignon, France

Rogelio Salinas-Gutiérrez
Center for Research in Mathematics (CIMAT), Guanajuato, México
e-mail: rsalinas@cimat.mx

Günter Schmidt
Saarland University, P.O. Box 151150, D-66041 Saarbrücken, Germany
e-mail: gs@itm.uni-sb.de

Oliver Schütze
Computer Science Department, CINVESTAV-IPN, Av. IPN 2508, Col. San Pedro
Zacatenco, 07360 Mexico City, México
e-mail: schuetze@cs.cinvestav.mx

Ignacio Segovia Domínguez
Center for Research in Mathematics (CIMAT), Guanajuato, México
e-mail: ijsegoviad@cimat.mx

Eduardo Segredo
Dpto. de Estadística, I.O. y Computación. Universidad de La Laguna, La Laguna,
38271, Santa Cruz de Tenerife, Spain
e-mail: esegredo@ull.es

Carlos Segura
Dpto. de Estadística, I.O. y Computación. Universidad de La Laguna, La Laguna,
38271, Santa Cruz de Tenerife, Spain
e-mail: csegura@ull.es

Mariette Sicard
UMR782 Génie et Microbiologie des Procédés Alimentaires. AgroParisTech,
INRA, F-78850 Thiverval-Grignon, France

Alexandru-Adrian Tantar
Computer Science and Communications Research Unit, University of Luxembourg
e-mail: alexandru.tantar@uni.lu

Emilia Tantar
Computer Science and Communications Research Unit, University of Luxembourg
e-mail: emilia.tantar@uni.lu

Marco Tomassini
Faculty of Business and Economics, Information System Department, University
of Lausanne, Switzerland
e-mail: Marco.Tomassini@unil.ch

Enrique R. Villa-Diharce
Center for Research in Mathematics (CIMAT), Guanajuato, México
e-mail: villadi@cimat.mx

Andrea Villagra
Emerging Technologies Laboratory, Universidad Nacional de la Patagonia Austral,
Argentine
e-mail: avillagra@uaco.unpa.edu.ar

Katrin Witting
University of Paderborn, Chair of Applied Mathematics, Warburger Str. 100,
D-33098 Paderborn, Germany
e-mail: witting@math.uni-paderborn.de

Pierre-Henri Wuillemin
LIP6-CNRS UMR7606, 75016 Paris, France

Part I
Foundations, Probability and Evolutionary Computation

Chapter 1
On the Foundations and the Applications of Evolutionary Computing

Pierre Del Moral, Alexandru-Adrian Tantar, and Emilia Tantar

Abstract. Genetic type particle methods are increasingly used to sample from complex high-dimensional distributions. They have found a wide range of applications in applied probability, Bayesian statistics, information theory, and engineering sciences. Understanding rigorously these new Monte Carlo simulation tools leads to fascinating mathematics related to Feynman-Kac path integral theory and their interacting particle interpretations. In this chapter, we provide an introduction to the stochastic modeling and the theoretical analysis of these particle algorithms. We also illustrate these methods through several applications.

1.1 Introduction

Most of population-based algorithms are described in terms of interacting samples evolving in some solution state space. The random samples are also termed solutions, particles, individuals or genotypes. Their time evolution mimics natural selection, physical adaptation, reinforced principles, or some social behavior. For a detailed discussion, and an overview of these classes of evolutionary computing models we refer the reader to the couple of books [8, 115], and references therein.

Pierre Del Moral
Centre INRIA Bordeaux Sud-Ouest, Institut de Mathématiques de Bordeaux,
Université de Bordeaux I, 351 cours de la Libération 33405 Talence cedex,
and Applied Mathematics Department, CMAP École Polyetchnique Paris, France
e-mail: `Pierre.Del-Moral@inria.fr`

Alexandru-Adrian Tantar · Emilia Tantar
Computer Science and Communications Research Unit, University of Luxembourg
e-mail: `{alexandru.tantar,emilia.tantar}@uni.lu`

E. Tantar et al. (Eds.): EVOLVE- A Bridge between Probability, SCI 447, pp. 3–89.
springerlink.com © Springer-Verlag Berlin Heidelberg 2013

Evolutionary type algorithms are very often presented with limited regard to their rigorous mathematical foundations, without any rigorous analysis of the underlying random evolutionary processes. Their performance often relies on intuition driven by numerical observations, so that evolutionary computing research sometimes falls in obscure inspired paradigms.

In this context, several questions arise: To what extent do we accept numerical evidence as a proof when dealing with complex systems of high risk as nuclear plants, health care management, security or defense systems? What is the sensitivity and the error those paradigms introduce as part of a solution's design? What is the impact and how uncertainties propagate w.r.t. time?

Evolutionary type algorithms can be interpreted in two different ways.

Firstly, population-based algorithms w.r.t. optimization problems can be viewed as a gradient type hole puncher in complex solution state spaces. In this context, the central idea is to use natural evolution mechanisms to improve step by step the population adaptation. For convex optimization problems, the performance and convergence of the algorithm follows from standard analysis of stochastic gradient models. For more complex optimization problems, we expect large population explorations to escape from local minima.

On the other hand, under certain regularity conditions, most of genetic type evolutionary computing algorithms converge towards some particular probability distributions. These target probability measures are often prescribed by distributions on path spaces w.r.t. a series of conditioning events. For regulation problems, and open loop optimal control problems, these two viewpoints can be encapsulated in a single mathematical framework [107, 108, 119].

Answering to all the questions provided above amounts to rigorously analyzing the convergence and the performance of these stochastic hole puncher gradients, or these genetic particle sampling models. The second and rather recent viewpoint is the central theme of this chapter.

We end this introduction with a brief discussion on the origins and the mathematical foundations of genetic type particle models.

Genetic type stochastic models are increasingly used to sample from complex high-dimensional distributions. As we mentioned above, they approximate, as the population size tends to infinty, a given target probability distributions by a large cloud of random samples termed particles. Practically, the particles evolve randomly around the space independently and to each particle is associated a positive potential function. Periodically we duplicate particles with high potentials at the expense of particles with low potentials which die. This intuitive genetic mutation-selection type mechanism appears in numerous applications ranging from nonlinear filtering [22, 36, 52, 45, 38, 68, 67, 69, 89, 118, 120], Bayesian statistics [29, 40, 76, 123], combinatorial counting [3], molecular and polymer simulation [90], rare events simulation [26, 27, 82], quantum Monte Carlo methods [6, 102, 125] and genetic algorithms [47, 48, 87, 104], among others.

From a mathematical point of view, these methods can be interpreted as stochastic numerical approximations of Feynman-Kac measures. These measures represent the distribution of the paths of a reference Markov process, weighted by a collection of potential functions. These functional models are natural mathematical extensions of the traditional change of probability measures, commonly used in importance sampling. The particle interpretation consists in evolving a population of particles mimicking natural evolution mechanisms. During the mutation stage, the particles evolve independently of one another, according to the same probability transitions as the ones of the reference Markov chain. During the selection stage, each particle evaluates the potential value of its location. The ones with small relative values are killed, while the ones with high relative values are multiplied. The corresponding genealogical tree occupation measure converges, as the population size tends to infinity, to the complete Feynman-Kac distribution on path space.

The origins of stochastic particle simulation certainly start with the seminal paper of N. Metropolis and S. Ulam [126] published in 1949. As explained by these two physicists in the introduction of their pioneering article, the Monte Carlo method is, "essentially, a statistical approach to the study of differential equations, or more generally, of integro-differential equations that occur in various branches of the natural sciences". The links between genetic type particle Monte Carlo models and quadratic type parabolic integro-differential equations have been developed in the beginning of 2000' in the series of articles on continuous time models [51, 52, 54].

The earlier works on heuristic type genetic particle schemes seem to have started in Los Alamos National Labs with works of M.N. Rosenbluth and A.W. Rosenbluth [143], and T.E. Harris and H. Kahn [94]. We also quote the work on artificial life of Nils Aall Barricelli at the Institute for Advanced Study in Princeton [10, 11]. In all of these works, the genetic Monte Carlo scheme is always presented as a natural heuristic resampling type algorithm to generate random population models, to sample molecular conformations, or to estimate high energy particle distributions, without a single convergence estimate to ensure the performance, nor the robustness of the Monte Carlo sampler.

Since the mid 90's, genetic particle algorithms have recorded a dramatic popularity increase due to the proliferation and wide accessibility of powerful computing resources. They are now extensively and routinely used in engineering, machine learning, statistics and physics under sometimes different names, such as: particle filters, bootstrap or genetic filters, population Monte Carlo methods, sequential Monte Carlo models, genetic search models, branching and multi-level splitting particle rare event simulations, condensation models, go-with-the winner, spawning models, walkers population reconfigurations, pruning-enrichment strategies, quantum and diffusion Monte Carlo, rejuvenation models, and many others.

The mathematical foundations, and the performance analysis of all of these discrete generation particle models are rather recent. The first rigorous study in this field seems to be the article published by the first author in 1996 on the applications of particle methods to nonlinear estimation problems [36]. This article provides the first proof of the unbiased property of particle likelihood approximation models (lemma 3 page 12); and adaptive resampling criteria w.r.t. the weight dispersions (see remark 1 on page p.4).

This article also presents the first convergence results for a new class of interacting particle filters, originally presented as heuristic Monte Carlo schemes in the beginning of the 1990's in three independent schools. A series of classified industrial Research Contracts on tracking and control developed between 1990 and 1993 by the P. Del Moral, J.C. Noyer, G. Rigal, and G. Salut [59, 60, 61, 62, 63, 64], and [23, 65, 66]. The first journal article presenting the heuristic of particle filters is the article by N.J. Gordon, D. Salmond and A.F.M. Smith [89], and the first conference article presenting the heuristic of particle filters is the article by G. Kitagawa [118].

For a more thorough discussion on these models, we refer the reader to [37, 52, 38, 58, 67], as well as in [8, 115, 9, 72, 5, 152], and references therein.

1.1.1   *From Evolutionary Computing to Particle Algorithms*

Besides a sustained research on evolutionary computing, theoretical support and convergence proofs were until recently regarded as non mandatory. While leading to significant results in practice [34, 8, 115], advances were only derived on intuition and empirical grounds. And while this can be sustained for a real world setup where finite time and resource constraints are imposed, understanding the different paradigms and afferent convergence properties demands an accurate description of the underlying mathematical models.

A first aspect to address is what evolutionary computing is applied for and what information is expected? As later detailed in this section, different classes of problems are considered, e.g. non-linear, non-convex, discrete or continuous, with one or multiple objectives to optimize, highly multimodal, ill-conditioned or with epistatic interactions defined, within dynamic environments or subject to stochastic perturbations (uncertainty). All are finally connected by assumptions implying (i) no asymptotic convergence proof, with no exact solution or reproducible (stochastic) behavior expected under finite time constraints, (ii) nonexistence of a polynomial time alternative approach, e.g. due to a combinatorial explosion of the search space, intractability or exponential increase of the number of local optima, and (iii) exploration (classically) ended with no explicit information on the distribution of the optima, only the best found solution being provided as a result.

At the opposite end, deterministic algorithms (not covered here), e.g. interval methods, branch-and-bound, provide an optimal solution within a finite time and with finite resources, nonetheless requiring an exponentially increasing time (as a function of instance size). With respect to the last assumption, while different evolutionary algorithms enclose by definition intrinsic support for estimating normalizing constants, observing ancestral and genealogical structures or convergence towards target distributions, in most cases this information is not regarded as relevant and discarded. And, while those aspects fall by excellence in the domain of particle algorithms, commonly referred to, for example, as particle filters, sequential, diffusion or quantum Monte Carlo, and with a focus for estimating distribution laws, evolutionary and particle algorithms, with marginal exceptions, follow identical conceptual lines.

From the probabilistic point of view, genetic type particle algorithms are a natural class of Monte Carlo methods for sampling complex high-dimensional probability distributions and estimating their normalizing constants. As we already mentioned, this class of algorithms approximate a given sequence of target probability measures by a large cloud of random samples termed particles (equivalent of individuals in evolutionary algorithms). The particles evolve randomly in the solution space (mutation and free exploration). A positive potential/fitness function is associated to each particle. Periodically, the particles with high potential value are duplicated at the expense of particles with low potentials which are discarded (selection and replacement).

An overview of the evolutionary computing domain is offered in the following, to no extent exhaustive, only in order to highlight connection points with particle algorithms. For the remainder of this section, let us consider a simplified scenario where, given an arbitrary deterministic, static black-box function, the optimal solution (or approximation of) is demanded. Having as sole assumption that the function can be sampled within the entire definition domain, i.e. with no other information on the nature of the function, continuous or discrete, and disregarding the encoding of solutions, e.g. fixed or variable size array of binary, integer or real values, Gray coding (reflected binary code), graphs, trees, cellular, messy, direct or indirect encoding, a straightforward approach would be to draw samples until some termination criterion is met. Except the simplification, this portrays the basic idea of a simple Monte Carlo algorithm. Extending this direction, a first axis of discussion leads to single-solution based exploration paradigms. Note that while this particular class is commonly referred to as local search algorithms this only relates to exploration being conducted by sampling from the neighborhood of a single solution, evolved in iterative manner, and does not automatically imply a limitation for the algorithm's exploration capabilities. A second part of the discussion focuses on algorithms that simultaneously evolve a set of independent solutions while ensuring a balance or trade-off between local and global exploration, later moving to hybrid and parallel aspects.

1.1.1.1   Local Search Algorithms

As a general classification, *direct* and *indirect* search methods can be considered [141]. The first class, also denominated as *zero order methods*, only relies on a direct sampling of the objective function, with no partial derivatives, i.e. no analytic or numeric gradient employed. Initially introduced in the work of Hooke and Jeeves [105], the *direct search* denomination offers an explicit delimitation from higher complexity methods: "*the phrase implies our preference, based on experience, for straightforward search strategies which employ no techniques of classical analysis* [...]". Examples include Hill-Climbing, Nelder-Mead [153, 135, 121], where the exploration is conducted by a set of perturbations applied to a randomly generated initial simplex, Solis and Wets [151], including self-adaptive mechanisms, Tabu Search [83, 84], Variable Neighborhood Search [131, 132], Guided

Local Search [161], Iterated Local Search [13, 124] or Simulated Annealing [117], developed as a generalization of Metropolis Monte Carlo [127]. The latter class can be further divided into *first* and *second order methods*, using the first, respectively second or higher order derivatives (analytic or numerical approximations of) for guiding the exploration process. Examples include steepest descent, Conjugate Gradient [100, 101, 77, 78, 139, 150] and second order methods like the Limited Memory BFGS [17, 79, 88, 149, 137, 140] or the Adaptive Simulated Annealing [112, 111, 109, 110, 113].

1.1.1.2 Set of Solutions Based Algorithms

A reference in the evolutionary computing domain, Genetic Algorithms (GAs), developed through the work of Holland [104], define a structure that inspired and set the bases for different other paradigms. The approach relies on several distinct stages as follows: (i) initialization – a set of initial solutions (chromosomes in the case of GAs) are randomly sampled, forming a population, (ii) selection – a subset of the best fit solutions is constructed, (iii) recombination and perturbation – new samples are drawn by applying several operators on the previously selected solutions, e.g. crossover and mutation operators, and (iv) replacement – least fit solutions in the initial population are replaced, the algorithm iterating steps (ii) to (iv) until a termination criterion is met.

While no in-depth details will be provided here, it may be worthwhile mentioning that different strategies and operators were proposed and analyzed for all stages of a genetic algorithm. Examples of mutation and recombination operators include (i) diversification oriented or mutation constructions [128] like bit-flip or swap operators, polynomial transforms, (non) uniform, Gaussian or Cauchy distribution based, as well as (ii) intensification, or crossover operators [97, 95, 98, 99, 96]. This latter class was extensively investigated, leading to operators with one or multiple cutting or intersection points, uniform, arithmetic, geometric, Wright's heuristic, linear BGA, α, β-blend, simulated binary crossover, fuzzy recombination or dynamic operators. At the same time, multiple offspring operators were studied, with only the best two offspring solutions out of all finally selected, e.g. linear or the min-max arithmetic operator. Examples of more advanced operators include higher or adjustable arity operators (multiple parent solutions), including global recombination, gene-pool recombination, linkage evolving operator or the m-tuple mating [70, 16, 148, 134].

From a mathematical point of view, disregarding recombination operators, these methods can be interpreted as stochastic numerical approximations of Feynman-Kac measures, representing the distribution of paths for a reference Markov process weighted by a collection of potential functions. These functional models are natural mathematical extensions of the traditional change of probability measures, commonly used in importance sampling. The particle interpretation, as a direct analogy to genetic algorithms, consists in evolving a population of particles that mimic natural evolution mechanisms. During the transition (mutation) stage, the particles evolve independently one of another, according to the same probability transitions as the ones of the reference Markov chain. During the selection stage, each particle evaluates the potential value of its location and the ones with small relative values are discarded while the ones with high relative values are multiplied. The corresponding genealogical tree occupation measure converges, as the population size tends to infinity, to the complete Feynman-Kac distribution on path space. As a direct analogy, particle algorithms also rely on transition operators (equivalent of mutation) although imposing to leave the initial distribution measure invariant. No direct equivalent of crossover operators exists however as a straightforward understanding and modeling of a recombination transition does not always make sense or is even possible and coherent in a simulation context. Analogously, different selection strategies were explored, including proportional selection, stochastic universal sampling, tournament, (linear, exponential) ranking, sigma scaling or Boltzmann selection, all with or without elitism or truncation. Additionally, replacement may consider the depletion of the worst, best or most similar individuals (crowding), replacement of parent(s) or of randomly selected individuals. While selection is also considered in particle algorithms, semantics may differ depending on the specific application area, e.g. being referred to as resampling, filtering, absorption, etc., and implicitly encloses replacement. Additional examples and applications are presented in Section 1.5, allowing for a direct analogy between evolutionary computing and particle algorithms.

Extensions of the classical genetic paradigm fostered different axes of study, leading to co-evolution and memetic algorithms [133, 136, 1], e.g. hybrid local vs global exploration strategies, Lamarckian evolution and Baldwin effect, meta and hyper-heuristics [18], cultural algorithms [142], differential evolution [157], swarm intelligence (ant colony, particle swarm, artificial immune systems, etc.) [15, 33, 116, 35], scatter search and path relinking [85, 93], genetic programming [122] (symbolic regression), or evolution strategies [7, 14], among many others. All and each of these paradigms finally led to intense research on, for example, different hybridization strategies at low-level, operator enclosed, or at high-level, as a sequence of heuristics or independently evolving parallel algorithms, different strategies in differential evolution, etc. Furthermore, different approaches like the Covariance Matrix Adaptation Evolution Strategies [92, 91], Estimation of Distribution Algorithms [138] or the Reactive Search [12] (sub-symbolic learning, adaptation and incremental model development) were introduced, exploiting landscape information in the form of second order model approximations, estimations of optima distribution or reinforcement based learning. As a side note, landscape studies developed as a standalone axis of research in an attempt to understand what correlation exists between specific features in an objective function's landscape or definition and the (non) efficiency of the different exploration strategies [159, 160, 154, 155, 162]. Extensive research was also conducted on (self) tuning and adaptive paradigms [103, 73, 71] with applications in ill-conditioned, dynamic and stochastic problems, including online problems or aspects as state dependency and decision making [106]. Additional axes, although out of scope and not detailed in this introduction, include multi-objective evolutionary computing algorithms [31, 30] where a set of best-compromise solutions have to be found (Pareto set and front in the solution, respectively objective space), and parallel models [2, 21], e.g. multi-start, islands and topological (a)synchronous information exchange models. As a converging trend, an affinity for including or exploiting aspects and techniques from probability and statistics is emerging, making that, except for correspondences with filtering algorithms and sampling, different analogies are possible with applications in tracking, non-linear estimation problems, signal processing or stochastic optimization [47, 48, 40, 38, 58, 67].

1.1.2   Outline of the Chapter

The remainder of this chapter includes a pedagogical introduction to the stochastic modeling and the theoretical analysis of interacting particle algorithms in an effort to shed new light on some interesting links between physical, engineering, statistical and mathematical domains that appear disconnected at first glance. Second, the mathematical concepts and models are now at a point where they provide a very natural and unifying mathematical basis for a large class of Monte Carlo algorithms. To simplify the presentation and to clarify the main ideas behind these stochastic models, we have chosen to restrict the contents of this chapter to finite or countable state space models, avoiding any measure theory irrelevancies. In this simplified framework, we develop a rigorous mathematical analysis only involving vector and matrix operations. We emphasize that all of these particle models and the associated convergence results can be extended to general state-space models, including path-space models and excursion spaces on abstract measurable state spaces. In Section 1.5 several application areas are presented and a detailed description of interacting particle algorithms is provided.

1.2   Basic Notation and Motivation

In this section, we provide some basic notation and some comments on the stochastic models presented in this chapter. First, we mention that probabilistic models are always defined in terms of measures, numerical functions, as well as operators on functions and measures. Besides the fact that measures on finite spaces can be seen as elementary functions and linear operators as simple matrices, in order to provide a rigorous presentation and to facilitate the extensions to more general models we have chosen to keep the probabilistic terminology and the corresponding notation.

Let E be a finite set equipped with a matrix $(Q(x,y))_{x,y \in E}$. A signed measure on a finite set E is a mapping $x \in E \mapsto \mu(x) \in \mathbb{R}$. For any subset $A \subset E$, and any numerical function $x \in E \mapsto f(x)$ we set

$$\mu(A) := \sum_{x \in A} \mu(x) = \sum_{x \in E} \mu(x)\,\mathbb{1}_A(x) \quad \text{and} \quad \mu(f) = \sum_{x \in E} \mu(x)f(x)$$

with the indicator function $\mathbb{1}_A$ of a subset A. The Dirac measure at some point $x \in E$ is the indicator function $\mathbb{1}_x : y \in E \mapsto \mathbb{1}_x(y)$ of the set $\{x\}$. In this slightly abusive notation, we have $\mu(A) = \mu(\mathbb{1}_A)$ and $\mathbb{1}_x(A) = \mathbb{1}_A(x)$. A probability measure is a non negative measure μ such that $\mu(E) = 1$. Given some nonnegative measure μ on E, sometimes we use the proportional relation and we write

$$v \propto \mu \quad \text{to define the probability measure } v(x) = \mu(x)/\textstyle\sum_{z \in E} \mu(z).$$

The empirical measure associated with a set of N states $(x^1, \ldots, x^N) \in E^N$ is the measure defined by

$$y \in E \mapsto \eta^N(y) := \frac{1}{N} \sum_{i=1}^{N} \mathbb{1}_{x^i}(y)$$

with $N \geq 1$. By construction, we have

$$\eta^N(f) = \sum_{y \in E} f(y)\eta^N(y) = \frac{1}{N} \sum_{i=1}^{N} f(x^i)$$

We also denote by $Q(f)$ and (μQ) the function $x \mapsto Q(f)(x)$ and the measure $y \mapsto (\mu Q)(y)$ defined below

$$Q(f)(x) := \sum_y Q(x,y)f(y) \quad \text{and} \quad (\mu Q)(y) := \sum_x \mu(x)Q(x,y) \qquad (1.1)$$

In this notation, reversing the summation order, we have $\mu(Q(f)) = (\mu Q)(f)$.

For instance, for finite ordered state spaces with cardinality $d \geq 1$ there is no loss of generality to suppose that $E = \{1, \ldots, d\}$. In this case, we can identify measures μ, matrices Q, and functions f by the conventional notation of vector calculus

$$\mu := [\mu(1), \ldots, \mu(d)] \qquad Q := \begin{pmatrix} Q(1,1) & \cdots & Q(1,d) \\ \vdots & \vdots & \vdots \\ Q(d,1) & \cdots & Q_1(d,d) \end{pmatrix} \qquad \mathbf{f} := \begin{pmatrix} f(1) \\ \vdots \\ f(d) \end{pmatrix}$$

In this situation, the formulae (1.1) coincide with the usual matrix operations, with the x-th entry $Q(f)(x)$ of the column vector Qf, and the y-th entry $(\mu Q)(y)$ of the line vector μQ.

Given a sequence of matrices $(Q_n(x,y))_{x,y \in E}$, indexed by the parameter $n \in \mathbb{N}$, we denote by $(Q_1 \ldots Q_n)$ the composition of the matrices Q_p, from $p = 1$ to $p = n$; that is, we have that

$$(Q_1 \ldots Q_n)(x_0, x_n) = \sum_{x_1, \ldots, x_{n-1} \in E} Q_1(x_0, x_1)Q_2(x_1, x_2) \ldots Q_n(x_{n-1}, x_n)$$

For time homogeneous matrices $Q_n = Q$, we set $Q^n = (Q_1 \ldots Q_n)$.

A Markov transition is a positive matrix $(M(x,y))_{x,y \in E}$ such that $\sum_y M(x,y) = 1$, for any $x \in E$. These matrices are sometimes called stochastic matrices in the literature on probability and Markov chains. We say that a measure $\mu(x)$ on E is reversible for a Markov transition $M(x,y)$ if we have for any states $x, y \in E$

$$\mu(x)M(x,y) = \mu(y)M(y,x)$$

We say that a probability measure $\mu(x)$ is invariant for the Markov transition $M(x,y)$ if we have for each $y \in E$

$$\mu(y) = \sum_x \mu(x)M(x,y)$$

Measures, matrices and functions are defined in the same way on more general measurable state spaces E under appropriate well known regularity conditions. We denote respectively by $\mathcal{M}(E)$, and $\mathcal{B}(E)$, the set of all finite signed measures on some measurable space $(E, \mathcal{E})$, and the Banach space of all bounded and measurable functions f equipped with the uniform norm $\|f\|$.

We let $\mu(f) = \int \mu(dx) f(x)$, be the Lebesgue integral of a function $f \in \mathcal{B}(E)$, with respect to a measure $\mu \in \mathcal{M}(E)$. We recall that a bounded integral operator M from a measurable space $(E, \mathcal{E})$ into an auxiliary measurable space $(F, \mathcal{F})$ is an operator $f \mapsto M(f)$ from $\mathcal{B}(F)$ into $\mathcal{B}(E)$ such that the functions $x \mapsto M(f)(x) :=$ $\int_F M(x,dy)f(y)$ are $\mathcal{E}$-measurable and bounded, for any $f \in \mathcal{B}(F)$. A Markov kernel is a positive and bounded integral operator M with $M(1) = 1$. Given a pair of bounded integral operators (M_1, M_2), we let $(M_1 M_2)$ the composition operator defined by $(M_1 M_2)(f) = M_1(M_2(f))$. For time homogenous state spaces, we denote by $M^m = M^{m-1}M = MM^{m-1}$ the m-th composition of a given bounded integral operator M, with $m \geq 1$.

We shall slightly abuse the notation and we denote by 0 and 1 the zero and the unit elements in the semi-rings $(\mathbb{R}, +, \times)$ and in the set of functions on some state space E. We recall that the gradient ∇f and the Hessian $\nabla^2 f$ of a smooth function $f : \theta = (\theta^i)_{1 \leq i \leq d} \in \mathbb{R}^d \mapsto f(\theta) \in \mathbb{R}$ are defined by the functions

$$\nabla f = \left(\frac{\partial f}{\partial \theta^1}, \frac{\partial f}{\partial \theta^2}, \dots, \frac{\partial f}{\partial \theta^d} \right) \quad \text{and} \quad \nabla^2 f = \begin{pmatrix} \frac{\partial^2 f}{\partial^2 \theta^1} & \frac{\partial^2 f}{\partial \theta^1 \partial \theta^2} & \cdots & \frac{\partial^2 f}{\partial \theta^1 \partial \theta^d} \\ \frac{\partial^2 f}{\partial \theta^2 \partial \theta^1} & \frac{\partial^2 f}{\partial^2 \partial \theta^2} & \cdots & \frac{\partial^2 f}{\partial \theta^2 \partial \theta^d} \\ \vdots & \vdots & \vdots \\ \frac{\partial^2 f}{\partial \theta^d \partial \theta^1} & \frac{\partial^2 f}{\partial \theta^d \partial \theta^2} & \cdots & \frac{\partial^2 f}{\partial^2 \theta^d} \end{pmatrix}$$

Given a $(d \times d')$ matrix M with random entries $M(i,j)$, we write $\mathbb{E}(M)$ the deterministic matrix with entries $\mathbb{E}(M(i,j))$. We also denote by $(\bullet)_+$, $(\bullet)_-$ and $\lfloor \bullet \rfloor$ respectively the positive, negative and integer part operations. The maximum and minimum operations are denoted respectively by $\vee$ and $\wedge$

$$a \vee b = \max(a,b) \quad \text{and} \quad a \wedge b = \min(a,b)$$

We also use the traditional conventions

$$\left(\sum_{\emptyset}, \prod_{\emptyset} \right) = (0, 1) \quad \text{and} \quad \left(\sup_{\emptyset}, \inf_{\emptyset} \right) = (-\infty, +\infty)$$

1.3   Genetic Particle Models

Genetic algorithms are often presented as a random search heuristic that mimics the process of evolution to generate useful solutions to complex optimization problems. The genetic evolution starts with a population of N candidate possible solutions $(\xi_0^1, \dots, \xi_0^N)$ randomly chosen w.r.t. some distribution $\eta_0(x)$ on some initial finite state space, say E_0, where the coordinates ξ_0^i are also called individuals or genotypes, with $1 \leq N$. In discrete generation models, the genetic evolution is decomposed into two main steps: the selection and the mutation transitions. During the selection-reproduction stage, multiple individuals in the current population $(\xi_n^1, \dots, \xi_n^N)$ at time $n \in \mathbb{N}$ are stochastically selected based on some problem dependent fitness function G_n that measure the quality of a solution on a given finite solution space E_n. In practice, we choose a random proportion B_n^i of an existing solution ξ_n^i in the current population with a mean value $\propto G_n(\xi_n^i)$ to breed a brand new generation of "improved" solutions $(\widehat{\xi}_n^1, \dots, \widehat{\xi}_n^N)$. During the mutation step, every selected individual $\widehat{\xi}_n^i$ mutates to a new solution $\xi_{n+1}^i = x$ randomly chosen with a distribution $M_{n+1}(\widehat{\xi}_n^i, x)$ on a possibly different finite solution space E_{n+1}. This generational random process is repeated until some desired termination condition has been reached.

An informal pseudocode description is provided in figure 1.

The question of why these genetic algorithms often succeed at generating high fitness solutions of complex practical problems is not really well understood. Sometimes some researchers say: *"If God uses this natural evolution procedures why I shouldn't use it to solve my problem?"*. More surprisingly, genetic type selection-mutation models are currently used in a variety of application domains, including numerical physics, biology, signal processing, Bayesian statistics, rare event simulation, uncertainty propagation in numerical codes, and many others. In Sequential Monte Carlo literature, the mutation and the selection steps are called the sampling and the resampling transition. In advanced signal processing, particle filters also coincide with these genetic models with mutation-selection stages given by the prediction-updating steps. In Diffusion Monte Carlo methods as well as in Quantum Monte Carlo methods, the mutation and the selection steps are interpreted as the free evolution of walkers and the reconfiguration of the population. In polymer chain simulations, the selection transition is often called pruning. Many other botanical names are given to the selection transition, including cloning, replenish, go with the winner, and many others. For a more thorough discussion on this question with rather detailed bibliographical references, we refer the reader to [52, 38].

One crucial comment is that the size of the population N should be a precision parameter, so that in some sense we solve the problem at hand when N tends to

---

Algorithm 1.1. Genetic algorithm pseudocode

{Fix some population size (precision of the algorithm) parameter N}

Initialization

$\xi_0 :=$ sample N particles, $(\xi_0^i)_{1 \leq i \leq N}$ randomly with some given law η_0.

for $k = 1$ to n **do**

    Selection

    for $i = 1$ to N **do**

      {For each particle}

$$\widehat{\xi}_{k-1}^i := \begin{cases} \xi_{k-1}^i, \text{ with probability } G_{k-1}(\xi_{k-1}^i)/\max_{1 \leq j \leq N} G_{k-1}(\xi_{k-1}^j) \quad (1) \\[2mm] \xi_{k-1}^i, \text{ a random variable with law } \sum_{i=1}^{N} \frac{G_{k-1}(\xi_{k-1}^i)}{\sum_{j=1}^{N} G_{k-1}(\xi_{k-1}^j)} \delta_{\xi_{k-1}^i}, \text{ otherwise } \quad (2). \end{cases}$$

    end for

    {We can replace the acceptance probability in the r.h.s. of (1) by the quantity $\varepsilon\, G_{k-1}(\xi_{k-1}^i)$, for any $\varepsilon \geq 0$, such that $\varepsilon\, \max_{1 \leq j \leq N} G_{k-1}(\xi_{k-1}^j) \leq 1$. If we choose $\varepsilon = 0$, we simply remove the line (1), so that the selection transition coincides with the proportional/roulette selection}

    Transition

    for $i = 1$ to N **do**

      {For each particle}

      $\xi_k^i := F_k(\widehat{\xi}_{k-1}^i, \omega_k^i)$,

      {$F_k(., \omega_k^i)$ designates the perturbation operator generating new candidate solutions. In other words, $\xi_k^i = x$ with probability $M_k(\widehat{\xi}_{k-1}^i, x)$.}

    end for

end for

---

infinity. In other words, when the computational resources $N \to \infty$ the genetic search model should increased its ability to find the desired solution. One way to understand these questions is to analyze the genealogical tree of a given population of individuals. If we interpret the genetic algorithm as a birth and death branching process, then we can trace back in time the whole ancestral line of the individual ξ_n^i at the n-th generation.

$$\xi_{0,n}^i \longleftarrow \xi_{1,n}^i \longleftarrow \cdots \longleftarrow \xi_{n-1,n}^i \longleftarrow \xi_{n,n}^i = \xi_n^i$$

The random state $\xi_{p,n}^i$ represents the ancestor of the individual ξ_n^i at the level p, with $0 \leq p \leq n$, and $1 \leq i \leq N$.

One could expect that this genealogical tree models have different asymptotic behaviors depending on their sampling and on the problem at hand. In fact, in terms of proportions and probability measures we don't have a lot of variability. The random occupation measure of the tree becomes more and more deterministic and we have the following convergence result

$$\lim_{N\to\infty} \frac{1}{N} \sum_{i=1}^{N} \mathbb{1}_{(\xi_{0,n}^i, \xi_{1,n}^i, \ldots, \xi_{n,n}^i)}(x_0, x_1, \ldots, x_n) = \frac{1}{\mathscr{Z}_n} \left\{ \prod_{0\leq p<n} G_p(x_p) \right\} \times \mathbb{P}_n(x_0, \ldots, x_n)$$

(1.2)

with some normalizing constant $\mathscr{Z}_n$, and the probability distribution of a Markov chain sequence

$$\mathbb{P}_n(x_0, \ldots, x_n) := \eta_0(x_0) M_1(x_0, x_1) \ldots M_n(x_{n-1}, x_n)$$

Furthermore, the product of the empirical population mean values of the fitness functions we used in the genetic evolution provides an unbiased estimate of the unknown normalizing constants

$$\mathscr{Z}_n^N := \prod_{0\leq p<n} \frac{1}{N} \sum_{i=1}^{N} G_p(\xi_p^i) \longrightarrow_{N\to\infty} \mathscr{Z}_n$$

These limiting probability measures in the r.h.s. of (1.2) are often called Feynman-Kac measures or Boltzmann-Gibbs distributions in physics and in the applied probability literature. Inversely, suppose that we have to sample from a Feynman-Kac probability measure on some product space and/or we need to compute their normalizing constants. Then, one particle sampling strategy is to run a genetic particle approximation model.

Besides the fact that these rather surprising theoretical results give some insight on the convergence of genetic algorithms and their range of applications, many questions remain to be answered: What is the rate of convergence in the estimates given above? Are they uniform w.r.t. the time parameter? Is it possible to quantify the law of a finite block of individuals? Do we have Central Limit Theorems and exponentially small sub-Gaussian deviation probabilities as in conventional Monte Carlo sampling? What is the interpretation of these limiting probability measures in practical situation and real world concrete problems? How to turn a given complex estimation problem into this probabilistic framework? In this chapter, we provide some answers to these natural questions. In Section 1.4, we provide a brief overview of the connections between abstract positive matrices and genetic type interacting particle models. This section should not be skipped since it contains a series of recipes on matrix models and their particle interpretations to be combined with one another and applied in the application domains discussed in Section 1.5. The displeasure practitioners may get when analyzing these matrix models and their particle interpretations will fade since the genetic type particle approximations of these quantities presented below will provide instantly a collection of powerful simulation tools for the numerical solution of the problem at hand. In Section 1.5, we discuss a series of application domains of genetic particle models, by no means exhaustive. Each application starts with an introduction connecting the results developed in earlier parts with the current description.

1.4   Positive Matrices and Particle Recipes

1.4.1   Positive Matrices and Measures

1.4.1.1   Description of the Models

Let E be a finite set. We consider a collection of matrices $Q_n := (Q_n(x,y))_{x,y \in E}$ with non negative entries $Q_n(x,y) \geq 0$. Given a probability measure η_0 on E, we denote by $\mathbb{Q}_n$ the measure on the product space $E_n := E^{n+1}$ defined for any sequence $(x_p)_{0 \leq p \leq n} \in E_n$ of length n by the following formula:

$$\mathbb{Q}_n(x_0, \ldots, x_n) \propto \eta_0(x_0)\, Q_1(x_0, x_1)\, Q_2(x_1, x_2)\, \cdots\, Q_n(x_{n-1}, x_n) \qquad (1.3)$$

When the matrices $Q_n(x,y)$ are such that $\sum_{y \in E} Q_n(x,y) = 1$, for any $x \in E$, we can interpret $Q_n(x,y)$ as the probability of the transition $X_{n-1} = x \rightsquigarrow X_n = y$ of a given Markov chain X_n. In this situation, we have

$$\mathbb{Q}_n(x_0, \ldots, x_n) = \mathrm{Proba}\left((X_0, \ldots, X_n) = (x_0, \ldots, x_n)\right) \qquad (1.4)$$

Moreover, if we set

$$M_n(x,y) = \mathrm{Proba}\left(X_n = y \mid X_{n-1} = x\right) \quad \text{and} \quad \eta_0(x) = \mathrm{Proba}\left(X_0 = x\right)$$

then we find that

$$\mathbb{Q}_n(x_0, \ldots, x_n)$$

$$= \mathbb{P}_n(x_0, \ldots, x_n)$$

$$:= \mathrm{Proba}\left((X_0, \ldots, X_n) = (x_0, \ldots, x_n)\right)$$

$$= \mathrm{Proba}\left(X_0 = x_0\right)\mathrm{Proba}\left(X_1 = x_1 \mid X_0 = x_0\right) \ldots \mathrm{Proba}\left(X_n = y \mid X_{n-1} = x\right)$$

$$= \eta_0(x_0)M_1(x_0, x_1) \ldots M_n(x_{n-1}, x_n)$$

In a variety of applications, we want to approximate the integral type mean values of functions f_n on the product space E_n

$$\mathbb{Q}_n(f_n) = \sum_{x_0, \ldots, x_n} \mathbb{Q}_n(x_0, \ldots, x_n) f_n(x_0, \ldots, x_n) \qquad (1.5)$$

as well as their normalizing constants

$$\mathscr{Z}_n := \sum_{x_0, \ldots, x_n} \eta_0(x_0)\, Q_1(x_0, x_1)\, Q_2(x_1, x_2)\, \cdots\, Q_n(x_{n-1}, x_n) \qquad (1.6)$$

Reducing a bit our initial objective, sometimes we only want to approximate the final time marginals

$$\eta_n(x_n) := \sum_{x_0 \dots x_{n-1}} \mathbb{Q}_n(x_0, \dots, x_{n-1}, x_n) \quad \text{and} \quad \gamma_n(x_n) := \mathcal{Z}_n \times \eta_n(x_n)$$

$$(\Rightarrow \gamma_n(1) = \mathcal{Z}_n) \tag{1.7}$$

1.4.1.2   Path Space Models

From the pure mathematical point of view, for path space models these marginal measure models are equivalent to the model defined in (1.3). More precisely, if we set

$$\forall n \geq 0 \qquad x_n := (x_{0,n}, x_{1,n}, \dots, x_{n-1,n}, x_{n,n}) \in E_n$$

then we find that

$$
\begin{aligned}
\mathbb{Q}_n(x_n) &= \mathbb{Q}_n(x_{0,n}, x_{1,n}, \dots, x_{n-1,n}, x_{n,n}) \\
&\propto \eta_0(x_{0,n}) \, Q_1(x_{0,n}, x_{1,n}) \, Q_2(x_{1,n}, x_{2,n}) \dots Q_n(x_{n-1,n-1}, x_{n,n}) \\
&\propto \sum_{x_{n-1} \in E_{n-1}} \mathbb{Q}_{n-1}(x_{n-1}) \, \mathcal{Q}_n(x_{n-1}, x_n)
\end{aligned}
$$

with the matrices

$$\mathcal{Q}_n(x_{n-1}, x_n) := \mathbb{1}_{x_{n-1}}(x_{0,n}, x_{1,n}, \dots, x_{n-1,n}) \times Q_n(x_{n-1,n}, x_{n,n}) \tag{1.8}$$

This implies that $\mathbb{Q}_n(x_n)$ is the n-th marginal of the extended measure on the product of the path spaces $\prod_{0 \leq p \leq n} E_p$ defined by

$$\mathbb{Q}_n^{(\text{path})}(x_0, \dots, x_n) \propto \eta_0(x_0) \mathcal{Q}_1(x_0, x_1) \mathcal{Q}_2(x_1, x_2) \dots \mathcal{Q}_n(x_{n-1}, x_n)$$

for any $x_p \in E_p$, with $0 \leq p \leq n$.

In the case of Markov transitions $Q_n = M_n$ discussed in (1.4), we have

$$
\begin{aligned}
\mathcal{Q}_n(x_{n-1}, x_n) &= \mathcal{M}_n(x_{n-1}, x_n) \\
&:= \mathbb{1}_{x_{n-1}}(x_{0,n}, x_{1,n}, \dots, x_{n-1,n}) \times M_n(x_{n-1,n}, x_{n,n})
\end{aligned}
$$

In other words, $\mathcal{M}_n(x_{n-1}, x_n)$ is the Markov transition of the historical process

$$\mathscr{X}_n = (X_0, \dots, X_n)$$

of the Markov chain X_n with transitions M_n; that is, we have that

$$\text{Proba}\left(\mathscr{X}_n = x_n \mid \mathscr{X}_{n-1} = x_{n-1}\right)$$

$$\text{Proba}\left((X_0,\ldots,X_n) = (x_{0,n},\ldots,x_{n,n}) \mid (X_0,\ldots,X_{n-1}) = (x_{0,n-1},\ldots,x_{n-1,n-1})\right)$$

$$= \mathbb{1}_{x_{n-1}}(x_{0,n},x_{1,n},\ldots,x_{n-1,n}) \times \text{Proba}\left(X_n = x_{n,n} \mid X_{n-1} = x_{n-1,n}\right)$$

$$= \mathbb{1}_{x_{n-1}}(x_{0,n},x_{1,n},\ldots,x_{n-1,n}) \times M_n(x_{n-1,n},x_{n,n})$$

Finally, in this situation we observe that

$$\mathbb{Q}_n(x_n) = \mathbb{Q}_n(x_{0,n},x_{1,n},\ldots,x_{n-1,n},x_{n,n}) = \mathbb{P}_n(x_{0,n},x_{1,n},\ldots,x_{n-1,n},x_{n,n}) = \mathbb{P}_n(x_n)$$

with

$$\mathbb{P}_n(x_{0,n},x_{1,n},\ldots,x_{n-1,n},x_{n,n}) = \eta_0(d_{0,n})M_1(x_{0,n},x_{1,n})\ldots M_n(x_{n-1,n},x_{n,n})$$

Another useful state space enlargement allows to work on "transition type" state spaces E^2. These models are defined as follows. For any time $n \geq 0$, we set

$$x_n := (x_{n-1,n},x_{n,n}) \in E^2 \quad \text{and} \quad \mathscr{Q}_n(x_{n-1},x_n) := \mathbb{1}_{x_{n-1,n-1}}(x_{n-1,n}) \, Q_n(x_{n-1,n},x_{n,n})$$

We also use the convention $\eta_0(x_0) = \eta_0(x_{-1,0},x_{0,0}) = \eta_0(x_{0,0})$, for $n = 0$. In this notation, for any sequence x_n with $x_{n-1,n} = x_{n-1,n-1}$, we have

$$\eta_0(x_0)\mathscr{Q}_1(x_0,x_1)\mathscr{Q}_2(x_1,x_2)\ldots\mathscr{Q}_n(x_{n-1},x_n)$$

$$\propto \eta_0(x_{0,0})Q_1(x_{0,0},x_{1,1})Q_2(x_{1,1},x_{2,2})\ldots Q_n(x_{n-1,n-1},x_{n,n})$$

1.4.2   Interacting Particle Models

1.4.2.1   Genetic Population Evolution

This section is concerned with the design of a genetic particle approximation of the measures η_n introduced in (1.7). One universal way to associate a genetic population evolution model to positive matrices $Q_n(x_{n-1},x_n)$ is to use the decomposition

$$Q_n(x_{n-1},x_n) = G_{n-1}(x_{n-1}) \times M_n(x_{n-1},x_n) \tag{1.9}$$

with the Markov transition M_n and the potential function G_{n-1}

$$M_n(x_{n-1},x_n) := \frac{Q_n(x_{n-1},x_n)}{\sum_{x_n} Q_n(x_{n-1},x_n)} \quad \text{and} \quad G_{n-1}(x_{n-1}) = \sum_{x_n} Q_n(x_{n-1},x_n) \tag{1.10}$$

We let X_n be a Markov chain with initial distribution η_0 and Markov transitions M_n. By the definition of the measure $\mathbb{Q}_n$, the integral type formula (1.5) has the following probabilistic interpretation

$$\mathbb{Q}_n(f_n) \propto \sum_{x_1,\ldots,x_n} f_n(x_0,\ldots,x_n) \left\{ \prod_{0 \leq p < n} G_p(x_p) \right\} \times \left\{ \eta_0(x_0) \prod_{1 \leq p \leq n} M_p(x_{p-1},x_p) \right\}$$

$$\propto \mathbb{E}\left[f_n(X_0,\ldots,X_n) \left\{ \prod_{0 \leq p < n} G_p(X_p) \right\} \right] \tag{1.11}$$

with the normalizing constant

$$\mathscr{Z}_n := \mathbb{E}\left[\prod_{0 \leq p < n} G_p(X_p) \right]$$

For unit potential functions $G_n(x) = 1$, the model resumes to the Markov transitions model $Q_n = M_n$ discussed in (1.4). In this context, we clearly have that $\mathscr{Z}_n = 1$ and

$$\mathbb{Q}_n(f_n) = \mathbb{P}_n(f_n) = \mathbb{E}[f_n(X_0,\ldots,X_n)]$$

In this situation, we can approximate these expectation by sampling N independent copies X_n^i of the Markov chain X_n and using the traditional Monte Carlo empirical estimates

$$\mathbb{P}_n^N(f_n) := \frac{1}{N} \sum_{i=1}^N f_n(X_0^i,\ldots,X_n^i)$$

In more general situations, the potential functions G_n may change radically the probability mass distributions of the measures $\mathbb{Q}_n$. For instance, for indicator functions $G_n = 1_A$, we have

$$\mathscr{Z}_n := \mathbb{E}\left[\prod_{0 \leq p < n} G_p(X_p) \right] = \mathrm{Proba}(X_p \in A, \forall 0 \leq p < n)$$

and

$$\mathbb{Q}_n = \mathrm{Law}((X_0,\ldots,X_n) \mid X_p \in A, \forall 0 \leq p < n)$$

as soon as $\mathscr{Z}_n > 0$. In this situation, the probability measure $\mathbb{Q}_n$ only charges random trajectories that remains in the set A, for any time $0 \leq p < n$. This situation is discussed in some details in section 1.5.1.1 dedicated to particle absorption models. We already mention that in this case we often have

$$\mathrm{Proba}(X_p \in A, \forall 0 \leq p < n) \to_{n \uparrow \infty} 0$$

so that it becomes more and more unlikely that a random sample copy of X_n remains in the set A for all times $0 \leq p < n$ during a large horizon n. when all the N independent copies X_n^i have left the desired A, we have

$$\frac{1}{N} \sum_{i=1}^{N} f_n(X_0^i, \ldots, X_n^i) \prod_{0 \leq p < n} G_p(X_p^i) = 0$$

To avoid this "technical" problem, we use a genetic type acceptance rejection sampling scheme. To avoid some unnecessary technical discussion, we further assume that the functions G_n take values in $]0,1]$. As we shall see, this condition can be removed. Practically, we approximate the desired target measures $\eta_n(x_n)$ by a large cloud of N random samples also termed particles ξ_n^i, with $1 \leq i \leq N$; that is, at any time $n \geq 0$, we have that

$$\eta_n^N(x) := \frac{1}{N} \sum_{i=1}^{N} \mathbb{1}_{\xi_n^i}(x) \longrightarrow_{N \to \infty} \eta_n(x) \tag{1.12}$$

The particle system $\xi_n := (\xi_n^i)_{1 \leq i \leq N}$ is a simple genetic sampling model combing a mutation transition and a selection transition

$$\xi_n := (\xi_n^i)_{1 \leq i \leq N} \xrightarrow{\text{selection}} \widehat{\xi}_n := (\widehat{\xi}_n^i)_{1 \leq i \leq N} \xrightarrow{\text{mutation}} \xi_{n+1} \tag{1.13}$$

During the mutation transition, every individual performs a local random move according to the Markov transition M_n. During the selection step, every individual evaluates its potential value $G_n(\xi_n^i)$, with $1 \leq i \leq N$. For every index i, with a probability $G_n(\xi_n^i)$, we set $\widehat{\xi}_n^i = \xi_n^i$, otherwise we replace ξ_n^i by a fresh new individual $\widehat{\xi}_n^i = \xi_n^j$ randomly chosen in the whole population with a probability $\propto G_n(\xi_n^j)$.

For more general potential functions, we can replace the acceptance rate $G_n(\xi_n^i)$ of any individual ξ_n^i by any acceptance rate of the form $\varepsilon_n(\xi_n)G_n(\xi_n^i)$ with some $\varepsilon_n(\xi_n) \geq 0$ that may depend on the whole population and s.t. $\varepsilon_n(\xi_n)G_n(\xi_n^i) \in [0,1]$. For instance, we can chose $\varepsilon_n = 0$ or $\varepsilon_n(\xi_n) = 1/\max_{1 \leq i \leq N} G_n(\xi_n^i)$. In the first case, the selection transition is often called the "proportional selection". In the second case, we notice that the best fit individuals are always accepted. In all cases, the acceptance probability of an individual ξ_n^i is proportional to $G_n(\xi_n^i)$. In the further development of this chapter, we write that the acceptance probability is $\propto G_n(\xi_n^i)$. We end this section with a second important remark:

Let us suppose that Q_n has the following form

$$Q_n(x_{n-1}, x_n) = K_n(x_{n-1}, x_n) \times H_n(x_n)$$

for some Markov transition K_n and some potential function H_n. In this situation, the decomposition (1.10) is met with

$$M_n(x_{n-1}, x_n) := \frac{K_n(x_{n-1}, x_n)H_n(x_n)}{\sum_{x_n} K_n(x_{n-1}, x_n)H_n(x_n)} \quad \text{and} \quad G_{n-1}(x_{n-1}) = \sum_{x_n} K_n(x_{n-1}, x_n)H_n(x_n)$$

In practice, the numerical solving of the sum in the r.h.s. equation and the computational cost of sampling a mutation $x_{n-1} \rightsquigarrow x_n$ with the transition M_n can be prohibitive. One way to solve these problems is to sample a sequence of transitions

$\xi_{n-1}^i \rightsquigarrow \zeta_n^{i,j}$, with $1 \le j \le N'$ using the Markov transition K_n so that for any $1 \le i \le N$ we have

$$K_n^{N'}(\xi_{n-1}^i, x_n) = \frac{1}{N'} \sum_{i=1}^{N'} \mathbb{1}_{\zeta_n^{i,j}}(x_n) \simeq_{N'\uparrow\infty} K_n(\xi_{n-1}^i, x_n)$$

We let $M_n^{N'}$ and $G_{n-1}^{N'}$ be the Markov transitions and the potential functions defined as M_n and G_{n-1}, replacing K_n by $K_n^{N'}$ in the above equations; that is, we have

$$M_n^{N'}(\xi_{n-1}^i, x_n) := \sum_{j=1}^{N'} \frac{H_n(\zeta_n^{i,j})}{\sum_{k=1}^{N'} H_n(\zeta_n^{i,k})} \mathbb{1}_{\zeta_n^{i,j}}(x_n) \quad \text{and} \quad G_{n-1}^{N'}(\xi_{n-1}^i) = \frac{1}{N'} \sum_{j=1}^{N'} H_n(\zeta_n^{i,j})$$

Another strategy is to observe that the measures (1.3) are given by

$$\mathbb{Q}_n(x_0,\ldots,x_n) \propto \left[\eta_0(x_0) \, Q_1'(x_0,x_1) \, Q_2'(x_1,x_2) \, \ldots \, Q_n'(x_{n-1},x_n) \right] \, H_n(x_n)$$

with

$$Q_n'(x_{n-1},x_n) = H_{n-1}(x_{n-1}) \, K_n(x_{n-1},x_n)$$

and the convention $H_0 = 1$, for $n = 1$. In this interpretation, we approximate the measures

$$\mathbb{Q}_n'(x_0,\ldots,x_n) \propto \eta_0(x_0) \, Q_1'(x_0,x_1) \, Q_2'(x_1,x_2) \, \ldots \, Q_n'(x_{n-1},x_n)$$

using the particle occupation measures defined above (with mutation transitions K_n and selection fitness functions H_n) weighted by the function $H_n(x_n)$

$$\sum_{i=1}^{N} \frac{H_n(\xi_n^i)}{\sum_{j=1}^{N} H_n(\xi_n^j)} \mathbb{1}_{\xi_n^i}(x) \longrightarrow_{N\to\infty} \eta_n(x)$$

1.4.2.2   Particle Normalizing Constants

In this section, we present an unbiased particle approximation of the normalizing constants $\mathscr{Z}_n$ introduced in (1.6). Using the decomposition

$$\mathbb{Q}_n(x_0,\ldots,x_n) = \frac{\mathscr{Z}_{n-1}}{\mathscr{Z}_n} \, \mathbb{Q}_{n-1}(x_0,\ldots,x_{n-1}) \, Q_n(x_{n-1},x_n)$$

we find the following matrix formulae

$$\eta_n = \frac{\mathscr{Z}_{n-1}}{\mathscr{Z}_n} \, \eta_{n-1} Q_n \implies \frac{\mathscr{Z}_n}{\mathscr{Z}_{n-1}} = \eta_{n-1} Q_n(1) = \eta_{n-1}(G_{n-1}) \quad \text{and} \quad \gamma_n = \gamma_{n-1} Q_n$$

Now, it is also easily checked that

$$\gamma_n(1) = \gamma_{n-1} Q_n(1) = \gamma_{n-1}(1) \, \eta_{n-1} Q_n(1) = \cdots = \prod_{0 \le p < n} \eta_p Q_{p+1}(1) = \prod_{0 \le p < n} \eta_p(G_p)$$

Mimicking the r.h.s. multiplicative formula, an N-particle approximation of the normalizing constants $\mathscr{Z}_n$ is given by the following unbiased estimates

$$\mathscr{Z}_n^N := \prod_{0 \le p < n} \eta_p^N(G_p) \longrightarrow_{N \to \infty} \mathscr{Z}_n = \prod_{0 \le p < n} \eta_p(G_p)$$

and for any $x \in E$

$$\gamma_n^N(x) := \mathscr{Z}_n^N \, \eta_n^N(x) \longrightarrow_{N \to \infty} \gamma_n(x) := \mathscr{Z}_n \, \eta_n(x) \qquad (1.14)$$

Furthermore, the particle estimate $\gamma_n^N(x)$ is unbiased in the sense that for any $x \in E$, and for any $n \ge 0$, we have

$$\mathbb{E}\left(\gamma_n^N(x)\right) = \gamma_n(x)$$

The proof of this property is not so obvious. To our knowledge, in the context of nonlinear filtering this property has first been proved in [36]. See also [52] and [38] for more general models.

1.4.3   *Genealogical and Ancestral Structures*

1.4.3.1   **Genealogical Trees**

The aim of this section is to use the genealogical tree structure of the genetic population model defined in the previous sections to approximate the measures $\mathbb{Q}_n$ defined in (1.3).

Running back in time, we can trace back the complete ancestral lines of the individuals and the time evolution of the genealogical tree model associated with the genetic algorithm described above. For instance, a realization of that tree for $N = 3$ individuals and $n = 4$ iterations is given by

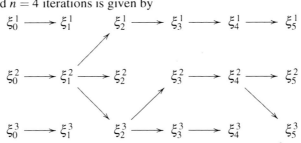

One way to encode the ancestral line of a current individual, say ξ_5^2, is to write $\xi_{p,5}^2$ with the level index $0 \le p \le 5$ of the ancestor; with the convention $\xi_{5,5}^2 = \xi_5^2$, for $p = 5$. In this notation, we obtain $N = 3$ ancestral lines, and the line associated with the i-th current individual ξ_n^i is the random vector in the product space E_n given by

$$\left(\xi_{0,n}^i, \xi_{1,n}^i, \xi_{2,n}^i, \ldots, \xi_{n-1,n}^i, \xi_{n,n}^i\right)$$

An N-particle approximation of $\mathbb{Q}_n$ is given by the occupation measure of these ancestral lines

$$\frac{1}{N} \sum_{i=1}^{N} \mathbb{1}_{(\xi_{0,n}^i, \xi_{1,n}^i, \ldots, \xi_{n,n}^i)}(x_0, \ldots, x_n) \longrightarrow_{N \to \infty} \mathbb{Q}_n(x_0, \ldots, x_n) \qquad (1.15)$$

Furthermore, the evolution model of the genealogical N ancestral lines

$$\xi_n^i := (\xi_{0,n}^i, \xi_{1,n}^i, \ldots, \xi_{n,n}^i)$$

with $1 \le i \le N$, coincides with the genetic model defined in (1.13) on product state spaces $E_n = \mathcal{E}_n$ with matrices $\mathcal{Q}_n(x_{n-1}, x_n)$ defined in (1.8). In the path space notation used in (1.8), the convergence result (1.15) takes the following form

$$\frac{1}{N} \sum_{i=1}^{N} \mathbb{1}_{\xi_n^i}(x_n) \longrightarrow_{N \to \infty} \mathbb{Q}_n(x_n)$$

for any $x_n := (x_{0,n}, x_{1,n}, \ldots, x_{n-1,n}, x_{n,n}) \in \mathcal{E}_n$; and this property coincides with (1.12).

Once again, this state space enlargement property is not really obvious. To our knowledge, this property has first been proved in [53].

1.4.3.2   Complete Ancestral Trees

To simplify the presentation, sometimes we denote by Ξ_n the sequence of complete genealogical trees $\xi_p := (\xi_p^i)_{1 \le i \le N}$, from the origin $p = 0$, up to time $p = n$; that is, we set

$$\Xi_n = (\Xi_{0,n}, \Xi_{1,n}, \ldots, \Xi_{n,n}) := (\xi_0, \ldots, \xi_n) \in \prod_{0 \le p \le n} E_p^N$$

with

$$\Xi_{p,n} = (\Xi_{p,n}^i)_{1 \le i \le N} = (\xi_p^i)_{1 \le i \le N} \quad \text{and} \quad \Xi_{p,n}^i := \xi_p^i := (\xi_{0,p}^i, \xi_{1,p}^i, \ldots, \xi_{p,p}^i)$$

for any $0 \le p \le n$.

Combining this observation with the exchangeability of the particle system and with the unbiased property of the unnormalized measures (1.14) discussed above, we conclude that for any function f_n on the product space $\mathcal{E}_n$

$$\mathbb{E}\left(\eta_n^N(f_n)\, \overline{\mathcal{Z}}_n^N \right) = \mathbb{E}\left(f_n\left(\xi_n^1\right) \overline{\mathcal{Z}}_n^N \right) = \mathbb{Q}_n(f_n) \quad \text{with} \quad \overline{\mathcal{Z}}_n^N := \mathcal{Z}_n^N / \mathcal{Z}_n$$

and with the first ancestral line $\xi_n^1 = \left(\xi_{0,n}^1, \xi_{1,n}^1, \ldots, \xi_{n,n}^1 \right)$ of the genealogical tree. In other words, we have that

$$\mathbb{E}\left(\overline{\mathcal{Z}}_n^N \mid \xi_n^1 = x_n \right) \times \mathrm{Proba}(\xi_n^1 = x_n) = \mathbb{Q}_n(x_n)$$

for any $x_n := (x_{0,n}, x_{1,n}, \ldots, x_{n,n}) \in E_n$. In addition, the measure defined by

$$\mathbb{T}_n^N(F_n) := \mathbb{E}\left(F_n(\Xi_n)\, \overline{\mathscr{Z}}_n^N\right) \tag{1.16}$$

for any function F_n on the product space $\prod_{0 \leq p \leq n} E_p^N$, is a probability measure whose ξ_n^i-marginals on E_n coincide with $\mathbb{Q}_n$, for any $1 \leq i \leq N$.

1.4.4   Complete Genealogical Tree Model

The aim of this section is to use the genealogical tree structure of the genetic population model defined in the previous sections to approximate the measures $\mathbb{Q}_n$ defined in (1.3).

1.4.4.1   Ancestral Lines Occupation Measures

The complete ancestral tree of the genetic model is the set of all the populations of individuals ξ_p, from the origin $p = 0$, up to the final time horizon $p = n$. Notice that these systems contain all the information about the genetic evolution, including the ancestral lines that have disappeared. A basic convergence estimate is the following

$$\frac{1}{N} \sum_{i=1}^N \mathbb{1}_{(\xi_0^i, \xi_1^i, \ldots, \xi_n^i)}(x_0, \ldots, x_n)$$

$$\longrightarrow_{N \to \infty} \eta_0(x_0) \times K_{1,\eta_0}(x_0, x_1) \times K_{2,\eta_1}(x_1, x_2) \times \cdots \times K_{n,\eta_{n-1}}(x_{n-1}, x_n)$$

with the stochastic matrices

$$K_{n,\eta_{n-1}}(x,y) = G_{n-1}(x)\, M_n(x,y) + (1 - G_{n-1}(x)) \sum_z \frac{\eta_{n-1}(z) G_{n-1}(z)}{\eta_{n-1}(G_{n-1})}\, M_n(z,y)$$

It is instructive to notice that the selection-mutation Markov transition $\xi_{n-1} \rightsquigarrow \xi_n$ defined in (1.13) is given by

$$\mathrm{Proba}\left(\xi_n = (x^1, \ldots, x^N) \mid \xi_{n-1}\right) := \prod_{1 \leq i \leq N} K_{n,\eta_{n-1}^N}(\xi_{n-1}^i, x^i)$$

We recall that $\eta_{n-1}^N(x) := \frac{1}{N} \sum_{i=1}^N \mathbb{1}_{\xi_{n-1}^i}(x)$ stands for the occupation measures of the genetic population at time $(n-1)$ so that the probability $K_{n,\eta_{n-1}^N}(\xi_{n-1}^i, x^i)$ that an individual $\xi_n^i = x^i$ is given by

$$G_{n-1}(\xi_{n-1}^i)\, M_n(\xi_{n-1}^i, x^i) + (1 - G_{n-1}(\xi_{n-1}^i)) \sum_{j=1}^N \frac{G_{n-1}(\xi_{n-1}^j)}{\sum_{k=1}^N G_{n-1}(\xi_{n-1}^k)}\, M_n(\xi_{n-1}^j, x^i)$$

Without altering the above convergence results, we can replace the fitness function G_n in the selection transition of the genetic model defined in section 1.4.2 by any function of the form $\varepsilon_n G_n$, for any constant $\varepsilon_n \geq 0$ s.t. $\varepsilon_n G_n \in [0,1]$. The selection transition associated with the choice $\varepsilon = 0$ is often called a simple selection, a proportional selection, as well as a multinomial branching model. In this situation, we have the following convergence result

$$\frac{1}{N} \sum_{i=1}^{N} \mathbb{1}_{(\xi_0^i, \xi_1^i, \ldots, \xi_n^i)}(x_0, \ldots, x_n) \longrightarrow_{N \to \infty} \eta_0(x_0) \times \eta_1(x_1) \times \cdots \times \eta_n(x_n) \qquad (1.17)$$

1.4.4.2   Backward Markov Chain Model

Using the matrix formulae given above, we observe that

$$\frac{\eta_{n-1}Q_n(x_n)}{\eta_n(x_n)} \times \frac{\eta_{n-2}Q_{n-1}(x_{n-1})}{\eta_{n-1}(x_{n-1})} \times \cdots \times \frac{\eta_0 Q_1(x_1)}{\eta_1(x_1)} = \frac{\mathscr{L}_n}{\mathscr{L}_{n-1}} \times \frac{\mathscr{L}_{n-1}}{\mathscr{L}_{n-2}} \times \cdots \times \frac{\mathscr{L}_1}{\mathscr{L}_0} = \mathscr{L}_n$$

From this observation, we readily prove the following backward representation of $\mathbb{Q}_n$

$$\mathbb{Q}_n(x_0, \ldots, x_n)$$

$$= \eta_n(x_n) \times \frac{\eta_{n-1}(x_{n-1})Q_n(x_{n-1}, x_n)}{\eta_{n-1}Q_n(x_n)} \cdots \frac{\eta_1(x_1)Q_2(x_1, x_2)}{\eta_1 Q_2(x_2)} \times \frac{\eta_0(x_0)Q_1(x_0, x_1)}{\eta_0 Q_1(x_1)}$$

$$:= \eta_n(x_n) \times \quad Q_{n,\eta_{n-1}}^\star(x_n, x_{n-1}) \quad \cdots \quad Q_{2,\eta_1}^\star(x_2, x_1) \times Q_{1,\eta_0}^\star(x_1, x_0)$$

$$(1.18)$$

with the time reversal Markov matrices $Q_{n,\eta_{n-1}}^\star(x_n, x_{n-1})$ defined below

$$Q_{n,\eta_{n-1}}^\star(x_n, x_{n-1}) = \frac{\eta_{n-1}(x_{n-1})Q_n(x_{n-1}, x_n)}{\eta_{n-1}Q_n(x_n)}$$

Mimicking formula (1.18) an alternative particle approximation of the measures $\mathbb{Q}_n$ by the following estimate

$$\mathbb{Q}_n^N(x_0, \ldots, x_n) \quad = \quad \eta_n^N(x_n) \times Q_{n,\eta_{n-1}^N}^\star(x_n, x_{n-1}) \cdots Q_{2,\eta_1^N}^\star(x_2, x_1) \times Q_{1,\eta_0^N}^\star(x_1, x_0)$$

$$\longrightarrow_{N \uparrow \infty} \mathbb{Q}_n(x_0, \ldots, x_n) \qquad (1.19)$$

with the time reversal random matrices $Q_{n,\eta_{n-1}^N}^\star(x_n, x_{n-1})$ defined by

$$Q_{n,\eta_{n-1}^N}^\star(x_n, x_{n-1}) = \frac{\eta_{n-1}^N(x_{n-1})Q_n(x_{n-1}, x_n)}{\eta_{n-1}^N Q_n(x_n)} = \sum_{i=1}^{N} \frac{Q_n(\xi_{n-1}^i, x_n)}{\sum_{j=1}^{N} Q_n(\xi_{n-1}^j, x_n)} \mathbb{1}_{\xi_{n-1}^i}(x_{n-1})$$

Once again, for any function f_n on the product space E_n, we have the following unbiased property

$$\mathbb{E}\left(\overline{\mathscr{Z}}_n^N \, \mathbb{Q}_n^N(f_n)\right) = \mathbb{E}\left(\overline{\mathscr{Z}}_n^N \, f_n(\zeta_n)\right) = \mathbb{Q}_n(f_n)$$

where $\zeta_n := (\zeta_{n,n}, \zeta_{n-1,n}, \ldots, \zeta_{1,n}, \zeta_{0,n})$ stands for a backward Markov chain (as well as a complete ancestral line) with distribution $\mathbb{Q}_n^N$. In other words, if we set

$$x_n := (x_{0,n}, x_{1,n}, \ldots, x_{n-1,n}, x_{n,n}) \in E_n \quad \text{with} \quad \overleftarrow{x}_n = (x_{n,n}, x_{n-1,n}, \ldots, x_{1,n}, x_{0,n})$$

then we have that

$$\mathbb{E}\left(\overline{\mathscr{Z}}_n^N \mid \zeta_n = \overleftarrow{x}_n\right) \times \mathrm{Proba}(\zeta_n = \overleftarrow{x}_n) = \mathbb{Q}_n(x_n)$$

In addition, the measures $\mathbb{A}_n^N$ defined by

$$\mathbb{A}_n^N(F_n) := \mathbb{E}\left(F_n(\Xi_n, \zeta_n) \, \overline{\mathscr{Z}}_n^N\right)$$

for any function F_n on the product space $\left\{\left[\prod_{0 \leq p \leq n} E_p^N\right] \times E_n\right\}$, is a probability measure whose ζ_n-marginals on E_n coincide with the measure $\mathbb{Q}_n$, for any $1 \leq i \leq N$.

1.4.5 Particle Derivation and Conditioning Principles

1.4.5.1 Particle Derivation Models

Besides the fact that the computation of $\mathbb{Q}_n$ and $\mathscr{Z}_n$ comes from a specific estimation problem, in some application areas, the distribution $\eta_0^{(\theta)}$, as well as the matrices $Q_n^{(\theta)}(x, y)$ also depend on some parameter $\theta \in \mathbb{R}^d$, and we want to estimate the gradient $\nabla\mathbb{Q}_n^{(\theta)}$ and $\nabla\mathscr{Z}_n^{(\theta)}$ of the corresponding functions $\theta \mapsto \mathbb{Q}_n^{(\theta)}$ and $\theta \mapsto \mathscr{Z}_n^{(\theta)}$. The computation of these quantities is again connected to the integral type computation w.r.t. the measure $\mathbb{Q}_n^{(\theta)}$, with the following easy to check formulae

$$\nabla\log\mathscr{Z}_n^{(\theta)} = \mathbb{Q}_n^{(\theta)}(\Lambda_n^{(\theta)}) \quad \text{and} \quad \nabla\log\mathbb{Q}_n^{(\theta)} = \Lambda_n^{(\theta)} - \mathbb{Q}_n^{(\theta)}(\Lambda_n^{(\theta)}) \tag{1.20}$$

with the gradient

$$\Lambda_n^{(\theta)} := \nabla\mathbb{L}_n^{(\theta)} \text{ of the additive functional } \mathbb{L}_n^{(\theta)}(x_0, \ldots, x_n) := \sum_{p=0}^{n} \log Q_p^{(\theta)}(x_{p-1}, x_p)$$

In the above display, we have used the convention $Q_0(x_{-1}, x_0) = \eta_0(x_0)$, for $p = 0$. We also have the correlation representation of the Hessian functions

$$\nabla^2 \log \mathscr{Z}_n^{(\theta)} = \mathbb{Q}_n^{(\theta)} \left[\left(\Lambda_n^{(\theta)} - \mathbb{Q}_n^{(\theta)}(\Lambda_n^{(\theta)}) \right)' \left(\Lambda_n^{(\theta)} - \mathbb{Q}_n^{(\theta)}(\Lambda_n^{(\theta)}) \right) \right] \qquad (1.21)$$

and for any f_n on E_n

$$\begin{aligned}
\nabla^2 \mathbb{Q}_n^{(\theta)}(f_n) &= \mathbb{Q}_n^{(\theta)} \left[f_n \left(\Lambda_n^{(\theta)} - \mathbb{Q}_n^{(\theta)}(\Lambda_n^{(\theta)}) \right)' \left(\Lambda_n^{(\theta)} - \mathbb{Q}_n^{(\theta)}(\Lambda_n^{(\theta)}) \right) \right] \\
&\quad - \mathbb{Q}_n^{(\theta)} \left[f_n \left(\nabla^2 \mathbb{L}_n^{(\theta)} - \mathbb{Q}_n^{(\theta)}(\nabla^2 \mathbb{L}_n^{(\theta)}) \right) \right]
\end{aligned} \qquad (1.22)$$

The physical interpretations or the engineering meaning of these rather abstract mathematical objects depend on the application domain they are thought for.

Next, we design particle approximations of these derivative models. We denote by $\eta_n^{(\theta,N)}$ the occupation measures associated with a genetic particle model with mutation transitions and fitness potential function defined by

$$M_n^{(\theta)}(x,y) := Q_n^{(\theta)}(x,y) / \sum_z Q_n^{(\theta)}(x,z) \quad \text{and} \quad G_n^{\theta}(x) := \sum_z Q_{n+1}^{(\theta)}(x,z)$$

We also denote by $\mathbb{Q}_n^{(\theta,N)}$ the random measures on path space defined by

$$\begin{aligned}
\mathbb{Q}_n^{(\theta,N)}(x_0,\dots,x_n) &:= \eta_n^{(\theta,N)}(x_n) \times Q^\star_{n,\eta_{n-1}^{(\theta,N)}}(x_n,x_{n-1}) \cdots Q^\star_{2,\eta_1^{(\theta,N)}}(x_2,x_1) \\
&\quad \times Q^\star_{1,\eta_0^{(\theta,N)}}(x_1,x_0)
\end{aligned} \qquad (1.23)$$

and the corresponding particle normalizing constants

$$\mathscr{Z}_n^{(\theta,N)} := \prod_{0 \leq p < n} \eta_p^{(\theta,N)} \left(G_p^{\theta} \right)$$

Mimicking the derivation formulae (1.20) we define the particle derivation of the logarithm of the normalizing constants by

$$\nabla_N \log \mathscr{Z}_n^{(\theta)} := \mathbb{Q}_n^{(\theta,N)}(\Lambda_n^{(\theta)}) \longrightarrow_{N \uparrow \infty} \nabla \log \mathscr{Z}_n^{(\theta)}$$

In the same vein, the particle derivation of the measure $\mathbb{Q}_n^{(\theta)}$ is defined by the following correlation formulae

$$\nabla_N \mathbb{Q}_n^{(\theta)}(f_n) := \mathbb{Q}^{(\theta,N)} \left(f_n \left[\Lambda_n^{(\theta)} - \mathbb{Q}_n^{(N,\theta)}(\Lambda_n^{(\theta)}) \right] \right) \longrightarrow_{N \uparrow \infty} \nabla \mathbb{Q}_n^{(\theta)}(f_n)$$

for any function f_n on E_n with the additive functional

$$\Lambda_n^{(\theta)}(x_0,\dots,x_n) := \sum_{p=0}^n \nabla \log Q_p^{(\theta)}(x_{p-1},x_p)$$

and the convention $Q_0(x_{-1},x_0) = \eta_0(x_0)$ for $p = 0$.

Analogously, we define the particle Hessian functions $\nabla_N^2 \log \mathscr{Z}_n^{(\theta)}$ and $\nabla_N^2 \mathbb{Q}_n^{(\theta)}(f_n)$ replacing in (1.21) and (1.22) the measures $\mathbb{Q}_n^{(\theta)}$ by $\mathbb{Q}_n^{(\theta,N)}$.

1.4.5.2 Particle Conditioning Models

This section is concerned with some conditional distributions of the measure $\mathbb{Q}_n$ and their particle approximations. First, we observe that

$$\mathbb{Q}_p^{(n)}(x_0,\ldots,x_p) := \sum_{x_{p+1},\ldots,x_n} \mathbb{Q}_n(x_0,\ldots,x_p,x_{p+1},\ldots,x_n)$$

$$= \eta_{p|n}(x_p)\, Q_{p,\eta_{p-1}}^{\star}(x_p,x_{p-1}) \cdots Q_{1,\eta_0}^{\star}(x_1,x_0) \qquad (1.24)$$

with the p-th marginal $\eta_{p|n}$ of the measure $\mathbb{Q}_n$ defined by the matrix formula

$$\eta_{p|n} := \eta_n Q_{n,\eta_{n-1}}^{\star} \cdots Q_{p+1,\eta_p}^{\star}$$

This clearly implies that

$$\mathbb{Q}_{n|p}((x_{p+1},\ldots,x_n) \mid (x_0,\ldots,x_p))$$

$$:= \mathbb{Q}_n(x_0,\ldots,x_p,x_{p+1},\ldots,x_n)/\mathbb{Q}_p^{(n)}(x_0,\ldots,x_p)$$

$$= \frac{1}{\eta_{p|n}(x_p)}\, \eta_n(x_n)Q_{n,\eta_{n-1}}^{\star}(x_n,x_{n-1}) \cdots Q_{p+1,\eta_p}^{\star}(x_{p+1},x_p)$$

The distributions $\eta_{p|n}(x_p)$ and $\mathbb{Q}_{n|p}((x_{p+1},\ldots,x_n) \mid (x_0,\ldots,x_p)) = \mathbb{Q}_{n|p}((x_{p+1},\ldots, x_n) \mid x_p)$ can be approximated using the complete ancestral tree and replacing in the above formulae the measures η_p by their particle approximations η_p^N, with $0 \le p \le n$:

$$\eta_{p|n}^N := \eta_n^N Q_{n,\eta_{n-1}^N}^{\star} \cdots Q_{p+1,\eta_p^N}^{\star} \qquad (1.25)$$

and

$$\mathbb{Q}_{n|p}^N((x_{p+1},\ldots,x_n) \mid x_p) = \frac{1}{\eta_{p|n}^N(x_p)}\, \eta_n^N(x_n)Q_{n,\eta_{n-1}^N}^{\star}(x_n,x_{n-1}) \cdots Q_{p+1,\eta_p^N}^{\star}(x_{p+1},x_p)$$

We end this section with a probabilistic interpretation of these mathematical objects. First, using (1.9) we find the following decomposition

$$\mathbb{Q}_n(x_0,\ldots,x_n) = \frac{1}{\mathscr{Z}_n} \left\{ \prod_{0 \le p < n} G_p(x_p) \right\} \mathbb{P}_n(x_0,\ldots,x_n)$$

with the distribution $\mathbb{P}_n(x_0,\ldots,x_n)$ of the Markov chain sequence with initial condition $\text{Law}(X_0) = \eta_0$ and Markov transitions M_n

$$\mathbb{P}_n(x_0,\ldots,x_n) = \text{Proba}((X_0,\ldots,X_n) = (x_0,\ldots,x_n))$$
$$= \eta_0(x_0)M_1(x_0,x_1)M_2(x_1,x_2)\cdots M_n(x_{n-1},x_n)$$

In this interpretation, we have

$$Q_p^{(n)}(x_0,\ldots,x_p) := \frac{\mathscr{Z}_{n|p}(x_p)}{\mathscr{Z}_n} \left\{\prod_{0\leq q<p} G_q(x_q)\right\} \mathbb{P}_p(x_0,\ldots,x_p)$$

$$= \frac{\mathscr{Z}_{n|p}(x_p)}{\mathscr{Z}_{n|p}} \mathbb{Q}_p(x_0,\ldots,x_p)$$

$$= \frac{\mathscr{Z}_{n|p}(x_p)}{\mathscr{Z}_{n|p}} \eta_p(x_p) \left[Q^\star_{p,\eta_{p-1}}(x_p,x_{p-1})\cdots Q^\star_{1,\eta_0}(x_1,x_0)\right]$$

with $\mathscr{Z}_{n|p}$ and $\mathscr{Z}_{n|p}(x_p)$ defined by

$$\mathscr{Z}_{n|p} = \mathscr{Z}_n/\mathscr{Z}_p = \prod_{p\leq q<n} \eta_q(G_q) \ \text{ and } \ \mathscr{Z}_{n|p}(x_p) := \mathbb{E}\left(\left\{\prod_{p\leq q<n} G_q(x_q)\right\} | X_p = x_p\right)$$

Using (1.24), this implies that

$$\eta_{p|n}(x_p) = \frac{1}{\mathscr{Z}_{n|p}} \, \mathscr{Z}_{n|p}(x_p) \, \eta_p(x_p) \tag{1.26}$$

and

$$\mathbb{Q}_{n|p}((x_{p+1},\ldots,x_n) \mid x_p) = \frac{1}{\mathscr{Z}_{n|p}(x_p)} \left\{\prod_{p\leq q<n} G_q(x_q)\right\} \mathbb{P}_{n|p}((x_{p+1},\ldots,x_n)|x_p)$$

with the conditional distribution

$$\mathbb{P}_{n|p}((x_{p+1},\ldots,x_n)|x_p) = \text{Proba}((X_{p+1},\ldots,X_n) = (x_{p+1},\ldots,x_n)|X_p = x_p)$$
$$= M_{p+1}(x_p,x_{p+1})\cdots M_n(x_{n-1},x_n)$$

Combining (1.25) and (1.26) we also have the following particle approximations

$$\eta_p^N(x) \, \mathscr{Z}_{n|p}^N(x) := \left\{\prod_{p\leq q<n} \eta_q^N(G_q)\right\} \eta_{p|n}^N(x) \simeq_{N\uparrow\infty} \eta_p(x) \mathscr{Z}_{n|p}(x)$$

1.5   Some Application Domains

1.5.1   Particle Absorption Models

1.5.1.1   Random Walks Confined in a Set

We consider a symmetric random walk X_n on the integers $\mathbb{Z}$ starting at the origin $X_0 = 0$. More formally, we take independent random variables U_n, where $\mathbb{P}(U_n = 1) = \mathbb{P}(U_n = -1) = 1/2$ and we set $X_n = X_0 + \sum_{1 \leq p \leq n} U_p$. We fix $A = \{-a+1, -a+2, ..., a-1\}$, with $a \in \mathbb{N}$. We want to compute the conditional distributions

$$\text{Law}\left((X_0, \ldots, X_n) \mid \forall 0 \leq p \leq n, \ X_p \in A\right) \tag{1.27}$$

as well as the quantities

$$\mathscr{Z}_n := \mathbb{P}\left(\forall 0 \leq p < n, \ X_p \in A\right)$$

This problem can be solved by simulation using the following particle algorithm. We start with N particles at the origin denoted by $\xi_0^i = 0$, with $i = 1, \ldots, N$. Each of them evolve $\xi_0^i \rightsquigarrow \xi_1^i$ according to one transition of the random walk; more formally, we sample N independent copies $(U_1^i)_{1 \leq i \leq N}$ of the random variables U_1, and we set $\xi_1^i = \xi_0^i + U_1^i$. We denote

$$\eta_1^N(\mathbb{1}_A) = \frac{1}{N} \sum_{1 \leq i \leq N} \mathbb{1}_A(\xi_1^i) = \frac{1}{N} \text{Card}\left\{1 \leq i \leq N \ : \ \xi_1^i \in A\right\}$$

the proportion of points ξ_1^i in the set A. We define from the sample population $\left(\xi_1^i\right)_{1 \leq i \leq N}$ a new population of N individuals $\left(\widehat{\xi}_1^i\right)_{1 \leq i \leq N}$ as follows. For each $i = 1, \ldots, N$, we perform the following operation: if $\xi_1^i \in A$, we set $\widehat{\xi}_1^i = \xi_1^i$. If $\xi_1^i \notin A$, we pick randomly an individual $\widetilde{\xi}_1^i$ among those ξ_1^j in the set A and we set $\widehat{\xi}_1^i = \widetilde{\xi}_1^i$. In other words, individuals within A do not move, while the individuals outside A are replaced by a randomly chosen individual among those in the set A. It may happen that all individuals ξ_1^i are outside of the set A. In this case, the algorithm stops and we set $\tau^N = 1$ to report the time of this event. If the algorithm has not stopped, we have a new configuration $\left(\widehat{\xi}_1^i\right)_{1 \leq i \leq N}$ of N individuals in the set A. We evolve $\widehat{\xi}_1^i \rightsquigarrow \xi_2^i$ according to one transition of the random walk; that is we sample N independent copies $(U_2^i)_{1 \leq i \leq N}$ of the random variables U_2, we set $\xi_2^i = \widehat{\xi}_1^i + U_2^i$ and we define

$$\eta_2^N(\mathbb{1}_A) = \frac{1}{N} \sum_{1 \leq i \leq N} \mathbb{1}_A(\xi_2^i) = \frac{1}{N} \text{Card}\left\{1 \leq i \leq N \ : \ \xi_2^i \in A\right\}.$$

As before, we define from the sample population $\left(\xi_2^i\right)_{1 \leq i \leq N}$ a new population of N individuals $\left(\widehat{\xi}_2^i\right)_{1 \leq i \leq N}$: individuals within A do not move, while the individuals outside the desired set are replaced by a randomly chosen individual among those in the set A. If all individuals ξ_2^i fall are outside of the set A, we set $\tau^N = 2$. Iterating this stochastic process, for every time n $(< \tau^N)$, we define a sequence of genetic type populations

$$\xi_n := \left(\xi_n^i\right)_{1 \le i \le N} \in \mathbb{Z}^N \xrightarrow{\text{selection}} \widehat{\xi}_n := \left(\widehat{\xi}_n^i\right)_{1 \le i \le N} \in \mathbb{Z}^N \xrightarrow{\text{mutation}} \xi_{n+1} \in \mathbb{Z}^N \tag{1.28}$$

This stochastic algorithm can be interpreted as a genetic type model with mutation transitions given by the one of a symmetric random walk and an acceptance-rejection selection type transition associated with the potential indicator type function $\mathbb{1}_A$. Several estimates can be extracted from this interacting sampling algorithm.

First, we mention that the stopping time τ^N tends to infinity as the size of the population $N \to \infty$. More precisely, the probability that the algorithm stops at a given time n tends to zero exponentially fast, as N tends to infinity. More interestingly, the product of the proportions of surviving particles at each time step

$$\mathscr{Z}_n^N := \prod_{0 \le p < n} \eta_p^N(\mathbb{1}_A)$$

is asymptotically a consistent estimate of the quantity $P_n(A)$ and it is unbiased; that is we have

$$\lim_{N \to \infty} \mathscr{Z}_n^N = \mathscr{Z}_n \quad \text{and} \quad \mathbb{E}\left(\mathscr{Z}_n^N\right) = \mathscr{Z}_n \tag{1.29}$$

The convergence on the l.h.s. is an almost sure asymptotic convergence. It can be made precise by non asymptotic estimates including non asymptotic variance estimates and more refined exponential type deviations. If we interpret the selection transition as a birth and death process, then the important notion of the ancestral line of a current individual arises. More precisely, when a particle $\widehat{\xi}_{n-1}^i \longrightarrow \xi_n^i$ evolves to a new location ξ_n^i, we can interpret $\widehat{\xi}_{n-1}^i$ as the parent of ξ_n^i. Looking backwards in time and recalling that the particle $\widehat{\xi}_{n-1}^i$ has selected a site ξ_{n-1}^j in the configuration at time $(n-1)$, we can interpret this site ξ_{n-1}^j as the parent of $\widehat{\xi}_{n-1}^i$ and therefore as the ancestor $\xi_{n-1,n}^i$ at level $(n-1)$ of ξ_n^i. Running back in time we can construct the whole ancestral line

$$\xi_{0,n}^i \longleftarrow \xi_{1,n}^i \longleftarrow \cdots \longleftarrow \xi_{n-1,n}^i \longleftarrow \xi_{n,n}^i = \xi_n^i \tag{1.30}$$

of each current individual. The occupation measures of the corresponding N-genealogical tree model converge as $N \to \infty$ to the conditional distribution (1.27). In a sense to be given, for any function f on the set $\mathbb{Z}^{n+1}$, we have the convergence, as $N \to \infty$,

$$\lim_{N \to \infty} \frac{1}{N} \sum_{i=1}^{N} f(\xi_{0,n}^i, \xi_{1,n}^i, \ldots, \xi_{n,n}^i) \, \mathbb{1}_{\tau^N > n} = \mathbb{E}\left(f(X_0, \ldots, X_n) \mid \forall 0 \le p < n, \, X_p \in A\right) \tag{1.31}$$

This convergence result can be refined in various directions. For instance, we can prove that the ancestral lines are "almost" independent with a common distribution given by the limiting conditional distribution. This is often called the propagation of chaos property in applied probability. It refers to the fact that the initial population consists of independent and identically distributed random variables and that this property "propagates" approximately despite the introduction of interactions. Many other results can be derived including the fluctuations and the exponential concentration of the occupation measures of the genealogical tree around the limiting conditional distribution.

Besides the fact that the particle model approximate the (rare event) probabilities (1.29) and the conditional distributions (1.31) in path spaces, it also contains some information about the top of the spectrum of the matrix Q defined below

$$\forall (x,y) \in \{-a, -a+1, ..., a-1, a\} \qquad Q(x,y) := G(x)\, M(x,y)$$

with

$$G(x) := \mathbb{1}_A(x) \quad \text{and} \quad M(x,y) = \frac{1}{2}\, \mathbb{1}_{x-1}(y) + \frac{1}{2}\, \mathbb{1}_{x+1}(y)$$

Indeed, if we consider λ to be the top eigenvalue of Q and we denote by h the corresponding eigenvector s.t. $\sum_x h(x) = 1$, then we have

$$\lim_{N \to \infty} \lim_{n \to \infty} \frac{1}{n} \sum_{0 \le p \le n} \log \eta_p^N(\mathbb{1}_A) = \log \lambda$$

In addition, the value $h(x)$ coincides with the long time proportion of visits of the algorithm to the state x. In other words, $h(x)$ can be interpreted as the limiting distribution of the individuals within the set A; that is

$$\lim_{N,n \to \infty} \frac{1}{n} \sum_{0 \le p \le n} \frac{1}{N} \sum_{1 \le i \le N} \mathbb{1}_x(\widehat{\xi}_n^i)\, \mathbb{1}_{\tau^N > n} = h(x) = \lim_{N,n \to \infty} \frac{1}{N} \sum_{1 \le i \le N} \mathbb{1}_x(\widehat{\xi}_n^i)\, \mathbb{1}_{\tau^N > n}$$

The particle approximation model discussed above is far from unique. Many other interacting sampling strategies can be introduced by a simple change of probability measure. For instance, we can replace the mutation or the free evolution of the individuals in the previous algorithm by local moves restricted to the desired set A. These mutation type transitions $\widehat{\xi}_{n-1} \rightsquigarrow \xi_n$ can also be seen as transitions of a simple random walk on $\mathbb{Z}$ reflected at the boundaries of the set A. By construction all the individuals ξ_n^i at any time horizon n and for any index $i = 1, ..., N$ are in the desired set A.

The corresponding selection transition $\xi_n \rightsquigarrow \widehat{\xi}_n$ is now defined as follows: each individual $\xi_n^i = x$ on the boundary $x \in \partial A = \{-a+1, (a-1)\}$ of the set A has a probability $G(x) := 1/2$ to stay in A, while the other individuals ξ_n^i (which are in the set A) have a probability $G(x) = 1$ to stay in A. The population $\widehat{\xi}_n$ is now defined as follows. For every index i, with a probability $G(\xi_n^i)$, we set $\widehat{\xi}_n^i = \xi_n^i$, otherwise we replace ξ_n^i by a new individual $\widehat{\xi}_n^i = \xi_n^j$ randomly chosen in the whole population with a probability proportional to $G(\xi_n^j)$. If we now write $\eta_n^N(G) = \frac{1}{N} \sum_{1 \leq i \leq N} G(\xi_n^i)$, all the previous particle approximation results (corresponding to $G(x) = \mathbb{1}_A(x)$) remain valid for this new particle algorithm.

1.5.1.2   Feynman-Kac Model

The sampling techniques described in section 1.5.1.1 are far from being restricted to random walks models confined to a set. These strategies apply to a variety of application areas including computational physics, nonlinear filtering, biology, as well as rare event analysis. From the pure mathematical point of view, they correspond to interacting particle approximation models of Feynman-Kac measures in path spaces.

To introduce these models, we recall that the conditional distributions discussed in (1.27) can be represented in terms of the distributions of the free path evolution

$$\mathbb{P}_n(x_0, \ldots, x_n) = \mathrm{Proba}\left((X_0, \ldots, X_n) = (x_0, \ldots, x_n)\right)$$
$$= \mathbb{1}_0(x_0)\, M_1(x_0, x_1) \, \ldots \, M_n(x_{n-1}, x_n) \qquad (1.32)$$

of the simple random walk starting at the origin with elementary transitions given by the matrix $M_n := (M_n(x, y))_{x,y \in \mathbb{Z}}$ with entries given by

$$M_n(x, y) := \frac{1}{2}\, \mathbb{1}_{x-1}(y) + \frac{1}{2}\, \mathbb{1}_{x+1}(y)$$

More formally, if we set

$$\mathbb{Q}_n(x_0, \ldots, x_n) := \mathrm{Proba}\left((X_0, \ldots, X_n) = (x_0, \ldots, x_n) \mid \forall 0 \leq p < n, \, X_p \in A\right)$$

then we have

$$\mathbb{Q}_n(x_0, \ldots, x_n) = \frac{1}{\mathscr{Z}_n} \left\{ \prod_{0 \leq p < n} G_p(x_p) \right\} \mathbb{P}_n(x_0, \ldots, x_n) \qquad (1.33)$$

with the indicator potential functions $G_n(x) = \mathbb{1}_A(x)$ and $\mathbb{P}_n(x_0, \ldots, x_n)$ being the distribution of a free path of length n of the symmetric random walk. In (1.33), $\mathscr{Z}_n$ is the normalizing constant given by

$$\mathscr{Z}_n = \mathbb{P}\left(\forall 0 \leq p < n, \, X_p \in A\right) = \mathbb{E}\left(\prod_{0 \leq p < n} G_p(X_p) \right)$$

These path integration type models are called Feynman-Kac measures in reference to Feynman path integral formulation of quantum mechanics where the classical notion of a single deterministic trajectory for a system is replaced by a sum over all possible trajectories weighted by the contributions of all the histories in configuration space.

1.5.1.3   A Killed Markov Chain

The Feynman-Kac measures presented in (1.33) can be regarded as the distribution of the paths of a Markov particle evolving using the Markov transitions M_n in an environment with absorbing obstacles related to potential functions G_n, and starting with some initial distribution $\mathrm{Law}(X_0) = \eta_0$ with $\eta_0(x_0) = \mathbb{1}_0(x_0)$ in (1.32). To be more precise, we consider an auxiliary coffin or cemetery state c and we set $E_c = E \cup \{c\}$. We define an E_c-valued Markov chain X_n^c with two separate killing/exploration transitions:

$$X_n^c \xrightarrow{\text{killing}} \widehat{X}_n^c \xrightarrow{\text{exploration}} X_{n+1}^c \tag{1.34}$$

This killing/exploration mechanism are defined as follows:

- **Killing:** If $X_n^c = c$, we set $\widehat{X}_n^c = c$. Otherwise the particle X_n^c is still alive. In this case, with a probability $G_n(X_n^c)$, it remains in the same site so that $\widehat{X}_n^c = X_n^c$, and with a probability $1 - G_n(X_n^c)$ it is killed and we set $\widehat{X}_n^c = c$.
- **Exploration:** Once a particle has been killed, it can not be brought back to life so if $\widehat{X}_n^c = c$ then we set $\widehat{X}_p^c = X_p = c$ for any $p > n$. Otherwise, the particle $\widehat{X}_n^c \in E$ evolves to a new location $X_{n+1}^c = x$ in E randomly chosen according to the distribution $M_{n+1}(X_n^c, x)$.

In this physical interpretation, the measure $\mathbb{Q}_n$ represent the conditional distributions of the paths of a non absorbed Markov particle. To see this claim, we denote by T the time at which the particle has been killed

$$T = \inf\{n \geq 0 \,;\, \widehat{X}_n^c = c\}$$

By construction, we have

$$\mathrm{Proba}(T > n - 1)$$

$$= \mathrm{Proba}(\widehat{X}_0^c \in E, \ldots, \widehat{X}_{n-1}^c \in E)$$
$$= \int_{E_n} \eta_0(dx_0)\, G_0(x_0)\, M_1(x_0, dx_1) \ldots M_{n-1}(x_{n-2}, dx_{n-1}) G_{n-1}(x_{n-1})$$

$$= \mathbb{E}\left(\prod_{p=0}^{n-1} G_p(X_p)\right)$$

This also shows that the normalizing constants $\mathscr{Z}_n$ represent respectively the probability for the particle to be alive at time $n - 1$. In other words, we have that

$$\mathscr{Z}_n = \mathrm{Proba}(T > n - 1)$$

Similar arguments yield that the distribution of a particle conditional of being alive at time $n - 1$ is

$$\mathbb{Q}_n(x_0, \ldots, x_n) = \mathrm{Proba}\left((X_0^c, \ldots, X_n^c) = (x_0, \ldots, x_n) \mid T > n - 1\right)$$

Using (1.18) we also have the following backward representation of $\mathbb{Q}_n$

$$\mathbb{Q}_n(x_0,\ldots,x_n) = \eta_n(x_n) \times Q^\star_{n,\eta_{n-1}}(x_n,x_{n-1}) \cdots Q^\star_{2,\eta_1}(x_2,x_1) \times Q^\star_{1,\eta_0}(x_1,x_0)$$

(1.35)

with the time reversal Markov matrices $Q^\star_{n,\eta_{n-1}}(x_n,x_{n-1})$ defined below

$$Q^\star_{n,\eta_{n-1}}(x_n,x_{n-1}) = \frac{\eta_{n-1}(x_{n-1})Q_n(x_{n-1},x_n)}{\eta_{n-1}Q_n(x_n)}$$

1.5.1.4   A Particle Sampling Model

The particle sampling technique of any distribution $\mathbb{Q}_n$ associated with some Markov transition M_n and some sequence of $[0,1]$-valued potential function G_n on some (countable) state space E is defined as before in terms of a genetic type algorithm with M_n-mutations and G_n-selection type transitions. More precisely, at every time step n, we sample the mutation-selection transitions as follows: during the mutation step, every individual performs a local random move according to the Markov transition M_n. During the selection step, every individual evaluates its potential value $G_n(\xi^i_n)$, with $1 \leq i \leq N$. For every index i, with a probability $G_n(\xi^i_n)$, we set $\widehat{\xi}^i_n = \xi^i_n$, otherwise we replace ξ^i_n be a fresh new individual $\widehat{\xi}^i_n = \xi^j_n$ randomly chosen from the population with a probability proportional to $G_n(\xi^j_n)$.

As in the confinement model (discussed in the previous section), it may happen that all individuals ξ^i_n have a null potential value $G_n(\xi^i_n) = 0$, at some time period n. In this case, the algorithm stops and we set $\tau^N = n$ to report the time of this event. Under some rather weak regularity properties, we also mention that the stopping time τ^N tends to infinity as the size of the population $N \to \infty$.

For any time horizon n and any function f on the set E_n, we have

$$\lim_{N\to\infty} \frac{1}{N} \sum_{i=1}^N f(\xi^i_{0,n},\xi^i_{1,n},\ldots,\xi^i_{n,n}) \, \mathbb{1}_{\tau^N>n} = \sum_{x_0,\ldots,x_n} f(x_0,\ldots,x_n) \, \mathbb{Q}_n(x_0,\ldots,x_n) \quad (1.36)$$

Furthermore, the unbiased approximations of the normalizing constants $\mathscr{Z}_n$ are given by

$$\mathscr{Z}_n^N := \prod_{0 \leq p < n} \eta^N_p(G_p) \quad \text{with} \quad \forall n \in \mathbb{N} \quad \eta^N_n(G_n) := \frac{1}{N} \sum_{1 \leq i \leq N} G_n(\xi^i_n) \quad (1.37)$$

In addition, mimicking formula (1.41), an alternative particle approximation of the measures $\mathbb{Q}_n$ is defined, replacing the measures η_n by their particle approximations

$$\mathbb{Q}_n^N(x_0,\ldots,x_n) = \eta_n^N(x_n) \times Q^\star_{n,\eta^N_{n-1}}(x_n,x_{n-1}) \cdots Q^\star_{2,\eta^N_1}(x_2,x_1) \times Q^\star_{1,\eta^N_0}(x_1,x_0)$$

$$\to_{N\uparrow\infty} \mathbb{Q}_n(x_0,\ldots,x_n)$$

with the time reversal random matrices $Q^\star_{n,\eta^N_{n-1}}(x_n, x_{n-1})$ defined below

$$Q^\star_{n,\eta^N_{n-1}}(x_n, x_{n-1}) = \frac{\eta^N_{n-1}(x_{n-1})Q_n(x_{n-1}, x_n)}{\eta^N_{n-1}Q_n(x_n)} = \sum_{i=1}^N \frac{Q_n(\xi^i_{n-1}, x_n)}{\sum_{j=1}^N Q_n(\xi^j_{n-1}, x_n)} \, \mathbb{1}_{\xi^i_{n-1}}(x_{n-1})$$

For time homogeneous models $(G_n, M_n) = (G, M)$ associated with a reversible matrix M w.r.t. to some measure μ on E, i.e. $\mu(x)M(x,y) = \mu(y)M(y,x)$, the corresponding particle model also contains information about the top of the spectrum of the matrix Q defined through

$$\forall (x,y) \in E \qquad Q(x,y) := G(x)\,M(x,y)$$

More precisely, if we consider λ to be the top eigenvalue of Q in $\mathbb{L}_2(\mu)$ and we denote by h the corresponding eigenvector s.t. $\sum_x \mu(x)h(x) = 1$, then we have

$$\lim_{N\to\infty}\lim_{n\to\infty}\frac{1}{n}\sum_{0\le p\le n}\log \eta^N_p(G) = \log \lambda$$

as well as

$$\lim_{N,n\to\infty}\frac{1}{n}\sum_{0\le p\le n}\frac{1}{N}\sum_{1\le i\le N}\mathbb{1}_x(\widehat{\xi}^i_n)\,\mathbb{1}_{\tau^N>n} = \mu(x)h(x) = \lim_{N,n\to\infty}\frac{1}{N}\sum_{1\le i\le N}\mathbb{1}_x(\widehat{\xi}^i_n)\,\mathbb{1}_{\tau^N>n}$$

For further details on this subject, we refer the reader to [38, 39, 54] and references therein.

1.5.2 Signal Processing and Bayesian Inference

1.5.2.1 Nonlinear Filtering Problems

We discuss here the application of these particle model to filtering problems. Suppose that at every time step the state of the Markov chain X_n is partially observed according to the following schematic picture

$$\begin{array}{ccccc} X_0 & \longrightarrow & X_1 & \longrightarrow & X_2 & \longrightarrow & \ldots \\ \downarrow & & \downarrow & & \downarrow & & \\ Y_0 & & Y_1 & & Y_2 & & \ldots \end{array}$$

with some random variables Y_n whose values only depend on the current state of the chain

$$\mathrm{Proba}\,(Y_n = y_n \mid X_n = x_n) := G(x_n, y_n) \tag{1.38}$$

We consider the following pair of events

$$A_n(x) := \{(X_0, \ldots, X_n) = (x_0, \ldots, x_n)\} \text{ and } B_{n-1}(y) := \{(Y_0, \ldots, Y_{n-1}) = (y_0, \ldots, y_{n-1})\}$$

The filtering problem consists of computing the conditional distributions of the state variables $A_n(x)$ given the observations $B_n(y)$. By construction, given $A_n(x)$, the random variables Y_n are independent and identically distributed with a distribution given by

$$\mathrm{Proba}\,(B_{n-1}(y)\,|A_n(x)) = \prod_{0 \le p < n} G(x_p, y_p)$$

By direct application of Bayes' rule we have the following formula

$$\mathrm{Proba}\,(A_n(x) \cap B_{n-1}(y)) = \mathrm{Proba}\,(B_{n-1}(y)\,|A_n(x)) \times \mathrm{Proba}\,(A_n(x))$$

$$= \left\{ \prod_{0 \le p < n} G(x_p, y_p) \right\} \mathbb{P}_n(x_0, \ldots, x_n) \qquad (1.39)$$

with the distributions of the path sequence $(X_0, \ldots, X_n)$ given by

$$\mathbb{P}_n(x_0, \ldots, x_n) = \mathrm{Proba}\,(X_0 = x_0, \ldots, X_n = x_n)$$

from which we conclude that

$$\mathrm{Proba}\,(A_n(x)\,|\,B_{n-1}(y)) = \frac{1}{\mathscr{Z}_n(y)} \left\{ \prod_{0 \le p < n} G(x_p, y_p) \right\} \mathbb{P}_n(x_0, \ldots, x_n)$$

with the normalizing constants

$$\mathscr{Z}_n(y) := \mathrm{Proba}(B_{n-1}(y)) = \sum_{x_0, \ldots, x_n} \left\{ \prod_{0 \le p < n} G(x_p, y_p) \right\} \mathbb{P}_n(x_0, \ldots, x_n)$$

These Feynman-Kac formulae express the conditional distributions of the path sequence $(X_0, \ldots, X_n)$ as the distribution $\mathbb{P}_n(x_0, \ldots, x_n)$ of the signal paths $(X_0, \ldots, X_n) = (x_0, \ldots, x_n)$ weighted by the product of the likelihood functions $G(x_p, y_p)$ from the origin $p = 0$ up to time $p = n$.

If we fix the observation sequence $Y_n = y_n$ and set

$$G_n(x_n) := G(x_n, y_n)$$

then we find that these measures have exactly the same form as the one presented in (1.33). We can also rewrite these conditional distributions as follows

$$\mathbb{Q}_n(x_0, \ldots, x_n) = \mathrm{Proba}\,(A_n(x)\,|\,B_{n-1}(y))$$

$$\propto \left\{ \prod_{0 \le p < n} G_p(x_p) \right\} \underbrace{\left\{ \eta_0(x_0) \prod_{1 \le p \le n} M_p(x_{p-1}, x_p) \right\}}_{\mathbb{P}_n(x_0, \ldots, x_n)}$$

$$= \eta_0(x_0) Q_1(x_0, x_1) Q_2(x_1, x_2) \ldots Q_n(x_{n-1}, x_n) \qquad (1.40)$$

with the positive matrices $Q_n(x_{n-1}, x_n)$ defined for any $n \geq 1$ by

$$Q_n(x_{n-1}, x_n) := G_{n-1}(x_{n-1}) \, M_n(x_{n-1}, x_n)$$

The corresponding particle approximations defined in section 1.5.1.4 are often referred to as particle filters in signal processing and statistics [36, 37, 52, 38, 67]. These particle algorithms can also be used to approximate the log-likelihood functions using (1.37); that is the log-likelihood

$$L_n(y) := \log \mathscr{Z}_n(y)$$

is approximated using

$$L_n^N(y) := \log \mathscr{Z}_n^N(y) = \sum_{0 \leq p < n} \log \eta_p^N(G_p).$$

1.5.2.2   Smoothing Estimation Models

Smoothing problems consist of estimating some values of the signal X_p at some time p, given s series of observations $Y_q = y_q$, with $0 \leq q \leq n$, and $p \leq n$. One strategy is to estimate the whole signal path sequence $(X_0, \ldots, X_n)$ given the observations from the origin, up to the time horizon n. The conditional distributions on path space defined in section 1.5.2.1 can be estimated using three methods:

- the genealogical tree evolution of the particle filters;
- the particle backward Markovian interpretation of conditional distributions;
- the particle conditional distributions of the noise of the signal.

These three methods are described below.

- The genealogical tree evolution of the particle filters. To describe with some precision these models, let E be the finite state space of the signal, and let $E_n = E^{(n+1)}$. These N particle approximations on path spaces coincide with a simple genetic type evolution model with N path-valued particles

$$\xi_n^i := \left(\xi_{0,n}^i, \xi_{1,n}^i, \ldots, \xi_{n,n}^i \right) \quad \text{and} \quad \widehat{\xi}_n^i := \left(\widehat{\xi}_{0,n}^i, \widehat{\xi}_{1,n}^i, \ldots, \widehat{\xi}_{n,n}^i \right) \in E_n$$

During the selection stage, with a probability $G(\xi_{n,n}^i, y_n)$ every path-valued individual ξ_n^i stays in the same place $\widehat{\xi}_n^i = \xi_n^i$; otherwise, we replace ξ_n^i be a new individual $\widehat{\xi}_n^i = \xi_n^j$ randomly chosen among the individuals ξ_0^j with a probability proportional to its weight $G(\xi_{n,n}^i, y_n)$. This mechanism is intended to favor more likely signal path sequences. During the mutation transition, $\widehat{\xi}_n^i$ evolves randomly to a new path sequence

$$\xi_{n+1}^i = ((\xi_{0,n+1}^i, \ldots, \xi_{n,n+1}^i), \xi_{n+1,n+1}^i) = ((\widehat{\xi}_{0,n}^i, \ldots \ldots, \widehat{\xi}_{n,n}^i), \xi_{n+1,n+1}^i) \in E_{n+1}$$
$$= (E_n \times E)$$

If $\widehat{\xi}_n^i = x_n$, then $\xi_{n+1,n+1}^i$ is a random variable that takes the value x with the distribution $\mathrm{Proba}(X_{n+1} = x_{n+1} | X_n = x_n)$. As usual, for any function f on $E_n = E^{(n+1)}$ and any time horizon n, we have

$$\lim_{N \to \infty} \frac{1}{N} \sum_{i=1}^{N} f(\xi_{0,n}^i, \xi_{1,n}^i, \ldots, \xi_{n,n}^i) = \sum_{x_0,\ldots,x_n} \mathbb{Q}_n(x_0,\ldots,x_n) f(x_0,\ldots,x_n)$$

- Particle backward Markov models. An alternative approach is to use the backward representation (1.18) of the conditional distribution $\mathbb{Q}_n$ defined in (1.40)

$$\mathbb{Q}_n(x_0,\ldots,x_n) = \eta_n(x_n) \times Q_{n,\eta_{n-1}}^{\star}(x_n,x_{n-1}) \cdots Q_{2,\eta_1}^{\star}(x_2,x_1) \times Q_{1,\eta_0}^{\star}(x_1,x_0) \tag{1.41}$$

with the time reversal Markov matrices $Q_{n,\eta_{n-1}}^{\star}(x_n,x_{n-1})$ defined below:

$$Q_{n,\eta_{n-1}}^{\star}(x_n,x_{n-1}) = \frac{\eta_{n-1}(x_{n-1}) Q_n(x_{n-1},x_n)}{\eta_{n-1} Q_n(x_n)} = \frac{\eta_{n-1}(x_{n-1}) G_{n-1}(x_{n-1}) M_n(x_{n-1},x_n)}{\sum_x \eta_{n-1}(x) G_{n-1}(x) M_n(x,x_n)}$$

Replacing the measures η_n by their particle estimates η_n^N, we define the particle approximation of $\mathbb{Q}_n$ by setting

$$\begin{aligned}
\mathbb{Q}_n^N(x_0,\ldots,x_n) &= \eta_n^N(x_n) \times Q_{n,\eta_{n-1}^N}^{\star}(x_n,x_{n-1}) \cdots Q_{2,\eta_1^N}^{\star}(x_2,x_1) \times Q_{1,\eta_0^N}^{\star}(x_1,x_0) \\
&\to_{N \uparrow \infty} \mathbb{Q}_n(x_0,\ldots,x_n)
\end{aligned}$$

with the time reversal random matrices $Q_{n,\eta_{n-1}^N}^{\star}(x_n,x_{n-1})$ defined by

$$\begin{aligned}
Q_{n,\eta_{n-1}^N}^{\star}(x_n,x_{n-1}) &= \frac{\eta_{n-1}^N(x_{n-1}) Q_n(x_{n-1},x_n)}{\eta_{n-1}^N Q_n(x_n)} \\
&= \sum_{i=1}^{N} \frac{G_{n-1}(\xi_{n-1}^i,) M_n(\xi_{n-1}^i, x_n)}{\sum_{j=1}^{N} G_{n-1}(xi_{n-1}^j) M_n(\xi_{n-1}^j, x_n)} \mathbb{1}_{\xi_{n-1}^i}(x_{n-1})
\end{aligned}$$

- Particle approximations of the noise of the signal. We further assume that the signal process given by recursive equations on some finite state space E of the following form

$$X_n := F_n(X_{n-1}, U_n) \tag{1.42}$$

with some independent random variables U_n, and with distribution v_n independent of X_0 on the finite set $\mathscr{U}$. If we consider the following events

$$C_n(u) = \{(X_0, (U_0, \ldots, U_n)) = (x_0, (u_1, \ldots, u_n))\}$$

then we find that

$$\text{Proba}\left(C_n(u)\,|B_{n-1}(y)\right) = \frac{1}{\mathscr{Z}_n(y)} \left\{ \prod_{0 \le p < n} G(\mathscr{X}_p^{(x_0,u)}, y_p) \right\} \mathbb{P}_n(x_0,(u_1,\ldots,u_n))$$

(1.43)

where $\mathscr{X}_n^{(x_0,u)}$ stands for the solution of the discrete generation system (1.42) associated with a given realization $(u_n)_{n \ge 1}$ and some initial condition x_0. The function G is the likelihood function defined in (1.38).

In the semigroup formulation, $\mathscr{X}_n^{(x_0,u)}$ is a function of the initial state and the control sequence $(u_1,\ldots,u_n)$; that is, we have that $\mathscr{X}_n^{(x_0,u)} = \phi_n(x_0,(u_1,\ldots,u_n))$ for some function ϕ_n from $(E \times \mathscr{U}^n)$ into E. For any $n \ge 0$, we set

$$H_n(x_0,(u_1,\ldots,u_n)) := G(\phi_n(x_0,(u_1,\ldots,u_n)), y_n)$$

In this notation we have

$$\text{Proba}\left(C_n(u)\,|B_{n-1}(y)\right) = \frac{1}{\mathscr{Z}_n(y)} \left\{ \prod_{0 \le p < n} H_p(x_0,(u_1,\ldots,u_p)) \right\} \mathbb{P}_n(x_0,(u_1,\ldots,u_n))$$

As above, the N particle approximation of these probability measures on control sequences is again described by genetic evolution models with N path-valued particles

$$\xi_n^i := (\xi_{0,n}^i, \xi_{1,n}^i, \ldots, \xi_{n,n}^i)$$

$$\widehat{\xi}_n^i := (\widehat{\xi}_{0,n}^i, \widehat{\xi}_{1,n}^i, \ldots, \widehat{\xi}_{n,n}^i) \in E_n := (E \times \mathscr{U}^n)$$

During the selection stage, with a probability $H_n(\xi_n^i)$ every path-valued individual stays in the same place $\widehat{\xi}_n^i = \xi_n^i$; otherwise, we replace ξ_n^i by a new individual $\widehat{\xi}_n^i = \xi_n^j$ randomly chosen among the individuals ξ_0^j with a probability proportional to its weight $H_n(\xi_n^i)$. This mechanism is intended to favor more likely noise sequences w.r.t. the observations. During the mutation transition, to every selected signal-noise sequence $\widehat{\xi}_n^i$ we add randomly new possible values of the noise at time $(n+1)$; that is, we set

$$\xi_{n+1}^i = ((\xi_{0,n+1}^i,\ldots,\xi_{n,n+1}^i),\xi_{n+1,n+1}^i)$$

$$= ((\widehat{\xi}_{0,n}^i,\ldots\ldots,\widehat{\xi}_{n,n}^i),\xi_{n+1,n+1}^i) \in E_{n+1} = (E_n \times \mathscr{U})$$

(1.44)

where $\xi_{n+1,n+1}^i$ is a random variable with distribution ν_n on $\mathscr{U}$. Various asymptotic estimates can be derived. For instance, for any function f on $E_n = E^n$ and any time horizon n, we have

$$\lim_{N \to \infty} \frac{1}{N} \sum_{i=1}^{N} f(\xi_{0,n}^i, \xi_{1,n}^i, \ldots, \xi_{n,n}^i) = \sum_{x_0,\ldots,x_n} \mathbb{Q}_n(x_0,(u_1,\ldots,u_n))\, f(x_0,(u_1,\ldots,u_n))$$

(1.45)

In other words, the occupation measures of the genealogical tree evolution, like the one illustrated below for $(N,n) = (4,5)$

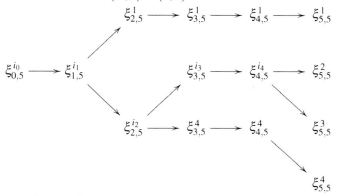

with any $i_0 \in \{1,2,3,4\}$, $i_2 \in \{2,3,4\}$, $i_3 \in \{2,3\}$, $i_4 \in \{2,3\}$, represent the conditional distribution of $(X_0, (U_1, U_2, U_3, U_4, U_5))$ w.r.t. the observations $(Y_0, Y_1, Y_2, Y_3, Y_4)$, in terms of the more likely initial condition $\xi_{0,5}^{i_0}$ and the four more likely signal-noise sequences $(\xi_{1,5}^i, \xi_{2,5}^i, \xi_{3,5}^i, \xi_{4,5}^i, \xi_{5,5}^i)_{i=1,2,3,4}$.

1.5.2.3   Approximate Bayesian Computation

Approximate Bayesian computation (*abbreviate ABC*) techniques are Bayesian inference methods currently used to evaluate posterior distributions without having to calculate likelihoods. For instance, in biology applications and more particularly in predictive bacteriology and food risk analysis, the observations of a kinetic biological complex system are given by counting bacteria individuals after successive dilutions of a food sample coming from an in vitro culture [74, 75, 80, 81]. Of course, this experimental observation process is often modeled by a series of Poisson type dependent random variables but the computation of the likelihood function often requires successive summations over the set of all the integers. In this situation likelihood functions are computationally intractable or very costly to estimate.

One of the central ideas of ABC methods is to replace the evaluation of the likelihood function by a simulation-based procedure of the observation process coupled with a numerical comparison between the observed and simulated data. This strategy is rather well known in particle filtering literature, see for instance [45, 46, 44]. In the same manner, these additional levels of simulation-based approximations can also be extended to compute the posterior distribution of fixed parameters in hidden Markov chain models. In signal processing literature, these ABC type particle models are sometimes called convolution particle filters, see for instance [19, 20, 144].

First, we notice that the transition probabilities of the signal-observation Markov chain $\mathscr{X}_n := (X_n, Y_n)$ are given by

$$\text{Proba}\left(\mathscr{X}_n = (x_n, y_n) \mid \mathscr{X}_n = (x_{n-1}, y_{n-1})\right) = M_n(x_{n-1}, x_n) \times G(x_n, y_n) \qquad (1.46)$$

with the likelihood function G defined in (1.38) and the Markov transitions of the chain X_n

$$\text{Proba}(X_n = x_n \mid X_n = x_{n-1}) = M_n(x_{n-1}, x_n)$$

Suppose that at every time step the state of the pair signal-observation Markov chain $\mathscr{X}_n := (X_n, Y_n)$ is partially observed according to the following schematic picture

$$\mathscr{X}_0 := \begin{bmatrix} X_0 \\ \downarrow \\ Y_0 \end{bmatrix} \longrightarrow \mathscr{X}_1 := \begin{bmatrix} X_1 \\ \downarrow \\ Y_1 \end{bmatrix} \longrightarrow \mathscr{X}_2 := \begin{bmatrix} X_2 \\ \downarrow \\ Y_2 \end{bmatrix} \longrightarrow \ldots$$
$$\downarrow \qquad\qquad \downarrow \qquad\qquad \downarrow$$
$$Y_0^\varepsilon \qquad\qquad Y_1^\varepsilon \qquad\qquad Y_2^\varepsilon \qquad \ldots$$

with some random variables Y_n^ε whose values only depend on the second component Y_n of the current state (X_n, Y_n) of the chain

$$\text{Proba}(Y_n^\varepsilon = y_n \mid (X_n, Y_n) = (x_n, z_n)) := G^\varepsilon(z_n, y_n)$$

We further assume that the likelihood function $G^\varepsilon(z, y)$ is a Markov transition indexed by some parameter $\varepsilon \in [0, 1]$ s.t. $\lim_{\varepsilon \to 0} G^\varepsilon(z, y) = \mathbb{1}_y(z)$. When the state space of the observation process is equipped with some neighborhood system, we can take

$$G^\varepsilon(z, y) = \frac{1}{\text{Card}(\mathcal{V}_\varepsilon(z))} \, \mathbb{1}_{\mathcal{V}_\varepsilon(z)}(y)$$

where $\mathcal{V}_\varepsilon(z)$ is a collection of neighborhoods of the point z s.t. $\mathcal{V}_\varepsilon(z) \to_{\varepsilon \to 0} \{z\}$. For instance, if the observation state space is equipped with some distance function d we can take $\mathcal{V}_\varepsilon(z) = \{y : d(z, y) \leq \varepsilon\}$. In this situation, given the the current state of the chain (X_n, Y_n), the observation Y_n^ε is randomly chosen in the set $\mathcal{V}_\varepsilon(Y_n)$.

Using (1.46) we prove that the distribution $\mathbf{P}_n((x_0, y_0), \ldots, (x_n, y_n))$ of the signal-observation paths

$$(\mathscr{X}_0, \ldots, \mathscr{X}_n) := ((X_0, Y_0), \ldots, (X_n, Y_n)) = ((x_0, y_0), \ldots, (x_n, y_n))$$

is given by

$$\mathbf{P}_n((x_0, y_0), \ldots, (x_n, y_n)) = \text{Proba}(A_n(x) \cap B_n(z))$$
$$= \left\{ \prod_{0 \leq p \leq n} G(x_p, z_p) \right\} \mathbb{P}_n(x_0, \ldots, x_n)$$

with the pair of events

$$A_n(x) := \{(X_0, \ldots, X_n) = (x_0, \ldots, x_n)\} \quad \text{and} \quad B_n(z) := \{(Y_0, \ldots, Y_n) = (z_0, \ldots, z_n)\}$$

and the distribution $\mathbb{P}_n(x_0, \ldots, x_n) := \text{Proba}(A_n(x))$ of the paths $(X_0, \ldots, X_n) = (x_0, \ldots, x_n)$. We consider the following events

$$\mathbf{A}_n((x,z)) := A_n(x) \cap B_n(z) \quad \text{and} \quad \mathbf{B}_n^{\varepsilon}(y) := \{(Y_0^{\varepsilon}, \ldots, Y_n^{\varepsilon}) = (y_0, \ldots, y_n)\}$$

As in section 1.5.2.1, the filtering problem defined above consists of computing the conditional distributions of the state variables $\mathbf{A}_n(x,z)$ given the observations $\mathbf{B}_n^{\varepsilon}(y)$. By construction, given $\mathbf{A}_n(x,z)$, the random variables Y_n^{ε} are independent and identically distributed with a distribution given by

$$\mathrm{Proba}\left(\mathbf{B}_n^{\varepsilon}(y) \,|\, \mathbf{A}_n(x,z)\right) = \prod_{0 \le p \le n} G^{\varepsilon}(z_p, y_p) \longrightarrow_{\varepsilon \to 0} \mathbb{1}_{(y_0,\ldots,y_n)}(z_0,\ldots,z_n)$$

from which we conclude that

$$\begin{aligned}
\mathrm{Proba}\left(\mathbf{A}_n(x,z) \,|\, \mathbf{B}_n^{\varepsilon}(y)\right) &= \frac{1}{\mathscr{L}_n^{\varepsilon}(y)} \left\{ \prod_{0 \le p \le n} G^{\varepsilon}(z_p, y_p) \right\} \mathbf{P}_n((x_0,z_0),\ldots,(x_n,z_n)) \\
&\to_{\varepsilon \downarrow 0} \mathrm{Proba}\left(A_n(x) \cap B_n(y)\right)
\end{aligned}$$

with the normalizing constants

$$\begin{aligned}
\mathscr{L}_n^{\varepsilon}(y) &:= \mathrm{Proba}(\mathbf{B}_n^{\varepsilon}(y)) = \sum_{(x_0,z_0),\ldots,(x_n,z_n)} \left\{ \prod_{0 \le p \le n} G^{\varepsilon}(z_p, y_p) \right\} \mathbf{P}_n((x_0,z_0),\ldots,(x_n,z_n)) \\
&\to_{\varepsilon \downarrow 0} \mathrm{Proba}(B_n(y))
\end{aligned}$$

As in section 1.5.2.1 these posterior distributions have exactly the same form as the one presented in (1.33). Notice that in this situation, at every time step n the stochastic model consists of N-particle samples $\xi_n^i := (\xi_n^{i,1}, \xi_n^{i,2})$ with a signal component $\xi_n^{i,1}$ and the corresponding observation component $\xi_n^{i,2}$, with $1 \le i \le N$. Given a series of observations $(y_n)_{n \ge 0}$, the conditional distributions defined above are approximated by the N-empirical measures of the particle model

$$\eta_n^N := \frac{1}{N} \sum_{i=1}^{N} \mathbb{1}_{(\xi_n^{i,1}, \xi_n^{i,2})} \longrightarrow_{N \to \infty} \mathrm{Proba}\left(\mathbf{A}_n(x,z) \,|\, \mathbf{B}_{n-1}^{\varepsilon}(y)\right)$$

and an unbiased estimate of the normalizing constants $\mathscr{L}_n^{\varepsilon}(y)$ is given by

$$\mathscr{L}_n^{\varepsilon,N}(y) := \prod_{0 \le p \le n} \frac{1}{N} \sum_{i=1}^{N} G^{\varepsilon}(\xi_p^{i,2}, y_p) \longrightarrow_{N \to \infty} \mathscr{L}_n^{\varepsilon}(y) = \mathrm{Proba}(\mathbf{B}_n^{\varepsilon}(y))$$

1.5.3   Interacting Kalman Filters

1.5.3.1   A Brief Introduction to Kalman Filters

We consider a $\mathbb{R}^{p+q}$-valued Markov chain (X_n, Y_n) defined by the recursive relations

$$\begin{cases} X_n = A_n X_{n-1} + B_n W_n, & n \geq 1 \\ Y_n = C_n X_n + D_n V_n, & n \geq 0 \end{cases} \tag{1.47}$$

for some $\mathbb{R}^{d_w}$ and $\mathbb{R}^{d_v}$-valued independent random sequences W_n and V_n, independent of X_0, and some matrices A_n, B_n, C_n, D_n with appropriate dimensions. We further assume that W_n and V_n are centered Gaussian random sequences with covariance matrices R_n^v, R_n^w and X_0 is a Gaussian random variable in $\mathbb{R}^p$ with a mean $\widehat{X}_0^-$ and covariance matrix $\widehat{P}_0^-$. In the further development of this section we shall denote by $\mathcal{N}(m,R)$ a Gaussian distribution in a d-dimensional space $\mathbb{R}^d$ with mean vector $m \in \mathbb{R}^d$ and covariance matrix $R \in \mathbb{R}^{d \times d}$

$$\mathcal{N}(m,R)(dx) = \frac{1}{(2\pi)^{d/2}\sqrt{|R|}} \exp\left[-2^{-1}(x-m)R^{-1}(x-m)'\right] dx$$

Using this notation, we have

$$\mathrm{Law}(X_n \,|\, Y_0, \ldots, Y_{n-1}) = \mathcal{N}(\widehat{X}_n^-, P_n^-) \quad \text{and} \quad \mathrm{Law}(X_n \mid Y_0, \ldots, Y_{n-1}, Y_n) = \mathcal{N}(\widehat{X}_n, P_n)$$

The synthesis of the conditional mean and covariance matrices is carried out using the traditional Kalman-Bucy recursive equations

$$\left(\widehat{X}_n^-, P_n^-\right) \xrightarrow{\text{updating}} \left(\widehat{X}_n, P_n\right) \xrightarrow{\text{prediction}} \left(\widehat{X}_{n+1}^-, P_{n+1}^-\right) \tag{1.48}$$

The updating and the prediction step are given below

[Updating] $\qquad \widehat{X}_n = \widehat{X}_n^- + \mathbf{G}_n\left(Y_n - C_n\widehat{X}_n^-\right) \quad \text{and} \quad P_n = P_n^- - \mathbf{G}_n C_n P_n^-$

with the gain matrix $\mathbf{G}_n = P_n^- C_n'(C_n P_n^- + D_n R_n^v D_n')^{-1}$, and

[Prediction] $\qquad \widehat{X}_{n+1}^- = A_{n+1}\widehat{X}_n \quad \text{and} \quad P_{n+1}^- = A_{n+1} P_n A_{n+1}' + B_{n+1} R_{n+1}^w B_{n+1}'$

Proof. The proof of the updating recursion equation is based on the fact that

$$\widehat{X}_n := \widehat{X}_n^- + \mathbf{G}_n\left(Y_n - \widehat{Y}_n^-\right) \quad \text{with} \quad \widehat{Y}_n^- = \mathbb{E}(Y_n | Y_0, \ldots, Y_{n-1}) = C_n\widehat{X}_n^-$$

Since $\mathbb{E}((X_n - \widehat{X}_n)(Y_n - \widehat{Y}_n^-)') = 0$, we find $\mathbb{E}((X_n - \widehat{X}_n^-)(Y_n - \widehat{Y}_n^-)') = \mathbf{G}_n\,\mathbb{E}((Y_n - \widehat{Y}_n^-)(Y_n - \widehat{Y}_n^-)')$, from which we find the gain matrix. Finally using the decomposition $X_n - \widehat{X}_n = (X_n - \widehat{X}_n^-) + (\widehat{X}_n^- - \widehat{X}_n)$ and by symmetry argument we conclude that

$$\begin{aligned} P_n &= P_n^- - \mathbb{E}((\widehat{X}_n^- - \widehat{X}_n)(\widehat{X}_n^- - \widehat{X}_n)') \\ &= P_n^- - \mathbf{G}_n\mathbb{E}((Y_n - \widehat{Y}_n^-)(Y_n - \widehat{Y}_n^-)')\mathbf{G}_n' = P_n^- - \mathbf{G}_n C_n P_n^- \end{aligned}$$

The proof of the prediction recursion is rather elementary. The first assertion is clear. The second one comes from the fact that

$$P_{n+1}^- = \mathbb{E}((A_{n+1}(X_n - \widehat{X}_n) + B_{n+1}W_{n+1})(A_{n+1}(X_n - \widehat{X}_n) + B_{n+1}W_{n+1})')$$
$$= A_{n+1} P_n A_{n+1}' + B_{n+1} R_{n+1}^w B_{n+1}'$$

It is also useful to observe that

$$\mathrm{Law}(Y_n \mid Y_0, \dots, Y_{n-1}) = \mathcal{N}(C_n \widehat{X}_n^-, C_n P_n^- C_n' + R_n^v)$$

We prove this claim using the fact that, given $(Y_0, \dots, Y_{n-1})$, the current observation takes the form

$$Y_n = C_n \tilde{X}_n + D_n V_n \text{ with some variable } \tilde{X}_n \text{ s.t. Law}\left(\tilde{X}_n \mid Y_0, \dots, Y_{n-1}\right) := \mathcal{N}(\widehat{X}_n^-, P_n^-).$$

We slight abuse the notation and we denote by $\mathcal{N}(m,R)(x)$ the density of a Gaussian distribution $\mathcal{N}(m,R)(dx) = \mathcal{N}(m,R)(x)dx$ w.r.t. the Lebesgue measure dx. In this notation, the density $p_n(y_0, \dots, y_n)$ of the random sequence of observation $(Y_0, \dots, Y_n)$ evaluated at the random observation path $(Y_0, \dots, Y_n)$ is given by

$$p_n(Y_0, \dots, Y_n) = \prod_{k=0}^n \mathcal{N}(C_k \widehat{X}_k^-, C_k P_k^- C_k' + R_k^v)(Y_k)$$

In Bayesian inference literature, this formula is sometimes written in the following form

$$p_n(Y_0, \dots, Y_n) = p_n(Y_n \mid Y_0, \dots, Y_{n-1}) \times p_{n-1}(Y_0, \dots, Y_{n-1}) = \prod_{k=0}^n p_k(Y_k \mid Y_0, \dots, Y_{k-1}).$$

1.5.3.2   Interacting Kalman Filters

We consider a Markov chain Θ_n taking values in some finite state space E, and a collection of matrices $A_n(\theta), B_n(\theta), C_n(\theta), D_n(\theta)$ indexed by $\theta \in E$, and of the same dimension as the matrices (A_n, B_n, C_n, D_n) introduced in (1.47)indexed We let (Θ_n, X_n, Y_n) be the $(E \times \mathbb{R}^{p+q})$-valued Markov chain defined by the same recursive relations as in (1.47)

$$\begin{cases} X_n = A_n(\Theta_n)X_{n-1} + B_n(\Theta_n)W_n, & n \geq 1 \\ Y_n = C_n(\Theta_n)X_n + D_n(\Theta_n)V_n, & n \geq 0 \end{cases} \quad (1.49)$$

Arguing as above, given a realization of the chain $\Theta = (\Theta_n)_{n \geq 0}$, we have

$$\mathrm{Law}(X_n \mid \Theta, Y_0, \dots, Y_{n-1}) = \mathcal{N}(\widehat{X}_n^{\Theta,-}, P_n^{\Theta,-})$$
$$\mathrm{Law}(X_n \mid \Theta, Y_0, \dots, Y_{n-1}, Y_n) = \mathcal{N}(\widehat{X}_n^\Theta, P_n^\Theta)$$

with some parameters $(\widehat{X}_n^{\Theta,-}, P_n^{\Theta,-})$ and $(\widehat{X}_n^\Theta, P_n^\Theta)$ that can be computed using the same Kalman recursions given above by replacing the matrices (A_n, B_n, C_n, D_n) by the matrices $(A_n(\Theta_n), B_n(\Theta_n), C_n(\Theta_n), D_n(\Theta_n))$. We observe that $(\widehat{X}_n^{\Theta,-}, P_n^{\Theta,-})$ only depends on the random sequence $(\Theta_0, \ldots, \Theta_n)$ so that

$$\mathcal{N}(C_n(\Theta_n)\widehat{X}_n^{\Theta,-}, C_n(\Theta_n) P_n^{\Theta,-} C_n'(\Theta_n) + R_n^v)(Y_n) := G_{n,Y_n}(\Theta_0, \ldots, \Theta_n) \qquad (1.50)$$

Therefore, the density $p_n((y_0, \ldots, y_n) \mid (\theta_0, \ldots, \theta_n))$ of the random sequence of observation $(Y_0, \ldots, Y_n)$ evaluated at the random observation path $(Y_0, \ldots, Y_n)$ and given a realization of the parameters $(\Theta_0, \ldots, \Theta_n) = (\theta_0, \ldots, \theta_n)$ is given by

$$p_n((Y_0, \ldots, Y_n) \mid (\theta_0, \ldots, \theta_n)) = \prod_{k=0}^{n} G_{k,Y_k}(\theta_0, \ldots, \theta_n) \qquad (1.51)$$

If we denote by $\mathbb{P}_n(\theta_0, \ldots, \theta_n)$ the probability measure of the sequence of random parameters $(\Theta_0, \ldots, \Theta_n)$, then using Bayes' rule we find that the probability measure

$$\mathbb{Q}_n(\theta_0, \ldots, \theta_n) := \frac{1}{\mathscr{Z}_{n,Y}} \left\{ \prod_{0 \leq k < n} G_{k,Y_k}(\theta_0, \ldots, \theta_n) \right\} \mathbb{P}_n(\theta_0, \ldots, \theta_n) \qquad (1.52)$$

(with some normalizing constant $\mathscr{Z}_{n,Y}$) coincides with the conditional distribution of the random sequence $(\Theta_0, \ldots, \Theta_n)$ given the observations $(Y_0, \ldots, Y_{n-1})$; that is, we have that

$$\mathbb{Q}_n = \mathrm{Law}((\Theta_0, \ldots, \Theta_n) \mid (Y_0, \ldots, Y_{n-1}))$$

The corresponding particle approximations on the set of sequences are often referred as particle methods in path space in signal processing literature and Bayesian inference studies (see for instance [38, 53, 67], and references therein).

1.5.4 Stochastic Optimization Algorithms

1.5.4.1 Interacting MCMC Models

We present now a genetic type particle strategy for sampling random states according to a sequence of probability measures on some finite state space E given by

$$\mu_n(x) = \frac{1}{\lambda(G_n)} G_n(x)\, \lambda(x) \quad \text{with} \quad G_n(x) := G_{n-1}(x) \times g_{n-1}(x) = \prod_{0 \leq p < n} g_p(x)$$

where $\lambda(x)$ is a probability measure and g_n is a collection of positive functions on E.

The interacting particle sampler of these measures is defined as follows. We start with a population of N independent individuals $\xi_0 := (\xi_0^i)_{1 \leq i \leq N}$ randomly chosen in E according to μ_0. We perform a selection transition $\xi_0 \rightsquigarrow \widehat{\xi}_0 := (\widehat{\xi}_0^i)_{1 \leq i \leq N}$ using the potential functions g_0. More precisely, every individual evaluates its potential value $g_0(\xi_0^i)$. For every index i, with a probability $g_0(\xi_0^i)$, we set $\widehat{\xi}_0^i = \xi_0^i$, otherwise we replace ξ_0^i by a new individual $\widehat{\xi}_0^i = \xi_0^j$ randomly chosen in the whole population with a probability proportional to $g_0(\xi_0^j)$. During the mutation transition $\widehat{\xi}_0 \leadsto \xi_1 := (\xi_1^i)_{1 \leq i \leq N}$, every selected individual $\widehat{\xi}_0^i$ performs a local random move $\widehat{\xi}_0^i \leadsto \xi_1^i$ (independently of one another) according to the Markov transition P_1 associated with an MCMC sampler with invariant measure μ_1. Then, we perform a selection transition $\xi_1 \leadsto \widehat{\xi}_1 := \left(\widehat{\xi}_1^i\right)_{1 \leq i \leq N}$ using the fitness functions g_1. After this selection stage we mutate each selected individual using the Markov transition P_2 associated with an MCMC sampler with invariant measure μ_2, and so on. Iterating these transitions, we define a simple genetic model with mutations transitions P_n and selection fitness functions g_n:

$$\xi_n := (\xi_n^i)_{1 \leq i \leq N} \in E^N \xrightarrow{\text{selection}} \widehat{\xi}_n := \left(\widehat{\xi}_n^i\right)_{1 \leq i \leq N} \in E^N \xrightarrow{\text{mutation}} \xi_{n+1} \in E^N$$

$$(1.53)$$

This algorithm belongs to the class of sequential Monte Carlo samplers proposed in [40]. Many convergence results can be established. For instance, under some weak regularity conditions we can show that for any $1 \leq q \leq N$, and any time horizon $n \geq 0$, the first q random samples $(\xi_n^i)_{1 \leq i \leq q}$ among N are almost independent and identically distributed with the desired target measure μ_n; that is, we have that

$$\sum_{x^1,\ldots,x^q} \left| \mathrm{Proba}\left(\xi_n^1 = x^1, \ldots, \xi_n^q = x^q\right) - \mu_n(x^1) \cdots \mu_n(x^q)\right| \leq c(n) \min\left(\frac{q^2}{N}, \sqrt{\frac{q}{N}}\right)$$

and some finite constant $c(n) < \infty$. We also have that for any $x \in E$ and any $n \geq 0$

$$\lim_{N \to \infty} \frac{1}{N} \sum_{1 \leq i \leq N} \mathbb{1}_{\xi_n^i}(x) = \mu_n(x) \quad \text{and} \quad \mathscr{Z}_n^N := \prod_{0 \leq p < n} \eta_p^N(g_p) \xrightarrow{N \to \infty} \mathscr{Z}_n$$

1.5.4.2   Interacting Monte Carlo Markov Chains

Suppose we want to compute the global minima of a given non negative cost function V on some finite state space E equipped with the counting measure $\lambda(x) := \frac{1}{\mathrm{Card}(E)}$. From the probabilistic point of view, this problem amounts to sampling random states according to the Boltzmann-Gibbs distributions associated with a large inverse temperature parameter β and given

$$\mu_\beta(x) := \frac{1}{\mathscr{Z}_\beta} e^{-\beta V(x)} \lambda(x) \quad \text{with} \quad \mathscr{Z}_\beta := \sum_x e^{-\beta V(x)} \lambda(x)$$

There is no loss of generality to assume that $\inf_x V(x) = 0$ and for any state $x \notin V_0 := V^{-1}(\{0\})$, $V(x) \geq \delta$ for some $\delta > 0$. It follows that we have

$$\mathrm{Card}(V_0) \leq \mathscr{Z}_\beta \leq \mathrm{Card}(V_0) + \mathrm{Card}(V_0^c)\, e^{-\beta\delta} \xrightarrow{\beta \uparrow \infty} \mathrm{Card}(V_0)$$

and therefore
$$\lim_{\beta \to \infty} \mu_\beta(x) = \mu_\infty(x) := \mathbb{1}_{V_0}(x)/\mathrm{Card}(V_0)$$

This simple observation shows that sampling according to μ_β is roughly equivalent to randomly sampling an unknown state variable with minimal cost. For very large state spaces, it is typically impossible to sample from μ_β directly. The celebrated simulated annealing algorithm to sample from μ_∞ consists of sampling approximately from a sequence of distributions μ_{β_n} where β_n is a non-decreasing sequence going to ∞. The rationale is that it is "easier" to sample from μ_β when β is small; if $\beta = 0$ then μ_0 is the uniform counting measure on E from which it is trivial to sample. For $\beta_n > 0$, we sample approximately from each intermediate distribution μ_{β_n} using Markov chain Monte Carlo (MCMC) sampling techniques; that is we select a transition matrix $M_{\beta_n} = \left(M_{\beta_n}(x,y)\right)_{x,y \in E}$ with left eigenvector μ_{β_n} associated with the eigenvalue 1, that is

$$\sum_x \mu_{\beta_n}(x) M_{\beta_n}(x,y) = \mu_{\beta_n}(y)$$

The probabilistic interpretation of the above equation is as follows: pick randomly a state x with distribution $\mu_{\beta_n}(x)$ and take a random transition $x \rightsquigarrow y$ from the distribution $M_{\beta_n}(x,y)$, then the probability of being at state y is again $\mu_{\beta_n}(y)$. The literature on MCMC methods discusses numerous choices of transitions M_{β_n} satisfying this property. The most famous is the Metropolis-Hastings transition associated to a symmetric transition matrix $K(x,y) = K(y,x)$ and defined by

$$M_{\beta_n}(x,y)$$

$$= K(x,y) \, \min\left(1, e^{-\beta_n(V(y)-V(x))}\right) + \left(1 - \sum_z K(x,z) \, \min\left(1, e^{-\beta_n(V(z)-V(x))}\right)\right) \mathbb{1}_x(y)$$

Using the fundamental ergodic theorem for regular Markov chains, starting from any initial state x_0, the n-th step of a run of the Markov chain with transitions M_{β_n} has a probability very close to $\mu_{\beta_n}(y)$ of being at the site y, for a large n. Practically, we select β_1 and we run the chain starting at $X_0 = x_0$ for a large enough number of runs n_1 such that the law of the state X_{n_1} is close to μ_{β_1}

$$X_0 = x_0 \xrightarrow{M_{\beta_1}} X_1 \xrightarrow{M_{\beta_1}} \ldots \xrightarrow{M_{\beta_1}} X_{n_1} \text{ with } n_1 \text{ large enough s.t. Law}(X_{n_1}) \simeq \mu_{\beta_1}$$

Notice that the choice of n_1 depends on β_1: the larger β_1 is, the "peakier" μ_{β_1} is and the larger n_1 is. When the chain is stabilized, we choose a $\beta_2 > \beta_1$ and we run the chain starting at X_{n_1} for a new large enough number of time steps n_2 such that the law of the state $X_{n_1+n_2}$ is close to μ_{β_2}

$$X_{n_1} \xrightarrow{M_{\beta_2}} X_{n_1+1} \xrightarrow{M_{\beta_2}} \ldots \xrightarrow{M_{\beta_2}} X_{n_1+n_2} \text{ with } n_2 \text{ large enough s.t. Law}(X_{n_1+n_2}) \simeq \mu_{\beta_2}$$

The theoretical "optimal" inverse temperature parameter ensuring convergence in some sense of the Markov chain to μ_∞ is logarithmic. This amounts to saying that we change by one unit the parameter β on every time interval with exponential length. This is unrealistic from a practical point of view.

We present now an alternative particle strategy for sampling random states according to the sequence of measures μ_{β_n} associated with a given non decreasing sequence of inverse temperature parameters β_n. We suppose that $\beta_0 = 0$ so that μ_{β_0} coincides with the uniform counting measure on the set E. We start with N independent individuals $\xi_0 := (\xi_0^i)_{1 \leq i \leq N}$ randomly chosen in E according to μ_{β_0}. We perform a selection transition $\xi_0 \rightsquigarrow \widehat{\xi}_0 := (\widehat{\xi}_0^i)_{1 \leq i \leq N}$ using the potential functions G_0 defined by

$$G_0(x) = \exp(-(\beta_1 - \beta_0)V(x))$$

In other words, every individual evaluates its potential value $G_0(\xi_0^i)$. For every index i, with a probability $G_0(\xi_0^i)$, we set $\widehat{\xi}_0^i = \xi_0^i$, otherwise we replace ξ_0^i by a new individual $\widehat{\xi}_0^i = \xi_0^j$ randomly chosen in the whole population with a probability proportional to $G_0(\xi_0^j)$. During the mutation step $\widehat{\xi}_0 \rightsquigarrow \xi_1 := (\xi_1^i)_{1 \leq i \leq N}$, every selected individual $\widehat{\xi}_0^i$ performs a local random move $\widehat{\xi}_0^i \rightsquigarrow \xi_1^i$ (independently of one another) according to the Markov transition M_{β_1}. Then, we perform another selection transition $\xi_1 \rightsquigarrow \widehat{\xi}_1 := (\widehat{\xi}_1^i)_{1 \leq i \leq N}$ using the fitness functions G_1 defined below:

$$G_1(x) = \exp(-(\beta_2 - \beta_1)V(x))$$

After this selection stage we mutate each selected individual using the Markov transition M_{β_2}, and so on. Iterating these transitions, we define a simple genetic model with mutation transitions M_{β_n} and selection fitness functions G_n:

$$\xi_n := (\xi_n^i)_{1 \leq i \leq N} \in E^N \xrightarrow{\text{selection}} \widehat{\xi}_n := (\widehat{\xi}_n^i)_{1 \leq i \leq N} \in E^N \xrightarrow{\text{mutation}} \xi_{n+1} \in E^N$$

$$(1.54)$$

This algorithm was first proposed in [40]. A variety of convergence results can be established for this algorithm. For instance, for any function f on E and any time horizon, we have

$$\lim_{N \to \infty} \frac{1}{N} \sum_{1 \leq i \leq N} f(\xi_n^i) = \sum_x \mu_{\beta_n}(x) f(x)$$

In addition, if we set $\eta_n^N(G_n) := \frac{1}{N} \sum_{1 \leq i \leq N} G_n(\xi_n^i)$, the unbiased N-particle approximation $\mathscr{Z}_{\beta_n}^N$ of the normalizing constants $\mathscr{Z}_{\beta_n}$ is given by

$$\mathscr{Z}_{\beta_n}^N := \prod_{0 \leq p < n} \eta_p^N(G_p) \xrightarrow{N \to \infty} \mathscr{Z}_{\beta_n}$$

1.5.4.3   Combinatorial Counting and Sampling

Suppose we want to compute the cardinality of a given subset A of some finite state space E equipped with the counting measure $\lambda(x) := \frac{1}{\text{Card}(E)}$. Once again, from a probabilistic point of view, this problem is equivalent to computing the normalizing constant of the following Boltzmann-Gibbs distribution

$$\mu_A(x) := \frac{1}{\mathscr{Z}_A}\ \mathbb{1}_A(x)\ \lambda(x) \quad \text{with} \quad \mathscr{Z}_A := \sum_x \mathbb{1}_A(x)\ \lambda(x)$$

To sample from μ_A and compute $\mathscr{Z}_A$, the idea consists of selecting a judicious sequence of decreasing subsets A_n in such a way that it is easy to sample states in A_n starting from the set A_{n-1}. We suppose that $A_0 = E$ so that μ_{A_0} coincides with the uniform counting measure on the set E. The algorithm is thus very similar to the one described previously for optimization. For any set A_n, we introduce an MCMC transition matrix $M_{A_n} = (M_{A_n}(x,y))_{x,y \in E}$ with left eigenvector μ_{A_n} associated with the eigenvalue 1, that is

$$\sum_x \mu_{A_n}(x) M_{A_n}(x,y) = \mu_{A_n}(y)$$

A simple Metropolis-Hasting type transition associated with a symmetric transition matrix $K(x,y) = K(y,x)$ is given by

$$M_{A_n}(x,y) = K(x,y)\ \mathbb{1}_{A_n}(y) + \left(1 - \sum_z K(x,z)\ \mathbb{1}_{A_n}(z)\right)\ \mathbb{1}_x(y)$$

The N-particle stochastic algorithm is defined as follows. We start with N independent random individuals $\xi_0 := \left(\xi_0^i\right)_{1 \leq i \leq N}$ randomly chosen in E with μ_{A_0}. We perform a selection transition $\xi_0 \rightsquigarrow \widehat{\xi}_0 := \left(\widehat{\xi}_0^i\right)_{1 \leq i \leq N}$ using the fitness functions $G_0 = \mathbb{1}_{A_1}$. In other words, every individual in the set A_1 stays in the same place $\widehat{\xi}_0^i = \xi_0^i$, otherwise we replace ξ_0^i by a fresh new individual $\widehat{\xi}_0^i = \xi_0^j$ randomly chosen among the individuals $\xi_0^j \in A_1$. When no individuals ξ_0^j are in the set A_1, the algorithm stops and we set $\tau^N = 0$. Assuming that $\tau^N > 0$, during the mutation step $\widehat{\xi}_0 \rightsquigarrow \xi_1 := \left(\xi_1^i\right)_{1 \leq i \leq N}$, every selected individual $\widehat{\xi}_0^i$ performs a local random move $\widehat{\xi}_0^i \rightsquigarrow \xi_1^i$ (independently of one another) in the set A_1 according to the Markov transition M_{A_1}. Then, we perform another selection transition $\xi_1 \rightsquigarrow \widehat{\xi}_1 := \left(\widehat{\xi}_1^i\right)_{1 \leq i \leq N}$ using the fitness functions $G_1 = \mathbb{1}_{A_2}$. When no individuals ξ_1^j are in the set A_2, the algorithm stops and we set $\tau^N = 1$. After this selection stage we mutate each selected individual using the Markov transition M_{A_2}, and so on. For any function f on E and any time horizon n, we have

$$\lim_{N \to \infty} \frac{1}{N} \sum_{1 \le i \le N} f(\xi_n^i) \mathbb{1}_{\tau^N > n} = \sum_x \mu_{A_n}(x) \, f(x)$$

In addition, if we set $\eta_n^N(G_n) := \frac{1}{N} \sum_{1 \le i \le N} G_n(\xi_n^i)$, the proportion of individuals in A_{n+1} after the n-th mutation, the unbiased N-particle approximation $\mathscr{Z}_{A_n}^N$ of the normalizing constants $\mathscr{Z}_{A_n}$ is given by

$$\mathscr{Z}_{A_n}^N := \prod_{0 \le p < n} \eta_p^N(G_p) \longrightarrow_{N \to +\infty} \mathscr{Z}_{A_n} = \mathrm{Card}(A_n)/\mathrm{Card}(E)$$

1.5.4.4   Genetic Search Algorithms

We consider an energy function or a cost criteria $V : x \in E \mapsto (x)$ on some finite state space E where we assume $\inf_x V(x) = 0$ without loss of generality. The objective is to find the global minima points $x^\star \in E$ s.t. $V(x^\star) = \inf_{x \in E} V(x)$. Let $V^\star$ denote the set of these points. We describe in Section 1.5.4.2 an interacting particle algorithm to solve this problem which relies on interacting simulated annealing type chains. We present here the more standard genetic algorithm with mutation and proportional selection.

To construct this algorithm, we introduce a collection of Markov transitions $M_n(x,y)$ from E into itself. This collection of transition matrices represents the probability $M_n(x,y)$ that a individual at site x evolves to a new state x during the n-th mutation transition.

The genetic algorithm with N individuals is defined as follows. We start with N independent random individuals $\xi_0 := (\xi_0^i)_{1 \le i \le N}$ randomly chosen in E with some distribution η_0. We perform a proportional type selection transition $\xi_0 \rightsquigarrow \widehat{\xi}_0 := (\widehat{\xi}_0^i)_{1 \le i \le N}$ using the potential functions $G_0(\xi_0^i) = \exp(-\beta_0 V(\xi_0^i))$, where $\beta_0 \ge 0$ is an inverse temperature parameter. In other words, with probability $G_0(\xi_0^i)$ every individual stays in the same place $\widehat{\xi}_0^i = \xi_0^i$; otherwise, we replace ξ_0^i by a new individual $\widehat{\xi}_0^i = \xi_0^j$ randomly chosen among the individuals ξ_0^j with a probability proportional to its weight $G_0(\xi_0^j)$. Formally, we set

$$\widehat{\xi}_0^i = \varepsilon_0^i \, \xi_0^i + \left(1 - \varepsilon_0^i\right) \widetilde{\xi}_0^i$$

where ε_0^i stands for a sequence of independent $\{0,1\}$-valued Bernoulli random variables with distributions

$$G_0(\xi_0^i) := \mathrm{Proba}\left(\varepsilon_0^i = 1 \mid \xi_0\right) = 1 - \mathrm{Proba}\left(\varepsilon_0^i = 0 \mid \xi_0\right)$$

and $\widetilde{\xi}_0 := \left(\widetilde{\xi}_0^i\right)_{1 \le i \le N}$ are independent, identically distributed and $\left\{\xi_0^j, \, 1 \le j \le N\right\}$-valued random variables with common distributions given for any index $1 \le i \le N$ by

$$\forall 1 \leq j \leq N \qquad \mathrm{Proba}\left(\widetilde{\xi}_0^i = \xi_0^j \mid \xi_0\right) = G_0(\xi_0^j)/\sum_{1\leq j\leq N} G_0(\xi_0^j)$$

During the mutation step $\widehat{\xi}_0 \leadsto \xi_1 := (\xi_1^i)_{1\leq i\leq N}$, every selected individual $\widehat{\xi}_0^i$ performs a local random move $\widehat{\xi}_0^i \leadsto \xi_1^i$ (independently of one another) according to the Markov transition M_1. Then, we perform another proportional type selection transition $\xi_1 \leadsto \widehat{\xi}_1 := \left(\widehat{\xi}_1^i\right)_{1\leq i\leq N}$ using the potential functions $G_1\left(\xi_1^i\right) = \exp\left(-\beta_1 V\left(\xi_1^i\right)\right)$, where $\beta_1 \geq 0$ is another inverse temperature parameter, and so on. We define in this way a sequence of genetic type populations $\xi_n, \widehat{\xi}_n$, as in (1.28) and the corresponding genealogical tree model (1.30) associated with the ancestral lines $(\xi_{p,n}^i)_{0\leq p\leq n}$ of every i-th individuals after the n-th mutation. In the same way, running back in time we have the whole ancestral line

$$\widehat{\xi}_{0,n}^i \longleftarrow \widehat{\xi}_{1,n}^i \longleftarrow \ldots \longleftarrow \widehat{\xi}_{n-1,n}^i \longleftarrow \widehat{\xi}_{n,n}^i = \widehat{\xi}_n^i \qquad (1.55)$$

of every i-th individual after the n-th selection.

For any function f on E_n and any time horizon n, we can prove that

$$\lim_{N\to\infty} \frac{1}{N}\sum_{i=1}^N f(\widehat{\xi}_{0,n}^i, \widehat{\xi}_{1,n}^i, \ldots, \widehat{\xi}_{n,n}^i) = \frac{\mathbb{E}\left(f_n(X_0,\ldots,X_n)\,\exp\left(-\sum_{0\leq p\leq n}\beta_p\,V(X_p)\right)\right)}{\mathbb{E}\left(\exp\left(-\sum_{0\leq p\leq n}\beta_p\,V(X_p)\right)\right)}$$

In other words, the proportion of paths $(\widehat{\xi}_{0,n}^i, \widehat{\xi}_{1,n}^i, \ldots, \widehat{\xi}_{n,n}^i)$ taking some value $(x_0,\ldots,x_n)$ is given by

$$\lim_{N\to\infty} \frac{1}{N}\sum_{i=1}^N \mathbb{1}_{(x_0,\ldots,x_n)}(\widehat{\xi}_{0,n}^i, \widehat{\xi}_{1,n}^i, \ldots, \widehat{\xi}_{n,n}^i) = \frac{1}{\mathscr{Z}_{n+1}}\,\exp\left(-\sum_{0\leq p\leq n}\beta_p\,V(x_p)\right)\,\mathbb{P}_n(x_0,\ldots,x_n)$$

with the probability of a free evolution path involving only mutation transitions

$$\mathbb{P}_n(x_0,\ldots,x_n) = \eta_0(x_0)M_1(x_0,x_1)\ldots M_n(x_{n-1},x_n)$$

where $\mathscr{Z}_{n+1}$ is a normalizing constant.

Suppose that every free evolution path has the same chance to be sampled, in the sense that

$$\mathbb{P}_n(x_0,\ldots,x_n) = \mathbb{P}_n(y_0,\ldots,y_n)$$

for any admissible pair of paths $(x_0,\ldots,x_n)$ and $(y_0,\ldots,y_n)$. This condition is satisfied if η_0 is the uniform counting measure on E and the mutation transitions $M_n(x,y)$ correspond to local random choices of the same number of neighbors, starting from any state x. In this case, for any admissible path $(x_0,\ldots,x_n)$ we have that

$$\lim_{N\to\infty} \frac{1}{N}\sum_{i=1}^N \mathbb{1}_{(x_0,\ldots,x_n)}(\widehat{\xi}_{0,n}^i, \widehat{\xi}_{1,n}^i, \ldots, \widehat{\xi}_{n,n}^i) = \frac{1}{\mathscr{Z}_n'}\,\exp\left(-\sum_{0\leq p\leq n}\beta_p\,V(x_p)\right)$$

for some normalizing constant $\mathscr{Z}'_n$. When the inverse temperature parameter β_p increases the r.h.s. probability mass quantity only charges admissible paths $(x_0, \ldots, x_n)$ that minimize the path potential function

$$\mathscr{V}_n(x_0, \ldots, x_n) = \inf_{(y_0, \ldots, y_n)} \sum_{0 \le p \le n} V(y_p)$$

In other words at low temperatures, the ancestral lines of the simple genetic model described above converge to the uniform measure on all the paths $(x_0, \ldots, x_n)$ of length n that minimize the energy function $\mathscr{V}_n$. For time homogenous mutation transitions associated with stochastic matrices $M_n(x, y) = M(x, y)$ satisfying the following condition for some integer $m \ge 1$ and any pair $(x, y) \in E^2$

$$M(x, x) > 0 \quad \text{and} \quad M^m(x, y) \ge \varepsilon M^m(x, z)$$

we also have the convergence result

$$\lim_{n \to \infty} \lim_{N \to \infty} \frac{1}{N} \sum_{i=1}^{N} \mathbb{1}_{V^\star}(\widehat{\xi}_n^i) = 1$$

as soon as $\beta_n = C \log(n + 1)$ for some constant C that depends on m and on the oscillations of the function V. This convergence result is also true for $\beta_n = C(n + 1)^\alpha$, with any $\alpha \in]0, 1[$, as soon as the above condition is met for $m = 1$.

Further details on these concentration properties can be found in [55]. Related convergence results for fixed population sizes can be found in [24]. To give a flavor of these results, let us suppose that the mutation transitions $M_n(x, y)$ also depend on the inverse temperature parameter and

$$M_n(x, y) \to_{n \to \infty} \mathbb{1}_x(y) \quad \text{as} \quad \beta_n \uparrow \infty$$

Intuitively speaking, the genetic mutations become rare transitions at low temperatures. In this situation, we can prove that there exists a "critical population size" $N^\star$ that depends on the energy function as well as on the free evolution model such that

$$\forall N \ge N^\star \quad \lim_{n \to \infty} \text{Proba}\left(\forall 1 \le i \le N \quad \widehat{\xi}_n^i \in V^\star\right) = 1$$

1.5.5   Analysis of Convergence under Uncertain Behavior

The following analysis focuses on a particular class of genetic type algorithms for which it is assumed that operators have a nonzero probability of erroneous or uncertain behavior. A direct example may be found in practice for distributed environments where remote nodes carry part of the steps of the algorithm and where nodes are prone to processing or communication errors and malicious behavior. Different questions arise in this context on the influence of erroneous (or abnormal) operation on the convergence of the algorithm.

In this work we do not concentrate on proving that we have a probability one in reaching the optimal solution when time goes to infinity, given a fixed population size, as these results are already present in the literature [145, 146], but rather on bounding the probability that the obtained results are within a certain error threshold. Another line of research is concerned with results using Feynman-Kac representations, focusing on the asymptotic stability and uniform convergence of genetic algorithms [47]. Note that in the following we will address convergence in finite spaces.

Finally, connections to dynamic optimization or in the presence of uncertainties could be made by considering noise or time dependent external factors as being an integrated part of how the operators function.

Let $(X_n)_{n \geq 0}$, $X_n \in E$, be a Markov chain, with E being an arbitrary space, for which a transition kernel is given as

$$M(x, dy) = \mathbb{P}(X_n \in dy | X_{n-1} = x).$$

Assumption A1: There exists ν, a probability measure over E, $\lambda > 0$, $m \geq 1$ s. t.

$$\mathbb{P}(X_m \in dx | X_0 = x_0) \geq \lambda \nu(dx)$$

Example 1.5.1. *Let* $E = \{x_1, x_2, \ldots, x_d\}$ *be a finite space and* $M(x, y) \geq \delta > 0$ *a Markov transition* $\left(M(x, y) \geq \delta d \times \frac{1}{d}\right)$. *Having* $\nu(y) = \frac{1}{d}$ *a uniform measure over* E *and by denoting with* $\lambda = \delta d$, *we obtain that*

$$M(x, y) = \mathbb{P}(X_1 = y | X_0 = x) \geq \lambda \nu(y).$$

Example 1.5.2. *Let* $E = \{x_1, x_2, \ldots, x_d\}$ *be a finite space and* $M^m(x, y)$ *a Markov transition involving the application of the M kernel m times, with* $M^m(x, y) \geq \delta > 0 \Leftrightarrow M^m(x, y) \geq \delta d \times \frac{1}{d}$. *By denoting as previously* $\lambda = \delta d$, *we obtain thus*

$$M^m(x, y) = \mathbb{P}(X_m = y | X_0 = x) \geq \lambda \ \nu(y).$$

Under **Assumption (A1)**, it is well known that there exists an unique probability measure π s.t. $\pi M = \pi$. This measure π is said to be an invariant measure.

Remark 1.5.3. *We further assume that E is finite and* **Assumption (A1)** *is met for* $m = 1$, *and some measure* ν *s.t.* $\nu(x) > 0$ *for any* $x \in E$. *We also let* π *be the invariant measure for the Markov chain of transition M. In this case, for any* $x \in E$ *we clearly have that:*

$$\pi(x) = \sum_y \pi(y) M(y, x) > 0, \forall x$$

The same goes for the case involving m successive transition steps (see Example 1.5.2.):

$$\pi(x) = \sum_y \pi(x) M^m(x, y) > 0, \forall x, \; as \; M^m(x, y) > 0, \forall x.$$

For more general state spaces E, for any measurable subset $A \subset E$ we have

$$\pi(A) = \int_A \pi(dx) = \int \pi(dx) M^m(x, A) > 0, \forall x, \; as \; M^m(x, A) > 0, \forall x.$$

Notation: We let $\pi^n = \frac{1}{n} \sum_{0 \le p < n} \delta_{X_p}$ be the occupation measure of the Markov chain $(X_p)_{p \ge 0}$ at time n, starting at some initial state, say $X_0 = x_0$. For any bounded measurable function f on E, we set

$$\pi^n(f) = \int f(x) \, \pi^n(dx) = \frac{1}{n} \sum_{0 \le p < n} f(X_p) \quad and \quad \pi(f) = \int f(x) \, \pi(dx)$$

In order to obtain stronger bounds we will base our further investigations on the result presented in [86], adjusted to our context, i.e. $\|f\| = 1$ and using the afore-mentioned notation.

Theorem 1.5.4. (Glynn and Ormoneit [86]) *Under the conditions of Assumption (A1), for any bounded measurable function f s.t. $\|f\| = 1$, and for any $n > 2m/(\lambda \varepsilon)$ and $\varepsilon > 0$, we have that*

$$\mathbb{P}\left(\pi^n(f) - \mathbb{E}(\pi^n(f))\right) \ge \varepsilon) \le e^{-\frac{\lambda^2 (n\varepsilon - 2m/\lambda)^2}{2nm^2}}$$

When considering the absolute value, it is implied moreover that

$$\mathbb{P}\left(|\pi^n(f) - \mathbb{E}(\pi^n(f))| \ge \varepsilon\right) \le 2 \times e^{-\frac{\lambda^2}{2m^2}\left(\varepsilon - \frac{2m}{\lambda n}\right)^2} \tag{1.56}$$

Example 1.5.5. *As a direct application of this result, for $m = 1$ and $\forall n > 2/(\lambda \varepsilon)$, we have that*

$$\mathbb{P}\left(\pi^n(f) - \mathbb{E}(\pi^n(f))\right) \ge \varepsilon) \le e^{-\frac{\lambda^2}{2n}\left(n\varepsilon - \frac{2}{\lambda}\right)^2} = e^{-\frac{\lambda^2 n}{2}\left(\varepsilon - \frac{2}{\lambda n}\right)^2}$$

We further assume that **Assumption (A1)** is met for some parameters $\lambda > 0$ and $m \ge 1$, and some probability measure ν on E. We notice that

$$\mathbb{E}\left(\frac{1}{n} \sum_{0 \le p < n} f(X_p)\right) = \frac{1}{n} \sum_{0 \le p < n} \mathbb{E}(f(X_p)) = \frac{1}{n} \sum_{0 \le p < n} M^p(f)(x_0).$$

On the other hand, under our assumptions it is well known that

$$\sup_{x_0, y_0} |M^p(f)(x_0) - M^p(f)(y_0)| \le c_1(m) \, e^{-c_2(m) \, p}$$

for any $p \geq 0$, and any measurable function f on E s.t. $\|f\| = 1$, for some nonnegative and finite constants $c_1(m)$ and $c_2(m)$ whose values only depend on the parameters λ and m. For a detailed proof of these inequalities we refer the reader to [38].

Thus, recalling that $\pi = \pi M^p$, for any $p \geq 0$, the following relations can be derived:

$$\mathbb{E}\left(\pi^n(f)\right) - \pi(f) = \frac{1}{n} \sum_{0 \leq p < n} [M^p(f)(x_0) - \pi(f)] = \frac{1}{n} \sum_{0 \leq p < n} [M^p(f)(x_0) - \pi M^p(f)]$$

so that

$$\left|\mathbb{E}\left(\pi^n(f)\right) - \pi(f)\right| = \frac{1}{n} \left| \sum_{0 \leq p < n} [M^p(f)(x_0) - \pi M^p(f)] \right|$$

$$\leq c_1(m) \times \frac{1}{n} \sum_{0 \leq p < n} e^{-c_2(m)\,p} \leq c_3(m)/n$$

for some constant

$$c_3(m) \leq c_1(m)/(1 - e^{-c_2(m)})$$

Considering these results with **Theorem** 1.5.4. one can conclude that

$$\mathbb{P}\left(|\pi^n(f) - \pi(f)| \geq \varepsilon\right) \leq \mathbb{P}\left(|\pi^n(f) - \mathbb{E}\left(\pi^n(f)\right)| + |\mathbb{E}\left(\pi^n(f)\right) - \pi(f)| \geq \varepsilon\right)$$

$$\leq \mathbb{P}\left(|\pi^n(f) - \mathbb{E}\left(\pi^n(f)\right)| \geq \varepsilon - c_3(m)/n\right)$$

$$\leq 2e^{-\frac{\lambda^2 n}{2}(\varepsilon - [c_3(m) + 2/\lambda]/n)^2}$$

$$(1.57)$$

for any $n \geq 1$ and any $\varepsilon > 0$ such that $\varepsilon > [c_3(m) + 2/\lambda]/n$. We summarize the above discussion with the following corollary:

Corollary 1.5.6. *Under the conditions of Assumption (A1), for any bounded measurable function f s.t. $\|f\| = 1$, any $\varepsilon > 0$ and any $n > [c_3(m) + 2/\lambda]/\varepsilon$, we have the exponential concentration inequality*

$$\mathbb{P}\left(|\pi^n(f) - \pi(f)| \geq \varepsilon\right) \leq 2e^{-\frac{\lambda^2 n}{2}(\varepsilon - [c_3(m) + 2/\lambda]/n)^2}$$

with some finite constant $c_3(m) \leq c_1(m)/(1 - e^{-c_2(m)})$.

1.5.5.1   Application in Optimization and Archive Models

As already mentioned, the existing results are mainly intended on the study of the behavior in an optimization environment, focusing on the limiting behavior [145, 146] as well as the limit probability distribution over populations as depicted in [147]. Further results, see for instance [130], deduce properties of the stationary distribution of the Markov chain associated with the evolutionary process by constructing a quotient chain associated with the original chain. The advantage offered by the herein depicted new result resides in the fact that it is based on an assumption that concerns the overall transitions, without being bounded to specific types of operators (mutation, selection) and without requiring any additional stronger assumptions. In fact, it provides the means of building specific transition operators that need only to satisfy the conditions from assumption (A1).

In order to apply the results presented in the previous section, in an optimization context, we consider a finite state space E for which $|E| = d$ and an objective function V having the set of optimal solutions defined as $V^* = \{x | V(x) = \inf(V)\} = \{x_1^*, x_2^*, \ldots, x_{d^*}^*\}$.

The aforementioned existing convergence results from [145, 146] were applied in an optimization context and considered the study of the behavior of the algorithm in limit, when time tends ot infinity. In the current case we consider the probability of deviation from the invariant measure to be bounded by a positive vlaue ε and establish bounds on these probability.

Let us also assume π an invariant measure such that $\forall i = \{1, 2, \ldots, d^*\}$, $\pi(x_i^*) > 0$ and $\pi(V^*) = \sum_{i=1}^{d^*} \pi(x_i^*)$. Let us further consider the optimization context modeled as

$$\begin{cases} X_n^* & = Argmin\{V(X_0), V(X_1), \ldots, V(X_n)\}, \\ V(X_n^*) & = \min\left(V(X_{n-1}^*), V(X_n)\right), \end{cases} \tag{1.58}$$

where $X_{n-1}^* \in V^*$ is the equivalent of having $\frac{1}{n} \sum_{i=0}^{n-1} \mathbb{1}_{V^*}(X_i) > 0$.

When applying the previously obtained results from equation (1.57), for $f = \mathbb{1}_{V^*}$ and by considering the measure π on V^*, the following holds:

$$\mathbb{P}\left(\left| \frac{1}{n} \sum_{i=0}^{n-1} \mathbb{1}_{V^*}(X_i) - \pi(V^*) \right| \geq \varepsilon \right) \leq 2e^{-\frac{\lambda^2 n}{2}(\varepsilon - [c_3(m) + 2/\lambda]/n)^2} \tag{1.59}$$

for any $n > [c_3(m) + 2/\lambda]/\varepsilon$. Adopting an opposite perspective, i.e. for the probability of having a deviation smaller than a given threshold, the following expression is derived:

$$\mathbb{P}\left(\left| \frac{1}{n} \sum_{i=0}^{n-1} \mathbb{1}_{V^*}(X_i) - \pi(V^*) \right| < \varepsilon \right) \geq 1 - 2e^{-\frac{\lambda^2 n}{2}(\varepsilon - [c_3(m) + 2/\lambda]/n)^2} \tag{1.60}$$

for any $n > [c_3(m) + 2/\lambda]/\varepsilon$. At the same time, without any loss of generality, we consider $\varepsilon = \varepsilon' \pi(V^*)$, with $\varepsilon' \in [0, 1[$, which leads to

$$\mathbb{P}\left(\left| \frac{1}{n} \sum_{i=0}^{n-1} \mathbb{1}_{V^*}(X_i) - \pi(V^*) \right| < \varepsilon \right) \leq \mathbb{P}\left(\frac{1}{n} \sum_{i=0}^{n-1} \mathbb{1}_{V^*}(X_i) \geq \pi(V^*) - \varepsilon \right) \tag{1.61}$$

where the right part of the expression can be rewritten as follows:

$$\mathbb{P}\left(\frac{1}{n}\sum_{i=0}^{n-1}\mathbb{1}_{V^*}(X_i) \geq \pi(V^*)(1-\varepsilon')\right) \geq 1 - 2 \times e^{-\frac{\lambda_*^2 n}{2}\left(\varepsilon'\pi(V^*)-[c_3(m)+2/\lambda]/n\right)^2}$$

$$(1.62)$$

for any $n > [c_3(m)+2/\lambda]/\varepsilon'\pi(V^*)$. Nonetheless, given that the following stands, in relation with the above result, we conclude the following

$$\mathbb{P}\left(\frac{1}{n}\sum_{i=0}^{n-1}\mathbb{1}_{V^*}(X_i) \geq \pi(V^*)(1-\varepsilon')\right) \leq \mathbb{P}\left(X_n^* \in V^*\right) \qquad (1.63)$$

and

$$\mathbb{P}\left(X_n^* \in V^*\right) \geq 1 - 2 \times e^{-\frac{\lambda_*^2 n}{2}\left(\varepsilon'\pi(V^*)-[c_3(m)+2/\lambda]/n\right)^2}$$

for any $0 < \varepsilon' < 1$ and any $n > [c_3(m)+2/\lambda]/\varepsilon'\pi(V^*)$. For instance, taking $\varepsilon' = 1/2$ we find that

$$n > 4\left(c_3(m)+2/\lambda\right)/\pi(V^*) \Rightarrow \pi(V^*)/2 - [c_3(m)+2/\lambda]/n > \pi(V^*)/4$$

From these observations, we obtain the following theorem:

Theorem 1.5.7. *Under the conditions of* **Assumption (A1)**, *we have that*

$$\forall n > 4\frac{c_3(m)+2/\lambda}{\pi(V^*)}, \qquad \mathbb{P}\left(X_n^* \in V^*\right) \geq 1 - 2e^{-n\,(\lambda\pi(V^*))^2/32}$$

The above result clearly shows that convergence is attained exponentially fast as $n \to \infty$. As a consideration for application in practice, if $\pi(V^*)$ or λ are close to zero, a large value is required for n, i.e. the algorithm needs a large number of iterations in order to converge. The current result reaches generality as it provides clear bounds on the probability that the evolutionary algorithm modeled as a Markov process, approaches the actual global optima of the optimization problem, without focusing on limit properties when time goes to infinity and without considering the absence of mutation/selection [129].

1.5.5.2 Bounds on Perturbed Processes

The current section aims at quantifying the error that a given stochastic perturbation has on the behavior of a genetic algorithm. Perturbations are considered to be induced, for example, as a result of external stochastic factors affecting the transition kernel and/or the selection kernel based on the use of a potential function. While the semantics of what exactly *perturbed behavior* means are widely open, e.g. some functional error, malicious behavior, etc., we will only consider that, with some known probability, the operators behave in some different manner than what is expected. In order to model this behavior we consider a genetic algorithm for which the transition kernel and the potential function are given by M_n, respectively G_n. A perturbed version of the algorithm, in the limits of the previous terms, is considered to be defined on M_n^ε and G_n^ε, where, with some fixed probability ε, the behavior of the transition kernel, for example, is different than what the M_n kernel models, while with probability $1 - \varepsilon$, the M_n kernel applies. An analogous definition is considered for G_n^ε, i.e. with probability ε the potential of a solution is given by G_n^ε, otherwise, with probability $1 - \varepsilon$, being given by G_n. Examples may be found in practice, e.g. algorithms executed across volatile resources or with the support of external, unreliable participants that offer or share computational power, and where, due to failures or malicious behavior, the way different operators act can not be ensured – a brief outline and discussion is offered by the end of this section.

A question that one may ask is, knowing that M_n and M_n^ε are comparable up to some constant, what impact on convergence does the M_n^ε transition kernel have, i.e. is there a significant difference between $\eta_n^{\varepsilon,N}(f)$ and $\eta_n^N(f)$, do the algorithms converge to similar or comparable results? A similar remark can be raised by observing the effect G_n^ε has on convergence. An analysis of both cases is presented in the following, within some assumptions on the relative difference of M_n^ε and M_n, respectively of G_n^ε and G_n.

We recall that the total variance distance is defined by

Definition 1.5.8. (Total variance distance)

$$\| \mu - v \|_{tv} = \sup_{f:\omega(f)\leq 1} |\mu(f) - v(f)|$$

Given a positive and bounded potential function G_n on E, we start by introducing the mappings $(\phi_n)_{n\geq 1}$, $(\psi_{G_n})_{n\geq 0}$, respectively $(\phi_n)_{n\geq 1}^\varepsilon$ and $(\psi_{G_n})_{n\geq 0}^\varepsilon$ from $P(E)$ into itself, with M_n being some Markov transition; ψ_{G_n} can be, for example, a Boltzmann-Gibbs mapping.

$$\begin{cases} \phi_{n+1}^\varepsilon(\eta) & = \Psi_{G_n^\varepsilon}(\eta)M_{n+1}^\varepsilon \\ \phi_{n+1}(\eta) & = \Psi_{G_n}(\eta)M_{n+1} \end{cases} \tag{1.64}$$

In order to quantify the difference in behavior of the two different variants, we would like to estimate $\phi_{n+1}^\varepsilon(\eta) - \phi_{n+1}(\eta)$, which can be further decomposed as

$$\begin{aligned} \phi_{n+1}^\varepsilon(\eta) - \phi_{n+1}(\eta) &= \Psi_{G_n^\varepsilon}(\eta)M_{n+1}^\varepsilon - \Psi_{G_n}(\eta)M_{n+1} \\ &= \Psi_{G_n^\varepsilon}(M_{n+1}^\varepsilon - M_{n+1}) + [\Psi_{G_n^\varepsilon}(\eta) - \Psi_{G_n}(\eta)]M_{n+1} \end{aligned} \tag{1.65}$$

By denoting with $v = \Psi_{G_n}(\eta)$, we notice that

$$\begin{aligned} \Psi_{G_n^\varepsilon}(\eta) - \Psi_{G_n}(\eta) &= [\Psi_{G_n^\varepsilon/G_n}(v) - v](f) \\ &= \frac{1}{v(G_n^\varepsilon/G_n)} v\left(\left(\frac{G_n^\varepsilon}{G_n}\right)[f - v(f)] \right) \end{aligned} \tag{1.66}$$

Next, we denote the oscillation of a function f by $\omega(f)$ and use it to define two working hypothesis, as described in the following:

Hypothesis H 1. *The Markovian transition kernels M_n^ε and M_n differ up to some constant c_1, the ε probability of having some alternative behavior than what is expected, and the oscillation of f*

$$\|M_n^\varepsilon(f) - M_n(f)\| \le c_1 \varepsilon \omega(f) \tag{1.67}$$

Example 1.5.9. *A simple example can be constructed by defining M_n^ε as $M_n^\varepsilon = \varepsilon_n K_n + (1 - \varepsilon_n) M_n$, with $\varepsilon = \sup_n(\varepsilon_n)$ and $K_n(x, dy) = \delta_n(dy)$, M_n Markov transitions. Following this rationale, the following relation can be inferred:*

$$M_n^\varepsilon(f)(x) = \varepsilon_n K_n(f)(x) + (1 - \varepsilon_n) M_n(f)(x)$$
$$M_n^\varepsilon(f)(x) - M_n(f)(x) = \varepsilon_n (K_n(f)(x) - M_n(f)(x))$$
$$|M_n^\varepsilon(f)(x) - M_n(f)(x)| \le \varepsilon_n \left| \iint K_n(x, dy) M_n(x, dz) (f(y) - f(z)) \right|$$

Nonetheless, as $f(y) - f(z) \le \omega(f)$, it directly follows that $|M_n^\varepsilon(f)(x) - M_n(f)(x)| \le \varepsilon_n \omega(f)$ which, when taking $\varepsilon = \sup_n \varepsilon_n$, leads to the following relation:

$$\sup_n |M_n^\varepsilon(f)(x) - M_n(f)(x)| = \|M_n^\varepsilon(f) - M_n(f)\| \le \varepsilon \omega(f).$$

Hypothesis H 2. *The difference between the potential functions is bounded, meaning:*

$$\|G_n^\varepsilon / G_n\| \le c_2 \varepsilon \tag{1.68}$$

which is equivalent to

$$1 - c_2 \varepsilon \le G_n^\varepsilon / G_n \le 1 + c_2 \varepsilon \tag{1.69}$$

Example 1.5.10. *As a direct example we can take $G_n^\varepsilon = e^{-V_n^\varepsilon}$ and $G_n = e^{-V_n}$, leading to $G_n^\varepsilon / G_n = e^{-(V_n^\varepsilon - V_n)}$. The expression can be rewritten to read $|G_n^\varepsilon / G_n - 1| = |e^{V_n - V_n^\varepsilon} - e^0|$ where the following stands:*

$$|e^{(V_n - V_n^\varepsilon)(x)} - e^0| \le |(V_n - V_n^\varepsilon)(x)| \times e^{\|V_n\| + \|V_n^\varepsilon\|}$$

Further, knowing that $\|V_n\| < \infty$ and that $\sup_\varepsilon \|V_n^\varepsilon\| < \infty$, there exists a constant v_1 such that $\|V - V^\varepsilon\| \le \varepsilon v_1$. Next, given that $e^{\|V_n\| + \|V_n^\varepsilon\|} \le e^{2\max(\|V_n\|, \sup_\varepsilon \|V_n^\varepsilon\|)}$, the following relation holds:

$$\|G_n^\varepsilon / G_n - 1\| \le \varepsilon c_2, \text{ where } c_2 \le v_1 e^{2\max(\|V_n\|, \sup_\varepsilon \|V_n^\varepsilon\|)}.$$

Example 1.5.11. *We adopt a different perspective, considering that* $V_n^\varepsilon = (1 - \varepsilon)V_n + \varepsilon W_n$, *or in a different form, that* $V_n^\varepsilon - V_n = \varepsilon(W_n - V_n)$. *Knowing that* $\sup_n \|W_n\| < \infty$ *and that* $\sup_n \|V_n\| < \infty$, *we obtain* $\|V_n^\varepsilon - V_n\| \leq \varepsilon c_2$, *with* $c_2 = \sup_n \|V_n\| + \sup_n \|W_n\|$. *In analogous manner we have the following:*

$$\|V_n^\varepsilon\| \leq (1-\varepsilon)\sup_n \|V_n\| + \varepsilon \sup_n \|W_n\| \leq \max(\sup_n \|V_n\|, \ \sup_n \|W_n\|).$$

From the second hypothesis **(H2)**, for any $\varepsilon < \frac{1}{2c_2}$ we obtain that

$$
\begin{aligned}
&[\Psi_{G_n^\varepsilon / G_n}(\nu)(f) - \nu](f) \leq \frac{c_2\varepsilon}{1 - c_2\varepsilon}\nu(|f - \nu(f)|) \leq \frac{c_2\varepsilon}{1 - c_2\varepsilon}, \\
&\|\Psi_{G_n^\varepsilon}(\eta) - \Psi_{G_n}(\eta)\|_{tv} = \sup_{f \,:\, \omega(f)\leq 1} |\Psi_{G_n^\varepsilon}(\eta)(f) - \Psi_{G_n}(\eta)(f)| \leq \frac{c_2\varepsilon}{1 - c_2\varepsilon},
\end{aligned}
\tag{1.70}
$$

Furthermore, equation (1.65), defining the difference between two mappings, can be rewritten as:

$$
\begin{aligned}
\|\phi_{n+1}^\varepsilon(\eta) - \phi_{n+1}(\eta)\|_{tv} &= c_1\varepsilon\omega(f) + \frac{c_2\varepsilon}{1 - c_2\varepsilon}\omega(M_n(f)) \\
&\leq c_1\varepsilon + \frac{c_2\varepsilon}{1 - c_2\varepsilon} \quad \text{as soon as } \omega(f) \leq 1 \\
&\leq (c_1 + 2c_2)\varepsilon \quad \text{for any } \varepsilon < 1/(2c_2).
\end{aligned}
\tag{1.71}
$$

In summary, we have proved the following technical lemma.

Lemma 1.5.12. *In the conditions defined by hypotheses **(H1)** and **(H2)**, for any* $\varepsilon < 1/(2c_2)$ *and any probability measure* η *we obtain that*

$$\|\phi_{n+1}^\varepsilon(\eta) - \phi_{n+1}(\eta)\|_{tv} \leq (c_1 + 2c_2)\varepsilon$$

Let (γ_n, η_n) be the Feynman-Kac model associated with the potential function G_n and the transition kernel M_n, representing the normalized, respectively unnormalized Feynman-Kac measures. At first we consider defining the sequential update of the flow of distributions $(\eta_n)_{n\geq 0}$ (standard case) and $(\eta_n^\varepsilon)_{n\geq 0}$ (perturbed variant):

$$\eta_n^\varepsilon = \phi_n^\varepsilon(\eta_{n-1}^\varepsilon),\tag{1.72}$$

where η_n is the measure associated with N independent samples of common law $\phi_n(\eta_{n-1})$. We will adopt the convention that for $p = 0$, $\eta_0 = \phi_0(\eta_{n-1}^\varepsilon)$.

An additional mapping notation is further considered $\phi_{p,n} = \phi_n \circ \ldots \circ \phi_{p+1}$. For $p = n$ we consider that $\phi_{p,n} = \phi_{n,n} = Id$, the identity mapping, and

$$\phi_{p,n}(\eta_p) = \eta_n.\tag{1.73}$$

For the normalized Feynman-Kac measure, we consider it as defined by

$$\gamma_p Q_{p,n} = \gamma_n \qquad (1.74)$$

with $Q_{p,n}$ having the following functional representation

$$Q_{p,n}(f)(x_p) = \mathbb{E}\left(f(X_n) \prod_{k=p}^{n-1} G_k(X_n) | X_p = x_p \right). \qquad (1.75)$$

The rationale behind introducing the additional mapping notation is to ease the description of the difference between the empirical measures in the presence of external stochastic factors and in the classical case, defined as follows:

$$\eta_n^\varepsilon - \eta_n = \sum_{p=0}^{n} [\phi_{p,n}(\eta_p^\varepsilon) - \phi_{p,n}(\phi_p(\eta_{p-1}^\varepsilon))] \qquad (1.76)$$

The proof of the above has been obtained using a telescoping sum decomposition.

Let us now consider the following regularity properties.

Hypothesis H. There exists some integer $m \geq 1$ and some parameter $\varepsilon > 0$ such that for any $p \geq 1$, any $(x,y) \in E^2$ and any measurable subset A we have that

$$M_{p+1},\dots,M_{p+m}(x,A) \geq \varepsilon \times M_{p+1},\dots,M_{p+m}(y,A)$$
$$g = \sup_p \sup_{x,y} \frac{G_p(x)}{G_p(y)} < \infty \qquad (1.77)$$

The following contraction inequalities are proved in [38, 42, 43, 49], see also [58] for a more recent development on these stability properties.

Theorem 1.5.13. *We assume (H) is met for some parameters (m,ε). In this situation, there exists some $\delta \in]0,1[$ such that for any probability measures (η, ν), and for any $p \leq n$, we have*

$$\|\phi_{p,n}(\eta) - \phi_{p,n}(\nu)\|_{tv} \leq c\,(1-\delta)^{(n-p)/m}\|\eta - \nu\|_{tv}. \qquad (1.78)$$

for some finite constant $c < \infty$ whose values do not depend the parameters (p,n,η,ν).

Replacing η with η_n^ε and ν with η_n, we find that

$$\|\eta_n^\varepsilon - \eta_n\|_{tv} \leq \sum_{p=0}^{n} c(1-\delta)^{\frac{n-p}{m}} \|\phi_p^\varepsilon(\eta_{p-1}^\varepsilon) - \phi_p(\eta_{p-1}^\varepsilon)\|_{tv}$$
$$\leq c\,(c_1 + 2c_2)\varepsilon \times \sum_{p \geq 0} (1-\delta)^{\frac{p}{m}} \qquad (1.79)$$

for any $\varepsilon < 1/(2c_2)$. The second implication comes from Lemma 1.5.12., under the conditions (H1) and (H2). We conclude that

$$\sup_{n \geq 0} \|\eta_n^\varepsilon - \eta_n\|_{tv} \leq c(\delta)\varepsilon \quad \text{with} \quad c(\delta) \leq c\,(c_1 + 2c_2)/(1 - (1-\delta)^{1/m}) \qquad (1.80)$$

We recall that we consider a population of N individuals and the occupation measures of the population for the two variants (standard and perturbed) are approximated as follows:

$$
\eta_n^{\varepsilon,N} = \frac{1}{N}\sum_{i=1}^{N}\delta_{\xi_n^{\varepsilon,i}}, \ with
\begin{cases}
M_n^{\varepsilon} \ mutation \\
G_n^{\varepsilon} \ selection
\end{cases}
$$
$$
\eta_n^{N} = \frac{1}{N}\sum_{i=1}^{N}\delta_{\xi_n^{i}}, \ with
\begin{cases}
M_n \ mutation \\
G_n \ selection
\end{cases}
\tag{1.81}
$$

Hypothesis $\mathbf{H}^{\varepsilon}$ There exists some integer $m \geq 1$ and some parameter $\varepsilon' > 0$ such that for any $p \geq 1$, any $(x,y) \in E^2$, and $\varepsilon > 0$, and any measurable subset A we have that

$$
\sup_{\varepsilon \geq 0}\sup_{p}\sup_{x,y} \frac{G_p^{\varepsilon}(x)}{G_p^{\varepsilon}(y)} < \infty
$$
$$
M_{p+1}^{\varepsilon},\ldots,M_{p+m}^{\varepsilon}(x,A) \geq \varepsilon'\ M_{p+1}^{\varepsilon},\ldots,M_{p+m}^{\varepsilon}(y,A)
\tag{1.82}
$$

Under the assumptions **(H)** and **($\mathbf{H}^{\varepsilon}$)**, based on the Proposition 2.9 from [52] (see also [43, 58]) it can be deduced that $(\forall p \geq 1)$ and $(\forall f : \omega(f) \leq 1)$

$$
\sup_{n \geq 0}\mathbb{E}\left(|\eta_n^{N}(f) - \eta_n(f)|^p\right)^{\frac{1}{p}} \leq c(p)/\sqrt{N}
$$
$$
\sup_{n \geq 0}\mathbb{E}\left(|\eta_n^{N,\varepsilon}(f) - \eta_n^{\varepsilon}(f)|^p\right)^{\frac{1}{p}} \leq c(p)/\sqrt{N}
\tag{1.83}
$$

for some finite constant $c(p)$. This implies that

$$
\mathbb{E}\left(|\eta_n^{N}(f) - \eta_n^{\varepsilon,N}(f)|^p\right)^{\frac{1}{p}} \leq \frac{2c(p)}{\sqrt{N}} + c(\delta)\varepsilon \leq c(p,\delta)\left(\frac{1}{\sqrt{N}} + \varepsilon\right)
\tag{1.84}
$$

for some finite constant $c(p,\delta) \leq \max\left(2c(p),c(\delta)\right)$.

Some exponential estimates can also be deduced using Bernstein-type martingales inequalities or alternatively by employing the Hoeffding's inequality [58, 57]. Under the assumptions **(H1)** and **(H2)** for any measurable function f, s.t. $\|f\| \leq 1$, for any $x > 0$ and any $N \geq 1$, when considering c_1 as being a finite constant related to *the bias* of the particle model and c_2 a constant related to the variance of the method, we have

$$\mathbb{P}\left((\eta_n^N - \eta_n)(f) \leq \frac{c_1}{N}(1 + x + \sqrt{x}) + \frac{c_2}{\sqrt{N}}\sqrt{x})\right) \geq 1 - e^{-x} \qquad (1.85)$$

$$\mathbb{P}\left(\left|\eta_n^N(f) - \eta_n(f)\right| \leq \frac{c_1}{N}(1 + x + \sqrt{x}) + \frac{c_2}{\sqrt{N}}\sqrt{x})\right) \geq 1 - 2e^{-x}. \qquad (1.86)$$

Next, by taking

$$(\eta_n^{N,\varepsilon} - \eta_n^N)(f) = (\eta_n^{N,\varepsilon} - \eta_n^\varepsilon)(f) + (\eta_n^\varepsilon - \eta_n)(f) + (\eta_n - \eta_n^N)(f)$$

where the middle term is bounded with respect to ε up to some constant b_1, i.e. $(\eta_n^\varepsilon - \eta_n)(f) \leq b_1\varepsilon$, and the relation in Equation (1.86) (with respect to the first and third terms), we obtain the following concentration inequalities

$$\mathbb{P}\left((\eta_n^{N,\varepsilon} - \eta_n^N)(f) \leq 2\left(\frac{c_1}{N}(1 + x + \sqrt{x}) + \frac{c_2}{\sqrt{N}}\sqrt{x}\right) + b_1\varepsilon\right) \geq 1 - 2e^{-x}$$

$$\mathbb{P}\left(\left|(\eta_n^{N,\varepsilon} - \eta_n^N)(f)\right| \leq 2\left(\frac{c_1}{N}(1 + x + \sqrt{x}) + \frac{c_2}{\sqrt{N}}\sqrt{x}\right) + b_1\varepsilon\right) \geq 1 - 4e^{-x}.$$

Direct applications of this result could find a way into, for example, distributed desktop computing. For a few notorious examples, one can refer to the Seti@Home (Search for Extraterrestrial Intelligence) [4], Leiden Classical (desktop computer grid dedicated to general classical dynamics) [158], Rosetta@Home (protein folding, design, and docking) [156] or MilkyWay@Home (highly accurate model of our galaxy) [32]. A common part all those projects have is the use of desktop resources offered by (anonymous) users all over the world. As underlying principles of how the workload is managed, the aspects hereafter need to be considered, within the limits of the herein results, i.e. genetic algorithms like structure:

- a high complexity of the problem to deal with, e.g. sampling a large conformational space or running a computationally intensive analysis over enormous amounts of data, surpassing the power provided by classical resources like clusters or even grids; a second important element is the use of spare computational cycles (or specified amount), allowing to give a meaning to otherwise wasted energy and computing power while simultaneously contributing to an advance of our knowledge on important scientific problems;
- the problem allows a decomposition into independent sub-tasks that can be independently processed – each of the participating users only needs to deal with such sub-tasks (locally installed clients) and does not see the complete picture the problem draws; as a straightforward example, this may be the equivalent of receiving a set of instances from a server (input data), processing and last, sending back the results.

While this (simplified) design has several advantages, it is not difficult to understand that it is also subject to several issues that direct to security or data integrity aspects. Errors, as a result of network or processing faults, may lead to data loss or corruption. At the same time, malicious behavior, while unlikely, can not be excluded – if reasons exist for malware and viruses, why would this particular context be any different? A question one may raise is how errors can be controlled or how results can be legitimated. While different approaches exist, the result presented in this section implies that an exponential decrease of the probability of having a difference above a given threshold can be obtained.

1.5.5.3   Weak Bounds for Behavior under Perturbation

We are modeling in the following a stochastic behavior that can occur for evolutionary algorithms. Let a candidate solution be modeled as a Markov chain $X_n = \begin{pmatrix} \varepsilon_n \\ Y_n \end{pmatrix}$ taking values in the product space $(\{0,1\} \times E)$. The parameter ε_n represents the stochastic factor, modeling the presence or absence of an uncertain behavior. We further assume that ε_n is a sequence of independent Bernoulli random variables with common law

$$\mathbb{P}(\varepsilon_n = 0) = (1 - \mathbb{P}(\varepsilon_n = 1)) = p$$

Let us now consider that the behavior involving no external stochastic factor and the absence of uncertainty is modeled through a value of zero attributed to the stochastic marker as $(\varepsilon_0, \dots, \varepsilon_n) = (0, \dots, 0)$. Given a realization $(\varepsilon_p)_{0 \le p \le n} = (u_p)_{0 \le p \le n}$, the second component Y_n forms a Markov chain with transitions M_{n+1,u_n} that depends on the parameter u_n, the initial random variable Y_0 is also distributed w.r.t. some probability measure η_{0,u_0} that depends on u_0.

We consider the space of possible values for the stochastic marker as Ω_n and let

$$\Omega_n^0 = \{\forall 0 \le p \le n \quad , \varepsilon_p = 0\}, \tag{1.87}$$

and $(\Omega_n^0)^C$ the complementary set. For a given function f, we use the following notation

$$f_n^{(0)}(y) = f_n \begin{pmatrix} 0 \\ y \end{pmatrix} \quad \text{and} \quad f_n^{(1)}(y) = f_n \begin{pmatrix} 1 \\ y \end{pmatrix}$$

We also set $G_n \begin{pmatrix} 0 \\ y \end{pmatrix} = G_n^{(0)}(y)$ and consider the uniform norm be given by $\|f\| = \sup_{u \in \{0,1\}, y \in E} \left| f \begin{pmatrix} u \\ y \end{pmatrix} \right|$.

Given the number of transitions that the algorithm is subject to, given by n, the normalized Feynman-Kac measure associated with the perturbed process behavior is given by

$$\gamma_n(f) = \mathbb{E} \left(f(X_n) \prod_{k=0}^{n-1} G_k(X_k) \right) \tag{1.88}$$

We let $\gamma_n^{(0)}$ be the Feynman-Kac measure defined by

$$\gamma_n^{(0)}\left(f_n^{(0)}\right) = \mathbb{E}\left(f_n^{(0)}(Y_n^{(0)})\prod_{k=0}^{n-1}G_n^{(0)}(Y_k^{(0)})\right) \tag{1.89}$$

where $Y_n^{(0)}$ stands for the Markov chain with transitions $M_{n+1,0}$ and initial distribution $\eta_{0,0}$.

The relation defined in equation (1.88) can be further decomposed in two cases according to the complete absence of external stochastic factors and the presence of perturbations.

$$\begin{aligned}\gamma_n(f) &= \mathbb{E}\left(f(X_n)\prod_{k=0}^{n-1}G_k(X_k)\mathbb{1}_{\Omega_n^0}\right) + \mathbb{E}\left(f(X_n)\prod_{k=0}^{n-1}G_k(X_k)\mathbb{1}_{(\Omega_n^0)^C}\right)\\ &= \gamma_n^{(0)}\left(f_n^{(0)}\right)\mathbb{P}(\Omega_n^0) + \mathbb{E}\left(f(X_n)\prod_{k=0}^{n-1}G_k(X_k)\mathbb{1}_{(\Omega_n^0)^C}\right).\end{aligned} \tag{1.90}$$

From equation 1.90 we can further derive

$$\begin{aligned}|\gamma_n(f) - \gamma_n^{(0)}\left(f_n^{(0)}\right)| &\leq |\gamma_n^{(0)}\left(f_n^{(0)}\right)|\left(1 - (1-p)^{n+1}\right) + \left|\mathbb{E}\left(f(X_n)\prod_{k=0}^{n-1}G_k(X_k)\mathbb{1}_{(\Omega_n^0)^C}\right)\right|\\ &\leq |\gamma_n^{(0)}\left(f_n^{(0)}\right)|\,|(1-p)^{n+1} - 1| + \|f\|\prod_{k=0}^{n-1}\|G_k\|(1 - \mathbb{P}(\Omega_n^0))\\ &\leq (1 - (1-p)^{n+1})\left(|\gamma_n^{(0)}\left(f_n^{(0)}\right)| + \|f\|\prod_{k=0}^{n-1}\|G_k\|\right)\end{aligned} \tag{1.91}$$

This implies that

$$|\gamma_n(f) - \gamma_n^{(0)}\left(f_n^{(0)}\right)| \leq (1 - (1-p)^{n+1})\,c(n)$$

for some constant $c(n) \leq 2\|f\|\prod_{k=0}^{n-1}\|G_k\|$.

In practice, when subject to a corrupted computing environment (e.g. involving malicious/cheating behavior or faults of the hardware material) this result provides a quantitative measure of the fault-tolerance accepted by the system. This can be useful in assesing the level of accuracy of the results.

1.5.6   Rare Events Stochastic Models

1.5.6.1   Calibration and Uncertainty Propagation

Modern computers are capable of simulating complex physical and engineering systems. Nevertheless, formalized mathematical models are rarely certain and error-free. For instance, the physical environment is often too complex to formalize perfectly, and all the different physical scales are difficult to capture with high precision. In addition, the reliability and the accuracy of computational approximation models often relies on complex calibration processes combined with the dispersion analysis of inputs and other sources of randomness.

Given some reference physical observation we would like to calibrate the model parameters so that the outputs simulated by some numerical code coincide with this reference data, or at least behave as much as possible as these physical observations.

In another context, with a successfully calibrated model one may be also interested in computing the probability that simulation outputs belong to some critical event; that is, to find the law of the input parameters and the sources of randomness leading to such events.

This couple of important issues can be formulated in terms of a classical input-output transformation

$$\underbrace{\text{Inputs} = I}_{\substack{\text{sources of randomness} \\ \text{uncertainty representations} \\ \text{tuning parameters} \\ \text{unknown kinetic parameters}}} \longrightarrow [\textbf{Black-box simulation model}] \longrightarrow \underbrace{\text{Outputs } O = C(I)}_{\substack{\text{physical, biological or forecasting predictions} \\ \text{partial differential equation profiles} \\ \text{mechanical forces} \\ \text{hydrodynamic profiles}}}$$

The prototype of question arising in practice is the following. We are given a desired domain, say $\mathcal{O}$, in the space of the outputs, and we want to estimate both the probability that the outputs fails into this set and the distribution of the inputs leading to these outputs; that is, we want to compute the following quantities

$$\text{Proba}(O \in \mathcal{O}) \quad \text{and} \quad \text{Law}(I \mid O \in \mathcal{O})$$

- The set $\mathcal{O}$ represents some critical event with a very small occurrence probability, say 10^{-9}. In this context, we are interested in computing rare event probabilities as well as the distribution of the random sources leading to this critical regime. The conditional distribution provides all the statistical information on the different contributions of the input parameters and the random sources on "the desired" critical rare event.
- The domain $\mathcal{O}$ is related to some distance-like criteria that measure the adequacy of some output profile with some reference data or some observations delivered by some sensors. In this context, we are interested in computing the chance that some collection of models may reach a given precision w.r.t. the physical data. Furthermore, we are also often interested in calibrating the numerical code with the selection of the most accurate input parameters that achieve a given precision w.r.t. the data.
- The couple of situations discussed above can be combined. For instance, we may be interested in computing the probability of a critical rare event given some observations, or reversely the law of some input parameters given some observations as well as some critical event.

Of course, the choice of the stochastic particle algorithm that solves these three questions is far from being unique. The design of an appropriate Monte Carlo simulation model strongly depends on the physical problem at hand, including the nature of the numerical code, the quantification of the sources of randomness, the specification of the inputs and the outputs profiles.

In the following we consider the case where some fixed critical level exists, given by a value h and the rare event of interest is the fact that the output value passes the maximal value h

The inputs are represented by a function like d-dimensional vector $I = (I^1, \ldots, I^d)$ and the set $\mathscr{O}$ represents a failure region. The outputs of the numerical code are represented by a function like d'-dimensional vector $O = (O^1, \ldots, O^{d'})$. In this situation, we are given some critical threshold value h and a random event of interest is the fact that the forces acting on the provided structure get above this maximal value; the corresponding probabilistic quantities of interest are given below

$$\text{Proba}\left(\sup_{1 \leq t \leq d'} |O^i| \geq h\right) \quad \text{and} \quad \text{Law}\left((I^1, \ldots, I^d) \mid \sup_{1 \leq t \leq d'} |O^i| \geq h\right)$$

Our next objective is to relate these questions to the probabilistic model of combinatorial counting and sampling presented in section 1.5.4.3. To this end, let us assume that the random input parameters I are distributed according to some probability measure λ on some finite or countable state space; that is we have that

$$\text{Proba}(I = x) = \lambda(x)$$

We also suppose that the input-output function is given by some mapping $C : x \mapsto C(x)$, and we set

$$A := \{x : C(x) \in \mathscr{O}\}$$

In this notation, the uncertainty propagation models presented above coincide with the ones discussed in section 1.5.4.3; that is, we have that

$$\text{Proba}(O \in \mathscr{O}) = \text{Proba}(I \in A)\, \lambda(A) := \sum_{x \in A} \lambda(x)$$

and

$$\text{Proba}(I = x \mid O \in \mathscr{O}) = \mu_A(x) := \frac{1}{\mathscr{Z}_A}\, \mathbb{1}_A(x)\, \lambda(x)$$

with the normalizing constant $\mathscr{Z}_A = \lambda(A)$.

In engineering literature, the multilevel genetic type splitting particle algorithms presented in section 1.5.4.3 are also called subset-simulation models. The central idea is to express a rather small failure probability as the product of not so small conditional probabilities:

$$\text{Proba}(I \in A) = \prod_{p=0}^{m} \text{Proba}(I \in A_{p+1} \mid I \in A_p) \tag{1.92}$$

The intermediate levels A_p are judiciously chosen failure regions s.t.

$$A_0 \supset A_1 \supset \ldots \supset A_m = A$$

In this interpretation, the computation of the small failure probability $\text{Proba}(I \in A)$ is now reduced to the computation of larger conditional probabilities $\text{Proba}(I \in A_{p+1} | I \in A_p)$. In the marine engineering problem discussed above, these failure regions are characterized by the choice of an increasing sequence of critical threshold values h_p; that is, if we set

$$I = (I^1, \ldots, I^d) = (x^1, \ldots, x^d) \mapsto O = C(I) = (C^1(I), \ldots, C^{d'}(I)) = (C^1(x), \ldots, C^{d'}(x))$$

then we have that

$$h_p \uparrow \quad \Rightarrow \quad A_p := \left\{ x = (x^1, \ldots, x^d) \mid \sup_{1 \le t \le d'} |C^t(x)| \ge h_p \right\} \downarrow$$

1.5.6.2   An Universal Particle Algorithm Based on Multilevel Splitting

In this section we introduce a generic particle simulation algorithm based on a multilevel splitting mechanism. A simple stochastic particle algorithm consists in propagating a population of N individuals representing potential solutions at each iteration. Hereafter we start by providing the main structure and pseudo-code of a generic multilevel splitting particle algorithm, depicted in Algorithm 1.2.

The algorithm is initialized with N random configurations chosen according to some distribution law ν_0 (in this case we consider it as uniform i.i.d. sampling) in the A_0 set. The algorithm considers a critical level to be reached, in order for the rare event to take place, denoted in the pseudo-code by c_n.

In the following p_1^N stands for the proportion of individuals that succeed to reach the level A_1, these individuals in A_1 being further selected for the next step of the algorithm. As we consider N as a fixed value, the notation will be simplified to p_1 instead of p_1^N. The rejected configurations are then randomly redistributed among the ones that passed into A_1 such that the number of individuals in the population remains constant, N. The following step consists on diversifying and enriching the population of solutions selected in A_1 during the first step. This is performed by applying the transition or perturbation operator F, leading to new candidate samples, while leaving the measure μ_A^1 invariant. In its most simple variants the transition operator can be seen as a mutation operator.

Each individual independently explores the space defined by A_1 following a local Markovian transition that leaves the measure μ_{A_1} invariant. As for the previous step, we denote by p_2 the proportion of individuals having succeeded to pass to the level A_2. We select afterwards the configurations having succeeded to pass at the second level A_2. The rejected configurations are again redistributed randomly among the previously selected ones. Each individual explores afterwards, in an independent manner, the space A_2 following a local Markovian transition that leaves the measure μ_{A_2} invariant. The process is reiterated until the last level n is reached.

---

[1] Note: h is the equivalent of h_m when $A = A_m$.

Algorithm 1.2. Multilevel splitting particle algorithm

$A_0 := E$, $h_i :=$ threshold levels, $j := 0$

Sample N particles, $(\xi_0^i)_{1 \leq i \leq N}$ from a given distribution μ_0 in A_0

$$\left\{ A_i = \left\{ x = (x^1, \ldots, x^d) \mid \sup_{1 \leq t \leq d'} |C^t(x)| \geq h_i \right\}, i \leq m+1 \text{ s.t. } h_{m+1} = h \text{ the fixed critical level.}^1 \right\}$$

while $\dfrac{1}{N} \displaystyle\sum_{i=1}^{N} \mathbb{1}_{A_{m+1}} (\xi_j^i) = 0$ **do**

$\quad \widehat{p}_{j+1}(A_{j+1} \mid A_j) := \dfrac{1}{N} \displaystyle\sum_{i=1}^{N} \mathbb{1}_{A_{j+1}} (\xi_j^i)$

$\quad \xi_j(A_{j+1}) := \{ \xi_j^l \mid 1 \leq l \leq N \ s.\, t.\ \xi_j^l \in A_{j+1} \}$

$\quad$ <u>Selection</u>
$\quad$ **for** $i = 1$ to N **do**
$$\widehat{\xi}_j^i := \begin{cases} \xi_j^i, \ if \ \xi_j^i \in A_{j+1} \\ sample \ randomly \ in \ the \ set \ \xi_j(A_{j+1}), \ otherwise. \end{cases}$$
$\quad$ **end for**

$\quad$ <u>Transition</u>
$\quad$ **for** $i = 1$ to N **do**
$$\xi_{j+1}^i := \begin{cases} F(\widehat{\xi}_j^i), \ if \ F(\widehat{\xi}_j^i) \in A_{j+1} \\ \widehat{\xi}_j^i, \quad otherwise. \end{cases}$$
$\quad$ **end for**

$\quad j := j+1$
end while

For the multilevel splitting, the result of this simulation can be explained as follows. When N is increased, the population of solutions obtained at each kth iteration is distributed according to the law of the variable X restricted to the set A_k. This is equivalent to saying, that at each iteration k, the variables $I_k^{(i,N)}$, $1 \leq i \leq N$, simulated on the A_k set are distributed approximately as a sequence of random variables $\bar{I}_k^{(i,N)}$, $1 \leq i \leq N$, independent and having the same law μ_{A_k}. This approximation can be further detailed in several forms. For example, we can have:

$$\left\| \text{Law} \left(I_k^{(1,N)}, \ldots, I_k^{(q,N)} \right) - \text{Law} \left(\bar{I}_k^{(1,N)}, \ldots, \bar{I}_k^{(q,N)} \right) \right\|_{tv} \leq \frac{q}{N} c(k)$$

for all $q \leq N$ and for a finite constant $c(k) < \infty$ which can be specified according to the parameters of the model. In the preceding equation, $\|P - Q\|_{tv}$ denotes the total variation distance between two probability measures P and Q. For any bounded function f and for any $\varepsilon > 0$, the following exponential error probabilities are verified:

$$\mathbb{P}\left(\left| \frac{1}{N} \sum_{i=1}^{N} f\left(I_k^{(i,N)} \right) - \mathbb{E}\left(f(I) \mid I \in A_k \right) \right| \geq \varepsilon \right) \leq c_1(k)\, e^{-\frac{N\varepsilon^2}{c_2(k)}}$$

Algorithm 1.3. Particle algorithm using acceptance-rejection

$A_0 := E$

$\hat{A} := \{A_1, A_2, \ldots, A_n\}$ {Constants specifying the fixed levels $(A_k)_{1 \le k \le n}$ }

Initialization

$\xi_0 :=$ sample N particles, $(\xi_0^i)_{1 \le i \le N}$ randomly of law η_0.

for $k = 1$ to n **do**

    {For each of the intermediate levels $(A_k)_{1 \le k \le n}$ }

    <u>Selection</u>

    for $i = 1$ to N **do**

        {For each particle}

$$\widehat{\xi}_{k-1}^i := \begin{cases} \xi_{k-1}^i, \; if \; \xi_{k-1}^i \in A_k \\ \xi_{k-1}^i, \; random \; variable \; of \; law \; \sum_{i=1}^{N} \frac{\mathbb{1}_{A_k}(\xi_{k-1}^i)}{\sum_{j=1}^{N} \mathbb{1}_{A_k}(\xi_{k-1}^j)} \delta_{\xi_{k-1}^i} \; otherwise. \end{cases}$$

    end for

    <u>Transition</u>

    for $i = 1$ to N **do**

        {For each particle}

$$\xi_k^i := \begin{cases} F(\widehat{\xi}_{k-1}^i), \; if \; F(\widehat{\xi}_{k-1}^i) \in A_k \\ \widehat{\xi}_{k-1}^i, \quad otherwise \end{cases}$$

        {F designates the perturbation operator generating new candidate solutions and ξ_k^i of law $M_k(\widehat{\xi}_{k-1}^i, \cdot)$ with a Markovian transition M_k leaving the measure μ_{A_k} invariant.}

    end for

end for

---

for a couple of finite constants $c_1(k), c_2(k) < \infty$ they may be specified depending on the model parameters.

Furthermore, the product of the proportions of success $\prod_{l=1}^{k} p_l^N$ is an unbiased estimator of the probability that the variable I is in A_k. Under certain regularity assumptions ([27]), the following convergence result can be proved (holds):

$$\mathbb{E}\left(\prod_{l=1}^{k} p_l^N\right) = \mathbb{P}(I \in A_k) \quad \text{and} \quad \mathbb{E}\left(\left[\prod_{l=1}^{k} p_l^N - \mathbb{P}(I \in A_k)\right]^2\right) \le \frac{c}{N} k \, \mathbb{P}(I \in A_k)^2,$$

for a finite universal constant $c < \infty$. Other convergence results, including estimates of error probability exponentially small are described in the book [38] and also in the articles [52, 41, 57].

1.5.6.3   Variants of Multilevel Splitting Simulation

In the description of the preceding algorithm, it is worth mentioning that the parameters (N, A_n) must be chosen judiciously so that at least one solution among the N ones is in A_n, otherwise the algorithm stops at the nth iteration. For this to be fulfilled we propose choosing a first level disparity $(A_n - A_{n-1})$ sufficiently low. Either case, when the number of proposed solutions N' increases, we can prove that all the levels are reachable and that the algorithm converges towards the desired solutions. It is also possible to chose the levels A_n adaptively online according to the proposed candidate samples. For example, one option consists in choosing the first level A_1 such that a given proportion of the current solutions (e.g. 80 percent) of the solutions proposed in A_0 reach the next level. These adaptive splitting algorithms can be recast in terms of sequential Monte Carlo models with adaptive resampling procedures. For a detailed discussion on these models with precise reference pointers we refer the reader to [41]. To our knowledge these adaptive resampling techniques were first introduced as an heuristic scheme in [36] (remark 1, section 2.1), see also [37]. These adaptive criteria for the choice of the levels were also discussed in three recent studies [26, 28, 27]. For a detailed theoretical analysis of these models, including central limit theorems and exponential cumulative ratios, we refer the reader to [41].

There are also other variants allowing the exploration of the search space according to these new data and solving this stopping problem. These techniques are more complex to describe and they will be detailed in the follow up. The main idea behind this is to create new candidate solutions until configurations in A_1 are reached. This step can also be catastrophic if the level A_1 is badly chosen.

We start by differentiating the two types of multilevel splitting considered in the following, according to the mechanism used in establishing the different splitting levels. The levels can be either fixed *a priori* or established adaptively at each iteration step by the threshold passed by a percentage of the sampled solutions. Also, two types of selection mechanisms are considered: the uniform selection and the acceptance-rejection selection.

1.5.6.4   Case Study

We illustrate these rather abstract models with a marine engineering problem we recently analyzed with Z. Guede from the French marine research institute (IFRE-MER). In this situation, we want to assess the reliability of an offshore structure, both at the design stage, as to validate the design choice, and in service for maintenance and inspection planning. The goal is to check whether the structure is able to withstand the loads from its environment for its entire planned lifetime, defined according to a physical criterion with respect to the structural response. The structural response is computed by a hydrodynamic numerical code with strong physical and geometrical non-linearity that lead to a complex failure region geometry. In this context, the input parameters are of different natures, some of them representing

---

Algorithm 1.4. Fixed levels particle algorithm

$A_0 := E$, $h_i :=$ threshold levels, $j := 0$

Sample N particles, $(\xi_0^i)_{1 \leq i \leq N}$ uniformly i.i.d. in A_0

$$\{A_i = \left\{ x = (x^1, \dots, x^d) \mid \sup_{1 \leq t \leq d'} |C^t(x)| \geq h_i \right\}, i \leq m+1 \text{ s.t. } h_{m+1} = h \text{ the fixed critical level}\}$$

while $\dfrac{1}{N} \sum_{i=1}^{N} \mathbb{1}_{A_{m+1}}(\xi_j^i) = 0$ **do**

$\qquad \widehat{p}_{j+1}(A_{j+1} \mid A_j) := \dfrac{1}{N} \sum_{i=1}^{N} \mathbb{1}_{A_{j+1}}(\xi_j^i)$

$\qquad \xi_j(A_{j+1}) := \{\xi_j^l \mid 1 \leq l \leq N \text{ s. t. } \xi_j^l \in A_{j+1}\}$

$\qquad$ **for** $i = 1$ to N **do**

$\qquad\qquad \widehat{\xi}_j^i := \begin{cases} \xi_j^i, & \text{if } \xi_j^i \in A_{j+1} \\ \xi_j^i \text{ a randomly chosen particle in the set } A_{j+1}, & \text{otherwise.} \end{cases}$

$\qquad$ **end for**

$\qquad$ **for** $i = 1$ to N **do**

$\qquad\qquad \xi_{j+1}^i := \begin{cases} F(\widehat{\xi}_j^i), & \text{if } F(\widehat{\xi}_j^i) \in A_{j+1} \\ \widehat{\xi}_j^i, & \text{otherwise.} \end{cases}$

$\qquad$ **end for**

$\qquad j := j+1$

end while

---

the spectral properties of wave mixtures, while the other ones represent temperature variations, waves periods and their direction. The outputs of the numerical code are represented by a function like $d^{\,prime}$-dimensional vector $O = (O^1, \dots, O^{d'})$ denoting the forces that act on the offshore structure surface at different time periods. In this situation, we are given some critical threshold value h for which a random event of interest is given by the fact that the forces acting on the offshore structure get above this maximal value.

In order to test the performance and the validity of the approaches in a general context, we employ as testbed the estimation of the evolution of a variable that follows a known law. The choice for χ^2 is due, among others, to its resemblance with the quadratic nature appearing in the real-life experiment proposed by IFREMER.

Let $U_1, U_2, \dots, U_k$ be k independent random variables following the same standard normal law, then the U variable is defined such that

$$\mathbb{P}(\max_{1 \leq i \leq n} X_i^2 \geq c) \tag{1.93}$$

$$\mathbb{P}(\max_{1 \leq i \leq n} X_i^2 \geq c) = 1 - (F_{X_1^2}^n(c)) \tag{1.94}$$

Algorithm 1.5. Adaptive particle algorithm

$A_0 := E, h := constant$ {Critical level}, $k_0 := 0, p := 1$
$\xi_0 := (\xi_0^i)_{1 \le i \le N}$ {N independent particles of law η_0}

while $k_{p-1} \le c$ **do**

$$k_p := \inf \left\{ k > k_{p-1} \ : \ \eta_{p-1}(\mathbb{1}_{A_k}) = \frac{1}{N} \sum_{i=1}^{N} \mathbb{1}_{A_k}(\xi_{p-1}^i) \le \alpha_p, \alpha_p \in \,]0,1[\right\}$$

$$\widehat{p}_p(A_{k_p}|A_{k_{p-1}}) = \frac{1}{N} \sum_{i=1}^{N} \mathbb{1}_{A_{k_p}}(\xi_{p-1}^i)$$

Selection
Select N particles $(\widehat{\xi}_{p-1}^i)_{1 \le i \le N}$ from the A_{k_p} set
{by using either an accept/reject technique or an uniform sampling}

Transition
for $i = 1$ to N **do**
ξ_p^i constructed by successively applying n_{p+1} Markovian elementary transitions M_{k_p} (of invariant measure η_{k_p}).
end for

$p := p + 1$
end while

$$U = \max_{1 \le i \le k} U_i^2 \tag{1.95}$$

follows a law χ^2 with one degree of freedom. For experimental purposes, we consider a number of $k = 2048$ random variables, as this number of degrees of freedom is considered also relevant for the practical IFREMER application.

$$\mathbb{P}\left(\max_{1 \le i \le k} U_i^2 \ge c\right) = 1 - \mathbb{P}(\max_{1 \le i \le k} U_i^2 < c)$$
$$= 1 - F_{U_1^2}(c)^n = 1 - \left(1 - 2\mathbb{P}\left(U_1 \ge \sqrt{c}\right)\right)^n \tag{1.96}$$

From this we obtain in fact the value of the distribution function of the variable U^n, which is denoted in the follow up by $F_{U_1^2}(c)$, where c represents a real value corresponding to a given critical level. Furthermore, the computation of the distribution function for a given level c and a variable $X = U_i^2, 1 \le i \le k$ can be done by using:

$$F_X(c) = \left(\frac{\gamma\left(\frac{c}{2}, \frac{1}{2}\right)}{\Gamma\left(\frac{1}{2}\right)}\right)^n = \left[\frac{1}{\sqrt{\pi}} \int_0^{\frac{c}{2}} \frac{e^{-t}}{\sqrt{t}} dt\right]^n \tag{1.97}$$

where γ is the lower incomplete gamma function. We can also use the following estimations in the formula (1.96)

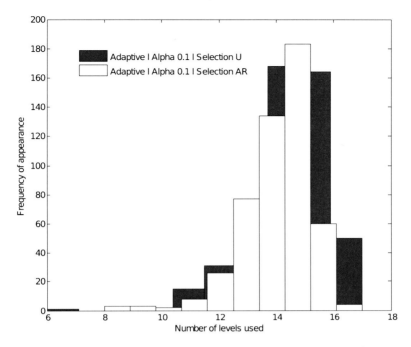

Fig. 1.1 Histogram depicting the number of levels for the adaptive algorithm with α set to 0.1

$$\frac{1}{\sqrt{2\pi}} \, \frac{1}{c+1/c} \, e^{-\frac{c^2}{2}} \leq \mathbb{P}\left(U_1 \geq c\right) \leq \frac{1}{\sqrt{2\pi}} \, \frac{1}{c} \, e^{-\frac{c^2}{2}}$$

A proof of these analytic estimations can be found in the book [56].

In order to prepare the calibration for the practical problem and prepare the testing environment we first studied the distribution of the number of levels employed for the adaptive method, as depicted in Figure 1.1. In concordance with the results, the number of chosen levels for the adaptive method was set to 15, this being given by the central tendency of the number of levels for the two adaptive cases.

The next step considered the analysis of the algorithms' evolution according to the theoretical estimate, obtained as described in Equation 1.96. Figure 1.2 depicts on the ordinate axis the distribution of the values obtained for the probability of passing the intermediate levels, while the abscissa represents the values of the system's response, i.e. from 15.0 to 52.2513 (critical level). The evolution of the adaptive algorithms (where the intermediate levels vary among two different executions) is approximated by a least squares method applied on the entire set of obtained values (cloud of points). The approximation in this latter case is done by estimating the average and standard deviation for a normal density. A comparison of the algorithms' evolution is illustrated in Figure 1.2 successively focusing closer to the critical level region. It is thus possible to evaluate the stalling of the algorithm using fixed levels as compared with the curve of the theoretical

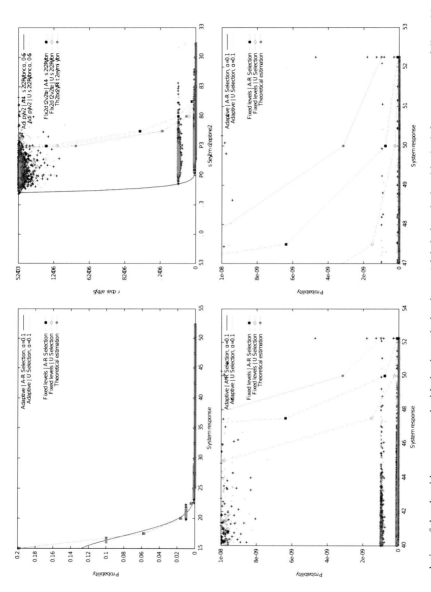

Fig. 1.2 The evolution of the algorithms compared with the theoretical estimate (global view depicted in the left upper corner, followed by successive zooms). Fixed levels algorithm: averaged maximal values per level. Adaptive algorithm: least squares approximation on the set of all the values.

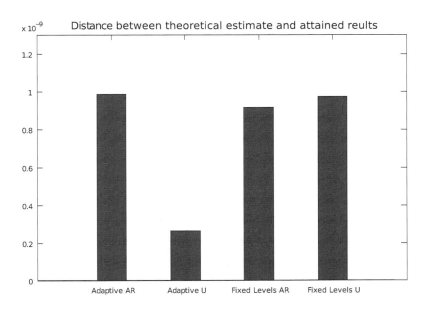

Fig. 1.3 The difference between the final probability attained by the four algorithm variants (averaged over 500 tests) and the theoretical estimate

estimate. Finally, one should note that adaptive variant whose averaged values follow the best the curve of the values estimated by Equation 1.96 is the one using the uniform selection. By considering the simulation of the normal variable described above, for comparative purposes, the adaptive method was compared with the fixed levels variant, by employing two types of selection. The comparison is done by considering the distance between the average of the final probability (obtained on 500 tests per algorithm) and the theoretical estimate of the final probability. As illustrated in Figure 1.3, the best results were obtained with the adaptive variant employing the uniform selection.

1.5.6.5   Multilevel Splitting Simulation

The analysis of rare events arise in various scientific areas including physics, biology, engineering science and financial mathematics. For instance in nuclear physics to study of the performance of a radiation source containment we are interested in computing the probability that a neutron particle emitted by the radiation source escapes from the containment before being absorbed and desintegrated by some obstacle. In biology they may represent an extinction probability of a given population evolution model. In engineering science these rare events are sometimes related to a catastrophic failure such as a buffer excedence in communication networks. Finally in financial mathematics they may represent a ruin process.

The random excursion model is defined in terms of some Markov chain $(X'_n)_{n \geq 0}$ taking values in some finite state space E'. We assume that the chain X'_n starts in some given subset $X_0 \in A \subset E'$ with a given distribution ν_0. We also let (B,C) be a pair of subsets (B,C) such that $A \cap C = \emptyset = B \cap C$. We also assume that the triplet (A,B,C) is chosen so that for any initial state $x \in A$ the chain X'_n hits one of the sets B or C in finite time.

We let T_A be the entrance time of X' into a given subset A; that is, we have

$$T_A = \inf\{n \geq 0 \ : \ X'_n \in A\}$$

One would like to estimate the probability that the chain hits B before C

$$\mathbb{P}(T_{B \cup C} < T_C) = \mathbb{P}(X'_{T_{B \cup C}} \in B) = \mathbb{E}(\mathbb{1}_B(X'_{T_{B \cup C}}))$$

and the law of the random excursion given the fact that it reached B before C

$$\mathrm{Law}(X'_t \ ; \ 0 \leq t \leq T_{B \cup C} \mid T_{B \cup C} < T_C) = \mathrm{Law}(X'_t \ ; \ 0 \leq t \leq T_{B \cup C} \mid X'_{T_{B \cup C}} \in B)$$

Of course we have implicitly assumed that $\mathbb{P}(T_{B \cup C} < T_C) > 0$ so that the conditional distributions are well defined.

In connection with the previous examples discussed in the early part of this section the rare level set B may represent the outside of the radiation containment, an undesired critical population size or buffer excedance as well as ruin level of a given company. The level set C is usually far from being rare but it corresponds to an almost sure event. For instance in the radiation containment model the set C represents the set of physical obstacles which hopefully absorb the radiation and avoid the particle to come out of the containment. In communication networks models the set C represents a recurrent and well behave buffer size level. In population models C is related to a natural fluctuation size level of the population evolution and in ruin processes it corresponds to a predicted gain or a desired equilibrium level.

During its excursion from A to $(B \cup C)$ the process passes through a decreasing sequence of level sets $\mathscr{B} = (B_n)_{n=0,\dots,m}$ with

$$A = B_0 \supset B_1 \supset \dots \supset B_m = B$$

Here again the splitting parameter m and the choice of the level sets $\mathscr{B}$ depends on the problem at hand.

This decomposition reflects the successive levels the stochastic process needs to cross before to enter into the relevant rare event. In other words the increasing levels behave as gateways from which the rare event is more and more likely to happen.

To clarify the presentation we shall slight abuse the notation and we write T_n instead of $T_{B_n \cup C}$ the entrance time of X into $B_n \cup C$. To capture the behavior of X between the different levels we introduce the excursion-valued Markov chain

$$X_n = (T_n, (X'_t \; ; \; T_{n-1} \leq t \leq T_n)) \in E = \cup_{p \leq q}(\{q\} \times (E')^{(q-p+1)}) \tag{1.98}$$

By a direct inspection we see that the random sequence of level-crossing times $(T_n)_{0 \leq n \leq m}$ is increasing and whenever $T_n < T_C$ the second component of X_n represents the excursion of the process X' between the successive levels B_{n-1} and B_n so that T_n can be alternatively be defined by the inductive formulae

$$T_n = \inf\{T_{n-1} \leq t \; : \; X'_t \in B_n \cup C\}$$

Under our assumptions we also observe that these entrance times are finite and

$$(T_{B \cup C} < T_C) = (T_m < T_C) = (T_1 < T_C, \ldots, T_m < T_C)$$

One simple way to check whether or not a random path has succeeded to reach the desired n-th level is to consider the potential functions G_n on E defined for each $n \in \{0, \ldots, m\}$ and $x = (x_q)_{p \leq q \leq r} \in (E')^{(r-p+1)}$ by

$$G_n(t, x) = \mathbb{1}_{B_n}(x_r) \tag{1.99}$$

In this notation we have for each $n \leq m$

$$(T_n < T_C) = (T_1 < T_C, \ldots, T_n < T_C) = (G_1(X_1) = 1 \ldots, G_n(X_n) = 1)$$

and

$$(X_0, \ldots, X_n)$$

$$= ((0, X'_0), (T_1, (X'_t \; ; \; 0 \leq t \leq T_1)), \ldots, (T_n, (X'_t \; ; \; T_{n-1} \leq t \leq T_n)))$$

In we write $[X'_t \; ; \; 0 \leq t \leq T_n]$ instead of $(X_0, \ldots, X_n)$ the sequence of excursions of X' between the levels, then for any $n \leq m$ and any function f_n on the product space E_n we have

$$\mathbb{E}_{v_0}\left(f_n(X_0, \ldots, X_n) \prod_{p=1}^{n} G_p(X_p)\right) = \mathbb{E}_{v_0}\left(f_n([X'_t \; ; \; 0 \leq t \leq T_n]) \mathbb{1}_{T_n < T_C}\right)$$

We denote by $\mathbb{P}_n$ the law of the excursion-valued Markov chain from the origin $p = 0$, up to the time $p = n$

$$\mathbb{P}_n(x_0, \ldots, x_n) = \mathbb{P}(X_0 = x_0, \ldots, X_n = x_n)$$

If we set

$$\mathbb{Q}_n(x_0, \ldots, x_n) = \frac{1}{\mathcal{Z}_n}\left\{\prod_{0 \leq p < n} G_p(x_p)\right\} \mathbb{P}_n(x_0, \ldots, x_n)$$

with the unit potential function $G_0 = 1$ then we have

$$\mathbb{Q}_n = \text{Law}\left(\left[X_t'\,;\,0 \le t \le T_n\right] \mid T_{n-1} < T_C\right) \quad \text{and} \quad \mathcal{Z}_n = \mathbb{P}\left(T_{n-1} < T_C\right)$$

Once again, these measures have exactly the same form as the one presented in (1.33). The corresponding particle approximations are often referred as multilevel splitting particle methods or sequential Monte Carlo samplers in the literature on rare event simulation (see for instance [25, 26, 50, 114], and references therein).

Acknowledgements. The second and third author acknowledge that their work has been supported by the National Research Fund, Luxembourg, and cofunded under the Marie Curie Actions of the European Commission (FP7-COFUND).

References

1. Ackley, D., Littman, M.: A case for lamarckian evolution. Artifical Life III: SFI studies in the sciences of complexity XVII, 3–10 (1993)
2. Alba, E., Luque, G.: Performance of Distributed GAs on DNA Fragment Assembly. In: Parallel Evolutionary Computations, pp. 97–116. Springer (2006)
3. Aldous, D., Vazirani, U.: Go with the winners algorithms. In: Proc. 35th Symp. Foundations of Computer Sci., pp. 492–501 (1994)
4. Anderson, D.P., Cobb, J., Korpela, E., Lebofsky, M., Werthimer, D.: SETI@home: an experiment in public-resource computing. Commun. ACM 45(11), 56–61 (2002)
5. Ashlock, D.A.: Evolutionary computation for modeling and optimization. Springer (2006)
6. Assaraf, R., Caffarel, M., Khelif, A.: Diffusion Monte Carlo methods with a fixed number of walkers. Phys. Rev. E 61, 4566–4575 (2000)
7. Bäck, T., Hoffmeister, F., Schwefel, H.P.: A survey of evolution strategies. In: Proceedings of the Fourth International Conference on Genetic Algorithms, pp. 2–9. Morgan Kaufmann (1991)
8. Bäck, T., Fogel, D.B., Michalewicz, Z.: Handbook of Evolutionary Computation. IOP Publishing Ltd., Bristol (1997)
9. Bäck, T., Hammel, U., Schwefel, H.P.: Evolutionary computation: comments on the history and current state. IEEE Trans. Evolutionary Computation 1(1), 3–17 (1997)
10. Barricelli, N.A.: Esempi numerici di processi di evoluzione. Methodos, 45–68 (1954)
11. Barricelli, N.A.: Symbiogenetic evolution processes realized by artificial methods. Methodos 9(35-36), 143–182 (1957)
12. Battiti, R., Brunato, M., Mascia, F.: Reactive Search and Intelligent Optimization. In: Operations Research/Computer Science Interfaces. Springer (2008) doi:10.1007/978-0-387-09624-7
13. Baum, E.B.: Towards practical 'neural' computation for combinatorial optimization problems. In: AIP Conference Proceedings 151 on Neural Networks for Computing, pp. 53–58. American Institute of Physics Inc., Woodbury (1987), http://dl.acm.org/citation.cfm?id=24140.24150
14. Belew, R.K., Booker, L.B. (eds.): Proceedings of the 4th International Conference on Genetic Algorithms. Morgan Kaufmann, San Diego (1991)
15. Bonabeau, E., Dorigo, M., Theraulaz, G.: Swarm intelligence: from natural to artificial systems. Oxford University Press, Inc., New York (1999)

16. Bremermann, H.J., Rogson, M., Salaff, S.: Global Properties of Evolution Processes. In: Pattee, H.H., Edlsack, E.A., Fein, L., Callahan, A.B. (eds.) Natural Automata and Useful Simulations, pp. 3–41. Spartan Books, Washington, DC (1966)

17. Broyden, C.G.: The Convergence of a Class of Double-rank Minimization Algorithms: 2. The New Algorithm. IMA Journal of Applied Mathematics 6(3), 222–231 (1970), `abstract/6/3/222, doi:10.1093/imamat/6.3.222`

18. Burke, E.K., Hyde, M., Kendall, G., Ochoa, G., Ozcan, E., Woodward, J.R.: A classification of hyper-heuristic approaches. Handbook of Metaheuristics 146, 1–21 (2010), `http://www.springerlink.com/index/XXM7126130381913.pdf`

19. Campillo, F., Rossi, V.: Convolution particle filtering for parameter estimation in general state-space models. In: Proceedings of the 45th IEEE Conference on Decision and Control, San Diego, USA (2006)

20. Campillo, F., Rossi, V.: Convolution filter based methods for parameter estimation in general state-space models. IEEE Transactions on Aerospace and Electronic Systems 45(3), 1063–1071 (2009)

21. Cantu-Paz, E.: A survey of parallel genetic algorithms. Calculateurs Paralleles Reseaux et Systems Repartis 10(2), 141–171 (1998), `http://ieeexplore.ieee.org/lpdocs/epic03/wrapper.htm?arnumber=879173`

22. Carpenter, J., Clifford, P., Fearnhead, P.: An improved particle filter for non-linear problems. IEE Proceedings F 146, 2–7 (1999)

23. Carvalho, H., Del Moral, P., Monin, A., Salut, G.: Optimal Non-linear Filtering in GPS/INS Integration. IEEE-Trans. on Aerospace and Electronic Systems 33(3), 835–850 (1997)

24. Cerf, R.: Asymptotic convergence of genetic algorithms. Adv. Appl. Probab. 30, 521–550 (1998)

25. Cérou, F., Del Moral, P., LeGland, F., Lezaud, P.: Limit Theorems for multilevel splitting algorithms in the simulation of rare events (preliminary version). In: Kuhl, M.E., Steiger, N.M., Armstrong, F.B., Joines, J.A. (eds.) Proceedings of the 2005 Winter Simulation Conference (2005)

26. Cérou, F., Del Moral, P., LeGland, F., Lezaud, P.: ALEA Lat. Am. J. Probab. Math. Stat. 1, 181–203 (2006)

27. Cérou, F., Del Moral, P., Guyader, A.: A non asymptotic variance theorem for unnormalized Feynman-Kac particle models. Technical Report HAL-INRIA RR-6716 (2008), Annales de l'Institut H. Poincaré, Série: Probabilités(B) 47(3) (2011)

28. Cérou, F., Del Moral, P., Furon, T., Guyader, A.: Rare event simulation for a static distribution. Research Report RR-6792, INRIA (2009)

29. Chopin, N.: A sequential particle filter method for static models. Biometrika 89, 539–552 (2002)

30. Coello Coello, C.: List of references on evolutionary multiobjective optimization, `http://www.lania.mx/~ccoello/EMOObib.html`

31. Coello Coello, C., Van Veldhuizen, D., Lamont, G.: Evolutionary Algorithms for Solving Multi-Objective Problems. In: Genetic Algorithms and Evolutionary Computation, vol. 5. Kluwer Academic Publishers, Boston (2002)

32. Cole, N., Desell, T., Lombraña González, D., Fernández de Vega, F., Magdon-Ismail, M., Newberg, H., Szymanski, B., Varela, C.: Evolutionary Algorithms on Volunteer Computing Platforms: The MilkyWay@Home Project. In: de Vega, F.F., Cantú-Paz, E. (eds.) Parallel and Distributed Computational Intelligence. SCI, vol. 269, pp. 63–90. Springer, Heidelberg (2010)

33. Colorni, A., Dorigo, M., Maniezzo, V.: Distributed Optimization by Ant Colonies. In: European Conference on Artificial Life, pp. 134–142 (1991)

34. Di Chio, C., Brabazon, A., Di Caro, G.A., Drechsler, R., Farooq, M., Grahl, J., Greenfield, G., Prins, C., Romero, J., Squillero, G., Tarantino, E., Tettamanzi, A.G.B., Urquhart, N., Uyar, A.Ş. (eds.): EvoApplications 2011, Part II. LNCS, vol. 6625. Springer, Heidelberg (2011)

35. De Castro, L.N., Timmis, J.: Artificial Immune Systems: A New Computational Intelligence Approach. Springer (2002),
    `http://books.google.com/books?hl=en&lr=&id=aMFP7p8DtaQC&`
    `oi=fnd&pg=PA1&dq=Artificial+immune+systems+a+new+`
    `computational+intelligence+approach&ots=zHjlTG5TiP&`
    `sig=VKMxGqTe4FhtUai-ET3wdQ2mJ78`

36. Del Moral, P.: Non Linear Filtering: Interacting Particle Solution. Markov Processes and Related Fields 2(4), 555–580 (1996)

37. Del Moral, P.: Measure Valued Processes and Interacting Particle Systems. Application to Non Linear Filtering Problems. Annals of Applied Probability 8(2), 438–495 (1998)

38. Del Moral, P.: Feynman-Kac Formulae: Genealogical and Interacting Particle Systems with Applications. Springer, New York (2004)

39. Del Moral, P., Doucet, A.: Particle motions in absorbing medium with hard and soft obstacles. Stochastic Anal. Appl. 22, 1175–1207 (2004)

40. Del Moral, P., Doucet, A., Jasra, A.: Sequential Monte Carlo samplers. J. Royal Statist. Soc. B 68, 411–436 (2006)

41. Del Moral, P., Doucet, A., Jasra, A.: On Adaptive Resampling Procedures for Sequential Monte Carlo Methods. Research Report INRIA (HAL-INRIA RR-6700), 46p. (October 2008); In: Bernoulli 18(1), 252–278 (2012)

42. Del Moral, P., Guionnet, A.: On the stability of measure valued processes with applications to filtering. C. R. Acad. Sci. Paris Sér. I Math. 329, 429–434 (1999)

43. Del Moral, P., Guionnet, A.: On the stability of interacting processes with applications to filtering and genetic algorithms. Annales de l'Institut Henri Poincaré 37(2), 155–194 (2001)

44. Del Moral, P., Jacod, J.: Interacting Particle Filtering With Discrete Observations. In: Doucet, A., de Freitas, J.F.G., Gordon, N.J. (eds.) Sequential Monte Carlo Methods in Practice. Statistics for Engineering and Information Science, pp. 43–77. Springer (2001)

45. Del Moral, P., Jacod, J., Protter, P.: The Monte-Carlo Method for filtering with discrete-time observations. Probability Theory and Related Fields 120, 346–368 (2001)

46. Del Moral, P., Jacod, J.: The Monte-Carlo Method for filtering with discrete time observations. Central Limit Theorems. In: Lyons, T.J., Salisbury, T.S. (eds.) The Fields Institute Communications, Numerical Methods and Stochastics. American Mathematical Society (2002)

47. Del Moral, P., Kallel, L., Rowe, J.: Modeling genetic algorithms with interacting particle systems. Revista de Matematica, Teoria y Aplicaciones 8(2) (July 2001)

48. Del Moral, P., Miclo, L.: Asymptotic Results for Genetic Algorithms with Applications to Non Linear Estimation. In: Naudts, B., Kallel, L. (eds.) Proceedings Second EvoNet Summer School on Theoretical Aspects of Evolutionary Computing. Natural Computing. Springer (2000)

49. Del Moral, P., Miclo, L.: On the Stability of Non Linear Semigroup of Feynman-Kac Type. Annales de la Faculté des Sciences de Toulouse 11(2), (2002)

50. Del Moral, P., Lezaud, P.: Branching and interacting particle interpretation of rare event probabilities. In: Blom, H., Lygeros, J. (eds.) Stochastic Hybrid Systems: Theory and Safety Critical Applications. Springer, Heidelberg (2006)

51. Del Moral, P., Miclo, L.: A Moran particle system approximation of Feynman-Kac formulae. Stochastic Processes and their Applications 86, 193–216 (2000)
52. Del Moral, P., Miclo, L.: Branching and interacting particle systems approximations of Feynman-Kac formulae with applications to non linear filtering. In: Azéma, J., Emery, M., Ledoux, M., Yor, M. (eds.) Séminaire de Probabilités XXXIV. Lecture Notes in Mathematics, vol. 1729, pp. 1–145. Springer (2000)
53. Del Moral, P., Miclo, L.: Genealogies and Increasing Propagations of Chaos for Feynman-Kac and Genetic Models. Annals of Applied Probability 11(4), 1166–1198 (2001)
54. Del Moral, P., Miclo, L.: Particle approximations of Lyapunov exponents connected to Schrödinger operators and Feynman-Kac semigroups. ESAIM: Probability and Statistics 7, 171–208 (2003)
55. Del Moral, P., Miclo, L.: Annealed Feynman-Kac models. Comm. Math. Phys. 235, 191–214 (2003)
56. Del Moral, P., Rémillard, B., Rubenthaler, S.: Introduction aux Probabilités. Ellipses Edition (2006)
57. Del Moral, P., Rio, E.: Concentration inequalities for mean field particle models. Technical report HAL-INRIA RR-6901 (2009). Annals of Applied Probability 21(3), 1017–1052 (2011)
58. Del Moral, P., Hu, P., Wu, L.: On the Concentration Properties of Interacting Particle Processes. Foundations and Trends in Machine Learning 3(3-4), 225–389 (2012)
59. Del Moral, P., Rigal, G., Salut, G.: Estimation and nonlinear optimal control: An unified framework for particle solutions LAAS-CNRS, Toulouse, Research Report no. 91137, DRET-DIGILOG- LAAS/CNRS contract (April 1991)
60. Del Moral, P., Rigal, G., Salut, G.: Nonlinear and non Gaussian particle filters applied to inertial platform repositioning. LAAS-CNRS, Toulouse, Research Report no. 92207, STCAN/DIGILOG-LAAS/CNRS Convention STCAN no. A.91.77.013, 94p. (September 1991)
61. Del Moral, P., Rigal, G., Salut, G.: Estimation and nonlinear optimal control: Particle resolution in filtering and estimation. Experimental results. Convention DRET no. 89.34.553.00.470.75.01, Research report no.2, 54p. (January 1992)
62. Del Moral, P., Rigal, G., Salut, G.: Estimation and nonlinear optimal control: Particle resolution in filtering and estimation. Theoretical results Convention DRET no. 89.34.553.00.470.75.01, Research report no.3, 123p. (October 1992)
63. Del Moral, P., Noyer, J.-C., Rigal, G., Salut, G.: Particle filters in radar signal processing: detection, estimation and air targets recognition. LAAS-CNRS, Toulouse, Research Report no. 92495 (December 1992)
64. Del Moral, P., Rigal, G., Salut, G.: Estimation and nonlinear optimal control: Particle resolution in filtering and estimation. Studies on: Filtering, optimal control, and maximum likelihood estimation. Convention DRET no. 89.34.553.00.470.75.01. Research report no.4, 210p. (January 1993)
65. Del Moral, P., Noyer, J.C., Rigal, G., Salut, G.: Traitement non-linéaire du signal par réseau particulaire: Application RADAR. In: Proceedings XIV Colloque GRETSI, Traitement du Signal et des Images, Juan les Pins, France, pp. 399–402 (September 1993)
66. Del Moral, P., Noyer, J.C., Salut, G.: Resolution particulaire et traitement non linéaire du signal: Application radar/sonar. Revue du Traitement du Signal (Septembre 1995)
67. Doucet, A., de Freitas, J.F.G., Gordon, N.J. (eds.): Sequential Monte Carlo Methods in Practice. Springer, New York (2001)

68. Doucet, A., Godsill, S.J., Andrieu, C.: On sequential Monte Carlo sampling methods for Bayesian filtering. Statistics and Computing 10, 197–208 (2000)
69. Doucet, A., Johansen, A.M.: A tutorial on particle filtering and smoothing: fifteen years later. In: Crisan, D., Rozovsky, B. (eds.) Handbook of Nonlinear Filtering. Cambridge University Press (2009)
70. Eiben, A.E., Bäck, T.: Empirical investigation of multiparent recombination operators in evolution strategies. Evolutionary Computation 5(3), 347–365 (1997)
71. Eiben, A.E., Hinterding, R., Hinterding, A.E.E.R., Michalewicz, Z.: Parameter control in evolutionary algorithms. IEEE Transactions on Evolutionary Computation 3, 124–141 (2000)
72. Eiben, A.E., Smith, J.E.: Introduction to Evolutionary Computing. Springer (2003)
73. Eiben, A., Schut, M.: New ways to calibrate evolutionary algorithms. In: Advances in Metaheuristics for Hard Optimization. Natural Computing, pp. 153–177. Springer (2008), http://dblp.uni-trier.de/db/conf/ncs/metaheuristics2008.html#EibenS08
74. Ellouze, M., Gauchi, J.P., Augustin, J.C.: Global sensitivity analysis applied to a contamination assessment model of Listeria monocytogenes in cold smoked salmon at consumption. Risk Anal. 30, 841–852 (2010)
75. Ellouze, M., Gauchi, J.P., Augustin, J.C.: Use of global sensitivity analysis in quantitative microbial risk assessment: Application to the evaluation of a biological time temperature integrator as a quality and safety indicator for cold smoked salmon. In: Food Microbiol. (2010), doi:10.1016/j.fm.2010.05.022
76. Fearnhead, P.: Computational methods for complex stochastic systems: A review of some alternatives to MCMC. Statistics and Computing 18, 151–171 (2008)
77. Fletcher, R., Powell, M.: A rapidly convergent descent method for minimization. Computer Journal 6, 163–168 (1963)
78. Fletcher, R., Reeves, C.: Function minimization by conjugate gradients. Computer Journal 7, 149–154 (1964)
79. Fletcher, R.: A new approach to variable metric algorithms. The Computer Journal 13(3), 317–322 (1970), http://comjnl.oxfordjournals.org/cgi/content/abstract/13/3/317, doi:10.1093/comjnl/13.3.317
80. Gauchi, J.P., Vila, J.P., Coroller, L.: New prediction confidence intervals and bands in the nonlinear regression model: Application to the predictive modelling in food. Communications in Statistics, Simulation and Computation 39(2), 322–330 (2009)
81. Gauchi, J.P., Bidot, C., Augustin, J.C., Vila, J.P.: Identification of complex microbiological dynamic system by nonlinear filtering. In: 6th Int. Conference on Predictive Modelling in Foods, Washington DC (2009)
82. Glasserman, P., Heidelberger, P., Shahabuddin, P., Zajic, T.: Multilevel splitting for estimating rare event probabilities. Operations Research 47, 585–600 (1999)
83. Glover, F.: Heuristics for integer programming using surrogate constraints. Decision Sciences 8(1), 156–166 (1977), http://dx.doi.org/10.1111/j.1540-5915.1977.tb01074.x, doi:10.1111/j.1540-5915.1977.tb01074.x
84. Glover, F.: Future paths for integer programming and links to artificial intelligence. Comput. Oper. Res. 13(5), 533–549 (1986), http://dx.doi.org/10.1016/0305-05488690048-1, doi:10.1016/0305-0548(86)90048-1
85. Glover, F.: A template for scatter search and path relinking. In: Hao et al. [93], pp. 1–51 (1997)

86. Glynn, P.W., Ormoneit, D.: Hoeffding's inequality for uniformly ergodic Markov chains. Statist. Probab. Lett. 56(2), 143–146 (2002)
87. Goldberg, D.E.: Genetic Algorithms in Search, Optimization and Machine Learning. Addison-Wesley, Reading (1989)
88. Goldfarb, D.: A family of variable metric updates derived by variational means. Mathematics of Computation 24, 23–26 (1970)
89. Gordon, N.J., Salmond, D., Smith, A.F.M.: A novel approach to state estimation to nonlinear non-Gaussian state estimation. IEE Proceedings F 40, 107–113 (1993)
90. Grassberger, P.: Pruned-enriched Rosenbluth method: Simulations of θ polymers of chain length up to 1 000 000. Phys. Rev. E, 3682–3693 (1997)
91. Hansen, N., Müller, S.D., Koumoutsakos, P.: Reducing the time complexity of the derandomized evolution strategy with covariance matrix adaptation (CMA-ES). Evol. Comput. 11(1), 1–18 (2003),
    http://dx.doi.org/10.1162/106365603321828970,
    doi:10.1162/106365603321828970
92. Hansen, N., Ostermeier, A., Gawelczyk, A.: On the adaptation of arbitrary normal mutation distributions in evolution strategies: The generating set adaptation. In: Proceedings of the 6th International Conference on Genetic Algorithms, pp. 57–64. Morgan Kaufmann Publishers Inc., San Francisco (1995),
    http://dl.acm.org/citation.cfm?id=645514.657936
93. Hao, J.-K., Lutton, E., Ronald, E., Schoenauer, M., Snyers, D. (eds.): AE 1997. LNCS, vol. 1363. Springer, Heidelberg (1998)
94. Harris, T.E., Kahn, H.: Estimation of particle transmission by random sampling. Natl. Bur. Stand. Appl. Math. Ser. 12, 27–30 (1951)
95. Herrera, F., Lozano, M.: Heuristic Crossovers for Real-Coded Genetic Algorithms Based on Fuzzy Connectives. In: Ebeling, W., Rechenberg, I., Voigt, H.-M., Schwefel, H.-P. (eds.) PPSN 1996. LNCS, vol. 1141, pp. 336–345. Springer, Heidelberg (1996), http://www.springerlink.com/content/y42m98n165872533, doi:10.1007/3-540-61723-X_998
96. Herrera, F., Lozano, M., Sánchez, A.M.: A taxonomy for the crossover operator for real-coded genetic algorithms: An experimental study. Int. J. Intell. Syst. 18(3), 309–338 (2003), http://dx.doi.org/10.1002/int.10091, doi:10.1002/int.10091
97. Herrera, F., Lozano, M., Verdegay, J.: Fuzzy connective based crossover operators to model genetic algorithms population diversity. Tech. Rep. DECSAI-95110. University of Granada, Spain (1995)
98. Herrera, F., Lozano, M., Verdegay, J.: Dynamic and heuristic fuzzy connectives-based crossover operators for controlling the diversity and convergence of real-coded genetic algorithms. Int. J. Intell. Syst. 11, 1013–1041 (1996)
99. Herrera, F., Lozano, M., Verdegay, J.: Fuzzy connectives based crossover operators to model genetic algorithms population diversity. Fuzzy Set. Syst. 92(1), 21–30 (1997), doi:10.1016/S0165-0114(96)00179-0
100. Hestenes, M., Stiefel, E.: Methods of conjugate gradients for solving linear systems. J. Research NBS 49(6), 409–436 (1952)
101. Hestenes, M.R.: Iterative methods for solving linear equations. Report 52-9, NAML (1951); reprinted in J. Optimiz. Theory App. 11, 323–334 (1973)
102. Hetherington, J.H.: Observations on the Statistical Iteration of Matrices. Phys. Rev. A. 30, 2713–2719 (1984)

103. Hinterding, R., Michalewicz, Z., Eiben, A.E.: Adaptation in Evolutionary Computation: A Survey. In: Proceedings of the 4th IEEE International Conference on Evolutionary Computation, pp. 65–69 (1997),
    http://ieeexplore.ieee.org/lpdocs/epic03/
    wrapper.htm?arnumber=592270
104. Holland, J.H.: Adaptation in Natural and Artificial Systems. University of Michigan Press, Ann Arbor (1975)
105. Hooke, R., Jeeves, T.: Direct search solution of numerical and statistical problems. Journal of the ACM 8(2), 212–229 (1961),
    doi:http://doi.acm.org/10.1145/321062.321069
106. Horn, J.: Multicriteria decision making and evolutionary computation. In: Handbook of Evolutionary Computation, Institute of Physics Publishing, London (1997)
107. Ikonen, E., Del Moral, P., Najim, K.: A genealogical decision tree solution to optimal control problems. In: IFAC Workshop on Advanced Fuzzy/Neural Control, Oulu, Finland, pp. 169–174 (2004)
108. Ikonen, E., Najim, K., Del Moral, P.: Application of genealogical decision trees for open-loop tracking control. In: Proceedings of the16th IFAC World Congress, Prague, Czech (2005)
109. Ingber, L.: Adaptive simulated annealing (asa), global optimization c-code. Tech. rep. Caltech Alumni Association (1993)
110. Ingber, L.: Simulated annealing: Practice versus theory. Math. Comput. Model. 18(11), 29–57 (1993)
111. Ingber, L.: Adaptive simulated annealing (asa): Lessons learned. Control and Cybern. 25, 33–54 (1996)
112. Ingber, L.: Adaptive simulated annealing (asa) and path-integral (pathint) algorithms: Generic tools for complex systems. Tech. rep. Chicago, IL (2001)
113. Ingber, L., Rosen, B.: Genetic algorithms and very fast simulated reannealing: A comparison. Math. Comput. Model. 16(11), 87–100 (1992)
114. Johansen, A.M., Del Moral, P., Doucet, A.: Sequential Monte Carlo Samplers for Rare Events. In: Proceedings of 6th International Workshop on Rare Event Simulation, Bamberg, Germany (2006)
115. Jong, K.A.D.: Evolutionary computation - a unified approach. MIT Press (2006)
116. Kennedy, J., Eberhart, R.: Particle swarm optimization. In: Proceedings of IEEE International Conference on Neural Networks, vol. 4, pp. 1942–1948 (1995),
    doi:10.1109/ICNN.1995.488968
117. Kirkpatrick, S., Gelatt, C., Vecchi, M.: Optimization by simulated annealing. Science 220(4598), 671–680 (1983),
    citeseer.ist.psu.edu/kirkpatrick83optimization.html
118. Kitagawa, G.: Monte Carlo filter and smoother for non-Gaussian nonlinear state space models. J. Comp. Graph. Statist. 5, 1–25 (1996)
119. Kolokoltsov, V.N., Maslov, V.P.: Idempotent analysis and its applications. Mathematics and its Applications, vol. 401. Kluwer Academic Publishers Group, Dordrecht (1997); Translation of Idempotent analysis and its application in optimal control, Russian, Nauka Moscow (1994); translated by Nazaikinskii, V. E. With an appendix by Pierre Del Moral: Maslov Optimization Theory: Optimality Versus Randomness, pp. 243–302
120. Künsch, H.R.: State-space and hidden Markov models. In: Barndorff-Nielsen, O.E., Cox, D.R., Kluppelberg, C. (eds.) Complex Stochastic Systems, pp. 109–173. CRC Press (2001)
121. Lagarias, J., Reeds, J., Wright, M., Wright, P.: Convergence properties of the Nelder-Mead simplex algorithm in low dimensions. SIAM J. Optimiz. 9, 112–147 (1998)

122. Langdon, W., Poli, R.: Foundations of Genetic Programming, vol. 5. Springer (2002),
     `http://discovery.ucl.ac.uk/124583/`
123. Liu, J.S.: Monte Carlo Strategies in Scientific Computing. Springer, New York (2001)
124. Martin, O., Otto, S.W., Felten, E.W.: Large-step markov chains for the traveling sales-
     man problem. Complex Systems 5, 299–326 (1991)
125. Melik-Alaverdian, V., Nightingale, M.P.: Quantum Monte Carlo methods in statistical
     mechanics. Internat. J. of Modern Phys. C. 10, 1409–1418 (1999)
126. Metropolis, N., Ulam, S.: The Monte Carlo Method. Journal of the American Statistical
     Association 44(247), 335–341 (1949)
127. Metropolis, N., Rosenbluth, A., Rosenbluth, M., Teller, A., Teller, E.: Equation of state
     calculations by fast computing machines. J. Chem. Phys. 21(6), 1087–1092 (1953),
     `http://link.aip.org/link/?JCP/21/1087/1`, doi:10.1063/1.1699114
128. Michalewicz, Z.: Genetic algorithms + data structures = evolution programs, 2nd, ex-
     tended edn. Springer-Verlag New York, Inc., New York (1994)
129. Mitavskiy, B., Rowe, J.: An Extension of Geiringer's Theorem for a Wide Class of
     Evolutionary Search Algorithms. Evolutionary Computation 14(1), 87–118 (2006)
130. Mitavskiy, B., Rowe, J., Wright, A., Schmitt, L.: Quotients of Markov chains and
     asymptotic properties of the stationary distribution of the Markov chain associated to an
     evolutionary algorithm. Genetic Programming and Evolvable Machines 9(2), 109–123
     (2008)
131. Mladenović, N.: A variable neighborhood algorithm – a new metaheuristics for combi-
     natorial optimization. In: Abstracts of Papers Presented at Optimization Days, Montreal
     (1995)
132. Mladenović, N., Hansen, P.: Variable neighborhood search. Comput. Oper. Res. 24(11),
     1097–1100 (1997),
     `http://dx.doi.org/10.1016/S0305-05489700031-2`,
     doi:10.1016/S0305-0548(97)00031-2
133. Moscato, P.: Memetic algorithms: a short introduction. In: Corne, D., Dorigo, M.,
     Glover, F., Dasgupta, D., Moscato, P., Poli, R., Price, K. (eds.) New Ideas in Opti-
     mization, pp. 219–234. McGraw-Hill Ltd., UK (1999),
     `http://dl.acm.org/citation.cfm?id=329055.329078`
134. Mühlenbein, H., Schlierkamp-Voosen, D.: Analysis of selection, mutation and recom-
     bination in genetic algorithms. Evolution and Biocomputation, 142–168 (1995)
135. Nelder, J.A., Mead, R.: A simplex method for function minimization. Comput. J. 7(4),
     308–313 (1965), `http://comjnl.oxfordjournals.org/cgi/content/`,
     `abstract/7/4/308`, doi:10.1093/comjnl/7.4.308
136. Neri, F., Cotta, C., Moscato, P.: Handbook of Memetic Algorithms. SCI. Springer
     (2011), `http://books.google.lu/books?id=uop6UvKu8q4C`
137. Nocedal, J.: Updating quasi-newton matrices with limited storage. Math. Com-
     put. 35(151), 773–782 (1980), `http://www.jstor.org/stable/2006193`
138. Pelikan, M., Goldberg, D.E., Lobo, F.G.: A survey of optimization by build-
     ing and using probabilistic models. Comput. Optim. Appl. 21(1), 5–20 (2002),
     `http://dx.doi.org/10.1023/A:1013500812258`,
     doi:10.1023/A:1013500812258
139. Polak, E., Ribière, G.: Note sur la convergence des méthodes de directions conjuguées.
     Revue Française d'informatique et de Recherche Opérationnelle 16, 35–43 (1969)
140. Powell, M.: On the Convergence of the Variable Metric Algorithm. Journal of the Insti-
     tute of Mathematics and its Applications 7, 21–36 (1971)
141. Rao, S., Shanta, C.: Numerical Methods: With Program in Basic, Fortan, Pascal & C++.
     Orient Blackswan (2004)

142. Reynolds, R.G., Sverdlik, W.: Problem solving using cultural algorithms. In: International Conference on Evolutionary Computation, pp. 645–650 (1994)
143. Rosenbluth, M.N., Rosenbluth, A.W.: Monte-Carlo calculations of the average extension of macromolecular chains. J. Chem. Phys. 23, 356–359 (1955)
144. Vila, J.-P., Rossi, V.: Nonlinear filtering in discret time: A particle convolution approach. Biostatistic Group of Montpellier, Technical Report 04-03 (2004), http://vrossi.free.fr/recherche.html
145. Rudolph, G.: Convergence of Evolutionary Algorithms in General Search Spaces. In: International Conference on Evolutionary Computation, pp. 50–54 (1996)
146. Rudolph, G.: Finite Markov Chain Results in Evolutionary Computation: A Tour d'Horizon. Fundam. Inform. 35(1-4), 67–89 (1998)
147. Schmitt, F., Rothlauf, F.: On the Importance of the Second Largest Eigenvalue on the Convergence Rate of Genetic Algorithms. In: Beyer, H., Cantu-Paz, E., Goldberg, D., Parmee, Spector, L., Whitley, D. (eds.) Proceedings of the Genetic and Evolutionary Computation Conference (GECCO 2001), pp. 559–564. Morgan Kaufmann Publishers, San Francisco (2001)
148. Schwefel, H.P., Rudolph, G.: Contemporary Evolution Strategies. In: Morán, F., Merelo, J.J., Moreno, A., Chacon, P. (eds.) ECAL 1995. LNCS, vol. 929, pp. 893–907. Springer, Heidelberg (1995)
149. Shanno, D.: Conditioning of quasi-newton methods for function minimization. Math. Comput. 24(111), 647–656 (1970)
150. Shewchuk, J.: An introduction to the conjugate gradient method without the agonizing pain. Tech. rep., Carnegie Mellon University, Pittsburgh, Pittsburgh, PA, USA (1994), http://portal.acm.org/citation.cfm?id=865018
151. Solis, F., Wets, R.B.: Minimization by random search techniques. Math. Oper. Res. 6, 19–30 (1981)
152. Spears, W.M., Jong, K.A.D., Ba, T., Fogel, D.B., Garis, H.D.: An overview of evolutionary computation. Evolutionary Computation 667(1), 442–459 (1993), http://www.springerlink.com/index/Y03055H012777681.pdf
153. Spendley, W., Hext, G., Himsworth, F.: Sequential application of simplex designs in optimisation and evolutionary operation. Technometrics 4(4), 441–461 (1962)
154. Stadler, P.: Towards a theory of landscapes. In: Lopéz-Peña, R., Capovilla, R., García-Pelayo, R., Waelbroeck, H., Zertuche, F. (eds.) Complex Systems and Binary Networks, vol. 461, pp. 77–163. Springer, Berlin (1995)
155. Stadler, P., Flamm, C.: Barrier trees on poset-valued landscapes. Genet. Program. Evol. M. 4(1), 7–20 (2003), http://dblp.uni-trier.de/db/journals/gpem/gpem4.html%5c#StadlerF03
156. Stewart, C.A., Mueller, M.S., Lingwall, M.: Progress Towards Petascale Applications in Biology: Status in 2006. In: Lehner, W., Meyer, N., Streit, A., Stewart, C. (eds.) Euro-Par Workshops 2006. LNCS, vol. 4375, pp. 289–303. Springer, Heidelberg (2007), http://dl.acm.org/citation.cfm?id=1765606.1765638
157. Storn, R., Price, K.: Differential evolution – a simple and efficient heuristic for global optimization over continuous spaces. J. of Global Optimization 11(4), 341–359 (1997), http://dx.doi.org/10.1023/A:1008202821328, doi:10.1023/A:1008202821328
158. Surhone, L.M., Tennoe, M.T., Henssonow, S.F.: Leiden Classical. VDM Verlag Dr. Mueller AG & Company Kg (2010)
159. Tantar, E., Dhaenens, C., Figueira, J.R., Talbi, E.G.: A priori landscape analysis in guiding interactive multi-objective metaheuristics. In: IEEE Congress on Evolutionary Computation, pp. 4104–4111 (2008)

160. Tantar, E., Schuetze, O., Figueira, J.R., Coello, C.A.C., Talbi, E.G.: Computing and selecting epsilon-efficient solutions of 0,1-knapsack problems. In: Multiple Criteria Decision Making for Sustainable Energy and Transportation Systems. Lecture Notes in Econom. and Math. Systems, vol. 634, pp. 379–387 (2010)

161. Tsang, E., Voudouris, C.: Fast local search and guided local search and their application to British Telecom's workforce scheduling problem. Oper. Res. Lett. 20(3), 119–127 (1997), http://www.sciencedirect.com/science/, article/pii/S0167637796000429, doi:10.1016/s0167-6377(96)00042-9

162. Weinberger, E.: Correlated and uncorrelated fitness landscapes and how to tell the difference. Biol. Cybern. 63, 325–336 (1990), http://dx.doi.org/10.1007/BF00202749, doi:10.1007/BF00202749

Chapter 2
Incorporating Regular Vines in Estimation of Distribution Algorithms

Rogelio Salinas-Gutiérrez, Arturo Hernández-Aguirre, and
Enrique R. Villa-Diharce

Abstract. This chapter presents the incorporation and use of regular vines into Estimation of Distribution Algorithms for solving numerical optimization problems. Several kinds of statistical dependencies among continuous variables can be taken into account by using regular vines. This work presents a procedure for selecting the most important dependencies in EDAs by truncating regular vines. Moreover, this chapter also shows how the use of mutual information in the learning of graphical models implies a natural way of employing copula functions.

2.1 Introduction

Nowadays, optimization methods have been recognized as important tools for finding optimal solutions in several fields, such as Computer Science, Statistics, Artificial Intelligence, Operations Research, among others. Optimization problems have been studied and solved with different proposals. Some of these proposals, named *metaheuristics*, have been designed for solving hard optimization problems. Although this kind of algorithms does not guarantee global optimal, in practice, they usually find good solutions in a reasonable period of time. The *Evolutionary Computation* (EC) is a field of artificial intelligence that consists of metaheuristic techniques for solving optimization problems. These metaheuristics use principles of Darwin's theory and they are also known as *Evolutionary Algorithms* (EAs). Each iteration in an EA involves a competitive *selection* that chooses the best solutions. The solutions with highest fitness are *crossed over* for creating new solutions. Some individuals of the new population can be mutated in order to preserve diversity in the solutions. In this way, the genetic operators crossover and mutation are used for giving *variation* to the set of solutions.

Rogelio Salinas-Gutiérrez · Arturo Hernández-Aguirre · Enrique R. Villa-Diharce
Center for Research in Mathematics (CIMAT), Guanajuato, México
e-mail: {rsalinas,artha,villadi}@cimat.mx

E. Tantar et al. (Eds.): EVOLVE- A Bridge between Probability, SCI 447, pp. 91–121.
springerlink.com   © Springer-Verlag Berlin Heidelberg 2013

   Estimation of Distribution Algorithms (EDAs) are a class of evolutionary optimization techniques that employ probabilistic models as a representation of the relationships between variables in the population. This recent paradigm in EC does not use genetic operators such as crossover and mutation. The goal in EDAs is to model the dependencies in the best individuals and transfer them into the next population. EDAs generate the new population by sampling from the probabilistic model of promissory individuals. These evolutionary optimization techniques have used several probabilistic models. For this reason, there are a number of EDAs for discrete and continuous domains. Some of these probabilistic models are based on Bayesian and Markov networks. Other EDAs have used Gaussian assumptions, such as Gaussian kernels, Gaussian mixture models and the multivariate Gaussian distribution.

   On the other hand, in different areas such as finance, climate, oceanography, hydrology, geodesy, and reliability, researchers have used probabilistic models that separate marginal distributions from the dependence structure. This has allowed more flexibility for modeling multivariate data, without the need of restricting the marginal distributions. The way in which this can be done is by means of the *copula functions*. In the last years, copula functions have became an important option for modeling multivariate data.

   One motivation for this research is the fact that EDAs have the capacity of *explicitly* taking into account dependencies among variables in optimization problems. This characteristic, along with the possibility of transferring dependencies into the next generation of new solutions, have received much attention from the EC community. An important goal of modeling dependencies among variables in an EDA is to learn the structure of the optimization problem [9, 25]. The learning of the problem structure by means of a probabilistic model can help ensure an efficient optimization behavior.

   Although EDAs have been investigated for discrete and continuous domains, the works and contributions for continuous domains are mainly based on the multivariate Gaussian distribution. The assumption of modeling dependencies under this probabilistic model can not be realistic for some optimization problems. This observation gives an opportunity, and a motivation, for proposing new continuous EDAs. Another motivation is related to the growing use of copula functions for getting flexible multivariate distributions. This is because of the important contributions that copula theory has had in many research and application works.

   The goal of this chapter is to present regular vines and show how they can be incorporated into continuous EDAs. Regular vines are graphical models that represent multivariate distributions using bivariate and conditional bivariate copula functions. In particular, two subset of regular vines known as Canonical vines and D-vines are adapted for optimizing several benchmark functions.

2.2   Estimation of Distribution Algorithms

The use of probabilistic models for searching and generating promising solutions is a recent paradigm in EC. Algorithms based on this principle have been called

Estimation of Distribution Algorithms (EDAs) [38], Probabilistic Model Building Genetic Algorithms (PMBGAs) [45], and also as Iterated Density Estimation Algorithms (IDEAs) [37]. In this chapter, the class of EAs that employs probabilistic models as a representation of the relationships among variables is identified as EDAs.

Similar to Genetic Algorithms (GAs), EDAs are population based. However, this new class of EAs does not use genetic operators such as crossover and mutation for generating new individuals. Instead, in EDAs, the new individuals are sampled from a probability distribution. Therefore, the goal in EDAs is to take into account the dependencies of the best solutions and using them for generating the next population. A pseudocode for EDAs is shown in Algorithm 3.1. The use of the estimated model in step 4 allows one to explicitly take into account the dependencies between decision variables and their structure. Step 5 shows the possibility of incorporating the dependencies among the variables into the new population, which greatly modifies the performance of an EDA.

Algorithm 2.1. Pseudocode for EDAs

1: Initialize the generation counter $t \longleftarrow 0$
   Generate the initial population $\mathscr{P}_0$ with N individuals at random.
2: Evaluate population $\mathscr{P}_t$ using the cost function.
3: Select a subset $\mathscr{S}_t$ from $\mathscr{P}_t$ according to the selection method.
4: Estimate a probabilistic model $\mathscr{M}_t$ from $\mathscr{S}_t$.
5: Generate the new population $\mathscr{P}_{t+1}$ by sampling from the model $\mathscr{M}_t$
   Assign $t \longleftarrow t + 1$.
6: If stopping criteria are not reached go to step 2.

The main advantage of using probabilistic distributions in evolutionary algorithms is that the interrelations among the variables of a population are *explicitly* modeled. Thus, according to step 4 of Algorithm 3.1, the estimation of a probabilistic model is an important procedure in EDAs. Therefore, ever since EDAs were introduced in the field of EC, many researchers have been interested in proposing and enhancing new probabilistic models. A number of EDAs have been proposed for optimization problems in discrete and continuous domains. EDAs can be classified according to the complexity of their probabilistic model used to learn the interactions between the variables.

It is known that EDAs models explicitly the dependencies among variables for solving optimization problems. This capacity is not presented in other EAs such as the GA. Some representative works in this direction are the papers presented by Heinz Mühlenbein [41, 42, 43]. These studies make an important contribution for considering a new and promissory class of EAs within the EC community.

The research in EDAs has been motivated by their capacity of taking into account the interactions between variables. This source of knowledge has been called *linkage information* and was investigated by different authors for extending simple GAs to process interrelations, i.e., *building blocks* [37]. Furthermore, it is possible to make theoretical analysis of the evolutive proccess in EDAs [24].

EDAs have become a growing field into the EC community. Nowadays the works in EDAs are presented in the three most important conferences of EC: *Genetic and Evolutionary Computation Conference* (GECCO), *Congress on Evolutionary Computation* (CEC) and *Parallel Problem Solving from Nature* (PPSN). According to [28], the EDA track for the GECCO and CEC appeared in 2005. The publication of papers in several journals and the publication of books such as [38, 48], the presentation of works in many other conferences, along with other academic activities like seminars and workshops, give evidence that the research on EDAs is an active research area in EC.

EDAs can be classified as univariate, bivariate or multivariate according to the complexity of their probabilistic model used to learn the interactions among the variables. The univariate EDAs consider all the variables as independent, for instance, the Univariate Marginal Distribution Algorithm (UMDA) [41, 36], the Population Based Incremental Learning (PBIL) [4], and the compact Genetic Algorithm (cGA)[27]. The bivariate EDAs take into account dependencies between pairs of variables and a few examples are the Bivariate Marginal Distribution Algorithm (BMDA) [47], Mutual Information Maximizing Input Clustering (MIMIC) [15, 36], and Dependency-Trees [5]. Many univariate and bivariate discrete EDAs have been extended to continuous domains by using Gaussian probabilistic models.

For multiple dependencies in discrete domain the EDAs have used probabilistic models such as the Polytree Approximation of Distribution Algorithm (PADA) [57], Estimation of Bayesian Network Algorithm (EBNA) [18, 35] and Bayesian Optimization Algorithm (BOA) [46]. For real-valued (continuous) multivariate variables the EDAs have used mostly multivariate Gaussian distributions and some examples are the Estimation of Multivariate Normal Algorithm (EMNA) and Estimation of Gaussian Network Algorithm (EGNA) [36]. The EDA AMaLGaM [10] and the algorithm CMA-ES [29] are also based on the multivariate Gaussian distribution. Both algorithms modify the estimated covariance matrix in order to improve the convergence rate towards the optimum. Currently they are the state of the art in real-valued optimization.

Although there have been many applications and studies related to discrete EDAs, it is interesting to design new EDAs for continuous domains. More specifically, it is important to *design multivariate distributions that represent satisfactorily dependencies among continuous variables and that are relatively easy to estimate and to sample*.

To the best of our knowledge the theses presented by [6] and [2] are the first attempts to incorporate a multivariate Gaussian copula function in EDAs. Since then, other related works have been published. These papers present EDAs based on (1) the Gaussian copula function [64], and (2) Archimedean copula functions with a fixed dependence parameter [61, 63, 13, 62]. Unlike the previous papers that use

Archimedean copula functions, the works of [20] and [22] present a way of estimating the copula parameter. All these works, with exception of [20], only use multivariate copula functions to model the dependence structure among decision variables and do not employ graphical models. On the other hand, in [52, 53, 54], we have proposed the use of the maximum likelihood method and copula entropies in order to (1) estimate the copula parameters, and (2) build a graphical model which establishes the most important dependencies between variables. One recent contribution of our research work [54], has been the incorporation of a procedure for selecting the most adequate copula function.

2.3   Copula Functions

Copula functions are suitable tools in statistics for modeling dependencies, not necessary linear, in several random variables. Copula theory was introduced by [28] to separate the effect of dependence from the effect of marginal distributions in a joint distribution. Although copula functions can model linear and nonlinear dependencies, they have been barely used in computer science applications where nonlinear dependencies are common and need to be represented. In this section, we provide an introduction to the copula theory and present several copula functions.

Copula functions have been widely used in economics and finance [5, 17, 21, 29, 60]. More recently copula functions have been used in other fields such as climate [26], oceanography [10], hydrology [12], geodesy [2], reliability [18], evolutionary computation [52, 53, 54] and engineering [26]. By using copula theory, a joint distribution can be built with a copula function and, possibly, several different marginal distributions. Copula theory has been used also for modeling multivariate distributions in *unsupervised learning* problems such as image segmentation [11, 19] and retrieval tasks [39, 49, 58]. In [30], the bivariate Archimedean copula functions Ali-Mikhail-Haq, Clayton, Frank and Gumbel are used for unsupervised classification. These copulas are well defined for two variables but when extended to three or more variables several complications arise, preventing their generalization and applicability. Some of these complications are (1) the copula parameter is the same for all pairs, and (2) it is not possible to model separately the dependence among all pairs of variables. For the Gaussian copula however, there exist a simple "general formula" for any number of variables. The research works by [50, 51] introduce the use of Gaussian copula in supervised classification, and compares an independent probabilistic classifier with a copula-based probabilistic classifier.

Definition 2.1. A copula function is a joint distribution function of standard uniform random variables. That is,

$$C(u_1, \ldots, u_d) = P[U_1 \leq u_1, \ldots, U_d \leq u_d] \ ,$$

where $U_i \sim U(0,1)$ for $i = 1, \ldots, d$.

As a consequence of Definition 2.1, the copula density for continuous random variables can be calculated as:

$$c(u_1, \ldots, u_d) = \frac{\partial^d C(u_1, \ldots, u_d)}{\partial u_1 \cdots \partial u_d} \quad . \tag{2.1}$$

The interested reader is referred to [14, 20, 29] for a more formal definition of copula function. The following result, known as Sklar's theorem, gives the relevance and practical utility to copula functions.

Theorem 2.1 (Sklar). *Let F be a d-dimensional distribution function with marginals* $F_1, F_2, \ldots, F_d$*, then there exists a copula C such that for all x in* $\overline{\mathbb{R}}^d$*,*

$$F(x_1, x_2, \ldots, x_d) = C(F_1(x_1), F_2(x_2), \ldots, F_d(x_d)) \quad ,$$

where $\overline{\mathbb{R}}$ *denotes the extended real line* $[-\infty, \infty]$*. If* $F_1(x_1)$*,* $F_2(x_2)$*,* $\ldots$*,* $F_d(x_d)$ *are all continuous, then C is unique. Otherwise, C is uniquely determined on* $Ran(F_1) \times Ran(F_2) \times \cdots \times Ran(F_d)$*, where Ran stands for the range.*

According to Theorem 3.1 and using the chain rule for differentiating composite functions along with (2.1), *any* d-dimensional density f can be represented as

$$f(x_1, \ldots, x_d) = \prod_{i=1}^{d} f_i(x_i) \cdot c(F_1(x_1), \ldots, F_d(x_d)) \quad , \tag{2.2}$$

where c is the density of the copula C, and $f_i(x_i)$ is the marginal density of variable x_i. Equation (2.2) shows that the dependence structure is modeled by the copula function. This expression separates any joint density function into the product of copula density and marginal densities. This contrasts with the usual way to model multivariate distributions, which suffers from the restriction that the marginal distributions are usually of the same type. The separation between marginal distributions and a dependence structure explains the modeling flexibility given by copula functions.

2.3.1   The Gaussian Copula

An important parametric family is the multivariate Gaussian copula. This copula function along with the multivariate Student's copula are members of the elliptical copulas.

Definition 2.2. The copula associated to the multivariate standard normal distribution is called Gaussian copula.

According to Definition 2.2 and Theorem 3.1, if the d-dimensional distribution of a random vector $(Z_1, \ldots, Z_d)$ is a joint standard normal distribution, then the associated Gaussian copula has the following expression:

$$C(\Phi(z_1), \ldots, \Phi(z_d); \Sigma) = \int_{-\infty}^{z_1} \cdots \int_{-\infty}^{z_d} \frac{e^{-\frac{1}{2}t'\Sigma^{-1}t}}{(2\pi)^{(n/2)}|\Sigma|^{1/2}} dt_d \cdots dt_1 \quad , \tag{2.3}$$

or equivalently,

$$C(u_1, \ldots, u_d; \Sigma) = \int_{-\infty}^{\Phi^{-1}(u_1)} \cdots \int_{-\infty}^{\Phi^{-1}(u_d)} \frac{e^{-\frac{1}{2}t'\Sigma^{-1}t}}{(2\pi)^{(n/2)}|\Sigma|^{1/2}} dt_d \cdots dt_1 \quad , \tag{2.4}$$

where Φ is the cumulative distribution function of the marginal standard normal distribution and Σ is a symmetric matrix with main diagonal of ones. The elements outside the main diagonal of matrix Σ are the pairwise correlations ρ_{ij} between variables Z_i and Z_j, for $i, j = 1, \ldots, d$ and $i \neq j$. It can be noticed that a d-dimensional standard normal distribution has mean vector zero and a correlation matrix Σ with $d(d-1)/2$ parameters.

The dependence parameters ρ_{ij} of a d-dimensional Gaussian copula can be estimated using the maximum likelihood method. To do so, the steps of Algorithm 2.2 can be followed. Algorithm 2.2 is based on reference [5].

Algorithm 2.2. Pseudocode for estimating Gaussian copula parameters

---

1: Transform values of each variable u_j by calculating $z_j = \Phi^{-1}(u_j)$, for $j = 1, \ldots, d$, where Φ is the cumulative standard normal distribution function.
2: Build the sample data matrix $\mathbf{z} = \{(z_{1i}, z_{2i}, \ldots, z_{di})\}_{i=1}^{n}$.
3: Estimate the correlation matrix $\widehat{\Sigma}$ using pseudo observations $\mathbf{z}_i = (z_{1i}, z_{2i}, \ldots, z_{di})$ and the formula

$$\widehat{\Sigma} = \frac{1}{n} \sum_{i=1}^{n} \mathbf{z}_i' \mathbf{z}_i \quad .$$

Due to Equation (2.2), the d-dimensional Gaussian copula density can be calculated as:

$$\begin{aligned}
c(\Phi(z_1), \ldots, \Phi(z_d); \Sigma) &= \frac{\frac{1}{(2\pi)^{(d/2)}|\Sigma|^{1/2}} e^{-\frac{1}{2}z'\Sigma^{-1}z}}{\prod_{i=1}^{d} \frac{1}{(2\pi)^{1/2}} e^{-\frac{1}{2}z_i^2}} \\
&= \frac{1}{|\Sigma|^{1/2}} e^{-\frac{1}{2}z'(\Sigma^{-1}-I)z} \quad .
\end{aligned} \tag{2.5}$$

Given that a Gaussian copula is also a distribution function, it is possible to simulate data from it. The main steps are the following: once a correlation matrix Σ is specified, a data set can be generated from a joint standard normal distribution. The next step consists of transforming this data set using the cumulative distribution function Φ. Algorithm 2.3 and Fig. 2.1 illustrate the sampling procedure for different correlations.

Algorithm 2.3. Pseudocode for generating data from a Gaussian copula

---

1: Simulate observations $(z_1, \ldots, z_d)$ from a joint standard normal distribution with correlation matrix Σ.

2: Calculate $u_i = \Phi(z_i)$ where Φ is the cumulative standard normal distribution function, for $i = 1, \ldots, d$.

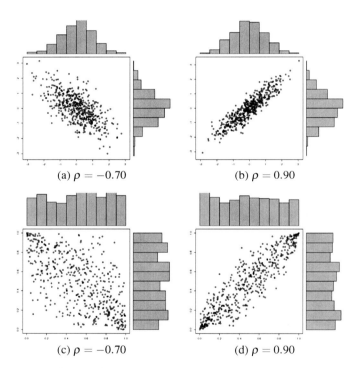

Fig. 2.1 A sample of 500 points from a standard normal distribution (top) and the corresponding sample for a Gaussian copula (bottom) with (a) a negative dependence, and (b) a positive dependence. Histograms show the marginal distribution for each variable.

Figure 2.1 (a) shows a sample drawn from a bivariate standard normal distribution with correlation $\rho = -0.70$ (step 1, Algorithm 2.3). The histogram on the vertical axis and the histogram on the horizontal axis illustrate that both marginals are univariate standard normal distributions. This data set is used to obtain a sample from a Gaussian copula, as shown in Fig. 2.1 (c) (step 2, Algorithm 2.3). Both histograms illustrate that marginals are uniform, according to Definition 2.1. In order to appreciate how the correlation parameter modifies the dependence structure, Fig. 2.1 (b) and Fig. 2.1 (d) show the corresponding information with $\rho = 0.90$.

An important result for parametric bivariate copulas is explained through the following theorem, which relates the dependence parameter θ of a copula and Kendall's τ.

Theorem 2.2. *Let X and Y be continuous random variables whose copula is C. Then the population version of Kendall's tau for X and Y is given by*

$$\tau(X,Y) = 4 \int_0^1 \int_0^1 C(u,v;\theta)dC(u,v;\theta) - 1 \ , \tag{2.6}$$

where $u = F_X(x)$ and $v = F_Y(y)$.

For a bivariate Gaussian copula, Equation (2.6) can be written as

$$\tau = \frac{2}{\pi}\arcsin(\rho) \ , \tag{2.7}$$

where $\rho = \theta$. As a consequence, the data sets in Fig. 2.1 (a)-(top,bottom) have the same concordance value, measured in Kendall's τ. A similar statement can be said for Fig. 2.1 (b)-(top,bottom).

In order to appreciate Gaussian dependence between non Gaussian marginals, Fig. 2.2 shows a scatter plot with data drawn from a joint distribution with marginals Beta and dependence structure modeled by a bivariate Gaussian copula.

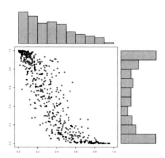

Fig. 2.2 A sample of 500 points that has been generated with a Gaussian copula with parameter $\theta = -0.9$ and marginal distributions Beta with parameters $(1,2)$ (histogram on the horizontal axis) and $(0.5, 0.5)$ (histogram on the vertical axis).

Given that is well established how to estimate correlation matrices, evaluate densities, and calculate integrals for the multidimensional normal distribution, the Gaussian copula function is relatively easy to implement.

2.4   Regular Vines

Vines is a class of undirected graphs for representing high dimensional probability distributions. These kind of graphs use bivariate and conditional bivariate copula functions.

According to [33], a vine on d variables is a set of nested trees, where the edges of the tree j are the nodes of the tree $j+1$, for $j=1,\ldots,d-2$, and each tree has the maximum number of edges. We illustrate the concept of a vine in the following example.

Example 2.1. Let (X_1,X_2,X_3) be a three dimensional random vector with a joint density function $f(x_1,x_2,x_3)$. A well known factorization for the trivariate density is given by the expression

$$f(x_1,x_2,x_3) = f(x_1) \cdot f(x_2|x_1) \cdot f(x_3|x_1,x_2) \ . \tag{2.8}$$

From the copula theory and (2.2), the joint density can also be factorized as

$$f(x_1,x_2,x_3) = f(x_1) \cdot f(x_2) \cdot f(x_3) \cdot c(u_1,u_2,u_3) \ . \tag{2.9}$$

However, once again by means of (2.2), we can decompose the conditional distributions into bivariate copulas and marginal densities

$$f(x_2|x_1) = \frac{f(x_1,x_2)}{f(x_1)} = c(u_1,u_2) \cdot f(x_2) \ , \tag{2.10}$$

$$\begin{aligned} f(x_3|x_1,x_2) &= \frac{f(x_1,x_3|x_2) \cdot f(x_2)}{f(x_1|x_2) \cdot f(x_2)} \\ &= \frac{c(u_1,u_3|u_2) \cdot f(x_1|x_2) \cdot f(x_3|x_2)}{f(x_1|x_2)} \\ &= c(u_1,u_3|u_2) \cdot c(u_2,u_3) \cdot f(x_3) \ . \end{aligned} \tag{2.11}$$

Inserting expressions (2.10) and (2.11) into (2.8) gives

$$f(\mathbf{x}) = c(u_1,u_2) \cdot c(u_2,u_3) \cdot c(u_1,u_3|u_2) \cdot \prod_{k=1}^{3} f(x_k) \ . \tag{2.12}$$

The pair copula decomposition in (2.12) can be represented by a graphical structure. Figure 2.3 (a) shows a graph with 6 nodes and 6 edges. Figure 2.3 (b) shows a vine with 5 nodes, 2 trees and 3 edges. The contents of each node in the vine represents the indexes of the random variables. For example, the node with number two is related to random variables (X_2,U_2). Two edges in the vine are associated to marginal bivariate copulas, whereas one edge is associated to a conditional bivariate copula.

By comparing (2.9) and (2.12), we see that a trivariate copula density can be built using only bivariate copulas as building blocks:

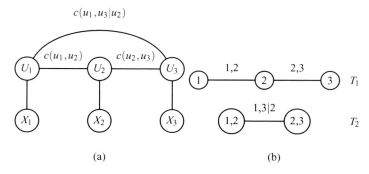

Fig. 2.3 (a) An undirected graphical model. (b) A typical vine representation. Both graphs refer to the trivariate density function (2.12).

$$c(u_1, u_2, u_3) = \underbrace{c(u_1, u_2) \cdot c(u_2, u_3)}_{\text{marginals}} \cdot \underbrace{c(u_1, u_3 | u_2)}_{\text{conditional}} \ . \tag{2.13}$$

From Fig. 2.3 (b), it can be noticed that the edges of the first tree T_1 are the nodes of the second tree T_2 and each tree has the maximum number of edges, i.e., two and one edges respectively. Moreover, from (2.13), it can be seen that the tree T_1 is related to the marginal bivariate copulas and the nested tree T_2 is related to the conditional bivariate copula. ◀

Several comments can be said from the exposition of Example 2.1. For example, the vine representation for the joint density function is not unique. There are six different permutations for the indexes of the variables, but only three permutations give different factorizations.

Vines give a way of extending bivariate copula functions to higher dimensions. By selecting an adequate set of bivariate copula functions, it is possible to design new d-dimensional copulas. Moreover, vines can be easily adapted to higher dimensions.

Finally, besides vines are graphical representations of pair copula decompositions, they can provide a more flexible representation of the joint distribution.

In this work, we are interested in using a special subset of vines as probabilistic models in EDAs. We refer to *regular vines* and present its formal definition [34].

Definition 2.3 (Regular vine). $\mathcal{V}$ is a regular vine on d elements if

1. $\mathcal{V} = (T_1, \ldots, T_{d-1})$, where T_i is a tree[1] for all $i = 1, \ldots, d - 1$.
2. T_1 is a connected tree with nodes $N_1 = \{1, \ldots, d\}$ and edges E_1. For $i = 2, \ldots, d - 1$, $T_i = (N_i, E_i)$ is a connected tree with nodes $N_i = E_{i-1}$.

---

[1] A tree, as defined in [34], can be considered as a forest of trees. A tree in which all nodes are connected is termed as a *connected tree*.

3. For $i = 2, \ldots, d-1$, if $\{a,b\} \in E_i$, then $\#a\triangle b = 2$, where $\triangle$ denotes the symmetric difference. In other words, if a and b are nodes of T_i connected by an edge in T_i, where $a = \{a_1, a_2\}$ and $b = \{b_1, b_2\}$, then exactly one of the a_i equals one of the b_i. This condition is called the *proximity condition*.

The first and second properties in Definition 2.3 refer to vines. Third property [7, 8] refers to the proximity condition in a regular vine, since it expresses the fact that two edges in tree j are joined by an edge in tree $j+1$ only if these edges share a common node, $j = 1, \ldots, d-2$.

Two families of regular vines are the *D-vine* and the *canonical vine (C-vine)*[2]. These special cases of regular vines impose additional restrictions and are characterized by minimal and maximal degrees of nodes in the trees.

Definition 2.4 (C-vine, D-vine). A regular vine is called a

1. **Canonical** or **C-vine** if each tree T_i has a unique node of degree $d-i$. The node with maximal degree in T_1 is the *root*.
2. **D-vine** if each node in T_1 has a degree of at most 2.

Examples of C-vines and D-vines on 4 nodes are shown in Figures 2.4 and 2.5 respectively.

Figure 2.5 shows a D-vine on four variables. The tree T_1 is built on marginal pairwise variables, whereas the tree T_2 takes into account conditional pairwise variables. Observe how tree T_2 is built on tree T_1. The last tree, T_3, involves variables U_1 and U_4 conditioned on variables U_2, and U_3. Every tree, except tree T_1, has associated conditional bivariate distributions. References [1, 34] provide formal definitions of regular vines and illustrative information.

From a theoretical point of view, it is possible to model any d-dimensional dependence structure by means of regular vines and bivariate copulas. However, for practical purposes, it is not necessary to build the complete C-vine or the complete D-vine. For example, a *truncated* regular vine can be adequate for modeling a d-dimensional distribution if it preserves as much information as possible.

Before presenting the implementation of regular vines into an EDA, we provide some theoretical relationships between multivariate distributions and their associated copula functions.

2.4.1   Copula Entropy and Mutual Information

In the EDA literature [38], the Kullback-Leibler divergence has been used as a measure of the difference between two probability distributions. This divergence is an information measure between two distributions. It is always non-negative for any two distributions, and is zero if and only if the distributions are identical. Hence, the

---

[2] D-vines were originally called *drawable vines*, while canonical vines owe their name to the fact that they are the most natural for sampling [34].

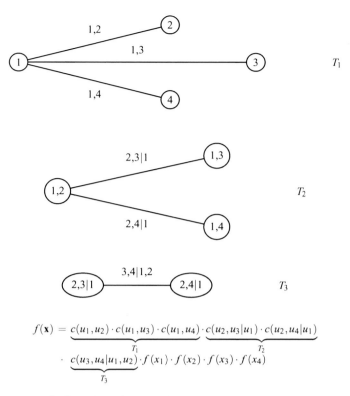

Fig. 2.4 Example of a four-dimensional C-vine

Kullback-Leibler divergence can be interpreted as a measure of the dissimilarity of two distributions. Below, we prove an important relationship between the Kullback-Leibler divergence and copula functions.

Proposition 2.1. *Let f and g be two d-dimensional density functions with marginal densities f_i and g_i, respectively for $i = 1, \ldots, d$. Then, the Kullback-Leibler divergence between multivariate densities f and g is given by the expression,*

$$D_{KL}(f\|g) = \sum_{i=1}^{d} D_{KL}(f_i\|g_i) + D_{KL}(c_f\|c_g) \quad,$$

where c_f and c_g are the associated copula functions for multivariate densities f and g.

Proof. We start the proof by using the definition of the Kullback-Leibler divergence and (2.2)

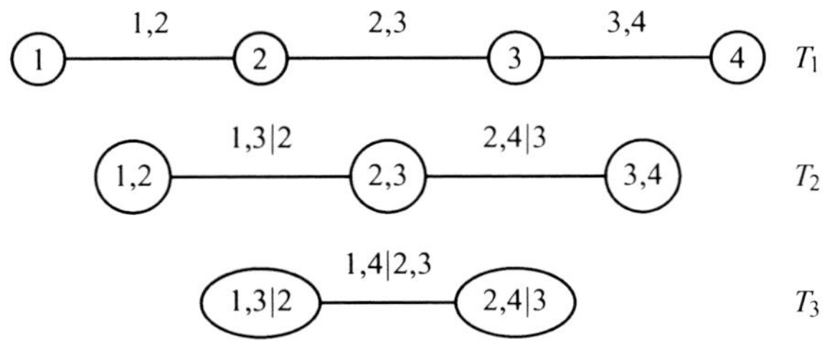

$$f(\mathbf{x}) = \underbrace{c(u_1,u_2) \cdot c(u_2,u_3) \cdot c(u_3,u_4)}_{T_1} \cdot \underbrace{c(u_1,u_3|u_2) \cdot c(u_2,u_4|u_3)}_{T_2}$$
$$\cdot \underbrace{c(u_1,u_4|u_2,u_3)}_{T_3} \cdot f(x_1) \cdot f(x_2) \cdot f(x_3) \cdot f(x_4)$$

Fig. 2.5 Example of a four-dimensional D-vine

$$D_{KL}(f\|g) = E_{f(\mathbf{x})}\left[\log \frac{f(\mathbf{x})}{g(\mathbf{x})}\right]$$
$$= E_{f(\mathbf{x})}\left[\log \frac{\prod_{i=1}^{d} f_i \cdot c_f}{\prod_{i=1}^{d} g_i \cdot c_g}\right]$$
$$= \sum_{i=1}^{d} E_{f(\mathbf{x})}\left[\log \frac{f_i}{g_i}\right] + E_{f(\mathbf{x})}\left[\log \frac{c_f}{c_g}\right]$$
$$= \sum_{i=1}^{d} D_{KL}(f_i\|g_i) + E_{f(\mathbf{x})}\left[\log \frac{c_f}{c_g}\right] .$$

It is known that $u_i = F_i(x_i)$ for $i = 1,\ldots,d$. By using a change of variables we have

$$E_{f(\mathbf{x})}\left[\log \frac{c_f}{c_g}\right] = \int f(\mathbf{x}) \log\left(\frac{c_f}{c_g}\right) d\mathbf{x}$$
$$= \int \prod_{i=1}^{d} f_i \cdot c_f \log\left(\frac{c_f}{c_g}\right) \frac{1}{\prod_{i=1}^{d} f_i} d\mathbf{u}$$
$$= D_{KL}(c_f\|c_g) .$$

Therefore,

$$D_{KL}(f\|g) = \sum_{i=1}^{d} D_{KL}(f_i\|g_i) + D_{KL}(c_f\|c_g) . \qquad \square$$

Proposition 2.1 gives an encouragement for using copula functions. When a probabilistic model is proposed for a multivariate data set, the Kullback-Leibler divergence between the unknown density of the data set and the proposed density model depends on the selection of marginal densities and the copula function. Under the assumption that marginal densities are well selected, the proposed density differs only from the unknown density in the term related to the dependence among variables.

Proposition 2.2. *Let f be a d-dimensional density function with marginal densities f_i for $i = 1, \ldots, d$. Then, the Kullback-Leibler divergence between the multivariate density f and the product of marginal densities $\prod_{i=1}^{d} f_i$ is given by the expression,*

$$D_{KL}\left(f \| \prod_{i=1}^{d} f_i\right) = -H(U_1, \ldots, U_d) \ ,$$

where $H(U_1, \ldots, U_d)$ is the entropy of the associated copula function for the multivariate density f.

Proof. Using Proposition 2.1

$$D_{KL}\left(f \| \prod_{i=1}^{d} f_i\right) = \sum_{i=1}^{d} D_{KL}(f_i \| f_i) + D_{KL}(c_f \| 1)$$
$$= D_{KL}(c_f \| 1) \ .$$

But,

$$D_{KL}(c_f \| 1) = \int c_f \log\left(\frac{c_f}{1}\right) d\mathbf{u}$$
$$= -H(U_1, \ldots, U_d) \ .$$

Thus,

$$D_{KL}\left(f \| \prod_{i=1}^{d} f_i\right) = -H(U_1, \ldots, U_d) \ . \qquad \qquad \square$$

For the particular case of a two-dimensional random vector (X_1, X_2), it is known that the Kullback-Leibler divergence between the bivariate density f and the product of marginal densities $f_1 \cdot f_2$ is equal to the mutual information between variables X_1 and X_2. In this sense, Proposition 2.2 gives a theoretical support for the connection between mutual information and the entropy of a bivariate copula function presented in [14].

The following result was presented in [31] and shows the relationship among the entropies of marginal densities, the entropy of the original distribution, and the entropy of the associated copula function.

Proposition 2.3. *Let f be a d-dimensional density function with marginal densities f_i for $i = 1, \ldots, d$. Then, the entropy of the associated copula function is given by,*

$$H(U_1, \ldots, U_d) = H(X_1, \ldots, X_d) - \sum_{i=1}^{d} H(X_i) \ .$$

Proof. We first calculate the Kullback-Leibler divergence

$$D_{KL}\left(f \| \prod_{i=1}^{d} f_i\right) = E_{f(\mathbf{x})}\left[\log \frac{f(\mathbf{x})}{\prod_{i=1}^{d} f_i}\right]$$

$$= E_{f(\mathbf{x})}\left[\log f(\mathbf{x})\right] - E_{f(\mathbf{x})}\left[\log \prod_{i=1}^{d} f_i\right]$$

$$= -H(X_1, \ldots, X_d) + \sum_{i=1}^{d} H(X_i) \ .$$

By using the result of Proposition 2.2, we complete the proof.   □

From information theory, it is known that the sum of marginal entropies is greater or equal than the joint entropy. As a consequence, Proposition 2.3 states that the entropy of a copula function is non positive. Moreover, all the information about dependencies between variables can be measured by the copula entropy.

We present in the next proposition, two results for bivariate and trivariate dependence structures.

Proposition 2.4. *Let X_1, X_2 and X_3 be continuous random variables with joint and marginal densities. The following expressions hold for the mutual information and the conditional mutual information,*

$$I(X_1, X_2) = I(U_1, U_2) \ , \tag{2.14}$$

$$I(X_1, X_2 | X_3) = I(U_1, U_2 | U_3) \ , \tag{2.15}$$

where U_1, U_2 and U_3 are the variables of the corresponding copula function.

Proof. By using Proposition 2.2 and the fact that marginal densities are uniform for copula functions,

$$\begin{aligned} I(X_1, X_2) &= D_{KL}(f \| f_1 \cdot f_2) \\ &= D_{KL}(c_f \| 1) \\ &= \int c_f \log(c_f) \, d\mathbf{u} \\ &= I(U_1, U_2) \ . \end{aligned}$$

Similarly,

$$I(X_1, X_2 | X_3) = D_{KL}\left(f || f_{1|3} \cdot f_{2|3} \cdot f_3\right)$$
$$= D_{KL}\left(c_f || c(u_1, u_3) \cdot c(u_2, u_3)\right)$$
$$= \int c_f \log\left(\frac{c_f}{c(u_1, u_3) \cdot c(u_2, u_3)}\right) d\mathbf{u}$$
$$= I(U_1, U_2 | U_3) \ . \qquad \Box$$

Thus, by (2.14) and (2.15), the information about marginal and conditional dependencies between *any* two variables, can be measured by the marginal and conditional mutual information of the corresponding copula functions. It is not difficult to prove that, for calculating the marginal and conditional mutual information, we can use copula entropies:

$$I(X_1, X_2) = -H(U_1, U_2) \ , \qquad (2.16)$$

$$I(X_1, X_2 | X_3) = -H(U_1, U_2, U_3) + H(U_1, U_3) + H(U_2, U_3) \ . \qquad (2.17)$$

The results in Proposition 2.4 show that, under the assumption that marginal distributions are well fitted, the source of information about the dependence between variables can be calculated by using only copula functions. These results along with (2.16) and (2.17) will be used for constructing the graphical structure of regular vines in EDAs.

The relationship between the entropy of a bivariate copula function and the marginal mutual information of two variables, (2.16), has both theoretical and practical importance: 1) the mutual information is given by the copula function *regardless* of the marginal distributions, and 2) the estimation of the copula entropy can be more accurate than the estimation of mutual information because the copula domain is *always* bounded and standardized.

An important consequence for bivariate Gaussian copulas is that, by definition, its entropy is equal to the negative of mutual information of two variables with standard joint Gaussian distribution

$$H(U_1, U_2) = \frac{1}{2}\log(1 - \rho^2) \ , \qquad (2.18)$$

where ρ is the correlation parameter.

The result in Proposition 2.2 implies that the entropy of a trivariate Gaussian copula is given by:

$$H(U_1, U_2, U_3) = \frac{1}{2}\log(1 + 2\rho_{12}\rho_{13}\rho_{23} - \rho_{12}^2 - \rho_{13}^2 - \rho_{23}^2) \ . \qquad (2.19)$$

We illustrate how the selection of the first two trees in a C-vine and a D-vine can modify the amount of information for the corresponding truncated regular vine.

Example 2.2 (Truncated C-vine). Consider a four-dimensional distribution $f(\mathbf{x})$ with an associated copula function. A full C-vine for the four-dimensional distribution is given by

$$f(\mathbf{x}) = \underbrace{c(u_1, u_2) \cdot c(u_1, u_3) \cdot c(u_1, u_4)}_{T_1} \cdot \underbrace{c(u_2, u_3|u_1) \cdot c(u_2, u_4|u_1)}_{T_2}$$
$$\cdot \underbrace{c(u_3, u_4|u_1, u_2)}_{T_3} \cdot f(x_1) \cdot f(x_2) \cdot f(x_3) \cdot f(x_4) \quad . \tag{2.20}$$

A graphical representation of (2.20) can be seen in Fig. 2.4. If we employ only the first two trees of the C-vine, the truncated model is given by

$$f_{T_1, T_2}(\mathbf{x}) = \underbrace{c(u_1, u_2) \cdot c(u_1, u_3) \cdot c(u_1, u_4)}_{T_1} \cdot \underbrace{c(u_2, u_3|u_1) \cdot c(u_2, u_4|u_1)}_{T_2}$$
$$\cdot f(x_1) \cdot f(x_2) \cdot f(x_3) \cdot f(x_4) \quad . \tag{2.21}$$

The loss of information given by the truncated model can be measured through the Kullback-Leibler divergence between (2.20) and (2.21):

$$D_{KL}(f\|f_{T_1, T_2}) = - \underbrace{\sum_{i=2}^{4} I(X_1, X_i)}_{T_1} - \underbrace{\sum_{j=3}^{4} I(X_2, X_j|X_1)}_{T_2}$$

$$+ \sum_{k=1}^{4} H(X_k) - H(X_1, X_2, X_3, X_4) \quad . \tag{2.22}$$

The terms in (2.22) related to marginal and conditional mutual information can be calculated by using (2.16) and (2.17). Moreover, according to the result in Proposition 2.3, the last two terms of the Kullback-Leibler divergence (2.22) can be substituted by the joint copula entropy.

If the Gaussian copula is assumed as the associated copula function, then, by using (2.18) and (2.19), the Kullback-Leibler divergence (2.22) can be written as

$$D_{KL}(f\|f_{T_1, T_2}) = \underbrace{\frac{1}{2} \sum_{i=2}^{4} \log\left(1 - \rho_{1i}^2\right)}_{T_1} + \underbrace{\frac{1}{2} \sum_{j=3}^{4} \log\left(1 - \rho_{2j|1}^2\right)}_{T_2}$$

$$- \frac{1}{2} \log(|\Sigma|) \quad . \tag{2.23}$$

where ρ_{1i} and $\rho_{2j|1}$ are the parameters of the marginal and conditional Gaussian copulas $c(u_1, u_i)$ and $c(u_2, u_j|u_1)$, respectively. Details of the Gaussian copula and its correlation matrix Σ can be seen in Sec. 2.3.1.

For a particular Gaussian copula with correlation matrix given by

$$\Sigma = \begin{bmatrix} 1 & 0.61 & 0.62 & 0.39 \\ 0.61 & 1 & 0.47 & 0.50 \\ 0.62 & 0.47 & 1 & 0.49 \\ 0.39 & 0.50 & 0.49 & 1 \end{bmatrix} , \qquad (2.24)$$

the Kullback-Leibler divergence (2.23) is

$$D_{KL}\left(f\|f_{T_1,T_2}\right) = \underbrace{(-0.557762)}_{T_1 \text{ with } X_1 \text{ as root}} + \underbrace{(-0.080097)}_{T_2 \text{ with } X_2 \text{ as root}} - (-0.6900143)$$
$$= 0.052155 . \qquad (2.25)$$

The loss of information can be modified if we considered other variables as roots. For example, if variables X_2 and X_3 are the roots for the first tree T_1 and the second tree T_2, respectively, then the corresponding Kullback-Leibler divergence is

$$D_{KL}\left(f\|f_{T_1,T_2}\right) = \underbrace{(-0.501336)}_{T_1 \text{ with } X_2 \text{ as root}} + \underbrace{(-0.18778)}_{T_2 \text{ with } X_3 \text{ as root}} - (-0.6900143)$$
$$= 0.000898 . \qquad (2.26)$$

For minimizing the loss of information, the above results suggest that an adequate truncated C-vine must select the tree T_1 with the greatest sum of marginal mutual information and then, conditioned to the root of tree T_1, select the tree T_2 with the greatest sum of conditional mutual information. With this idea, the minimum of the Kullback-Leibler divergence is gotten when variables X_1 and X_4 are the roots of the first tree T_1 and the second tree T_2, respectively:

$$D_{KL}\left(f\|f_{T_1,T_2}\right) = \underbrace{(-0.557762)}_{T_1 \text{ with } X_1 \text{ as root}} + \underbrace{(-0.131869)}_{T_2 \text{ with } X_4 \text{ as root}} - (-0.6900143)$$
$$= 0.000383 . \qquad (2.27)$$

◀

Example 2.3 (Truncated D-vine). Similar to the Example 2.2, we consider a four-dimensional distribution $f(\mathbf{x})$ with an associated copula function. A full D-vine for the four-dimensional distribution is given by

$$f(\mathbf{x}) = \underbrace{c(u_1, u_2) \cdot c(u_2, u_3) \cdot c(u_3, u_4)}_{T_1} \cdot \underbrace{c(u_1, u_3|u_2) \cdot c(u_2, u_4|u_3)}_{T_2}$$
$$\cdot \underbrace{c(u_1, u_4|u_2, u_3)}_{T_3} \cdot f(x_1) \cdot f(x_2) \cdot f(x_3) \cdot f(x_4) . \qquad (2.28)$$

A graphical representation of (2.28) can be seen in Fig. 2.5. If we employ only the first two trees of the D-vine, the truncated model is given by

$$f_{T_1,T_2}(\mathbf{x}) = \underbrace{c(u_1,u_2) \cdot c(u_2,u_3) \cdot c(u_3,u_4)}_{T_1} \cdot \underbrace{c(u_1,u_3|u_2) \cdot c(u_2,u_4|u_3)}_{T_2}$$
$$\cdot \, f(x_1) \cdot f(x_2) \cdot f(x_3) \cdot f(x_4) \quad . \tag{2.29}$$

The loss of information given by the truncated model can be measured through the Kullback-Leibler divergence between (2.28) and (2.29):

$$D_{KL}(f||f_{T_1,T_2}) = -\underbrace{\sum_{i=1}^{3} I(X_i, X_{i+1})}_{T_1} - \underbrace{\sum_{j=1}^{2} I(X_j, X_{j+2}|X_{j+1})}_{T_2}$$
$$- H(U_1, U_2, U_3, U_4) \quad . \tag{2.30}$$

For the permutation of variables (X_1, X_2, X_3, X_4) in the first tree T_1, and the particular Gaussian copula with correlation matrix given by (2.24), the Kullback-Leibler divergence (2.30) is

$$D_{KL}(f||f_{T_1,T_2}) = \underbrace{(-0.494779)}_{T_1 \text{ based on } (X_1,X_2,X_3,X_4)} + \underbrace{(-0.194337)}_{T_2} - (-0.6900143)$$
$$= 0.000898 \quad . \tag{2.31}$$

The loss of information can be modified if we considered other permutation of variables for the first tree T_1. For example, if variables X_1, X_2, X_4 and X_3 are a permutation of variables for the first tree T_1, then the corresponding Kullback-Leibler divergence is

$$D_{KL}(f||f_{T_1,T_2}) = \underbrace{(-0.513812)}_{T_1 \text{ based on } (X_1,X_2,X_4,X_3)} + \underbrace{(-0.054242)}_{T_2} - (-0.6900143)$$
$$= 0.12196 \quad . \tag{2.32}$$

In order to minimize the loss of information, it is necessary to consider all possible permutations of variables for the first tree. By doing so, an optimal permutation is given by variables X_2, X_1, X_4 and X_3:

$$D_{KL}(f||f_{T_1,T_2}) = \underbrace{(-0.452468)}_{T_1 \text{ based on } (X_1,X_2,X_4,X_3)} + \underbrace{(-0.237164)}_{T_2} - (-0.6900143)$$
$$= 0.000383 \quad . \tag{2.33}$$

◀

As can be seen, for minimizing the loss of information with a truncated D-vine, there is no a straightforward procedure as the C-vine case. However, for practical purposes, it is enough to select an adequate permutation of variables in order to define the tree T_1 with the greatest information as possible.

In general, by truncating the C-vine or the D-vine some piece of information can be lost. A motivation for truncating the C-vine or the D-vine is to reduce the complexity of the model and reduce the number of conditional bivariate copulas. However, it is convenient a procedure for choosing an adequate permutation without a huge loss of information.

2.5   EDAs Based on Regular Vines

In order to show how a multivariate probabilistic model based on regular vines can be used in EDAs, we propose the use of truncated C-vine and D-vine models.

2.5.1   Description of the C-Vine EDA

As shown in Example 2.2, there is a direct way of proposing a C-vine model with only two trees. We define a class of density functions based on a truncated C-vine with only two trees, T_1 and T_2:

$$f_{\text{C-vine}}(\mathbf{x}) = \prod_{i=2}^{d} c\left(u_{\beta_1}, u_{\beta_i}\right) \prod_{j=3}^{d} c\left(u_{\beta_2}, u_{\beta_j} | u_{\beta_1}\right) \prod_{k=1}^{d} f(x_k) \ , \qquad (2.34)$$

where $\boldsymbol{\beta} = (\beta_1, \ldots, \beta_d)$ is a permutation of the integers between 1 and d. Therefore, the d-dimensional density $f_{\text{C-vine}}(\mathbf{x})$ defined in (2.34) is an approximation to a multivariate density based on a full C-vine. Then, the goal is to choose the first two elements of the permutation $\boldsymbol{\beta}$ that minimizes the Kullback-Leibler divergence between the true density function $f(\mathbf{x})$ and the proposed density function $f_{\text{C-vine}}(\mathbf{x})$:

$$D_{KL}\left(f \| f_{\text{C-vine}}\right) = -\sum_{i=2}^{d} I(X_{\beta_1}, X_{\beta_i}) - \sum_{j=3}^{d} I(X_{\beta_2}, X_{\beta_j} | X_{\beta_1})$$
$$- H(\mathbf{U}) \ . \qquad (2.35)$$

The last term in the divergence (2.35) does not depend on $\boldsymbol{\beta}$. Therefore, minimizing the Kullback-Leibler is equivalent to maximizing

$$J_{\text{C-vine}}(\mathbf{X}) = \sum_{i=2}^{d} I(X_{\beta_1}, X_{\beta_i}) + \sum_{j=3}^{d} I(X_{\beta_2}, X_{\beta_j} | X_{\beta_1}) \ . \qquad (2.36)$$

The optimal permutation $\boldsymbol{\beta}$ is the one that produces the highest value for (2.36). A straightforward procedure, based on copula entropies (2.16) and (2.17), is given by Algorithm 2.4, which shows how to find the first two elements of the permutation $\boldsymbol{\beta}$.

Algorithm 2.4. Algorithm to pick the roots of the first two trees of a C-vine

1: Find $\beta_1 = \arg\min_i \sum_{j=2}^{d} \widehat{H}(U_i, U_j)$, where $\widehat{H}()$ is an estimation of the bivariate copula entropy among variables $u_i = F_{X_i}(x_i)$, and $u_j = F_{X_j}(x_j)$.

2: Find $\beta_2 = \arg\min_i \sum_{j=3}^{d} \widehat{H}(U_{\beta_1}, U_i, U_j) - \widehat{H}(U_{\beta_1}, U_i) - \widehat{H}(U_{\beta_1}, U_j)$, where $\widehat{H}()$ is an estimation of the corresponding bivariate and trivariate copula entropies.

If Gaussian copula is used for modeling depencies among variables, the trivariate and bivariate copula entropies of Algorithm 2.4 can be calculated by using (2.18) and (2.19).

2.5.2   Description of the D-Vine EDA

We define a class of density functions based on a truncated D-vine with only two trees, T_1 and T_2:

$$f_{\text{D-vine}}(\mathbf{x}) = \prod_{i=1}^{d-1} c\left(u_{\gamma_i}, u_{\gamma_{i+1}}\right) \prod_{j=1}^{d-2} c\left(u_{\gamma_j}, u_{\gamma_{j+2}} | u_{\gamma_{j+1}}\right) \prod_{k=1}^{d} f(x_k) \;, \qquad (2.37)$$

where $\boldsymbol{\gamma} = (\gamma_1, \ldots, \gamma_d)$ is a permutation of the integers between 1 and d. Therefore, the d-dimensional density $f_{\text{D-vine}}(\mathbf{x})$ defined in (2.37) is composed by the product of marginal densities and a copula density given by a D-vine with only two trees. Then, the goal is to choose a permutation $\boldsymbol{\gamma} = (\gamma_1, \ldots, \gamma_d)$ that minimizes the Kullback-Leibler divergence between the true density function $f(\mathbf{x})$ and the proposed density function $f_{\text{D-vine}}(\mathbf{x})$:

$$D_{KL}\left(f \| f_{\text{D-vine}}\right) = -\sum_{i=1}^{d-1} I(X_{\gamma_i}, X_{\gamma_{i+1}}) - \sum_{j=1}^{d-2} I(X_{\gamma_j}, X_{\gamma_{j+2}} | X_{\gamma_{j+1}})$$
$$- H(\mathbf{U}) \;. \qquad (2.38)$$

The last term in the divergence (2.38) does not depend on $\boldsymbol{\gamma}$. Therefore, minimizing the Kullback-Leibler is equivalent to maximizing

$$J_{\text{D-vine}}(\mathbf{X}) = \sum_{i=1}^{d-1} I(X_{\gamma_i}, X_{\gamma_{i+1}}) + \sum_{j=1}^{d-2} I(X_{\gamma_j}, X_{\gamma_{j+2}} | X_{\gamma_{j+1}}) \;. \qquad (2.39)$$

The optimal permutation $\boldsymbol{\gamma}$ is the one that produces the highest marginal and conditional pairwise mutual information with respect to the true distribution. But due to computational efficiency reasons we propose a greedy algorithm based on (2.16) and (2.17). We first select the three variables with the smallest trivariate copula entropy and choose a random order to make a chain. The following variables of the permutation γ are chosen according to their bivariate copula entropy with respect to any of the variables in the ends of the chain. Algorithm 2.5 shows a straightforward greedy algorithm to find a permutation γ.

Algorithm 2.5. Greedy algorithm to pick a permutation γ in a D-vine

1: Find $(\gamma_{m-1}, \gamma_m, \gamma_{m+1}) = \arg\min_{j \neq k \neq l} \widehat{H}(U_j, U_k, U_l)$, where $\widehat{H}()$ is an estimation of the trivariate copula entropy among variables $u_j = F_{X_j}(x_j)$, $u_k = F_{X_k}(x_k)$, and $u_l = F_{X_l}(x_l)$.
2: Choose variables with the smallest bivariate copula entropy with respect to any of the ends of the chain. The constraint is to avoid a circular chain.
3: The order of the chain defines permutation γ.

If Gaussian copula is used for modeling depencies among variables, the trivariate and bivariate copula entropies of Algorithm 2.5 can be calculated by using (2.18) and (2.19).

2.5.3   Incorporating the Gaussian Copula

In [53], a regular vine is considered for the first time as a graphical model for designing a new EDA. This EDA is based on a D-vine and Gaussian copulas.

The probabilistic models C-vine and D-vine have been previously presented in Sec. 2.4. We summarize the proposed approach in Algorithm 2.6.

Algorithm 2.6. Pseudocode for estimating the C-vine (D-vine) model and generating a new population

1: **for** $i = 1 \to d$ **do**
2:     For each variable X_i, estimate its marginal distribution function $\widehat{F}_i$.
3:     Determine $U_i = \widehat{F}_i(X_i)$.
4: **end for**
5: Estimate the parameters of the Gaussian copula Σ using Algorithm 2.2.
6: Calculate all bivariate and trivariate copula entropies, (2.18) and (2.19).
7: Pick a permutation β (γ) for the graphical model using Algorithm 2.4 (Algorithm 2.5).
8: Simulate U_{β_1} (U_{γ_1}) from an uniform distribution $U(0,1)$.
9: Simulate U_{β_2} (U_{γ_2}) from the conditional Gaussian copula $C(U_{\beta_2}|U_{\beta_1})$ $(C(U_{\gamma_2}|U_{\gamma_1}))$.
10: **for** $i = 3 \to d$ **do**
11:     Simulate U_{β_i} (U_{γ_i}) from the conditional Gaussian copula $C(U_{\beta_i}|U_{\beta_1}, U_{\beta_2})$ $(C(U_{\gamma_i}|U_{\gamma_{i-2}}, U_{\gamma_{i-1}}))$.
12: **end for**
13: **for** $i = 1 \to d$ **do**
14:     Generate new data X_i by using the quasi-inverse $\widehat{F}_i^{-1}(U_i)$.
15: **end for**

2.5.3.1   Experiments

We use three algorithms in order to optimize five test problems. One of these algorithms is the Estimation on Multivariate Normal Algorithm (EMNA), the other two algorithms are the EDAs based on regular vines. They are represented by the following notation:

- C-vine $_{\text{Kernel}}^{\text{Gaussian}}$: A truncated C-vine model with Gaussian copula functions and Gaussian kernels as marginal distributions.
- D-vine $_{\text{Kernel}}^{\text{Gaussian}}$: A truncated D-vine model with Gaussian copula functions and Gaussian kernels as marginal distributions.

The test problems used in the experiments are the Ackley, Griewangk, Rastrigin, Rosenbrock, and Sphere functions. These test functions are described in Fig. 2.6. The benchmark test suite includes separable functions and non-separable functions, from which there are unimodal and multimodal functions. In addition, the search domains are symmetric and asymmetric. All test functions are scalable. We use test problems in 10 dimensions and different search domains. Each algorithm is run 20 times for each problem. The population size is 200. The maximum number of evaluations is 50,000. However, when convergence to a local minimum is detected the run is stopped. Any improvement less than 1×10^{-6} in 25 iterations is considered as convergence. The goal is to reach the optimum with an error less than 1×10^{-4}.

2.5.3.2   Numerical Results

In Table 2.1 we report the descriptive statistics for the fitness values reached in all the runs. For each algorithm, the minimum, median, mean, maximum, standard deviation and success rate are shown. The minimum (maximum) value reached is labelled best (worst). The success rate is the proportion of runs in which an algorithm found the global optimum.

2.5.3.3   Discussion

According to Table 2.1, all the algorithms have a better performance in symmetric domains. The asymmetric domains represent a serious difficult for the EDAs. An asymmetric domain does not include the global optimum, so the algorithms must move their populations in each iteration in order to get good solutions. However, according to the success rate indicator, the algorithm with a better performance in asymmetric domains is the C-vine EDA. The success rate for the Ackley function and the Sphere function is non zero for the C-vine. Moreover, at least in a descriptive comparison, the average fitness for the C-vine EDA in all the test functions is always less than the average fitness of the D-vine EDA and the EMNA.

On the other hand, the best value gotten by the EMNA in asymmetric domains for all the test functions is not better than the best ones of the EDAs based on regular

Description

Ackley

$$-20 \cdot \exp\left(-0.2\sqrt{\frac{1}{d} \cdot \Sigma_{i=1}^{d} x_i^2}\right) - \exp\left(\frac{1}{d} \cdot \Sigma_{i=1}^{d} \cos(2\pi x_i)\right) + 20 + \exp(1)$$

Symmetric domain: $\mathbf{x} \in [-10,10]^d$ Asymmetric domain: $\mathbf{x} \in [5,10]^d$

Properties: Multimodal, Non-separable   Global Minimum: $f(\mathbf{0}) = 0$

Griewangk

$$1 + \Sigma_{i=1}^{d} \frac{x_i^2}{4000} - \Pi_{i=1}^{d} \cos\left(\frac{x_i}{\sqrt{i}}\right)$$

Symmetric domain: $\mathbf{x} \in [-600,600]^d$ Asymmetric domain: $\mathbf{x} \in [300,600]^d$

Properties: Multimodal, Non-separable   Global Minimum: $f(\mathbf{0}) = 0$

Rastrigin

$$\Sigma_{i=1}^{d}(x_i^2 - 10\cos(2\pi x_i) + 10)$$

Symmetric domain: $\mathbf{x} \in [-5.12,5.12]^d$ Asymmetric domain: $\mathbf{x} \in [2.56,5.12]^d$

Properties: Multimodal, Separable   Global Minimum: $f(\mathbf{0}) = 0$

Rosenbrock

$$\Sigma_{i=1}^{d-1}[100 \cdot (x_{i+1} - x_i^2)^2 + (1 - x_i)^2]$$

Symmetric domain: $\mathbf{x} \in [-10,10]^d$ Asymmetric domain: $\mathbf{x} \in [5,10]^d$

Properties: Unimodal, Non-separable   Global Minimum: $f(\mathbf{1}) = 0$

Sphere Model

$$\Sigma_{i=1}^{d} x_i^2$$

Symmetric domain: $\mathbf{x} \in [-100,100]^d$ Asymmetric domain: $\mathbf{x} \in [50,100]^d$

Properties: Unimodal, Separable   Global Minimum: $f(\mathbf{0}) = 0$

Fig. 2.6 Names, mathematical definition, search domains, global minimum and properties of the test functions

vines. In general, although the fitness results for asymmetric domains are not good enough, this would mean that the EDAs based on regular have more capacity for moving their population than the EMNA.

For symmetric domains, the performance of the algorithms is very similar for all the test functions. It is well known that solving multimodal and non-separable functions is more difficult than solving unimodal and separable functions. However, according to the success rate, the three algorithms have a very good performance in the Ackley, Griewangk and Sphere functions. None of the three algorithms can find the global optimum for the Rastrigin and Rosenbrock functions. It is worth saying that symmetric domains with a global optimum at the center provide a favorable condition for search distributions such as a multivariate normal distribution. This is the case for all the test functions used in the experiments, except the Rosenbrock function. Even so, the performance for the EDAs based on regular vines is very similar to the EMNA in symmetric domains.

Table 2.1 Descriptive fitness results for all test functions

Algorithm	Best	Median	Mean	Worst	Std. deviation	Success rate
Ackley, asymmetric domain						
C-Vine $^{\text{Gaussian}}_{\text{Kernel}}$	4.93E-005	5.16E-001	1.24E+000	5.56E+000	1.67E+000	0.50
D-Vine $^{\text{Gaussian}}_{\text{Kernel}}$	2.82E+000	7.04E+000	6.88E+000	1.01E+001	1.52E+000	0.00
EMNA	9.55E+000	1.04E+001	1.06E+001	1.22E+001	6.80E-001	0.00
Ackley, symmetric domain						
C-Vine $^{\text{Gaussian}}_{\text{Kernel}}$	5.96E-005	8.15E-005	8.20E-005	9.88E-005	1.15E-005	1.00
D-Vine $^{\text{Gaussian}}_{\text{Kernel}}$	6.76E-005	8.82E-005	8.39E-005	9.87E-005	9.48E-006	1.00
EMNA	5.39E-005	9.01E-005	8.83E-005	1.53E-004	2.03E-005	0.95
Griewangk, asymmetric domain						
C-Vine $^{\text{Gaussian}}_{\text{Kernel}}$	1.97E-002	2.35E+000	5.84E+000	3.71E+001	9.11E+000	0.00
D-Vine $^{\text{Gaussian}}_{\text{Kernel}}$	2.26E+001	4.45E+001	4.58E+001	9.94E+001	1.84E+001	0.00
EMNA	3.31E+002	3.52E+002	3.55E+002	3.81E+002	1.30E+001	0.00
Griewangk, symmetric domain						
C-Vine $^{\text{Gaussian}}_{\text{Kernel}}$	2.57E-005	7.42E-004	5.65E-004	9.86E-003	2.19E-003	0.90
D-Vine $^{\text{Gaussian}}_{\text{Kernel}}$	3.01E-005	7.17E-005	3.83E-002	4.72E-001	1.21E-001	0.90
EMNA	5.90E-005	8.04E-005	9.64E-002	4.18E-001	1.58E-001	0.70
Rastrigin, asymmetric domain						
C-Vine $^{\text{Gaussian}}_{\text{Kernel}}$	5.07E+000	2.64E+001	2.85E+001	7.79E+001	1.51E+001	0.00
D-Vine $^{\text{Gaussian}}_{\text{Kernel}}$	1.49E+001	2.90E+001	3.02E+001	5.77E+001	1.22E+001	0.00
EMNA	3.18E+001	5.67E+001	5.78E+001	8.91E+001	1.57E+001	0.00
Rastrigin, symmetric domain						
C-Vine $^{\text{Gaussian}}_{\text{Kernel}}$	2.16E+001	3.06E+001	3.22E+001	4.29E+001	5.79E+000	0.00
D-Vine $^{\text{Gaussian}}_{\text{Kernel}}$	1.84E+001	2.86E+001	2.90E+001	3.73E+001	5.23E+000	0.00
EMNA	1.68E+001	2.82E+001	2.91E+001	3.73E+001	4.67E+000	0.00
Rosenbrock, asymmetric domain						
C-Vine $^{\text{Gaussian}}_{\text{Kernel}}$	1.04E+001	2.06E+003	8.27E+003	5.07E+004	1.42E+004	0.00
D-Vine $^{\text{Gaussian}}_{\text{Kernel}}$	1.43E+004	4.75E+004	5.77E+004	1.53E+005	3.53E+004	0.00
EMNA	6.53E+005	9.29E+005	9.20E+005	1.05E+006	8.76E+004	0.00
Rosenbrock, symmetric domain						
C-Vine $^{\text{Gaussian}}_{\text{Kernel}}$	6.59E+000	7.82E+000	7.68E+000	8.65E+000	5.49E-001	0.00
D-Vine $^{\text{Gaussian}}_{\text{Kernel}}$	6.84E+000	7.72E+000	8.12E+000	1.38E+001	1.51E+000	0.00
EMNA	6.83E+000	8.02E+000	8.13E+000	9.28E+000	6.00E-001	0.00
Sphere, asymmetric domain						
C-Vine $^{\text{Gaussian}}_{\text{Kernel}}$	3.13E-005	3.13E+002	4.99E+002	2.51E+003	6.70E+002	0.15
D-Vine $^{\text{Gaussian}}_{\text{Kernel}}$	2.26E+003	3.94E+003	4.38E+003	7.30E+003	1.50E+003	0.00
EMNA	3.54E+004	3.86E+004	3.84E+004	4.08E+004	1.66E+003	0.00
Sphere, symmetric domain						
C-Vine $^{\text{Gaussian}}_{\text{Kernel}}$	2.79E-005	6.11E-005	6.50E-005	9.70E-005	2.19E-005	1.00
D-Vine $^{\text{Gaussian}}_{\text{Kernel}}$	4.67E-005	7.24E-005	7.38E-005	9.65E-005	1.47E-005	1.00
EMNA	4.34E-005	7.50E-005	7.43E-005	9.88E-005	1.68E-005	1.00

2.6 Conclusions

This chapter has presented elements and methods from information theory, graphical models, and copula theory for designing multivariate distributions and applying them into optimization problems. Our approach has been to model the most important dependencies in the selected population and to estimate their corresponding multivariate distribution. Theoretical results concerning measures such as entropy and mutual information along with copula functions have been provided for designing truncated C-vine and D-vine.

The incorporation of copula functions into continuous EDAs has been shown in this work. Besides, it has also shown how the structure of a probabilistic model can be learnt by taking into account the dependence among variables, regardless of the behavior of marginal distributions. The EDAs presented in this chapter use the copula entropy as a measure of dependence and integrate copula functions into graphical models. From a theoretical point of view, this provides the following advantages: (1) the most important dependencies are represented by the graphical model, (2) dependencies can be linear or nonlinear, (3) any joint distribution can be factorized by copula functions of lower order, and (4) marginal distributions can be selected separately.

According to our experiments, the performance of the copula based EDA strongly depends on the selected marginal distributions, copula functions and the graphical model. This suggests that dependencies between decision variables must be modeled adequately in order to get good solutions.

Finally, the presented methods for learning probabilistic models and using copula functions can be applied to other problems not necessarily related to optimization. For example, in [51, 50] Gaussian copulas have been applied to classification problems.

Acknowledgements. The first author acknowledges support from the National Council of Science and Technology of México (CONACyT) through a scholarship to pursue graduate studies in the Department of Computer Science at the Center for Research in Mathematics.

References

1. Aas, K., Czado, C., Frigessi, A., Bekken, H.: Pair-copula constructions of multiple dependence. Insurance: Mathematics and Economics 44(2), 182–198 (2009)
2. Arderí-García, R.J.: Algoritmo con Estimación de Distribuciones con Cópula Gaussiana. Universidad de La Habana, La Habana, Cuba (June 2007), Bachelor's thesis. in Spanish
3. T. Bacigál and M. Komorníková. Fitting Archimedean copulas to bivariate geodetic data. In A. Rizzi and M. Vichi, editors, *Compstat 2006 Proceedings in Computational Statistics*, pages 649–656, Heidelberg, Germany, 2006. Physica-Verlag HD.
4. Baluja, S.: Population-Based Incremental Learning: A Method for Integrating Genetic Search Based Function Optimization and Competitive Learning. Technical Report CMU-CS-94-163. Carnegie Mellon University, Pittsburgh, PA, USA (June 1994)
5. Baluja, S., Davies, S.: Using Optimal Dependency-Trees for Combinatorial Optimization: Learning the Structure of the Search Space. In: Fisher, D.H. (ed.) Proceedings of the Fourteenth International Conference on Machine Learning, pp. 30–38. Morgan Kaufmann (1997)
6. Barba-Moreno, S.E.: Una propuesta para EDAs no paramétricos. Master's thesis, Centro de Investigación en Matemáticas, Guanajuato, México (December 2007) (in Spanish)
7. Bedford, T., Cooke, R.M.: Probability Density Decomposition for Conditionally Dependent Random Variables Modeled by Vines. Annals of Mathematics and Artificial Intelligence 32(1), 245–268 (2001)
8. Bedford, T., Cooke, R.M.: Vines – A New Graphical Model for Dependent Random Variables. The Annals of Statistics 30(4), 1031–1068 (2002)

9. Bosman, P.A.N., Grahl, J.: Matching Inductive Search Bias and Problem Structure in Continuous Estimation of Distribution Algorithms. Technical Report 03/2005. University of Mannheim, Mannheim, Germany (2005)

10. Bosman, P.A.N., Grahl, J., Thierens, D.: Enhancing the Performance of Maximum–Likelihood Gaussian EDAs Using Anticipated Mean Shift. In: Rudolph, G., Jansen, T., Lucas, S., Poloni, C., Beume, N. (eds.) PPSN 2008. LNCS, vol. 5199, pp. 133–143. Springer, Heidelberg (2008)

11. Brunel, N., Pieczynski, W., Derrode, S.: Copulas in vectorial hidden markov chains for multicomponent image segmentation. In: ICASSP 2005: Proceedings of the 2005 IEEE International Conference on Acoustics, Speech and Signal Processing, pp. 717–720 (2005)

12. Cherubini, U., Luciano, E., Vecchiato, W.: Copula Methods in Finance. Wiley, Chichester (2004)

13. Cuesta-Infante, A., Santana, R., Hidalgo, J.I., Bielza, C., Larrañaga, P.: Bivariate empirical and n-variate Archimedean copulas in Estimation of Distribution Algorithms. In: WCCI 2010 IEEE World Congress on Computational Intelligence, pp. 1355–1362 (July 2010)

14. Davy, M., Doucet, A.: Copulas: a new insight into positive time-frequency distributions. IEEE Signal Processing Letters 10(7), 215–218 (2003)

15. De Bonet, J.S., Isbell, C.L., Viola, P.: MIMIC: Finding Optima by Estimating Probability Densities. In: Advances in Neural Information Processing Systems, vol. 9, pp. 424–430. The MIT Press (1997)

16. De-Waal, D.J., Van-Gelder, P.H.A.J.M.: Modelling of extreme wave heights and periods through copulas. Extremes 8(4), 345–356 (2005)

17. Dowd, K.: Copulas in Macroeconomics. Journal of International and Global Economic Studies 1(1), 1–26 (2008)

18. Etxeberria, R., Larrañaga, P.: Global optimization with Bayesian networks. In: Ochoa, A., Soto, M., Santana, R. (eds.) Second International Symposium on Artificial Intelligence. Adaptive Systems, CIMAF 1999, La Habana, pp. 332–339. Academia (1999)

19. Flitti, F., Collet, C., Joannic-Chardin, A.: Unsupervised Multiband Image Segmentation using Hidden Markov Quadtree and Copulas. In: IEEE International Conference on Image Processing, Genova, Italy (September 2005)

20. Flores de la Fuente, E.: EDAs con Funciones de cópula. Master's thesis, Centro de Investigación en Matemáticas, Guanajuato, México (September 2009) (in Spanish)

21. Frees, E.W., Valdez, E.A.: Understanding relationships using copulas. North American Actuarial Journal 2(1), 1–25 (1998)

22. Gao, Y.: Multivariate Estimation of Distribution Algorithm with Laplace Transform Archimedean Copula. In: Hu, W., Li, X. (eds.) 2009 International Conference on Information Engineering and Computer Science, ICIECS 2009, Wuhan, China (December 2009)

23. Genest, C., Favre, A.C.: Everything You Always Wanted to Know about Copula Modeling but Were Afraid to Ask. Journal of Hydrologic Engineering 12(4), 347–368 (2007)

24. González, C.: Contributions on Theoretical Aspects of Estimation of Distribution Algorithms. PhD thesis. University of the Basque Country, Donostia-San Sebastián, Spain (November 2005)

25. Grahl, J., Minner, S., Bosman, P.A.N.: Learning structure illuminates black boxes – an introduction into Estimation of Distribution Algorithms. Technical Report 10/2006, University of Mannheim, Mannheim, Germany (2006)

26. Grigoriu, M.: Multivariate distributions with specified marginals: Applications to Wind Engineering. Journal of Engineering Mechanics 133(2), 174–184 (2007)

27. Harik, G., Lobo, F.G., Goldberg, D.E.: The Compact Genetic Algorithm. In: Proceedings of the IEEE Conference on Evolutionary Computation, pp. 523–528 (1998)
28. Höns, R.: Estimation of Distribution Algorithm and Minimum Relative Entropy. PhD thesis. University of Bonn, Bonn, Germany (2005)
29. Igel, C., Suttorp, T., Hansen, N.: A computational efficient covariance matrix update and a (1+1)-CMA for evolution strategies. In: Proceedings of the 8th Annual Conference on Genetic and Evolutionary Computation, GECCO 2006, pp. 453–460. ACM (2006)
30. Jajuga, K., Papla, D.: Copula Functions in Model Based Clustering. In: Spiliopoulou, M., Kruse, R., Borgelt, C., Nürnberger, A., Gaul, W. (eds.) From Data and Information Analysis to Knowledge Engineering. Studies in Classification, Data Analysis, and Knowledge Organization, pp. 606–613. Springer, Heidelberg (2006)
31. Jenison, R.L., Reale, R.A.: The Shape of Neural Dependence. Neural Computation 16, 665–672 (2004)
32. Joe, H.: Multivariate models and dependence concepts. Chapman and Hall, Boca Raton (1997)
33. Kurowicka, D., Cooke, R.: The vine copula method for representing high dimensional dependent distributions: application to continuous belief nets. In: Yücesan, E., Chen, C.-H., Snowdon, J.L., Charnes, J.M. (eds.) Proceedings of the 2002 Winter Simulation Conference, pp. 270–278 (2002)
34. Kurowicka, D., Cooke, R.: Uncertainty Analysis. Series in Probability and Statistics. Wiley. Wiley (2006)
35. Larrañaga, P., Etxeberria, R., Lozano, J.A., Peña, J.M.: Combinatorial optimization by learning and simulation of Bayesian networks. In: Proceedings of the Sixteenth Conference on Uncertainty in Artificial Intelligence, pp. 343–352 (2000)
36. Larrañaga, P., Etxeberria, R., Lozano, J.A., Peña, J.M.: Optimization in continuous domains by learning and simulation of Gaussian networks. In: Proceedings of the Optimization by Building and Using Probabilistic Models OBUPM Workshop at the Genetic and Evolutionary Computation Conference GECCO 2000, pp. 201–204 (2000)
37. Larrañaga, P.: A Review on Estimation of Distribution Algorithms. In: Larrañaga and Lozano (eds.) [38], ch. 3, pp. 57–100 (2002)
38. Larrañaga, P., Lozano, J.A. (eds.): Estimation of Distribution Algorithms: A New Tool for Evolutionary Computation. Genetic Algorithms and Evolutionary Computation. Kluwer Academic Publishers (2002)
39. Mercier, G., Bouchemakh, L., Smara, Y.: The Use of Multidimensional Copulas to Describe Amplitude Distribution of Polarimetric SAR Data. In: IGARSS 2007 (2007)
40. Monjardin, P.E.: Análisis de dependencia en tiempo de falla. Master's thesis, Centro de Investigación en Matemáticas, Guanajuato, México (December 2007) (in Spanish)
41. Mühlenbein, H.: The Equation for Response to Selection and its Use for Prediction. Evolutionary Computation 5(3), 303–346 (1998)
42. Mühlenbein, H., Mahnig, T., Ochoa-Rodriguez, A.: Schemata, distributions and graphical models in evolutionary optimization. Journal of Heuristics 5, 215–247 (1999)
43. Mühlenbein, H., Paaß, G.: From Recombination of Genes to the Estimation of Distributions I. Binary Parameters. In: Ebeling, W., Rechenberg, I., Voigt, H.-M., Schwefel, H.-P. (eds.) PPSN 1996. LNCS, vol. 1141, pp. 178–187. Springer, Heidelberg (1996)
44. Nelsen, R.B.: An Introduction to Copulas, 2nd edn. Springer Series in Statistics. Springer (2006)
45. Pelikan, M., Goldberg, D.E., Lobo, F.G.: A Survey of Optimization by Building and Using Probabilistic Models. Computational Optimization and Applications 21(1), 5–20 (2002)

46. Pelikan, M., Goldberg, D.E., Cantú-Paz, E.: BOA: The Bayesian Optimization Algorithm. In: Banzhaf, W., Daida, J., Eiben, A.E., Garzon, M.H., Honavar, V., Jakiela, M., Smith, R.E. (eds.) Proceedings of the Genetic and Evolutionary Computation Conference, GECCO 1999, vol. 1, pp. 525–532. Morgan Kaufmann Publishers (1999)
47. Pelikan, M., Mühlenbein, H.: The Bivariate Marginal Distribution Algorithm. In: Roy, R., Furuhashi, T., Chawdhry, P.K. (eds.) Advances in Soft Computing - Engineering Design and Manufacturing, pp. 521–535. Springer, London (1999)
48. Pelikan, M., Sastry, K., Cantú-Paz, E. (eds.): Scalable Optimization via Probabilistic Modeling: From Algorithms to Applications. SCI, vol. 33. Springer (2006)
49. Sakji-Nsibi, S., Benazza-Benyahia, A.: Multivariate indexing of multichannel images based on the copula theory. In: IPTA 2008 (2008)
50. Salinas-Gutiérrez, R., Hernández-Aguirre, A., Rivera-Meraz, M.J.J., Villa-Diharce, E.R.: Supervised Probabilistic Classification Based on Gaussian Copulas. In: Sidorov, G., Hernández Aguirre, A., Reyes García, C.A. (eds.) MICAI 2010, Part II. LNCS(LNAI), vol. 6438, pp. 104–115. Springer, Heidelberg (2010)
51. Salinas-Gutiérrez, R., Hernández-Aguirre, A., Rivera-Meraz, M.J.J., Villa-Diharce, E.R.: Using Gaussian Copulas in Supervised Probabilistic Classification. In: Castillo, O., Kacprzyk, J., Pedrycz, W. (eds.) Soft Computing for Intelligent Control and Mobile Robotics. SCI, vol. 318, pp. 355–372. Springer, Heidelberg (2010)
52. Salinas-Gutiérrez, R., Hernández-Aguirre, A., Villa-Diharce, E.R.: Using Copulas in Estimation of Distribution Algorithms. In: Aguirre, A.H., Borja, R.M., Garciá, C.A.R. (eds.) MICAI 2009. LNCS(LNAI), vol. 5845, pp. 658–668. Springer, Heidelberg (2009)
53. Salinas-Gutiérrez, R., Hernández-Aguirre, A., Villa-Diharce, E.R.: D-vine EDA: a new Estimation of Distribution Algorithm based on Regular Vines. In: GECCO 2010: Proceedings of the 12th Annual Conference on Genetic and Evolutionary Computation, pp. 359–366. ACM, New York (2010)
54. Salinas-Gutiérrez, R., Hernández-Aguirre, A., Villa-Diharce, E.R.: Dependence Trees with Copula Selection for Continuous Estimation of Distribution Algorithms. In: GECCO 2011: Proceedings of the 13th Annual Conference on Genetic and Evolutionary Computation, pp. 585–592. ACM (2011); Estimation of Distribution Algorithms Track Papers
55. Schölzel, C., Friederichs, P.: Multivariate non-normally distributed random variables in climate research – introduction to the copula approach. Nonlinear Processes in Geophysics 15(5), 761–772 (2008)
56. Sklar, A.: Fonctions de répartition à n dimensions et leurs marges. Publications de l'Institut de Statistique de l'Université de Paris 8, 229–231 (1959)
57. Soto, M., Ochoa, A., Acid, S., de Campos, L.M.: Introducing the polytree approximation of distribution algorithm. In: Ochoa, A., Soto, M., Santana, R. (eds.) Second International Symposium on Artificial Intelligence. Adaptive Systems, CIMAF 1999, pp. 360–367. Academia, La Habana (1999)
58. Stitou, Y., Lasmar, N., Berthoumieu, Y.: Copulas based multivariate gamma modeling for texture classification. In: ICASSP 2009: Proceedings of the 2009 IEEE International Conference on Acoustics, Speech and Signal Processing, pp. 1045–1048. IEEE Computer Society, Washington, DC (2009)
59. Trivedi, P.K., Zimmer, D.M.: Copula Modeling: An Introduction for Practitioners. Foundations and Trends® in Econometrics. Now Publishers (2007)
60. Venter, G., Barnett, J., Kreps, R., Major, J.: Multivariate Copulas for Financial Modeling. Variance 1(1), 103–119 (2007)

61. Wang, L., Guo, X., Zeng, J., Hong, Y.: Using Gumbel Copula and Empirical Marginal Distribution in Estimation of Distribution Algorithm. In: Third International Workshop on Advanced Computational Intelligence, IWACI 2010, pp. 583–587. IEEE (August 2010)

62. Wang, L., Zeng, J., Hong, Y., Guo, X.: Copula Estimation of Distribution Algorithm Sampling from Clayton Copula. Journal of Computational Information Systems 6(7), 2431–2440 (2010)

63. Wang, L.F., Wang, Y.C., Zeng, J.C., Hong, Y.: An Estimation of Distribution Algorithm Based on Clayton Copula and Empirical Margins. In: Li, K., Li, X., Ma, S., Irwin, G.W. (eds.) LSMS 2010. CCIS, vol. 98, pp. 82–88. Springer, Heidelberg (2010)

64. Wang, L.F., Zeng, J.C.: Estimation of Distribution Algorithm Based on Copula Theory. In: Chen, Y.P. (ed.) Exploitation of Linkage Learning in Evolutionary Algorithms. Adaptation, Learning, and Optimization, vol. 3, pp. 139–162. Springer (2010)

Chapter 3
The Gaussian Polytree EDA with Copula Functions and Mutations

Ignacio Segovia Domínguez, Arturo Hernández Aguirre,
and Enrique Villa Diharce

Abstract. This chapter introduces the Gaussian Poly-Tree Estimation Distribution
Algorithm, and two extensions: i) with Gaussian copula functions, and ii) with local
optimizers. The new construction and simulation algorithms, and its application to
estimation of distribution algorithms with continuous Gaussian variables are also
introduced. The algorithm for the construction of the structure and for edge orientation is based on information theoretic concepts such as mutual information and
conditional mutual information. The three models are tested on a benchmark of 20
unimodal and multimodal functions. The version with copula function and mutations excels in most problems achieving near optimal success rate.

3.1   Introduction

The performance of function optimizers based on Estimation of Distribution Algorithms (EDA) relies on the model used to represent the structure and the dependencies of the data. Such model, simple or complex, is the joint probability distribution
(JPD) of the population; its construction takes place in step 4 of the algorithm 3.1.
An example of such model, frequently used in EDAs, is the multivariate normal
distribution. Other suitable models are graph based models, such as the chain or the
tree, depicted in Figure 3.1. Graph based models encode a JPD as the product of
conditional distributions.

How well does a graph based model approximate the true distribution of the
population ? That depends on a number of reasons, however, the approximation improves with the number of condition variables (parents) allowed to the nodes. For
instance, in the dependence tree all relations are bivariate (see Figure 3.1-a). In a
poly-tree (PT) the nodes may have more than one parent, Figure 3.1-b, with no link

Ignacio Segovia Domínguez · Arturo Hernández Aguirre · Enrique Villa Diharce
Center for Research in Mathematics
Guanajuato, México
e-mail: {ijsegoviad, artha, villadi}@cimat.mx

E. Tantar et al. (Eds.): EVOLVE- A Bridge between Probability, SCI 447, pp. 123–153.
springerlink.com   © Springer-Verlag Berlin Heidelberg 2013

Algorithm 3.1. Pseudocode for EDAs

1: assign $t \longleftarrow 0$
2: generate the initial population P_0 with N individuals at random
3: select a collection of M individuals S_t, with $M < N$, from P_t
4: estimate a probabilistic model $\mathcal{M}_t$ from S_t
5: generate the new population by sampling from the distribution of S_t.
6: assign $t \longleftarrow t + 1$
7: if stopping criterion is not reached go to step 2

---

between the parents. However, a Bayesian network, Figure 3.1-d, allows links between parents, and also several parents to the nodes. In general, the approximation can be improved up to some extent if more parents are permitted, and if there are edges between the parents. The models in Figure 3.1 can be ranked by their approximation capacity, as follows: $chain < dependencetree < polytree < BayesNet$. Indeed, it seems natural to apply very rich models to our problems. The trade-off here is the computational complexity of the construction algorithm, which for a Bayesian network is NP-complete [6]. There is, however, one graph based model, with relatively high approximation capacity and acceptable construction complexity ($\emptyset(N^3)$): the poly-tree (PT). A typical PT is shown in Figure 3.1-b.

This chapter seeks several goals: to introduce the Gaussian Poly-tree (GPT), that is, a poly-tree graphical model with Gaussian variables; to explain the construction algorithm; and to show the application of EDAs to function optimization problems. Notice again that the proposed GPT uses real variables. In short, the main algorithm and two variations are studied in this chapter:

1. **The Gaussian Poly-Tree (GPT).** Explains the construction of the graphical model, implementation details, and experiments. An edge is a conditional distribution modeled with conditional normal variables.
2. **The Gaussian Poly-tree with Gaussian copula.** The poly-tree adopts a Gaussian copula as a way to improve the dependency model between the variables. An edge is a conditional distribution modeled with a Gaussian copula function.
3. **The Gaussian Poly-tree with Gaussian copula + mutations.** EDAs with graphical models tend to stagnate. This variation shows how to generate individuals as mutations of the population and preserve the diversity for longer generations and improve the performance of the algorithm.

The construction algorithm of a graphical model needs two steps: 1) determination of the structure or "skeleton", and 2) determination of the parameters.

There are two main approaches to determine the structure:

1. Complexity measure. The process starts with a fully connected directed graph, or with a fully disconnected directed graph. While edges are added or removed an estimator of the likelihood of the graph assess the current quality at every iteration.
2. Dependence test. A dependency measure, such as the Persons chi-square, or an entropy based measure such as mutual information, is used to grow a tree

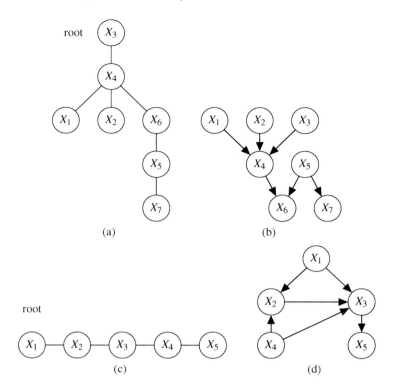

Fig. 3.1 Graphical models: (a) Dependence tree, (b) Poly-tree, (c) Chain, (d) Bayesian network

starting with the pair of nodes with the highest dependency value. Next node to be linked is chosen as that node with the highest dependency with the nodes already in the tree. The construction of GPTs is commonly made in two phases: 1) Create a dependence tree (undirected graph) which will serve as the basic skeleton, and 2) orient the edges according to the conditional dependencies of the data.

The parameters are computed once the structure is determined. For instance, marginal probabilities and conditional probabilities are computed from the data but according to the explicit dependencies denoted by the graph.

The proposed approach builds a GPT using a dependence test strategy with two phases. In phase one, a Chow and Liu algorithm (which uses mutual information as dependence measure) returns a dependence tree (which is an undirected graph). In phase two, the edges of that tree are then oriented (addition of arrow-heads) to create a GPT. The new orientation algorithm described in this chapter is based on mutual information (MI) and conditional mutual information (CMI). In order to compute the parameters the poly-tree is traversed from all the roots, level by level, downward to the lief, checking if a node has zero, one, or more parents. Depending on the node type, either the marginal mean and variance, or conditional mean and variance of a Normal distribution are calculated and stored.

This chapter is organized as follows: Section 3.2 reviews some works related to poly-trees. The next three sections describe the algorithms proposed. In Section 3.3 the Gaussian Poly-tree, the construction algorithm, and the simulation procedure are introduced. Then, the Poly-tree with Gaussian copula function, together with the construction and simulation algorithms are presented in Section 3.4. The last variation, the Poly-tree with Gaussian copula + mutations, is presented in Section 3.5. Section 3.6 presents all experimental results. Conclusions are given in Section 10.6.

3.2   Related Work

A poly-tree is a Directed Acyclic Graph (DAG) with only one undirected path between any two nodes [11],[16]. For discrete variables the poly-tree is well known for its applications to belief networks because inference algorithms run in linear time over them [23]. Acid and de Campos investigated the application of poly-trees to causal networks [1], [24]. More recently, Soto investigated the use of poly-tree models to approximate the population distribution, and designed the poly-tree approximation distribution algorithm, known as PADA [21]. The construction algorithm of PADA uses a dependency measure strategy that grows a poly-tree from a fully unconnected directed graph. For over-fitting control two threshold variables $\varepsilon_1, \varepsilon_2$ are used to filter out the (weak) dependencies. However, no recommendation about how to set those parameters is given in the PADA literature. Discrete poly-trees have been applied to classification problems.

A Gaussian Poly-Tree (GPT) represents a joint probability density function (JPDF) as the product of conditional normal distributions. The encoded JPDF is a multivariate normal distribution, as follows:

$$JPDF(X_1, X_2, \ldots, X_n) = \prod_{\forall i \in R} P(X_i) \prod_{\forall j \notin R} P(X_j | pa(X_j)), \qquad (3.1)$$

where (R) represents one or more root nodes of the poly-tree [19],[8]. The literature about poly-trees with continuous variables is small. Ouerd studied the properties of poly-tree based models and the construction of the poly-tree structure by measuring dependencies using the Chi^2 distribution [22]. Later, Ouerd proposed an approach for edge orientation [17]. Based on the previous work of Rebane and Pearl [25],[23], Ouerd at al. start with a dependence tree computed with the Chow & Liu algorithm [17]. Then they propose to orient the edges by traversing the dependence tree in a depth first search order. Articulation points and causal basins must be detected first. Their approach addresses four issues not completely solved by Rebane and Pearl. Two of them are: how to traverse the tree, and what to do with edges that are already oriented but need to be traversed again and may end up with another orientation. For edge orientation, their algorithm performs a marginal independence test on the parents X and Y of a node Z. If X and Y are independent then the node Z becomes a common child, with arrows pointing to Z, therefore Z is called a head to head node.

Another related model is the Gaussian network which has been widely studied. A few very relevant works are [19], Lauritzen [16], Whittaker [30], Castillo [4], and Joe [14].

3.3   The Gaussian Poly-Tree

The proposed approach starts with a dependence tree created with the Chow and Liu algorithm. Then the edges are oriented accordingly to the mutual information and conditional mutual information criteria. Mutual information (MI), measures bivariate dependencies between any pair of variables X_i and X_j, but conditional mutual information (CMI) can tell us if a third variable X_k may increase the mutual information of X_i and X_j. If $CMI(X_i, X_j | X_k) > MI(X_i, X_j))$ the node X_k gets X_i and X_j as parents. Otherwise X_i is the sole parent of X_j

3.3.1   Construction of the GPT

1. The Gaussian dependence tree. The first step to construct a Gaussian poly-tree is to construct a *Gaussian dependence tree*. The procedure is the same as that of the discrete Chow and Liu algorithm [7], but now we use Gaussian mutual information to estimate the dependencies. For Gaussian variables the mutual information is defined as:

$$MI(X,Y) = -\frac{1}{2}\log\left(1 - r_{x,y}^2\right).$$ (3.2)

The term $r_{x,y}$ is the Pearson's correlation coefficient which for Gaussian variables is defined as:

$$r_{x,y} = \frac{cov(x,y)}{\sigma_x \sigma_y}$$ (3.3)

This Gaussian dependence tree shares some properties with its discrete version: a) mutual information is the maximum likelihood estimator; and b) when the data comes from a tree-like distribution the Gaussian dependence tree is the best approximation.

2. Edge orientation. The goal of the edge orientation algorithm is to achieve the orientation principle [25]: if in a triplet $X - Z - Y$ the variables X and Y are independent, then Z is a head to head node, with X and Y as parents, as follows: $X \rightarrow Z \leftarrow Y$. Similarly, if in a triplet $X \rightarrow Z - Y$, the variables X and Y are

independent, then Z is a head to head node with X and Y as parents: $X \rightarrow Z \leftarrow Y$; otherwise Z is the parent of Y: $X \rightarrow Z \rightarrow Y$.

Proposed orientation based on information measures: for any triplet $X - Z - Y$, if $CMI(X,Y|Z) > MI(X,Y)$ then Z is a head to head node with X and Y as parents, as follows: $X \rightarrow Z \leftarrow Y$. The equation of CMI for Gaussian variables is the following.

$$CMI(X,Y|Z) = \frac{1}{2} \log \left[\frac{\sigma_x^2 \sigma_y^2 \sigma_z^2 \left(1 - r_{xz}^2\right)\left(1 - r_{yz}^2\right)}{|\Sigma_{xyz}|} \right] \qquad (3.4)$$

We shall prove that the proposed approach based on MI and CMI finds the correct orientation.

Proof. Assume three variables X, Y and Z. They can be connected in the four possible ways shown in the Figure 3.2. We wish to prove that the model M_4, head to head (a node with two parents), is the correct choice for $CMI(X,Y|Z) > MI(X,Y)$.

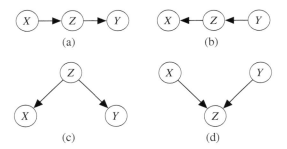

Fig. 3.2 The causal models that can be obtained with three variables X, Y y Z. (a) Model M_1. (b) Model M_2. (c) Model M_3. (d) Model M_4.

The quality of the causal models shown in the Figure 3.2 can be expressed by its log-likelihood. If the parents of any node X_i is the set of nodes $pa(X_i)$, the negative of the log-likelihood of a model M is [9]:

$$-ll(M) = \sum_{i=1}^{n} H(X_i|pa(X_i)) \qquad (3.5)$$

where $H(X_i|pa(X_i))$ is the conditional entropy of X_i given its parents $pa(X_i)$. It is well known that the causal models M_1, M_2 and M_3 are equivalent, or indistinguishable in probability [25]. The negative log-likelihood are the Equations 3.6, 3.7 and 3.8, respectively.

$$\begin{aligned}
-ll(M_1) &= H(X) + H(Z|X) + H(Y|Z) \\
&= H(X,Z) + H(Y,Z) - H(Z) \\
&\quad -H(X,Y,Z) + H(X,Y,Z) \\
&= H(X,Y,Z) + CMI(X,Y|Z)
\end{aligned} \tag{3.6}$$

$$\begin{aligned}
-ll(M_2) &= H(Z) + H(X|Z) + H(Y|Z) \\
&= H(X,Z) + H(Y,Z) - H(Z) \\
&\quad +H(X,Y,Z) - H(X,Y,Z) \\
&= H(X,Y,Z) + CMI(X,Y|Z)
\end{aligned} \tag{3.7}$$

$$\begin{aligned}
-ll(M_3) &= H(Y) + H(Z|Y) + H(X|Z) \\
&= H(X,Z) + H(Y,Z) - H(Z) \\
&\quad -H(X,Y,Z) + H(X,Y,Z) \\
&= H(X,Y,Z) + CMI(X,Y|Z)
\end{aligned} \tag{3.8}$$

For the head to head model (M_4), the negative of the log-likelihood is Equation 3.9.

$$\begin{aligned}
-ll(M_4) &= \quad H(X) + H(Y) + H(Z|X,Y) \\
&= H(X) + H(Y) + H(X,Y,Z) - H(X,Y) \\
&= \quad\quad H(X,Y,Z) + MI(X,Y)
\end{aligned} \tag{3.9}$$

The best model is that one with the smallest negative log-likelihood or smallest summation of conditional entropy. When is the negative log-likelihood of Model M_4 smaller than the log-likelihood of model M_1 or M_2 or M_3 ?

$$H(X,Y,Z) + MI(X,Y) < H(X,Y,Z) + CMI(X,Y|Z) \tag{3.10}$$

The answer is in Equation 3.10. When the conditional mutual information $CMI(X,Y|Z)$ is larger than $MI(X,Y)$ the model M_4 has smaller negative log-likelihood value, therefore, $M4$ is the correct choice.   □

The edge orientation principle runs on the depth first search algorithm [17], as follows: assume node A has nodes B, C, and D as candidate parents. There are 3 triplets to test: $B - A - C$, $B - A - D$ and $C - A - D$. For each triplet MI and CMI are calculated in the order they are found while traversing the tree. For instance, if $CMI(B,C|A) > MI(B,C)$ then the node A gets nodes B and C as parents. Then, for the next triplet, say $B - A - D$, the dependence measures are tested, and if it is the case then node A gets node D as parent, shown as another arrow.

3. Over-fitting control. The inequality $MI(X,Y) < CMI(X,Y|Z)$ could be made true due to the small biases of the data and thus creating false positive parents. As a rule, the larger the allowed number of parents the better the approximation. However, the PT gets over-fitted when such noisy nodes appear as vicious parents destroying the quality of the model. A computational inexpensive approach to filter the vicious nodes is based on a threshold value, but which value? We wish to know: how many times the CMI must be larger than the MI as to represent true parents? Which is a good threshold value?. Empirically we solve this question by randomly creating a huge database of triplet-vectors X, Y and Z (from random gaussian distributions) that made true the inequality

$CMI(X,Y|Z) > MI(X,Y)$. Within this large set there are two subsets: triplets that satisfy the condition, and the rest. A hypothesis test based on a non parametric bootstrap test was performed using the proposed data set. We found out that false parents are created in 95% of the cases when $\frac{CMI(X,Y|Z)}{MI(X,Y)} < 3$. Therefore the sought threshold value is 3. Thus, a head to head node is created whenever $\frac{CMI(X,Y|Z)}{MI(X,Y)} \geq 3$.

4. Selection. Truncation selection is commonly used in EDAs as a strategy to pick the best individuals from which the probability model is created. The Gaussian poly-tree EDA, however, uses a $\mu + \lambda$ approach. The whole population is used to create the poly-tree, then as many as $\lambda = 0.5\mu$ individuals are simulated from the GPT and stored along with the current parents. The new population is chosen by applying truncation selection to the new set.

3.3.2   Simulating Data from a Poly-Tree

The generation or simulation of new data from the PT is a somewhat cumbersome process which we describe for completeness purposes [19]. Assume we have sorted the nodes in an ancestral order, that is, if $X_j \rightarrow X_i$ then $i > j$. The conditional density function of a variable X_i given its parents can be expressed as:

$$f(X_i|pa(X_i)) = \mathcal{N}\left(\mu_i + \sum_{X_j \in pa(X_i)} \beta_{i,j}(X_j - \mu_j), \sigma_i^2\right) \qquad (3.11)$$

where μ_i is the unconditional mean of X_i, $\left(\mu_i + \sum_{X_j \in pa(X_i)} \beta_{i,j}(X_j - \mu_j)\right)$ is the conditional mean of X_i, σ_i^2 is the conditional variance of X_i given the values of parent nodes, and $\beta_{i,j}$ is the linear regression coefficient of X_j when X_i is regressed on all X_j such that X_j is a parent of X_i. Figure 3.3 shows the regression coefficients. When there is no edge between two variables, for instance X_1 and X_2, the regression coefficient $\beta_{2,1} = 0$.

The Equation 3.11 suggests to calculate the regression coefficients and simulate new data in the order they are found while traversing the tree. The GPT uniquely determines a non-singular multivariate normal distribution and vice versa, therefore, we can simulate samples from a Gaussian poly-tree using the distribution $\mathcal{N}(\mu_G, \Sigma_G)$. The unconditional mean vector is μ_G, and the covariance matrix Σ_G can be obtained from the precision matrix. Shachter and Kenley [27] described the general transformation from σ_i^2 and $(\beta_{i,j}|j < i)$ to the precision matrix $T = \Sigma_G^{-1}$. Algorithm 3.2 constructs the desired matrix.

Where:

$$t_i = \frac{1}{\sigma_i^2} \qquad (3.12)$$

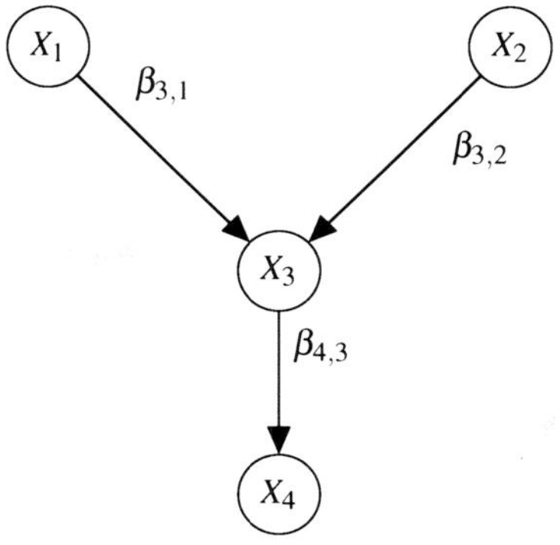

Fig. 3.3 Regression coefficients between variables in the Poly-tree

---

Algorithm 3.2. Building the precision matrix

Function: getPrecisionMatrix ($\sigma_1^2, \ldots, \sigma_n^2, \beta_1, \ldots, \beta_n$)
Results: Precision matrix T
$T_1 = t_1$
for $i = 2; i \leq n; i++$ **do**
$$T_i = \begin{pmatrix} T_{i-1} + t_i \beta_i \beta_i^t & -t_i \beta_i \\ -t_i \beta_i^t & t_i \end{pmatrix}$$
$T = T_n$

---

and

$$\beta_i = \begin{pmatrix} \beta_{i,1} \\ \vdots \\ \beta_{i,i-1} \end{pmatrix} \tag{3.13}$$

The covariance matrix is calculated from $\Sigma_G = T^{-1}$.

Ordinary least square (OLS) method can be used to find the regression coefficients. OLS has several important statistical properties. It is unbiased with expected value β. If the errors are independent, identically and normally distributed, the least square estimate is also the maximum likelihood estimate.

With the Algorithm 3.2 there is no need to traverse the tree but only read data from it. The ancestral order of the tree allows to run the algorithm, determine $\Sigma_G = T^{-1}$, and then sample new data from a multivariate Normal distribution.

3.4   The Gaussian Poly-Tree with Gaussian Copula Function

In this section the GPT adopts the Gaussian copula function and it is called the Gaussian Copula Poly-Tree (GCPT), and the EDA using GCPT is called the Gaussian Copula Poly-Tree Estimation of Distribution Algorithm (GCPT-EDA). The goal is to assess the impact of Gaussian copulas in optimization problems. The new construction algorithm and the simulation procedure are introduced. So far, the GPT represents pairs or triplets or quartets of variables where one node has one, two or three parents (larger tuples are possibly created depending on the data dependencies). Let us isolate any triplet from the PT, say one node Z with two parents X and Y, which are modeled with Gaussian distributions. This triplet has a joint density function factorized as $P(X) \times P(Y) \times P(Z|X,Y)$, which is also Gaussian. Since the conditional Gaussian distribution is still Gaussian, the variable Z is unimodal [14]. However, when the marginals of Z are not Gaussians the use of another joint density function, such as log-normal, exponential, etc., will not solve the problem because all of them have marginals of the same type. The solution is to use copula functions!. With the proper dependency parameter, the Gaussian copula function captures the data dependencies which need to be transfer to the new simulated data, however, the marginals need not be of Gaussian type. In this work, marginal distributions are approximated with Gaussian kernels.

3.4.1 Gaussian Copula Functions

A copula is an important tool in statistics for modeling multiple dependence in several variables. For this reason, copula functions have been widely used in some research and application areas such as finance [5, 29], climate [26], oceanography [10], hydrology [12], geodesy [2], and reliability [18].

   The copula concept was introduced by Sklar [28] to separate the effect of dependence from the effect of marginal distributions in a joint distribution.

Definition 3.1. A copula is a joint distribution function of standard uniform random variables. That is,

$$C(u_1,\ldots,u_n) = \Pr[U_1 \leq u_1,\ldots,U_n \leq u_n] \ ,$$

where $U_i \sim U(0,1)$ for $i = 1,\ldots,n$.

For a more formal definition of copula functions, the reader is referred to [14, 20]. The following result, known as Sklar's theorem, states how a copula function is related to a joint distribution function.

Theorem 3.1 (Sklar). *Let F be a n-dimensional distribution function with marginals $F_1, F_2, \ldots, F_n$, then there exists a copula C such that for all x in $\overline{\mathbb{R}}^n$,*

$$F(x_1,x_2,\ldots,x_n) = C(F_1(x_1),F_2(x_2),\ldots,F_n(x_n)) \ ,$$

where $\overline{\mathbb{R}}$ denotes the extended real line $[-\infty,\infty]$. If $F_1(x_1)$, $F_2(x_2)$, $\ldots,F_n(x_n)$ are all continuous, then C is unique. Otherwise, C is uniquely determined on $Ran(F_1) \times Ran(F_2) \times \cdots \times Ran(F_n)$, where Ran stands for the range.

According to Theorem 3.1, a n-dimensional density f can be represented as

$$f(x_1, x_2, \ldots, x_n) = c(F_1(x_1), F_2(x_2), \ldots, F_n(x_n)) \cdot \prod_{i=1}^{n} f_i(x_i) \ ,$$

where c is the density of the copula C and $f_i(x_i)$ is the marginal density of the variable x_i. The equation shows that the marginals f_i can belong to other and possibly different distributions. The dependence structure is given by the copula, and their product builds a multivariate distribution. This contrasts with the usual way to construct multivariate distributions, which suffers from the restriction that the margins are usually of the same type. The separation between marginal distributions and a dependence structure explains the modeling flexibility given by copulas.

Definition 3.2. The copula associated to the standard joint Gaussian distribution is called Gaussian copula.

Table 3.1 Gaussian copulas for 2 and 3 variables

2-dimensional Gaussian copula
Distribution: $C(u_1, u_2; \theta) = \Phi_G\left(\Phi^{-1}(u_1), \Phi^{-1}(u_2)\right)$, where Φ_G is the standard bivariate normal distribution with correlation θ Parameter: $\theta \in (-1, 1)$
3-dimensional Gaussian copula
$C(u_1, u_2, u_3; \theta) = \Phi_G\left(\Phi^{-1}(u_1), \Phi^{-1}(u_2), \Phi^{-1}(u_3)\right)$, where Φ_G is the standard trivariate normal distribution with correlation matrix $\theta = \begin{bmatrix} 1 & \theta_{12} & \theta_{13} \\ \theta_{12} & 1 & \theta_{23} \\ \theta_{13} & \theta_{23} & 1 \end{bmatrix}$ Parameters: $\theta_{ij} \in (-1, 1)$, $1 \le i < j \le 3$

Table 3.1 shows the defining equations of the bivariate and trivariate Gaussian copula.

For parametric bivariate copulas, the dependence parameter is related to Kendall's τ through the equation (see [20])

$$\tau(X_1, X_2) = 4 \int_0^1 \int_0^1 C(u_1, u_2; \theta) dC(u_1, u_2; \theta) - 1 \ . \tag{3.14}$$

For a bivariate Gaussian copula, Equation (3.14) can be written as

$$\tau = \frac{2}{\pi}\arcsin(\theta) \ . \tag{3.15}$$

Since Kendalls τ is easily estimated from the population using its well know formula, the nonparametric estimation of θ is known by solving Equation (3.15). Notice the parameter θ is all is needed to simulate data from the copulas shown in Table 3.1. (In fact, θ is the parameter required at step 5 of the Algorithm 3.4). For 2 or 3 variables there are several choices of copula functions, however only the Gaussian copula scales up without difficulty with the number of variables. Hence, to the best of our knowledge, only the Gaussian copula can be used to model multivariate distributions with any number of variables without any particular modification of the sampling algorithm.

Figure 3.4 shows some examples of random samples using a Gaussian copula and multimodal marginal distributions. In this, the Pearson correlation coefficient θ_{12} modify the dependence among X_1 and X_2.

3.4.2   Building the Gaussian Copula Poly-Tree and Data Simulation

The overall algorithm of the GCPT-EDA is listed in Algorithm 3.4, which includes our proposal to sample from the PT with Gaussian copula functions associated to the edges. The procedure to construct the skeleton of the Gaussian Poly-tree is the same described in Subsection 3.3.1, only a few steps are necessary. The marginal distribution of each variable X_i is approximated with Gaussian kernels in step 2, and the cumulative distribution function U_i of each X_i in step 3. In step 4 all variables X_i are mapped to Z_i using the inverse of the cumulative distribution U_i. In step 5 the correlation matrix between the variables in the domain of the variables Z_i is computed. The GPT is constructed also in the domain of Z_i at step 6.

Algorithm 3.4. The Gaussian Polytree EDA with Gaussian Copula

1: Pob_t = Population in the domain $X_1, X_2 \ldots X_d$
2: K_{X_i} = Gaussian kernel of each marginal in Pob_t
3: U_i = Cumulative Distribution Function of X_i using K_{X_i}; so $U_i \in [0,1]$
4: Z_i = Inverse Cumulative Distribution Function of $U_i \in [0,1]$ using $\mathcal{N}(0,1)$; so $Z_i \sim \mathcal{N}(0,1)$
5: Calculate the copula parameters $\theta_{i,j} = \rho_{i,j}$; where $\rho_{i,j}$ is the Pearson correlation
6: Learn a Gaussian polytree, $Poly^{gauss}$, over Z_i variables
7: Simulate $S_i \sim Poly^{gauss}$
8: V_i = Cumulative Distribution Function of S_i using $\mathcal{N}(0,1)$; so $V_i \in [0,1]$
9: Y_i = Inverse Cumulative Distribution Function of V_i using K_{X_i};
10: Pob_{t+1} = Best individuals in $X_1, X_2 \ldots X_d \cup Y_1, Y_2 \ldots Y_d$

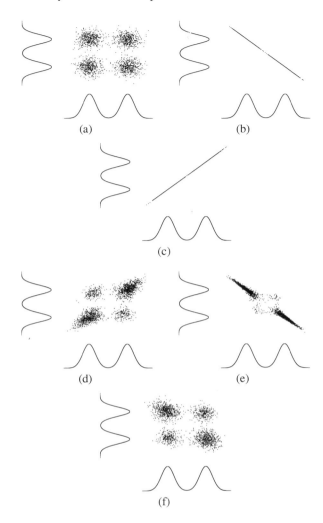

Fig. 3.4 Examples of dependence with Gaussian copula using multimodal marginal distributions. Each figure shows random samples on the (X_1, X_2) plane. (a) $\theta_{12} = 0$. (b) $\theta_{12} = -1$. (c) $\theta_{12} = 1$. (d) $\theta_{12} = 0.8$. (e) $\theta_{12} = -0.99$. (f) $\theta_{12} = -0.5$.

Once new data is simulated at step 7, each variable is evaluated in the Normal cumulative distribution function in step 8. Since each variable is Normal the evaluation maps the data to $[0, 1]$. Then, in step 9 it is ready to be send back to the marginals via the inverse cumulative distribution (previously generated using kernels).

3.5   Gaussian Poly-Trees with Gaussian Copula Functions + Mutations

In this section the Gaussian Poly-Tree with Copula function is extended with local mutations. The issue is the maintenance of population diversity which is necessary to keep exploration. A simple approach would be to use a new population and keeping the best individual, or to grow the population variance by some artificial method. In the proposed approach local mutations are used to create new individuals for the next population. The process is shown in the Figure 3.5. Assume the dimension of the problem is D.

In order to create a new individual do the following. Pick one individual at random from the population, call this the base element $\mathscr{B}$. Find D closer individuals to the base element. Compute the centroid of the $D \cup \mathscr{B}$ individuals and call it $\overline{\mathscr{C}}$. From the set of D individuals select the element which have best fitness value and call it $\mathscr{K}$. The new individual is $\mathscr{W} = \mathscr{K} + \left(\mathscr{K} - \overline{\mathscr{C}} \right)$.

For a population with size S, as many as 15% of S are mutated individuals, and 35% is generated from the Gaussian poly-tree with Gaussian copulas. Then truncation selection is applied as the normal $\mu + \lambda$ approach.

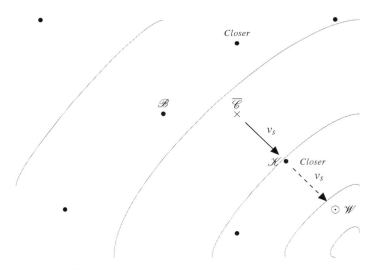

Fig. 3.5 Example of 2D local mutation. A new individual is created using the vector $v_s = \mathscr{K} - \overline{\mathscr{C}}$. The new individual, $\mathscr{W}$, improves the fitness according to level set.

3.6   Experiments

The Gaussian poly-tree EDA is an EDA as described in Algorithm 3.1 which uses a Gaussian Poly-tree to approximate the distribution of the selected set. It is tested with a total of 20 functions, 12 unimodal and 8 multimodal, which are described in Appendix A. The following parameters are used in all experiments except for the Experiment 1 below which is designed for a particular comparison. 30 independent runs are performed with the following initialization:

Number of runs: 30 .
Initialization. Asymmetric initialization for all problems, using the search space shown in Table 3.24 .
Population size. For a problem in D dimensions, the population is $2 \times (10(D^{0.7}) + 10)$ [3].
Stopping conditions. Maximum number of fitness function evaluations is reached: 3×10^5; or target error smaller than 1×10^{-10} for unimodal functions, and target error smaller than 1×10^{-6} for multimodal functions ; or no improving larger than 1×10^{-13} is detected after 30 generations when the mean of D standard deviations, one for each dimension, is less than 1×10^{-13}.

3.6.1   Experiment 1: Contrasting the Gaussian Poly-Tree with the Dependence Tree

The following experiment is designed to show how a richer graph based model may improve the performance of an EDA. In some ways, a poly-tree is a dependence tree with oriented edges. In Figure 3.1 is shown a poly-tree and a dependence tree with the same structure, the only difference is made by the nodes with several parents. This kind of edges provides the poly-tree with a richer representation capacity since they capture the conditional dependencies of the data. The problem to optimize is a Gaussian density function in ten dimensions. The optimum is shifted by 5 in every dimension, $\mathbf{X}^* = \mathbf{5}_{10 \times 1}$. The covariance matrix Σ_G of the Poly-tree in Figure 3.6 was created using the algorithm 3.2. The minimization problem is stated as follows:

$$Minimize \ f(\mathbf{X}) = -\mathcal{N}(\mathbf{X}; \mu_G, \Sigma_G) + \mathcal{N}(\mu_G; \mu_G, \Sigma_G) \qquad (3.16)$$

where $\mathbf{X}$ is the vector of decision variables, and $\mu_G = \mathbf{5}_{10 \times 1}$. So, the term $\mathcal{N}(\mu_G; \mu_G, \Sigma_G)$ is a constant offset that makes $f(\mathbf{X}^*) = 0.0$
The conditional variances are:

$$\left(\sigma_1^2, \cdots, \sigma_n^2\right)^t = (4, 14, 4, 6, 26, 9, 8, 4, 2, 12)^t \qquad (3.17)$$

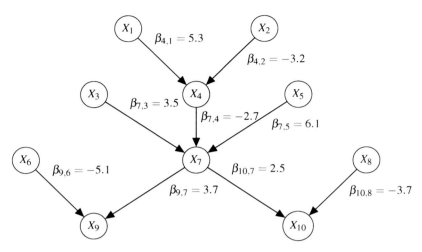

Fig. 3.6 The covariance matrix of experiment 1 is derived from this Gaussian Poly-tree

We performed an EDA using the proposed Gaussian Poly-tree vs the EDA with dependency tree. Table 3.2 reports the statistics of the best value found in 100 runs, and Table 3.3 shows the number of iterations required to reach the goal. The population size with $l = 10$ is $2 \times (10(l^{0.7}) + 10)$ [3], symmetric initialization is used for all the variables: $X_i \in [-5, 15]$. The stopping conditions is: maximum number of fitness function evaluations: 3×10^5; or target error smaller than 1×10^{-11}; or no improving larger than 1×10^{-13} is detected after 30 generations when the mean of the standard deviation of all the decision variables is less than 1×10^{-13}.

Table 3.2 Experiment 1: statistics for fitness value from 100 runs

Model	Best	Worst	Mean	Median	SD	Success (%)
Dependency tree	4.62E-12	2.65E-10	5.24E-11	3.58E-11	5.14E-11	22.0
Polytree	5.37E-12	2.82E-10	1.91E-11	9.08E-12	3.94E-11	72.0

Table 3.3 Experiment 1: statistics for number of function evaluations from 100 runs

Model	Best	Worst	Mean	Median	SD	Success (%)
Dependency tree	3.56E+04	1.69E+05	1.04E+05	1.16E+05	3.41E+04	22.0
Polytree	3.77E+04	2.62E+05	1.27E+05	1.27E+05	4.07E+04	72.0

The success rate using our poly-tree approach is definitely better than the EDA with dependency tree model. The number of evaluations shows a premature convergence in the EDA with dependency tree approach.

3.6.2   Experiment 2: Solving Unimodal Functions with the GPT-EDA

The success rate of the GPT-EDA out of 30 runs over 12 functions is shown in Table 3.10. The mean and standard deviation of the optimum values reached is shown in Table 3.5, the best value is reported between parenthesis. The mean and standard deviation of the number of fitness function evaluations is reported in Table 3.6, the best value is reported between parenthesis.

Table 3.4 Experiment 2. The GPT-EDA: success rate (%) on the unimodal problems in dimensions 4, 8, 10, 20, and 40. Target error= 1×10^{-10}

Alias / D	4	8	10	20	40
$\mathscr{F}_1$	100.0	100.0	100.0	100.0	100.0
$\mathscr{F}_2$	96.7	100.0	93.3	100.0	100.0
$\mathscr{F}_3$	90.0	93.3	83.3	100.0	96.7
$\mathscr{F}_4$	93.3	100.0	100.0	100.0	100.0
$\mathscr{F}_5$	90.0	100.0	93.3	96.7	100.0
$\mathscr{F}_6$	96.7	93.3	90.0	76.7	93.3
$\mathscr{F}_7$	96.7	100.0	93.3	86.7	60.0
$\mathscr{F}_8$	53.3	80.0	73.3	100.0	86.7
$\mathscr{F}_9$	56.7	100.0	96.7	100.0	100.0
$\mathscr{F}_{10}$	90.0	13.3	3.3	0.0	0.0
$\mathscr{F}_{11}$	100.0	0.0	0.0	0.0	0.0
$\mathscr{F}_{12}$	96.7	80.0	86.7	96.7	0.0

3.6.3   Experiment 3: Solving Multimodal Functions with the GPT-EDA

The success rate of the GPT-EDA out of 30 runs over 12 functions is shown in Table 3.7. The mean and standard deviation of the optimum values reached is shown in Table 3.8, the best value is reported between parenthesis. The mean and standard deviation of the number of fitness function evaluations is reported in Table 3.9, the best value is reported between parenthesis.

Table 3.5 Experiment 2. The GPT-EDA:optimum reached on the unimodal problems in dimensions 4, 8, 10, 20 and 40. In parenthesis the best value. Target error= 1×10^{-10}

Alias / D	4	8	10	20	40
$\mathscr{F}_1$	$(1.4236E-11)$	$(2.5384E-11)$	$(3.7539E-11)$	$(5.1387E-11)$	$(7.3703E-11)$
	$6.3E-11 \pm 2.3E-11$	$6.9E-11 \pm 2.1E-11$	$7.7E-11 \pm 1.6E-11$	$8.2E-11 \pm 1.3E-11$	$9E-11 \pm 6.7E-12$
$\mathscr{F}_2$	$(1.8246E-11)$	$(3.7246E-11)$	$(2.1298E-11)$	$(4.59E-11)$	$(5.9804E-11)$
	5.5 ± 30	$7.8E-11 \pm 1.6E-11$	$8.1E-05 \pm 0.00031$	$8.5E-11 \pm 1.2E-11$	$8.8E-11 \pm 1.2E-11$
$\mathscr{F}_3$	$(1.1589E-11)$	$(3.6939E-11)$	$(3.7055E-11)$	$(5.3446E-11)$	$(7.2089E-11)$
	0.019 ± 0.084	0.011 ± 0.046	0.075 ± 0.27	$8.4E-11 \pm 1.2E-11$	0.0022 ± 0.012
$\mathscr{F}_4$	$(6.2397E-12)$	$(3.4999E-11)$	$(4.0271E-11)$	$(5.1968E-11)$	$(5.7077E-11)$
	0.00031 ± 0.0015	$7.6E-11 \pm 1.9E-11$	$7.1E-11 \pm 1.5E-11$	$8.6E-11 \pm 1.2E-11$	$8.6E-11 \pm 1E-11$
$\mathscr{F}_5$	$(1.1234E-11)$	$(3.9127E-11)$	$(1.259E-11)$	$(6.4969E-11)$	$(6.483E-11)$
	0.0049 ± 0.021	$7.7E-11 \pm 1.8E-11$	0.043 ± 0.17	0.0022 ± 0.012	$9E-11 \pm 8.7E-12$
$\mathscr{F}_6$	$(1.0682E-11)$	$(2.5852E-11)$	$(4.2515E-11)$	$(5.8942E-11)$	$(6.9313E-11)$
	0.0013 ± 0.007	0.0016 ± 0.0087	0.0033 ± 0.013	0.043 ± 0.16	0.028 ± 0.16
$\mathscr{F}_7$	$(4.4846E-13)$	$(1.91E-12)$	$(3.646E-12)$	$(2.2199E-11)$	$(4.1786E-11)$
	$5.9E-11 \pm 9.6E-11$	$4.5E-11 \pm 2.8E-11$	$5.3E-07 \pm 2.9E-06$	$2.5E-05 \pm 0.00013$	0.00071 ± 0.0028
$\mathscr{F}_8$	(-5)	(-5)	(-5)	(-5)	(-5)
	-4.5 ± 1.1	-4.8 ± 0.47	-4.5 ± 1.4	$-5 \pm 1.2E-11$	-4.9 ± 0.23
$\mathscr{F}_9$	(-5)	(-5)	(-5)	(-5)	(-5)
	-4.8 ± 0.73	$-5 \pm 1.3E-11$	-5 ± 0.071	$-5 \pm 8.3E-12$	$-5 \pm 4.8E-12$
$\mathscr{F}_{10}$	$(1.3677E-11)$	$(6.147E-11)$	$(9.7902E-11)$	$(9.2597E-05)$	(0.15183)
	$6E-09 \pm 3.2E-08$	0.00013 ± 0.00039	0.00023 ± 0.00048	0.01 ± 0.01	0.25 ± 0.048
$\mathscr{F}_{11}$	(-16)	(-112)	(-208.8)	(-807.94)	(-2262)
	$-16 \pm 2.1E-11$	$-1.1E+02 \pm 4$	$-1.7E+02 \pm 25$	$-3.2E+02 \pm 3.2E+02$	$1.2E+03 \pm 2.2E+03$
$\mathscr{F}_{12}$	$(8.5561E-12)$	$(3.9845E-11)$	$(2.7323E-11)$	$(5.8239E-11)$	(0.25303)
	$1.8E-07 \pm 1E-06$	$9.5E-06 \pm 4.6E-05$	$1.8E-07 \pm 9.2E-07$	$7.8E-10 \pm 3.8E-09$	0.51 ± 0.16

3.6.4 Experiment 4: Solving Unimodal Functions with the GCPT-EDA

The success rate of the GCPT-EDA out of 30 runs over 12 functions is shown in Table 3.10. The mean and standard deviation of the optimum values reached is shown in Table 3.11, the best value is reported between parenthesis. The mean and standard deviation of the number of fitness function evaluations is reported in Table 3.12, the best value is reported between parenthesis.

Table 3.6 Experiment 2. GPT-EDA number of function evaluations on the unimodal problems in dimensions 4, 8, 10, 20 and 40. In parenthesis the best value. Target error= 1×10^{-10}

Alias / D	4	8	10	20	40
$\mathscr{F}_1$	(2404)	(5936)	(7860)	(19043)	(45044)
	$2.6E+03 \pm 90$	$6.3E+03 \pm 1.4E+02$	$8.2E+03 \pm 1.2E+02$	$1.9E+04 \pm 1.7E+02$	$4.6E+04 \pm 3.1E+02$
$\mathscr{F}_2$	(3329)	(7897)	(10500)	(24471)	(57628)
	$4.4E+03 \pm 4.8E+03$	$8.2E+03 \pm 1.2E+02$	$1.1E+04 \pm 2.5E+03$	$2.5E+04 \pm 2.7E+02$	$5.9E+04 \pm 5E+02$
$\mathscr{F}_3$	(3662)	(8904)	(11700)	(27967)	(66208)
	$4.9E+03 \pm 3.7E+03$	$9.9E+03 \pm 2.5E+03$	$1.4E+04 \pm 4E+03$	$2.9E+04 \pm 1.9E+02$	$6.8E+04 \pm 5.8E+03$
$\mathscr{F}_4$	(2811)	(6678)	(8760)	(19963)	(46760)
	$3.7E+03 \pm 2.3E+03$	$6.9E+03 \pm 1.4E+02$	$9E+03 \pm 1.4E+02$	$2E+04 \pm 2.5E+02$	$4.7E+04 \pm 3.6E+02$
$\mathscr{F}_5$	(3366)	(8480)	(11040)	(25667)	(60345)
	$5.3E+03 \pm 5.4E+03$	$8.7E+03 \pm 1.7E+02$	$1.2E+04 \pm 2.7E+03$	$2.7E+04 \pm 3.3E+03$	$6.1E+04 \pm 3E+02$
$\mathscr{F}_6$	(3292)	(7897)	(10500)	(25115)	(59201)
	$4.2E+03 \pm 3.6E+03$	$8.9E+03 \pm 2.2E+03$	$1.2E+04 \pm 2.9E+03$	$3E+04 \pm 7.4E+03$	$6.2E+04 \pm 7.5E+03$
$\mathscr{F}_7$	(1331)	(3180)	(4080)	(10671)	(26454)
	$2.2E+03 \pm 4.4E+03$	$3.4E+03 \pm 1.2E+02$	$2E+04 \pm 5.9E+04$	$5E+04 \pm 1E+05$	$1.4E+05 \pm 1.4E+05$
$\mathscr{F}_8$	(3255)	(7791)	(10200)	(23735)	(54482)
	$1.1E+04 \pm 8.9E+03$	$1E+04 \pm 3.6E+03$	$1.3E+04 \pm 4.7E+03$	$2.4E+04 \pm 2.6E+02$	$5.9E+04 \pm 1.1E+04$
$\mathscr{F}_9$	(5882)	(13727)	(17940)	(41307)	(95666)
	$1E+04 \pm 7.3E+03$	$1.4E+04 \pm 1.2E+02$	$1.9E+04 \pm 2.3E+03$	$4.2E+04 \pm 2.6E+02$	$9.7E+04 \pm 5.2E+02$
$\mathscr{F}_{10}$	(2663)	(14840)	(24300)	(2.4711E+05)	(3.0001E+05)
	$3.7E+03 \pm 2.8E+03$	$4.2E+04 \pm 1.4E+04$	$7.1E+04 \pm 1.2E+04$	$2.8E+05 \pm 1.1E+04$	$3E+05 \pm 0$
$\mathscr{F}_{11}$	(2774)	(68529)	(1.242E+05)	(3.0001E+05)	(3.0001E+05)
	$2.9E+03 \pm 92$	$1.2E+05 \pm 4.2E+04$	$2.2E+05 \pm 6.6E+04$	$3E+05 \pm 0$	$3E+05 \pm 0$
$\mathscr{F}_{12}$	(2885)	(14204)	(24300)	(1.4196E+05)	(3.0001E+05)
	$3.5E+03 \pm 1.6E+03$	$1.9E+04 \pm 9.3E+03$	$2.8E+04 \pm 5E+03$	$1.5E+05 \pm 1E+04$	$3E+05 \pm 0$

3.6.5   Experiment 5: Solving Multimodal Functions with the GCPT-EDA

The success rate of the GCPT-EDA out of 30 runs over 12 functions is shown in Table 3.13. The mean and standard deviation of the optimum values reached is shown in Table 3.14, the best value is reported between parenthesis. The mean and standard deviation of the number of fitness function evaluations is reported in Table 3.15, the best value is reported between parenthesis.

3.6.6   Experiment 6: Solving Unimodal Functions with the GCPT-EDA + Mutations

The success rate of the GCPT-EDA out of 30 runs over 12 functions is shown in Table 3.16. The mean and standard deviation of the optimum values reached is shown

Table 3.7 Experiment 3. GPT-EDA: success rate (%) on the Multimodal problems in dimensions 5, 10, 20, 30, and 50. Target error = 1×10^{-6}

Alias / D	5	10	20	30	50
$\mathscr{F}_{13}$	100.0	100.0	100.0	100.0	100.0
$\mathscr{F}_{14}$	0.0	100.0	100.0	100.0	100.0
$\mathscr{F}_{15}$	100.0	3.3	0.0	0.0	0.0
$\mathscr{F}_{16}$	0.0	0.0	0.0	0.0	0.0
$\mathscr{F}_{17}$	100.0	100.0	100.0	100.0	100.0
$\mathscr{F}_{18}$	100.0	100.0	100.0	100.0	100.0
$\mathscr{F}_{19}$	100.0	100.0	100.0	100.0	100.0
$\mathscr{F}_{20}$	100.0	100.0	100.0	100.0	100.0

Table 3.8 Experiment 3. GPT-EDA: optimum reached on the multimodal problems in dimensions 5, 10, 20, 30 and 50. In parenthesis the best value. Target error = 1×10^{-6}

Alias / D	5	10	20	30	50
$\mathscr{F}_{13}$	($2.7715E - 07$)	($5.5217E - 07$)	($8.092E - 07$)	($7.9579E - 07$)	($8.9916E - 07$)
	$8E - 07 \pm 1.9E - 07$	$8.6E - 07 \pm 1.2E - 07$	$9.2E - 07 \pm 4.9E - 08$	$9.3E - 07 \pm 4.9E - 08$	$9.6E - 07 \pm 3E - 08$
$\mathscr{F}_{14}$	(0.013266)	($3.5618E - 07$)	($5.0282E - 07$)	($6.4314E - 07$)	($6.3561E - 07$)
	0.036 ± 0.011	$7.9E - 07 \pm 1.8E - 07$	$8.4E - 07 \pm 1.1E - 07$	$8.6E - 07 \pm 9.5E - 08$	$9E - 07 \pm 8.4E - 08$
$\mathscr{F}_{15}$	($3.6526E - 07$)	($8.8942E - 07$)	(60.737)	(134.79)	(276.41)
	$6.6E - 07 \pm 1.6E - 07$	14 ± 4.5	76 ± 5.5	$1.5E + 02 \pm 7.3$	$3.2E + 02 \pm 13$
$\mathscr{F}_{16}$	(1.796)	(7.4973)	(17.736)	(27.76)	(47.637)
	2.6 ± 0.34	7.9 ± 0.17	18 ± 0.12	28 ± 0.066	48 ± 0.06
$\mathscr{F}_{17}$	(-0.5)	(-1)	(-2)	(-3)	(-5)
	$-0.5 \pm 2.1E - 07$	$-1 \pm 1.7E - 07$	$-2 \pm 1.2E - 07$	$-3 \pm 8.1E - 08$	$-5 \pm 7.1E - 08$
$\mathscr{F}_{18}$	($2.7767E - 07$)	($3.882E - 07$)	($6.4845E - 07$)	($5.7604E - 07$)	($7.5193E - 07$)
	$6.2E - 07 \pm 2.2E - 07$	$7.9E - 07 \pm 1.7E - 07$	$8.6E - 07 \pm 9.7E - 08$	$8.5E - 07 \pm 9.1E - 08$	$9E - 07 \pm 6.3E - 08$
$\mathscr{F}_{19}$	($3.3012E - 07$)	($3.0066E - 07$)	($5.1432E - 07$)	($5.8572E - 07$)	($6.7507E - 07$)
	$7E - 07 \pm 1.8E - 07$	$7.7E - 07 \pm 1.9E - 07$	$8.6E - 07 \pm 1.3E - 07$	$8.8E - 07 \pm 9.6E - 08$	$9E - 07 \pm 9E - 08$
$\mathscr{F}_{20}$	($1.6374E - 07$)	($3.0329E - 07$)	($4.8189E - 07$)	($6.1397E - 07$)	($7.359E - 07$)
	$6.5E - 07 \pm 2.3E - 07$	$6.9E - 07 \pm 1.9E - 07$	$8.1E - 07 \pm 1.4E - 07$	$8.8E - 07 \pm 9.2E - 08$	$9.1E - 07 \pm 7E - 08$

in Table 3.17, the best value is reported between parenthesis. The mean and standard deviation of the number of fitness function evaluations is reported in Table 3.18, the best value is reported between parenthesis.

Table 3.9 Experiment 3. GPT-EDA: number of function evaluations on the multimodal problems in dimensions 5, 10, 20, 30 and 50. In parenthesis the best value. Target error=1×10^{-6}

Alias / D	5	10	20	30	50
$\mathscr{F}_{13}$	(4141)	(9600)	(22355)	(36226)	(67319)
	$4.4E+03 \pm 1.1E+02$	$9.9E+03 \pm 1.4E+02$	$2.3E+04 \pm 1.7E+02$	$3.7E+04 \pm 2.6E+02$	$6.8E+04 \pm 4.1E+02$
$\mathscr{F}_{14}$	($3.0004E+05$)	(10740)	(16743)	(27376)	(50159)
	$3E+05 \pm 0$	$7.3E+04 \pm 8.6E+04$	$1.7E+04 \pm 3E+02$	$2.8E+04 \pm 3.1E+02$	$5.1E+04 \pm 3.4E+02$
$\mathscr{F}_{15}$	(53218)	($2.7174E+05$)	($3.0001E+05$)	($3.0007E+05$)	($3.0013E+05$)
	$9.9E+04 \pm 3.7E+04$	$3E+05 \pm 5.2E+03$	$3E+05 \pm 0$	$3E+05 \pm 0$	$3E+05 \pm 0$
$\mathscr{F}_{16}$	(19639)	(28560)	(55659)	(78942)	($1.3018E+05$)
	$4.3E+04 \pm 1.2E+04$	$3.9E+04 \pm 6.1E+03$	$5.8E+04 \pm 2.8E+03$	$8.2E+04 \pm 1.3E+03$	$1.3E+05 \pm 1.4E+03$
$\mathscr{F}_{17}$	(2091)	(5220)	(12511)	(20886)	(39764)
	$2.2E+03 \pm 80$	$5.4E+03 \pm 1.3E+02$	$1.3E+04 \pm 1.8E+02$	$2.1E+04 \pm 2.4E+02$	$4.1E+04 \pm 3.5E+02$
$\mathscr{F}_{18}$	(2009)	(4500)	(10487)	(17464)	(31844)
	$2.3E+03 \pm 1.1E+02$	$4.9E+03 \pm 1.8E+02$	$1.1E+04 \pm 2.5E+02$	$1.8E+04 \pm 2E+02$	$3.2E+04 \pm 3.5E+02$
$\mathscr{F}_{19}$	(2214)	(4860)	(11959)	(20414)	(38609)
	$2.6E+03 \pm 1.2E+02$	$5.6E+03 \pm 2.3E+02$	$1.3E+04 \pm 3.1E+02$	$2.1E+04 \pm 2.8E+02$	$3.9E+04 \pm 3.7E+02$
$\mathscr{F}_{20}$	(1886)	(4860)	(11591)	(19706)	(36794)
	$2.1E+03 \pm 1E+02$	$5.1E+03 \pm 1.3E+02$	$1.2E+04 \pm 2.1E+02$	$2E+04 \pm 1.8E+02$	$3.8E+04 \pm 4E+02$

Table 3.10 Experiment 4. GCPT-EDA: success rate (%) on the unimodal problems in dimensions 4, 8, 10, 20, and 40. Target error=1×10^{-10}

Alias / D	4	8	10	20	40
$\mathscr{F}_1$	100.0	100.0	100.0	100.0	100.0
$\mathscr{F}_2$	80.0	90.0	96.7	93.3	90.0
$\mathscr{F}_3$	83.3	86.7	86.7	70.0	86.7
$\mathscr{F}_4$	96.7	100.0	96.7	100.0	100.0
$\mathscr{F}_5$	80.0	86.7	83.3	93.3	100.0
$\mathscr{F}_6$	76.7	70.0	66.7	60.0	43.3
$\mathscr{F}_7$	100.0	100.0	96.7	53.3	23.3
$\mathscr{F}_8$	43.3	83.3	73.3	86.7	93.3
$\mathscr{F}_9$	40.0	90.0	76.7	100.0	100.0
$\mathscr{F}_{10}$	46.7	0.0	0.0	0.0	0.0
$\mathscr{F}_{11}$	80.0	0.0	0.0	0.0	0.0
$\mathscr{F}_{12}$	40.0	20.0	16.7	53.3	0.0

Table 3.11 Experiment 4. GCPT-EDA: optimum reached on the unimodal problems in dimensions 4, 8, 10, 20 and 40. In parenthesis the best value. Target error=1×10^{-10}

Alias / D	4	8	10	20	40
$\mathscr{F}_1$	($1.2394E-11$)	($3.4025E-11$)	($4.9352E-11$)	($4.5243E-11$)	($6.8435E-11$)
	$5.6E-11\pm2.4E-11$	$7.7E-11\pm1.5E-11$	$8.1E-11\pm1.4E-11$	$8.6E-11\pm1.4E-11$	$8.9E-11\pm8.7E-12$
$\mathscr{F}_2$	($1.1674E-11$)	($2.0639E-11$)	($2.7653E-11$)	($6.5835E-11$)	($6.372E-11$)
	0.14 ± 0.54	0.079 ± 0.3	0.00055 ± 0.003	0.33 ± 1.6	0.18 ± 0.72
$\mathscr{F}_3$	($1.1345E-11$)	($4.6005E-11$)	($4.3376E-11$)	($4.6178E-11$)	($6.2041E-11$)
	0.03 ± 0.14	0.15 ± 0.67	0.24 ± 0.89	0.19 ± 0.59	0.054 ± 0.27
$\mathscr{F}_4$	($1.097E-11$)	($1.8962E-11$)	($3.4219E-11$)	($3.6763E-11$)	($5.8411E-11$)
	$4.1E-05\pm0.00022$	$7.4E-11\pm2E-11$	0.073 ± 0.4	$8.4E-11\pm1.5E-11$	$8.8E-11\pm9.4E-12$
$\mathscr{F}_5$	($1.7992E-11$)	($2.7048E-11$)	($2.1638E-11$)	($6.2959E-11$)	($7.1218E-11$)
	0.12 ± 0.38	0.13 ± 0.45	0.096 ± 0.32	0.055 ± 0.21	$9E-11\pm8E-12$
$\mathscr{F}_6$	($1.4345E-11$)	($3.1124E-11$)	($4.4043E-11$)	($5.6644E-11$)	($6.4075E-11$)
	0.21 ± 0.68	0.068 ± 0.21	0.57 ± 1.4	0.34 ± 1.2	0.5 ± 1.3
$\mathscr{F}_7$	($4.6723E-13$)	($2.5019E-12$)	($9.757E-12$)	($2.7094E-11$)	($7.1496E-11$)
	$4.7E-11\pm2.8E-11$	$5.6E-11\pm2.5E-11$	$1E-10\pm2.2E-10$	$1.4E-06\pm7.6E-06$	$4E-05\pm0.00014$
$\mathscr{F}_8$	(-5)	(-5)	(-5)	(-5)	(-5)
	-4.9 ± 0.26	-4.8 ± 0.56	-4.6 ± 0.78	-4.8 ± 0.69	-5 ± 0.24
$\mathscr{F}_9$	(-5)	(-5)	(-5)	(-5)	(-5)
	-4.4 ± 0.89	-4.8 ± 0.82	-4.3 ± 1.6	$-5\pm8.1E-12$	$-5\pm5.3E-12$
$\mathscr{F}_{10}$	($2.7409E-11$)	($3.8446E-10$)	($1.9151E-05$)	(0.010311)	(0.12983)
	0.00018 ± 0.00058	0.02 ± 0.063	0.018 ± 0.064	0.078 ± 0.1	0.33 ± 0.14
$\mathscr{F}_{11}$	(-16)	(-111.97)	(-209.12)	(-1155.9)	(-2638.7)
	-16 ± 0.00039	$-1E+02\pm13$	$-1.6E+02\pm31$	$-2.5E+02\pm4.4E+02$	$1.1E+03\pm2.3E+03$
$\mathscr{F}_{12}$	($1.6444E-11$)	($3.1813E-11$)	($6.5982E-11$)	($4.3488E-11$)	(0.23668)
	0.0023 ± 0.0083	0.0022 ± 0.0063	0.0029 ± 0.014	0.00012 ± 0.00047	0.45 ± 0.12

3.6.7   Experiment 7: Solving Multimodal Functions with the GCPT-EDA + Mutations

The success rate of the GCPT-EDA + mutations out of 30 runs over 12 functions is shown in Table 3.19. The mean and standard deviation of the optimum values reached is shown in Table 3.20, the best value is reported between parenthesis. The mean and standard deviation of the number of fitness function evaluations is reported in Table 3.21, the best value is reported between parenthesis.

Table 3.12 Experiment 4. GCPT-EDA: number of function evaluations on the unimodal problems in dimensions 4, 8, 10, 20 and 40. In parenthesis the best value. Target error=1×10^{-10}

Alias / D	4	8	10	20	40
$\mathscr{F}_1$	(2663)	(6148)	(8100)	(18859)	(43471)
	$2.9E+03\pm1E+02$	$6.6E+03\pm1.5E+02$	$8.5E+03\pm2.1E+02$	$1.9E+04\pm2.2E+02$	$4.5E+04\pm4.1E+02$
$\mathscr{F}_2$	(3551)	(8109)	(10320)	(24931)	(57199)
	$8.2E+03\pm9.3E+03$	$9.6E+03\pm3.3E+03$	$1.1E+04\pm1.8E+03$	$2.7E+04\pm4.4E+03$	$6.1E+04\pm1E+04$
$\mathscr{F}_3$	(3847)	(9222)	(11880)	(28151)	(64635)
	$7.6E+03\pm7.8E+03$	$1.1E+04\pm3.8E+03$	$1.4E+04\pm4.1E+03$	$3.3E+04\pm7.5E+03$	$6.9E+04\pm9.4E+03$
$\mathscr{F}_4$	(3033)	(6943)	(8700)	(20331)	(46045)
	$4.1E+03\pm4.1E+03$	$7.3E+03\pm2.3E+02$	$9.6E+03\pm1.9E+03$	$2.1E+04\pm2.3E+02$	$4.7E+04\pm3.2E+02$
$\mathscr{F}_5$	(3810)	(8586)	(11160)	(25759)	(59058)
	$8.4E+03\pm9.3E+03$	$1E+04\pm3.6E+03$	$1.3E+04\pm4E+03$	$2.7E+04\pm4.2E+03$	$6E+04\pm3.6E+02$
$\mathscr{F}_6$	(3662)	(8427)	(10860)	(25299)	(59487)
	$8.9E+03\pm9.5E+03$	$1.2E+04\pm4.4E+03$	$1.5E+04\pm6.5E+03$	$3.4E+04\pm1.1E+04$	$7.7E+04\pm1.5E+04$
$\mathscr{F}_7$	(1405)	(3445)	(4620)	(12143)	(29600)
	$1.7E+03\pm1.2E+02$	$4.7E+03\pm4.7E+03$	$6.3E+03\pm6.4E+03$	$1.2E+05\pm1.3E+05$	$2.4E+05\pm1.2E+05$
$\mathscr{F}_8$	(3440)	(7897)	(9960)	(22723)	(52480)
	$1.5E+04\pm1.1E+04$	$1E+04\pm4E+03$	$1.3E+04\pm4.4E+03$	$2.6E+04\pm5.5E+03$	$5.5E+04\pm6.9E+03$
$\mathscr{F}_9$	(5956)	(13674)	(17760)	(40755)	(92949)
	$1.8E+04\pm1E+04$	$1.5E+04\pm2.7E+03$	$2.1E+04\pm4.7E+03$	$4.1E+04\pm3.5E+02$	$9.4E+04\pm4E+02$
$\mathscr{F}_{10}$	(2959)	(23850)	(50880)	(1.6606E+05)	(3.0001E+05)
	$1.1E+04\pm7.7E+03$	$4.1E+04\pm4.5E+03$	$5.7E+04\pm3.2E+03$	$1.7E+05\pm3.2E+03$	$3E+05\pm0$
$\mathscr{F}_{11}$	(2811)	(43460)	(61860)	(1.472E+05)	(3.0001E+05)
	$5.7E+03\pm5.8E+03$	$4.9E+04\pm3.6E+03$	$6.8E+04\pm3.2E+03$	$1.7E+05\pm1.1E+04$	$3E+05\pm0$
$\mathscr{F}_{12}$	(3144)	(14628)	(24840)	(1.3726E+05)	(3.0001E+05)
	$1.3E+04\pm9.1E+03$	$3.1E+04\pm8.9E+03$	$4.4E+04\pm9.6E+03$	$1.7E+05\pm2.2E+04$	$3E+05\pm0$

Table 3.13 Experiment 5. GCPT-EDA: success rate (%) on the multimodal problems in dimensions 5, 10, 20, 30, and 50. Target error=1×10^{-6}

Alias / D	5	10	20	30	50
$\mathscr{F}_{13}$	100.0	100.0	100.0	100.0	100.0
$\mathscr{F}_{14}$	80.0	96.7	100.0	100.0	100.0
$\mathscr{F}_{15}$	100.0	100.0	93.3	100.0	86.7
$\mathscr{F}_{16}$	0.0	0.0	0.0	0.0	0.0
$\mathscr{F}_{17}$	100.0	100.0	100.0	100.0	100.0
$\mathscr{F}_{18}$	100.0	100.0	100.0	100.0	100.0
$\mathscr{F}_{19}$	100.0	100.0	100.0	100.0	100.0
$\mathscr{F}_{20}$	100.0	100.0	100.0	100.0	100.0

Table 3.14 Experiment 5. GCPT-EDA: optimum reached on the multimodal problems in dimensions 5, 10, 20, 30 and 50. In parenthesis the best value. Target error=1×10^{-6}

Alias / D	5	10	20	30	50
$\mathcal{F}_{13}$	(4.5832E − 07)	(6.5954E − 07)	(7.7004E − 07)	(7.97E − 07)	(8.4945E − 07)
	8.1E − 07 ± 1.6E − 07	8.4E − 07 ± 8.8E − 08	9.1E − 07 ± 6.7E − 08	9.4E − 07 ± 4.7E − 08	9.5E − 07 ± 3.8E − 08
$\mathcal{F}_{14}$	(1.6288E − 07)	(2.7181E − 07)	(5.9879E − 07)	(6.6933E − 07)	(7.4417E − 07)
	0.0021 ± 0.0047	0.00025 ± 0.0014	8.5E − 07 ± 1E − 07	8.7E − 07 ± 8.6E − 08	9.2E − 07 ± 6.9E − 08
$\mathcal{F}_{15}$	(9.249E − 08)	(4.4719E − 07)	(5.1973E − 07)	(6.6918E − 07)	(6.7196E − 07)
	7E − 07 ± 2.1E − 07	7.5E − 07 ± 1.5E − 07	0.066 ± 0.25	8.8E − 07 ± 9.7E − 08	0.13 ± 0.34
$\mathcal{F}_{16}$	(0.44688)	(7.0943)	(16.916)	(27.272)	(47.019)
	2.5 ± 0.7	7.7 ± 0.34	18 ± 2.8	28 ± 0.44	48 ± 0.34
$\mathcal{F}_{17}$	(−0.5)	(−1)	(−2)	(−3)	(−5)
	−0.5 ± 2.2E − 07	−1 ± 1.5E − 07	−2 ± 1.1E − 07	−3 ± 1.2E − 07	−5 ± 7.7E − 08
$\mathcal{F}_{18}$	(3.9585E − 07)	(2.9783E − 07)	(6.601E − 07)	(6.312E − 07)	(7.4252E − 07)
	7E − 07 ± 2E − 07	7.5E − 07 ± 1.9E − 07	8.4E − 07 ± 9.9E − 08	8.7E − 07 ± 9.9E − 08	9.2E − 07 ± 6.4E − 08
$\mathcal{F}_{19}$	(9.1323E − 08)	(4.2477E − 07)	(5.1582E − 07)	(5.9602E − 07)	(6.6655E − 07)
	5.9E − 07 ± 2.3E − 07	7.5E − 07 ± 1.5E − 07	8.3E − 07 ± 1.3E − 07	8.8E − 07 ± 1E − 07	9.1E − 07 ± 9.1E − 08
$\mathcal{F}_{20}$	(1.2002E − 07)	(3.8777E − 07)	(3.6962E − 07)	(6.1903E − 07)	(8.2308E − 07)
	6.5E − 07 ± 2.4E − 07	7.5E − 07 ± 1.6E − 07	8.2E − 07 ± 1.5E − 07	8.8E − 07 ± 1.1E − 07	9.3E − 07 ± 5.3E − 08

Table 3.15 Experiment 5. GCPT-EDA: number of function evaluations on the multimodal problems in dimensions 5, 10, 20, 30 and 50. In parenthesis the best value. Target error=1×10^{-6}

Alias / D	5	10	20	30	50
$\mathcal{F}_{13}$	(4305)	(9480)	(21711)	(34810)	(64514)
	4.5E + 03 ± 1.2E + 02	9.9E + 03 ± 1.7E + 02	2.2E + 04 ± 2.5E + 02	3.6E + 04 ± 3.4E + 02	6.5E + 04 ± 3.2E + 02
$\mathcal{F}_{14}$	(19147)	(7800)	(16651)	(27022)	(49334)
	5.6E + 04 ± 2.9E + 04	9.6E + 03 ± 1.9E + 03	1.7E + 04 ± 2.4E + 02	2.7E + 04 ± 2.3E + 02	5E + 04 ± 3.7E + 02
$\mathcal{F}_{15}$	(4469)	(11940)	(31647)	(58882)	(1.188E + 05)
	6.3E + 03 ± 1E + 03	1.6E + 04 ± 2.5E + 03	4.1E + 04 ± 6.1E + 03	7.2E + 04 ± 9.8E + 03	1.5E + 05 ± 1.8E + 04
$\mathcal{F}_{16}$	(23698)	(22080)	(39927)	(61714)	(1.0461E + 05)
	4.7E + 04 ± 6.9E + 04	3.4E + 04 ± 5.3E + 03	5.4E + 04 ± 1E + 04	7.2E + 04 ± 4.7E + 03	1.2E + 05 ± 4.8E + 03
$\mathcal{F}_{17}$	(1927)	(4680)	(11039)	(18644)	(35474)
	2.1E + 03 ± 98	5E + 03 ± 1.4E + 02	1.2E + 04 ± 1.8E + 02	1.9E + 04 ± 2.1E + 02	3.6E + 04 ± 3.6E + 02
$\mathcal{F}_{18}$	(1804)	(4320)	(9935)	(16284)	(29534)
	2.1E + 03 ± 1.2E + 02	4.7E + 03 ± 1.5E + 02	1E + 04 ± 2.6E + 02	1.7E + 04 ± 3.1E + 02	3E + 04 ± 3.8E + 02
$\mathcal{F}_{19}$	(2050)	(5040)	(11959)	(19470)	(36464)
	2.3E + 03 ± 1.3E + 02	5.3E + 03 ± 1.4E + 02	1.2E + 04 ± 1.5E + 02	2E + 04 ± 2.2E + 02	3.7E + 04 ± 2.6E + 02
$\mathcal{F}_{20}$	(1927)	(4620)	(11223)	(18290)	(35309)
	2.1E + 03 ± 1.1E + 02	4.9E + 03 ± 1.4E + 02	1.2E + 04 ± 1.9E + 02	1.9E + 04 ± 3.5E + 02	3.6E + 04 ± 3.5E + 02

Table 3.16 Experiment 6. GCPT-EDA + mutations: success rate (%) on the unimodal problems in dimensions 4, 8, 10, 20, and 40. Target error=1×10^{-10}

Alias / D	4	8	10	20	40
$\mathscr{F}_1$	100.0	100.0	100.0	100.0	100.0
$\mathscr{F}_2$	100.0	100.0	100.0	100.0	100.0
$\mathscr{F}_3$	100.0	100.0	100.0	100.0	100.0
$\mathscr{F}_4$	100.0	100.0	100.0	100.0	100.0
$\mathscr{F}_5$	100.0	100.0	100.0	100.0	100.0
$\mathscr{F}_6$	100.0	100.0	100.0	100.0	100.0
$\mathscr{F}_7$	100.0	100.0	100.0	100.0	100.0
$\mathscr{F}_8$	100.0	100.0	100.0	100.0	100.0
$\mathscr{F}_9$	100.0	100.0	100.0	100.0	100.0
$\mathscr{F}_{10}$	100.0	100.0	100.0	100.0	0.0
$\mathscr{F}_{11}$	100.0	100.0	100.0	0.0	0.0
$\mathscr{F}_{12}$	100.0	100.0	100.0	100.0	0.0

Table 3.17 Experiment 6. GCPT-EDA + mutations: optimum reached on the unimodal problems in dimensions 4, 8, 10, 20 and 40. In parenthesis the best value. Target error=1×10^{-10}

Alias / D	4	8	10	20	40
$\mathscr{F}_1$	$(1.7911E-11)$ $6.4E-11 \pm 2.3E-11$	$(2.2897E-11)$ $7.2E-11 \pm 2E-11$	$(5.7751E-11)$ $8.1E-11 \pm 1.4E-11$	$(4.5888E-11)$ $8.3E-11 \pm 1.3E-11$	$(6.1861E-11)$ $9.1E-11 \pm 9.7E-12$
$\mathscr{F}_2$	$(8.9081E-12)$ $6.2E-11 \pm 2.6E-11$	$(1.4722E-11)$ $7.5E-11 \pm 2.3E-11$	$(2.6123E-11)$ $8.4E-11 \pm 1.5E-11$	$(6.525E-11)$ $8.4E-11 \pm 9.4E-12$	$(7.1892E-11)$ $9.1E-11 \pm 8E-12$
$\mathscr{F}_3$	$(8.3119E-12)$ $6.2E-11 \pm 2.4E-11$	$(3.4398E-11)$ $7.6E-11 \pm 1.8E-11$	$(4.8211E-11)$ $8.3E-11 \pm 1.2E-11$	$(5.478E-11)$ $8.5E-11 \pm 1.3E-11$	$(6.0675E-11)$ $8.8E-11 \pm 9.5E-12$
$\mathscr{F}_4$	$(2.7635E-11)$ $6.4E-11 \pm 2.3E-11$	$(3.2351E-11)$ $7.5E-11 \pm 1.8E-11$	$(4.7385E-11)$ $7.9E-11 \pm 1.2E-11$	$(4.3649E-11)$ $8.6E-11 \pm 1.3E-11$	$(7.2538E-11)$ $9.1E-11 \pm 7.3E-12$
$\mathscr{F}_5$	$(2.4112E-11)$ $7.4E-11 \pm 2.2E-11$	$(3.4796E-11)$ $7.8E-11 \pm 1.7E-11$	$(3.892E-11)$ $8E-11 \pm 1.6E-11$	$(6.6926E-11)$ $8.6E-11 \pm 8.8E-12$	$(7.053E-11)$ $9.3E-11 \pm 5.8E-12$
$\mathscr{F}_6$	$(4.7301E-12)$ $5.4E-11 \pm 2.7E-11$	$(1.8631E-11)$ $7E-11 \pm 2.2E-11$	$(4.5646E-11)$ $7.6E-11 \pm 1.4E-11$	$(5.3733E-11)$ $8.5E-11 \pm 1.2E-11$	$(6.9544E-11)$ $9.1E-11 \pm 7.1E-12$
$\mathscr{F}_7$	$(2.2497E-13)$ $3.9E-11 \pm 3.4E-11$	$(1.0451E-11)$ $5.2E-11 \pm 2.5E-11$	$(1.4424E-11)$ $6.1E-11 \pm 2.4E-11$	$(5.8304E-11)$ $8.3E-11 \pm 1.1E-11$	$(6.7912E-11)$ $9.1E-11 \pm 7.4E-12$
$\mathscr{F}_8$	(-5) $-5 \pm 2.3E-11$	(-5) $-5 \pm 1.7E-11$	(-5) $-5 \pm 2.1E-11$	(-5) $-5 \pm 1.2E-11$	(-5) $-5 \pm 8E-12$
$\mathscr{F}_9$	(-5) $-5 \pm 2E-11$	(-5) $-5 \pm 1E-11$	(-5) $-5 \pm 8.9E-12$	(-5) $-5 \pm 7.2E-12$	(-5) $-5 \pm 3.5E-12$
$\mathscr{F}_{10}$	$(2.3161E-11)$ $6.6E-11 \pm 2.3E-11$	$(6.694E-11)$ $8.7E-11 \pm 9.5E-12$	$(4.0397E-11)$ $8.5E-11 \pm 1.4E-11$	$(7.4611E-11)$ $9.4E-11 \pm 5.8E-12$	(2.4768) 18 ± 16
$\mathscr{F}_{11}$	(-16) $-16 \pm 2E-11$	(-112) $-1.1E+02 \pm 1.5E-11$	(-210) $-2.1E+02 \pm 2.4E-12$	(-1513.4) $-1.5E+03 \pm 8.7$	(-8842.7) $-2.8E+03 \pm 2.4E+03$
$\mathscr{F}_{12}$	$(1.715E-11)$ $6.5E-11 \pm 1.8E-11$	$(3.7743E-11)$ $8E-11 \pm 1.9E-11$	$(3.2669E-11)$ $7.8E-11 \pm 1.6E-11$	$(4.4259E-11)$ $8.9E-11 \pm 1.1E-11$	(1.1488) 67 ± 54

Table 3.18 Experiment 6. GCPT-EDA + mutations: number of function evaluations on the unimodal problems in dimensions 4, 8, 10, 20 and 40. In parenthesis the best value. Target error=1×10^{-10}

Alias / D	4	8	10	20	40
$\mathscr{F}_1$	(2811) $3.1E+03 \pm 1.1E+02$	(7882) $8.1E+03 \pm 1.7E+02$	(10140) $1.1E+04 \pm 2E+02$	(25107) $2.6E+04 \pm 2.6E+02$	(58629) $6E+04 \pm 4.4E+02$
$\mathscr{F}_2$	(3477) $3.8E+03 \pm 1.4E+02$	(9988) $1.1E+04 \pm 2.7E+02$	(13560) $1.4E+04 \pm 2.9E+02$	(33012) $3.4E+04 \pm 7E+02$	(75789) $8.1E+04 \pm 7.1E+03$
$\mathscr{F}_3$	(3958) $4.3E+03 \pm 1.5E+02$	(11392) $1.2E+04 \pm 3.3E+02$	(15600) $1.6E+04 \pm 2.5E+02$	(37197) $3.8E+04 \pm 7.9E+02$	(86657) $8.9E+04 \pm 4.7E+03$
$\mathscr{F}_4$	(3181) $3.6E+03 \pm 1.6E+02$	(8476) $8.9E+03 \pm 2.4E+02$	(10860) $1.2E+04 \pm 3E+02$	(26316) $2.7E+04 \pm 3.1E+02$	(61060) $6.2E+04 \pm 4.4E+02$
$\mathscr{F}_5$	(3773) $4.1E+03 \pm 2E+02$	(10420) $1.1E+04 \pm 3E+02$	(14220) $1.5E+04 \pm 3.1E+02$	(34314) $3.5E+04 \pm 4.7E+02$	(78220) $8E+04 \pm 1.9E+03$
$\mathscr{F}_6$	(3625) $3.9E+03 \pm 2.1E+02$	(10204) $1.1E+04 \pm 5.3E+02$	(13680) $1.4E+04 \pm 3.6E+02$	(33477) $3.5E+04 \pm 1.1E+03$	(79221) $9.2E+04 \pm 1.8E+04$
$\mathscr{F}_7$	(1294) $1.7E+03 \pm 1.5E+02$	(4372) $4.8E+03 \pm 2.1E+02$	(6000) $6.6E+03 \pm 4.4E+02$	(15342) $2E+04 \pm 4.3E+03$	(40754) $6.7E+04 \pm 2.7E+04$
$\mathscr{F}_8$	(2774) $3.1E+03 \pm 1.6E+02$	(8854) $9.4E+03 \pm 2.6E+02$	(12000) $1.3E+04 \pm 3.7E+02$	(30501) $3.1E+04 \pm 4.8E+02$	(69640) $7.4E+04 \pm 6.9E+03$
$\mathscr{F}_9$	(4994) $5.4E+03 \pm 1.7E+02$	(15928) $1.6E+04 \pm 3E+02$	(21420) $2.2E+04 \pm 4.4E+02$	(54309) $5.5E+04 \pm 4.3E+02$	($1.2455E+05$) $1.3E+05 \pm 5.3E+02$
$\mathscr{F}_{10}$	(2996) $3.3E+03 \pm 1.5E+02$	(16144) $1.8E+04 \pm 1.3E+03$	(30960) $3.4E+04 \pm 2.1E+03$	($2.2692E+05$) $2.6E+05 \pm 1.5E+04$	($3.0001E+05$) $3E+05 \pm 0$
$\mathscr{F}_{11}$	(3033) $3.3E+03 \pm 1.5E+02$	(12850) $2.4E+04 \pm 5.9E+03$	(63660) $9.7E+04 \pm 1.3E+04$	($3.0002E+05$) $3E+05 \pm 0$	($3.0001E+05$) $3E+05 \pm 0$
$\mathscr{F}_{12}$	(3218) $3.6E+03 \pm 2E+02$	(15334) $1.7E+04 \pm 8.7E+02$	(26760) $2.9E+04 \pm 1.3E+03$	($1.4694E+05$) $1.7E+05 \pm 1.1E+04$	($3.0001E+05$) $3E+05 \pm 0$

Table 3.19 Experiment 7. GCPT-EDA + mutations: success rate (%) on the multimodal problems in dimensions 5, 10, 20, 30, and 50. Target error=1×10^{-6}

Alias / D	5	10	20	30	50
$\mathscr{F}_{13}$	100.0	100.0	100.0	100.0	100.0
$\mathscr{F}_{14}$	6.7	80.0	100.0	100.0	100.0
$\mathscr{F}_{15}$	83.3	93.3	93.3	96.7	86.7
$\mathscr{F}_{16}$	96.7	100.0	0.0	0.0	0.0
$\mathscr{F}_{17}$	100.0	100.0	100.0	100.0	100.0
$\mathscr{F}_{18}$	100.0	100.0	100.0	100.0	100.0
$\mathscr{F}_{19}$	100.0	100.0	100.0	100.0	100.0
$\mathscr{F}_{20}$	100.0	100.0	100.0	100.0	100.0

Table 3.20 Experiment 7. GCPT-EDA + mutations: optimum reached on the multimodal problems in dimensions 5, 10, 20, 30 and 50. In parenthesis the best value. Target error=1×10^{-6}

Alias / D	5	10	20	30	50
$\mathscr{F}_{13}$	($4.5007E-07$)	($6.7356E-07$)	($7.8418E-07$)	($8.4756E-07$)	($8.9817E-07$)
	$8.1E-07 \pm 1.6E-07$	$9.1E-07 \pm 8.2E-08$	$9.3E-07 \pm 5.4E-08$	$9.6E-07 \pm 3.3E-08$	$9.6E-07 \pm 2.9E-08$
$\mathscr{F}_{14}$	($6.9561E-07$)	($3.4956E-07$)	($5.4599E-07$)	($6.0876E-07$)	($7.2706E-07$)
	0.016 ± 0.0098	0.0016 ± 0.0032	$8.4E-07 \pm 1.2E-07$	$8.8E-07 \pm 1E-07$	$9.1E-07 \pm 7.3E-08$
$\mathscr{F}_{15}$	($2.938E-07$)	($5.0189E-07$)	($5.7057E-07$)	($6.5462E-07$)	($7.71E-07$)
	0.2 ± 0.48	0.099 ± 0.4	0.066 ± 0.25	0.033 ± 0.18	0.13 ± 0.34
$\mathscr{F}_{16}$	($2.2295E-07$)	($6.508E-07$)	(10.758)	(23.693)	(45.175)
	0.01 ± 0.055	$9.4E-07 \pm 8.7E-08$	12 ± 0.27	24 ± 0.14	45 ± 0.18
$\mathscr{F}_{17}$	(-0.5)	(-1)	(-2)	(-3)	(-5)
	$-0.5 \pm 2.2E-07$	$-1 \pm 2E-07$	$-2 \pm 8.7E-08$	$-3 \pm 6.4E-08$	$-5 \pm 6E-08$
$\mathscr{F}_{18}$	($2.6329E-07$)	($4.0355E-07$)	($5.4575E-07$)	($6.7551E-07$)	($6.7427E-07$)
	$6.9E-07 \pm 2.1E-07$	$7.6E-07 \pm 1.7E-07$	$8.5E-07 \pm 1.2E-07$	$9E-07 \pm 8.2E-08$	$9.1E-07 \pm 8.3E-08$
$\mathscr{F}_{19}$	($3.8271E-08$)	($2.808E-07$)	($5.6814E-07$)	($6.6527E-07$)	($6.7601E-07$)
	$6.8E-07 \pm 2.3E-07$	$7.8E-07 \pm 1.7E-07$	$8.8E-07 \pm 9.2E-08$	$8.9E-07 \pm 9E-08$	$9.1E-07 \pm 7.6E-08$
$\mathscr{F}_{20}$	($1.9525E-07$)	($3.5566E-07$)	($5.5169E-07$)	($6.3609E-07$)	($7.6725E-07$)
	$6.6E-07 \pm 2.3E-07$	$7.8E-07 \pm 1.8E-07$	$8.3E-07 \pm 1.3E-07$	$8.8E-07 \pm 9.8E-08$	$9.1E-07 \pm 6.8E-08$

Table 3.21 Experiment 7. GCPT-EDA + mutations: number of function evaluations on the multimodal problems in dimensions 5, 10, 20, 30 and 50. In parenthesis the best value. Target error=1×10^{-6}

Alias / D	5	10	20	30	50
$\mathscr{F}_{13}$	(4744)	(12360)	(28827)	(46765)	(85985)
	$5.2E+03 \pm 1.6E+02$	$1.3E+04 \pm 1.9E+02$	$3E+04 \pm 3.4E+02$	$4.8E+04 \pm 4.3E+02$	$8.7E+04 \pm 6.6E+02$
$\mathscr{F}_{14}$	(10708)	(9600)	(22038)	(35936)	(65733)
	$3.9E+04 \pm 3E+04$	$1.4E+04 \pm 3.8E+03$	$2.3E+04 \pm 3.4E+02$	$3.7E+04 \pm 3.8E+02$	$6.7E+04 \pm 4.7E+02$
$\mathscr{F}_{15}$	(4744)	(15480)	(41568)	(77943)	($1.6367E+05$)
	$7.2E+03 \pm 1.6E+03$	$2.2E+04 \pm 3.4E+03$	$5.6E+04 \pm 7.8E+03$	$9.6E+04 \pm 1E+04$	$2E+05 \pm 2.2E+04$
$\mathscr{F}_{16}$	(5626)	(89520)	($3.0002E+05$)	($3.0012E+05$)	($3.0012E+05$)
	$1.7E+04 \pm 5.3E+04$	$1.4E+05 \pm 2.5E+04$	$3E+05 \pm 0$	$3E+05 \pm 0$	$3E+05 \pm 0$
$\mathscr{F}_{17}$	(1930)	(6000)	(14877)	(24869)	(47473)
	$2.4E+03 \pm 1.8E+02$	$6.3E+03 \pm 1.8E+02$	$1.5E+04 \pm 3.1E+02$	$2.6E+04 \pm 3.7E+02$	$4.9E+04 \pm 4.7E+02$
$\mathscr{F}_{18}$	(2266)	(5700)	(13296)	(21656)	(40003)
	$2.5E+03 \pm 1.1E+02$	$6E+03 \pm 1.8E+02$	$1.4E+04 \pm 2.8E+02$	$2.2E+04 \pm 3.5E+02$	$4.1E+04 \pm 4.2E+02$
$\mathscr{F}_{19}$	(2392)	(6240)	(15249)	(25464)	(48635)
	$2.7E+03 \pm 1.8E+02$	$6.6E+03 \pm 2.3E+02$	$1.6E+04 \pm 2.7E+02$	$2.6E+04 \pm 3.4E+02$	$4.9E+04 \pm 4.1E+02$
$\mathscr{F}_{20}$	(2182)	(5940)	(14877)	(24750)	(46975)
	$2.4E+03 \pm 1.2E+02$	$6.2E+03 \pm 1.8E+02$	$1.5E+04 \pm 2.8E+02$	$2.6E+04 \pm 3.3E+02$	$4.8E+04 \pm 5.2E+02$

Table 3.22 Set A: convex problems

Name	Alias	Definition
Sphere	$\mathscr{F}_1$	$\sum_{i=1}^{d} x_i^2$
Ellipsoid	$\mathscr{F}_2$	$\sum_{i=1}^{d} 10^{6\frac{i-1}{d-1}} x_i^2$
Cigar	$\mathscr{F}_3$	$x_1^2 + \sum_{i=2}^{d} 10^6 x_i^2$
Tablet	$\mathscr{F}_4$	$10^6 x_1^2 + \sum_{i=2}^{d} x_i^2$
Cigar Tablet	$\mathscr{F}_5$	$x_1^2 + \sum_{i=2}^{d-1} 10^4 x_i^2 + 10^8 x_d^2$
Two Axes	$\mathscr{F}_6$	$\sum_{i=1}^{[d/2]} 10^6 x_i^2 + \sum_{i=[d/2]}^{d} x_i^2$
Different Powers	$\mathscr{F}_7$	$\sum_{i=1}^{d} \lvert x_i \rvert^{2+10\frac{i-1}{d-i}}$
Parabolic Ridge	$\mathscr{F}_8$	$-x_1 + 100 \sum_{i=2}^{d} x_i^2$
Sharp Ridge	$\mathscr{F}_9$	$-x_1 + 100 \sqrt{\sum_{i=2}^{d} x_i^2}$
Schwefel 1.2	$\mathscr{F}_{10}$	$\sum_{i=1}^{d} \left(\sum_{j=1}^{i} x_j \right)^2$
Trid	$\mathscr{F}_{11}$	$\sum_{i=1}^{d} (x_i - 1)^2 - \sum_{i=2}^{d} x_i x_{i-1}$
Zakharov	$\mathscr{F}_{12}$	$\sum_{i=1}^{d} x_i^2 + \left(\sum_{i=1}^{d} 0.5 i x_i \right)^2 + \left(\sum_{i=1}^{d} 0.5 i x_i \right)^4$

Table 3.23 Set B: multimodal problems

Name	Alias	Definition
Ackley	$\mathscr{F}_{13}$	$-20 \exp \left(-0.2 \sqrt{\frac{1}{d} \sum_{i=1}^{d} x_i^2} \right)$
		$- \exp \left(\frac{1}{d} \sum_{i=1}^{d} \cos(2\pi x_i) \right) + 20 + e$
Griewangk	$\mathscr{F}_{14}$	$\sum_{i=1}^{d} \frac{x_i^2}{4000} - \prod_{i=1}^{d} \cos \left(\frac{x_i}{\sqrt{i}} \right) + 1$
Rastrigin	$\mathscr{F}_{15}$	$10d + \sum_{i=1}^{d} \left[x_i^2 - 10 \cos(2\pi x_i) \right]$
Rosenbrock	$\mathscr{F}_{16}$	$\sum_{i=1}^{d-1} \left[(1 - x_i)^2 + 100 \left(x_{i+1} - x_i^2 \right)^2 \right]$
Negative Cosine Mixture	$\mathscr{F}_{17}$	$\sum_{i=1}^{d} x_i^2 - 0.1 \sum_{i=1}^{d} \cos(5\pi x_i)$
Levy-Montalvo 1	$\mathscr{F}_{18}$	$\left(\frac{\pi}{d} \right) \left(10 \sin^2(\pi y_1) + (y_d - 1)^2 \right)$
with $y_i = 1 + \frac{1}{4}(x_i + 1)$		$+ \frac{\pi}{d} \sum_{i=1}^{d-1} (y_i - 1)^2 \left[1 + 10 \sin^2(\pi y_{i+1}) \right]$
Levy-Montalvo 2	$\mathscr{F}_{19}$	$0.1 \left(\sin^2(3\pi x_1) + (x_d - 1)^2 \left[1 + \sin^2(2\pi x_d) \right] \right)$
		$+ 0.1 \sum_{i=1}^{d-1} (x_i - 1)^2 \left[1 + \sin^2(3\pi x_{i+1}) \right]$
Levy 8	$\mathscr{F}_{20}$	$\sum_{i=1}^{d-1} (y_i - 1)^2 \left[1 + 10 \sin^2(\pi y_{i+1}) \right]$
with $y_i = 1 + \frac{1}{4}(x_i + 1)$		$\sin^2(\pi y_1) + (y_d - 1)^2$

Table 3.24 Search domain, global minimum and properties of test problems

Alias	Modes	Separability	Global Minimum	Domain
$\mathcal{F}_1$	Unimodal	Separable	$f(\mathbf{x}^*) = 0 : x_i = 0$	$x_i \in [-10,5]^d$
$\mathcal{F}_2$	Unimodal	Separable	$f(\mathbf{x}^*) = 0 : x_i = 0$	$x_i \in [-10,5]^d$
$\mathcal{F}_3$	Unimodal	Separable	$f(\mathbf{x}^*) = 0 : x_i = 0$	$x_i \in [-10,5]^d$
$\mathcal{F}_4$	Unimodal	Separable	$f(\mathbf{x}^*) = 0 : x_i = 0$	$x_i \in [-10,5]^d$
$\mathcal{F}_5$	Unimodal	Separable	$f(\mathbf{x}^*) = 0 : x_i = 0$	$x_i \in [-10,5]^d$
$\mathcal{F}_6$	Unimodal	Separable	$f(\mathbf{x}^*) = 0 : x_i = 0$	$x_i \in [-10,5]^d$
$\mathcal{F}_7$	Unimodal	Separable	$f(\mathbf{x}^*) = 0 : x_i = 0$	$x_i \in [-10,5]^d$
$\mathcal{F}_8$	Unimodal	Separable	$f(\mathbf{x}^*) = -5 : x_1 = 5, x_{i>1} = 0$	$x_i \in [-10,5]^d$
$\mathcal{F}_9$	Unimodal	Non-separable	$f(\mathbf{x}^*) = -5 : x_1 = 5, x_{i>1} = 0$	$x_i \in [-10,5]^d$
$\mathcal{F}_{10}$	Unimodal	Non-separable	$f(\mathbf{x}^*) = 0 : x_i = 0$	$x_i \in [-10,5]^d$
$\mathcal{F}_{11}$	Unimodal	Non-separable	$f(\mathbf{x}^*) = -\frac{d(d+4)(d-1)}{6} : x_i = i(d+1-i)$	$x_i \in [-d^2, d^2]^d$
$\mathcal{F}_{12}$	Unimodal	Non-separable	$f(\mathbf{x}^*) = 0 : x_i = 0$	$x_i \in [-10,5]^d$
$\mathcal{F}_{13}$	Multimodal	Non-Separable	$f(\mathbf{x}^*) = 0 : x_i = 0$	$x_i \in [-10,5]^d$
$\mathcal{F}_{14}$	Multimodal	Non-separable	$f(\mathbf{x}^*) = 0 : x_i = 0$	$x_i \in [-600,300]^d$
$\mathcal{F}_{15}$	Multimodal	Separable	$f(\mathbf{x}^*) = 0 : x_i = 0$	$x_i \in [-10,5]^d$
$\mathcal{F}_{16}$	Multimodal	Non-separable	$f(\mathbf{x}^*) = 0 : x_i = 1$	$x_i \in [-10,5]^d$
$\mathcal{F}_{17}$	Multimodal	Separable	$f(\mathbf{x}^*) = -0.1d : x_i = 0$	$x_i \in [-1,0.5]^d$
$\mathcal{F}_{18}$	Multimodal	Non-separable	$f(\mathbf{x}^*) = 0 : x_i = -1$	$x_i \in [-10,5]^d$
$\mathcal{F}_{19}$	Multimodal	Non-separable	$f(\mathbf{x}^*) = 0 : x_i = 1$	$x_i \in [-10,5]^d$
$\mathcal{F}_{20}$	Multimodal	Non-separable	$f(\mathbf{x}^*) = 0 : x_i = -1$	$x_i \in [-10,5]^d$

3.7   Conclusions

In this paper we described a new EDA based on Gaussian polytrees, then we
studied polytrees with copula functions, and poly-trees with copula functions plus
mutations. A polytree is a rich modeling structure that can be built with moderate computing costs. At the same time the Gaussian polytree is found to have a
good performance on the tested functions, mainly on the convex functions which
are solved without any assistance as other EDA algorithms which had convergence
problems [13] [3]. Comparing Tables 4, 10 and 16 which are the results on unimodal problems using the proposed approaches, the best results are obtained using
the poly-tree with gaussian copula + mutations. Also, the comparison of Tables 7,
13 and 19 for multimodal functions using the three approaches, the winner is again
the poly-tree with gaussian copula + mutations. For the multivariate problems, the
use of copula functions (Table 13) seems to slightly improve Table 7, however note
that the Rastrigin function ($\mathcal{F}_{15}$) could be solved up to the dimension 50 (which
is the largest value tested).Table 19 shows clearly that mutations mean a great advantage to the algorithm. In fact Tables 16 and 19 show almost perfect score of the

success rate for unimodal and multimodal functions solved with GCPT + mutations. The proposed sampling method favors diversity of the population since it is based on the covariance matrix of the parent nodes and the children nodes. Also the proposed selection strategy applies low selection pressure (recall the whole population is used to create the model) to the individuals because it is based on a $\mu + \lambda$ technique, therefore improving diversity and delaying convergence. In this comparison the three approaches had good results but note this is the first time the multimodal functions are solved with a graph based EDA and certainly mean a great improvement over the first results reported for some of these problems [15].

References

1. Acid, S., de Campos, L.M.: Approximations of Causal Networks by Polytrees: an Empirical Study. In: Bouchon-Meunier, B., Yager, R.R., Zadeh, L.A. (eds.) IPMU 1994. LNCS, vol. 945, pp. 149–158. Springer, Heidelberg (1995)
2. Bacigál, T., Komorníková, M.: Fitting archimedean copulas to bivariate geodetic data. In: Rizzi, A., Vichi, M. (eds.) Compstat 2006 Proceedings in Computational Statistics, pp. 649–656. Physica-Verlag HD, Heidelberg (2006)
3. Bosman, P.A.N., Grahl, J., Thierens, D.: Enhancing the performance of maximum-likelihood gaussian edas using anticipated mean shift. In: Proceedings of BNAIC 2008, the Twentieth Belgian-Dutch Artificial Intelligence Conference, pp. 285–286. BNVKI (2008)
4. Castillo, E., Gutiérrez, J.M., Hadi, A.S.: Expert Systems and Probabilistic Network Models. Springer (1997)
5. Cherubini, U., Luciano, E., Vecchiato, W.: Copula Methods in Finance. Wiley, Chichester (2004)
6. Chikering, D.M.: Learning bayesian networks is np-complete. In: Learning from Data: Artificial Intelligence and Statistics V, pp. 121–130. Springer (1996)
7. Chow, C.K., Liu, C.N.: Approximating discrete probability distributions with dependence trees. IEEE Transactions on Information Theory IT-14(3), 462–467 (1968)
8. Darwiche, A.: Modeling and Reasoning with Bayesian Networks. Cambridge University Press (2009)
9. Dasgupta, S.: Learning polytrees. In: Proceedings of the Fifteenth Conference Annual Conference on Uncertainty in Artificial Intelligence (UAI 1999), pp. 134–14. Morgan Kaufmann, San Francisco (1999)
10. De-Waal, D.J., Van-Gelder, P.H.A.J.M.: Modelling of extreme wave heights and periods through copulas. Extremes 8(4), 345–356 (2005)
11. Edwards, D.: Introduction to Graphical Modelling. Springer, Berlin (1995)
12. Genest, C., Favre, A.C.: Everything you always wanted to know about copula modeling but were afraid to ask. Journal of Hydrologic Engineering 12(4), 347–368 (2007)
13. Bosman, P.A.N., Grahl, J., Rothlauf, F.: The correlation-triggered adaptive variance scaling idea. In: GECCO 2006, Proceedings of the 8th Annual Conference on Genetic and Evolutionary Computation, pp. 397–404. ACM (2006)
14. Joe, H.: Multivariate models and dependence concepts. Chapman and Hall, London (1997)

15. Lozano, I.I.J.A., Larrañaga, P., Bengoetxea, E., Larrañaga, P., Bengoetxea, E.: Towards a New Evolutionary Computation: Advances on Estimation of Distributions Algorithms. STUDFUZZ. Springer (2006)
16. Lauritzen, S.L.: Graphical models. Clarendon Press (1996)
17. Oommen, B.J., Ouerd, M., Matwin, S.: A formal approach to using data distributions for building causal polytree structures. Information Sciences, an International Journal 168, 111–132 (2004)
18. Monjardin, P.E.: Análisis de dependencia en tiempo de falla. Master's thesis, Centro de Investigación en Matemáticas, Guanajuato, México (December 2007) (in Spanish)
19. Neapolitan, R.E.: Learning Bayesian Networks. Prentice Hall series in Artificial Intelligence (2004)
20. Nelsen, R.B.: An Introduction to Copulas, 2nd edn. Springer Series in Statistics. Springer (2006)
21. Ortiz, M.S.: Un estudio sobre los Algoritmos Evolutivos con Estimacion de Distribuciones basados en poliarboles y su costo de evaluacion. PhD thesis. Instituto de Cibernetica, Matematica y Fisica, La Habana, Cuba (2003)
22. Ouerd, M.: Learning in Belief Networks and its Application to Distributed Databases. PhD thesis. University of Ottawa, Ottawa, Ontario, Canada (2000)
23. Pearl, J.: Probabilistic Reasoning in Intelligent Systems: Networks of Plausible Inference. Morgan Kaufmann Publishers Inc., San Francisco (1988)
24. de Campos, L.M., Molina, J.M.R.: Using bayesian algorithms for learning causal networks in classification problems. In: Proceedings of the Fourth International Conference of Information Processing and Management of Uncertainty in Knowledge-Based Systems (IPMU), pp. 395–398 (1993)
25. Rebane, G., Pearl, J.: The recovery of causal poly-trees from statistical data. In: Proceedings, 3rd Workshop on Uncertainty in AI, Seattle, WA, pp. 222–228 (1987)
26. Schölzel, C., Friederichs, P.: Multivariate non-normally distributed random variables in climate research – introduction to the copula approach. Nonlinear Processes in Geophysics 15(5), 761–772 (2008)
27. Shachter, R., Kenley, C.: Gaussian influence diagrams. Management Science 35(5), 527–550 (1989)
28. Sklar, A.: Fonctions de répartition à n dimensions et leurs marges. Publications de l'Institut de Statistique de l'Université de Paris 8, 229–231 (1959)
29. Trivedi, P.K., Zimmer, D.M.: Copula Modeling: An Introduction for Practitioners. Foundations and Trends® in Econometrics, vol. 1. Now Publishers (2007)
30. Whittaker, J.: Graphical Models in Applied Multivariate Statistics. John Wiley & Sons (1990)

Appendix A
Test Function Definitions

Two sets of functions are used in the experiments: Set A of 12 convex functions are shown in Table 3.22, and Set B of 8 multimodal functions is shown in Table 3.23. Relevant information about the functions, such as the optimum vector, value of the function at that vector, and search domain is provided in Table 3.24.

Part II
Set Oriented Numerics

Chapter 4
On Quality Indicators for Black-Box Level Set Approximation

Michael T.M. Emmerich, André H. Deutz, and Johannes W. Kruisselbrink

Abstract. This chapter reviews indicators that can be used to compute the quality of approximations to level sets for black-box functions. Such problems occur, for instance, when finding sets of solutions to optimization problems or in solving non-linear equation systems. After defining and motivating level set problems from a decision theoretic perspective, we discuss quality indicators that could be used to measure how well a set of points approximates a level set. We review simple indicators based on distance, indicators from biodiversity, and propose novel indicators based on the concept of Hausdorff distance. We study properties of these indicators with respect to continuity, spread, and monotonicity and also discuss computational complexity. Moreover, we study the use of these indicators in a simple indicator-based evolutionary algorithm for level set approximation.

4.1   Introduction

In many problem settings one may ask for a diverse set of solutions that have a certain property fulfilled (not necessarily optimality) that can be verified by means of the evaluation of a black-box function. In this chapter we consider such problems. More precisely, we study the following class of problems:

Definition 4.1 (Black-box Level Set Problem). Given a black-box function $f : X \rightarrow \mathbb{R}$ and a target space $T \subseteq \mathbb{R}$, find all solutions in $L = \{x \in X | f(x) \in T\}$. Solutions in L will be termed *feasible solutions*.

For instance, T could be a singleton, in which case L is a level set. Other examples are superlevel sets (or sublevel sets), where T is the set of values above (below) a

Michael T.M. Emmerich · André H. Deutz · Johannes W. Kruisselbrink
LIACS, Niels Bohrweg 1, 2333-CA Leiden, NL
e-mail: {emmerich,deutz,jkruisse}@liacs.nl

E. Tantar et al. (Eds.): EVOLVE- A Bridge between Probability, SCI 447, pp. 157–185.
springerlink.com   © Springer-Verlag Berlin Heidelberg 2013

certain threshold. For simplicity, we refer to all these problems as *level set problems*. We restrict our attention to problems where for all possible inputs x the condition $f(x) \in T$ can be evaluated efficiently.

Examples for such level set problems are listed in the following:

1: Geometrical Modeling. Determine all points of an n-dimensional geometrical shape that is implicitly defined by an (in)equality.

2: System Identification. Find all inputs (possible causes, or parameters) x of a system (model) f such that the output (effect) is in T (measurement). These problems also arise in inverse design and fault diagnosis.

3: Constraint Based Design. Find all solutions of an inverse design optimization problem that score above a certain threshold τ (see also [18]); here the target set would be $T = [\tau, \infty)$.

As L might be infinite, instead of computing L we may compute a finite set A that represents L. The problem to find such a finite *approximation set A* will be termed *level set approximation problem*.

The assessment of (and search for) a finite approximation set should possibly be based on a preference relation on approximation sets. In many cases it is convenient to have a scalar quality indicator. By a quality indicator (QI) we mean a function that assigns to an approximation set a scalar value that can be interpreted as the fitness or quality of the approximation set.

Approximation sets for level sets should be as good as possible with respect to the following two properties (see below, **P1** and **P2**) and the QIs must distinguish the better approximations from the worse. That is, the QIs must be able to measure these properties or at least one of the two. In the latter case, ideally two indicators should be used one for each of the two properties.

P1: Representativeness. Distances of elements of the level set L to the approximation set A should be as small as possible.

P2: Feasibility. Symmetrically: Distances of elements of the approximation set A to the level set L should be as small as possible.

Quality indicators should be able to indicate when an approximation set A is better than an approximation set B with respect to **P1** and/or **P2**. In this paper we discuss different QIs that could be used for designing and studying approximation methods for level sets. In particular we are interested in unary QIs. That is, QIs that provide a quality value for each level set approximation. Moreover, the attention will be focused on QIs that do not require a priori knowledge of L, because they can be used for the design of indicator-based search heuristics (as in indicator-based multiobjective optimization) and for monitoring the progress of algorithms.

Example 4.1 (Starfish Level Set). Figure 4.1 shows, as a motivating example, a fifty point approximation set (the red dots) of an interesting geometric set; the fattened, Gielis starfish shape. The result is generated with the indicator-based evolutionary level set approximation algorithm (ELSA) using an ADI indicator with reference set $R = [-3,3]^2$, as it will be discussed later in this chapter. In order to define the starfish shape we introduce an auxiliary function $\alpha(x_1, x_2)$ which assigns to a point

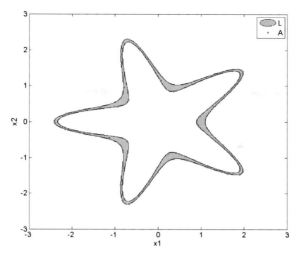

Fig. 4.1 A fifty point approximation of the fattened starfish level set

in the plane the angle between the positive direction of the X-axis and the halfline determined by $(0,0)^T$ and $(x_1,x_2)^T$. The fattened starfish level set L is defined as

$$L = \{(x_1,x_2) \in [-3,3]^2 \mid f(x_1,x_2) \leq 0.2\},$$

where

$$f(x_1,x_2) = g(x_1,x_2)^2 - x_1^2 - x_2^2$$

and

$$g(x_1,x_2) = \left(\left| \frac{1}{a} \cdot \cos\left(\frac{m \cdot \alpha(x_1,x_2)}{4} \right) \right|^{n_2} + \left| \frac{1}{b} \cdot \sin\left(\frac{m \cdot \alpha(x_1,x_2)}{4} \right) \right|^{n_3} \right)^{-\frac{1}{n_1}}.$$

This level set is obtained from the level set $\{(x_1,x_2) \mid f(x_1,x_2) \in T\}$, where $T = \{0\}$ by means of a 0.2-fattening (see Remark 4.1) and the superformula of Gielis [7] with instantiation $m = 5, n_1 = 2, n_2 = 7, n_3 = 7, a = 1, b = 1$ (which gives the level set a starfish shape).

This chapter starts with a discussion of related work in Section 4.2, followed by a study of design principles for quality indicators from a decision theoretic point of view in Section 4.3. In Section 4.4, different designs for QIs will be studied, including the discussion of computational aspects. Section 4.5, illustrates properties of the indicators in an empirical case study for approximating level sets. For this, a simple indicator-based evolutionary algorithm is introduced. Finally, the main findings and directions for future research are summarized in Section 4.6.

4.2   Related Work

For the numerical solution of level set approximation, continuation or homotopy methods are commonly used [1]. These methods start with a feasible solution on the optimal set and extend the set of feasible solutions by constructing new solutions in the neighborhood of given solutions using, for instance, predictor-corrector schemes. Continuation methods require the solution set to be connected and f to be differentiable.

Evolutionary methods and stochastic search techniques have hardly been applied for level set search, although they are quite popular in global optimization and for the approximation of Pareto fronts in multiobjective optimization. In fact, Pareto front approximation can be recasted as level set problem, given certain differentiability and convexity conditions [8, 11]. In the context of Pareto optimization promoting solution diversity in the search space can have an important role [15, 17].

Level set approximation problems and Pareto-front approximation problems have in common that in both cases finite approximation sets are searched for that should represent an a priori unknown solution set as good as possible. This is why it is interesting to look for analogous ways to solve these problems, in particular in settings that forbid the use of continuation methods (e.g., problems with disconnected level sets, discontinuities, etc.). The problems with non-connectedness of the Pareto front approximation by continuation methods have been addressed in [5, 12]. As opposed to hybridization in this chapter we mainly focus on indicator-based approaches. A first approach in the direction of evolutionary indicator-based level set approximation has been recently proposed with the NOAH algorithm [18]. NOAH has been designed for finding diverse sets of solutions that perform above a certain threshold. It is an indicator-based approach. That is, it selects populations based on a quality indicator that allows to compare different sets on the basis of diversity. NOAH uses the Solow Polasky diversity indicator [16], which will, among other indicators, be reviewed later in this chapter in the context of level set approximation.

Although closely related to diversity-oriented search as defined by Ulrich [18], in this chapter we argue that aiming for a maximally diverse set is not exactly the same as aiming for a good representation of a level set, which to a certain extent are complementary goals.

4.3   Decision Theoretic Motivation of Quality Indicators

This section considers the design of quality indicators for finite set approximations A to level sets L from the perspective of multiobjective decision theory. The guiding question in this discussion will be how the desirable properties P1 (representativeness) and P2 (feasibility) can be translated into preference relations, and eventually into scalar quality indicators that comply with these preference relations. For the sake of technical simplicity our discussion will be restricted to the following *regular level set approximation problems*:

Definition 4.2 (Regularity). A regular level set problem is a level set problem with the following properties: $X \subseteq \mathbb{R}^n$, $n > 0$. Thus the set X is equipped with a distance function $d : X \times X \to \mathbb{R}_0^+$ (inherited from $\mathbb{R}^n$), where $\mathbb{R}_0^+$ denotes the non-negative reals. Furthermore:

1. $L = \{x \in X \mid f(x) \in T\}$ is compact,
2. $\dim(X) = \dim(L)$,
3. for all $x \in X$, the local dimension of x is equal to $\dim(X)$,
4. for all $l \in L$, the local dimension of l is equal to $\dim(L)$.

Remark 4.1. In many cases, especially when T is a singleton (classical level set problems), L will not satisfy the condition $\dim(L) = \dim(X)$. In this case we often can take the ε-fattening of T (where the ε-fattening of a set $S \subset \mathbb{R}^n$ with $\varepsilon > 0$ is the set $\bigcup_{s \in S}\{x \in \mathbb{R}^n \mid d(x,s) \le \varepsilon\}$). Ideally we would like to take the ε-fattening of L, but in our approach we do not assume explicit *a priori* knowledge of L.

In the following we often refer to the notion of the distance of a point to a set:

Definition 4.3 (Distance of a Point to a Set). Let $S \subseteq \mathbb{R}^n$ and let $p \in \mathbb{R}^n$. The distance of the point p to the subset S is defined as $d(x,S) = \inf\{d(x,y) \mid y \in S\}$.

Note that most of our definitions are applicable in a more general context such as metric spaces or topological spaces, but for the sake of clarity and brevity we restrict our discussion to subsets of $\mathbb{R}^n$.

From a decision theoretic point of view, we may translate P1 and P2 into multi-objective optimization problems and state a preference relation which establishes a set-preference relation with respect of P1, and, respectively, P2. Set indicators can be discussed in the context of this preference relation. We will first introduce and discuss a preference relation for P1, then do the same for P2, and proceed with a discussion on the combination of these.

4.3.1   Pareto Order for Representativeness

In a level set problem we may view the representation of each $\ell \in L$ as an objective. That is, for each $\ell \in L$, the distance to the nearest neighbor in A should be minimized. This gives rise to a multiobjective optimization problem with an infinite number of objectives, on which we could establish an Edgeworth-Pareto order (or simply Pareto order) as follows:

Definition 4.4 (Pareto Order for Representativeness $\prec_R$). Given a level set problem with solution L, then an approximation set A Pareto dominates A' with respect to representativeness (in symbols $A \prec_R A'$), if and only if:

$$\forall \ell \in L : d(\ell, A) \le d(\ell, A') \text{ and}$$

$$\exists \ell \in L : d(\ell, A) < d(\ell, A').$$

We write $A \parallel_R A'$, iff $A \ne A'$ and $\neg(A \prec_R A')$ and $\neg(A' \prec_R A)$.

Note, that we do not demand solutions in A to be feasible. Two simple propositions on the feasible parts of approximation sets can be stated:

Proposition 4.1. *Let A, B, and L be subsets of $\mathbb{R}^n$ (or of a metric space) such that $A \subset B \subseteq L$. Furthermore let A be closed. Then, $B \prec_R A$.*

Proof. It is clear that $d(l, B) \leq d(l, A)$ for each l in L. Next we want to show that $\exists l_0 \in L$ such that $d(l_0, B) < d(l_0, A)$. Since $A \subset B$, it holds that $\exists b_0 \in B \setminus A$. For this b_0 it holds that $d(b_0, B) < d(b_0, A)$ (as $d(b_0, B) = 0$ and because A is a closed subset of L it holds $0 < d(b_0, A)$). $\square$

Proposition 4.2. *Let $A \subseteq L, B \subseteq L$, $A \setminus B \neq \emptyset$, $B \setminus A \neq \emptyset$, A and B are closed in L, and $L \subseteq \mathbb{R}^n$ (or L is a metric space). Then, $A \|_R B$.*

Proof. Since $A \setminus B \neq \emptyset$ we have $\exists a_0 \in A \setminus B$. For this a_0 it holds that $d(a_0, A) = 0$ and because B is a closed set we have $0 < d(a_0, B)$ (in other words, $\exists l_0 \in L$ such that $d(l_0, A) < d(l_0, B)$). By symmetry we can also find an $\tilde{l}_0 \in L$ such that $d(\tilde{l}_0, A) > d(\tilde{l}_0, B)$. Thus $A \|_R B$. $\square$

Remark 4.2. Proposition 4.1 basically tells us that if we restrict ourselves solely to feasible solution sets A and B, the set that includes the other should be preferred. This property was stated in the literature as *monotonicity in species* (cf. [19] and [18]).

Remark 4.3. Recall that a finite subset of a metric space is necessarily closed. Thus, Proposition 4.1 holds in case the set A is finite. Secondly, it is easy to construct counterexamples to the statement of Proposition 4.1 when the closedness condition for A does not hold: Let $A = \{1/n \,|\, n \in \mathbb{N}\}, B = A \cup \{0\}, L = [0, 1]$, then $B \preceq_R A$, but $B \prec_R A$ does not hold (where $B \preceq_R A$ means $\forall l \in L, d(l, B) \leq d(l, A)$).

Remark 4.4. Proposition 4.2 tells us that it is probably insufficient from a practical point of view to consider only $\prec_R$ as a preference relation, as it will be too coarse to establish a ranking between solutions, particularly in cases where sets of equal size need to be compared and all of these points belong to L.

Remark 4.5. The conditions of Proposition 4.2 are satisfied in case A and B are non-empty finite sets, $A \nsubseteq B$, $B \nsubseteq A$, and $L \subseteq \mathbb{R}^n$ (or L metric). If you drop either condition: A closed, B closed, then the Proposition 4.2 does not hold. For instance, take $C = \{1/n \,|\, n \in \mathbb{N}\} \cup \{1 - 1/n \,|\, n \in \mathbb{N}\}, L = [0, 1], A = C \cup \{0\}$, and $B = C \cup \{1\}$, then we have $\forall l \in L, d(l, A) = d(l, B)$.

4.3.2   Lorenz Order for Representativeness

As said, the relation $\prec_R$ provides us little guidance on how to distribute sets $A \subset L$. For instance, a strongly clustered set $A \subset L$ is incomparable to an evenly spread set $A' \subset L$ even if both sets are finite and of the same size.

Let us thus propose an extension to the order $\prec_R$. Instead of stating the problem as a Pareto optimization problem with each $\ell \in L$ giving rise to an individualized objective, $d(\ell, A) \to \min$, we consider these objectives as interchangeable. For instance, we consider an approximation set as equally good compared to a set for which some of the points the distances are incremented while compensated by a decrease by the same amount for some other points. In other words, we can take a set A and obtain a set B by moving one of the points of A further away from L and move another point by the same amount closer to L; in this case A and B are considered equally good (while in the Pareto order $A \parallel B$). However, if the increase is not compensated for, the new solution will be defined as being inferior to the original one (while in the Pareto order we may still $A \parallel B$). To put things into more concrete terms, let us introduce the following dominance relation:

Definition 4.5 (Lorenz Order on Representativeness $\prec_C$). Let $L \subseteq X$ be the level set we search for and let L be measurable[1] and of non-zero measure. For $A \subseteq X$, let

$$C(A,x) = \frac{1}{\lambda(L)} \lambda \{\ell \in L \mid d(\ell, A) \leq x\},$$

where λ denotes the Lebesgue measure and $x \in \mathbb{R}$. Then, we say A Lorenz dominates A' with respect to the distance distribution (in symbols $A \preceq_C A'$), if and only if

$$\forall x \in \mathbb{R} : C(A,x) \geq C(A',x).$$

Moreover, we define strict Lorenz dominance: $A \prec_C A'$, if and only if $A \preceq_C A'$ and $C(A, \_) \neq C(A', \_)$.

Remark 4.6. The function $C(A, \_)$ can be interpreted as a cumulative probability distribution function indexed by A. In this interpretation, $C(A,x)$ measures the probability that a randomly chosen point in L has a distance to its nearest neighbor in A that is smaller or equal than x.

Remark 4.7. The term Lorenz dominance refers to a concept introduced by Atkinson [2], who used it in economics to compare wealth distributions or Lorenz curves, named after W.O. Lorenz.

Example 4.2. In Figure 4.2, an example for a comparison with $\prec_C$ is displayed. The level set is $L = [0,3]$. Three approximations to this level set $A_b = \{1,2\}, A_r = \{\frac{1}{4}, \frac{3}{4}\}$, and $A_g = \{0,3\}$ are compared with each other. The picture on the left hand side displays the curves $C(A_b, \_), C(A_g, \_)$ and $C(A_r, \_)$. The curve dominance reveals that $A_b \prec_C A_r, A_b \prec_C A_g$. As $C(A_g, \_)$ and $C(A_r, \_)$ intersect, A_g and A_r are incomparable with respect to $\prec_C$.

Remark 4.8. An interesting aspect of this example is that the most diverse set, which is $A_g = \{0,3\}$, is not the set that best represents L according to $\prec_C$. This shows a subtle difference between the concept of representativeness and diversity. This contrast will be elaborated in more detail later in this section.

---

[1] In this context the Lebesgue measure on $\mathbb{R}^n$ would be the natural choice.

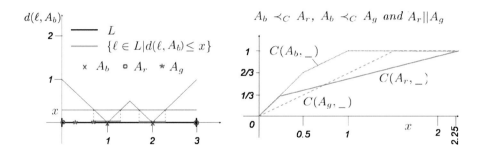

Fig. 4.2 Example for the order on cumulative distance distributions $\prec_C$

Proposition 4.3 (Compatibility $\prec_C$ with $\prec_R$). *Given a measurable level set L with dimension $\dim(L) > 0$ such the local dimension in each $l \in L$ is equal to $\dim(L)$, then for all subsets A, A' of L it holds that: $A \prec_R A' \Rightarrow A \prec_C A'$, but in general not the converse.*

Proof. $A \prec_R A' \Rightarrow A \preceq_C A'$ holds, since by definition of $\prec_R$ for all $\ell \in L$, $d(\ell, A) \leq d(\ell, A')$ we get that the Lebesgue measure $\lambda \{\ell \in L \,|\, d(\ell, A) \leq x\}$ will be *bigger or equal* for any given x compared to $\lambda \{\ell \in L \,|\, d(\ell, A') \leq x\}$.

For the strictness proof we can argue as follows. The fact that $A \prec_R A'$ implies that $\exists l_0 \in L$ such that $d(l_0, A) < d(l_0, A')$. Define $p := d(l_0, A') - d(l_0, A)$. Let $x := \frac{1}{2}p + d(l_0, A)$. Since $d(.,A)$ and $d(.,A')$ are continuous, we can find a $\delta_1 > 0$ such that $\forall l \in B_{\delta_1}(l_0)$, $d(l, A) < x$ and a $\delta_2 > 0$ such that $\forall l \in B_{\delta_2}(l_0)$, $d(l, A') > x$. Taking $\delta = \min\{\delta_1, \delta_2\}$ we get $\forall l \in B_\delta$, $d(l, A) < x < d(l, A')$. By our assumption that the local dimension is equal to the global dimension, we get $\lambda \{\ell \in L \,|\, d(\ell, A) \leq x\}$ is strictly bigger than $\lambda \{\ell \in L \,|\, d(\ell, A') \leq x\}$.

As a simple counterexample we refer to A_b and A_r in Example 4.2. In this case A_b is incomparable to A_r with respect to $\prec_R$ and $A_b \prec_C A_r$. $\square$

In the context of orders this means that $\prec_C$ is an extension of $\prec_R$.

4.3.3   Unary Indicators for Representativeness

To extend $\prec_R$ and $\prec_C$ to a total order we could introduce unary indicators for measuring representativeness, for instance the average distance (ADI):

$$\mathrm{ADI}(L, A) := \int_{\ell \in L} d(\ell, A) d\ell,$$

or the maximum distance of L to A

$$\mathrm{MDI}(L, A) := \max_{\ell \in L}\{d(\ell, A)\}.$$

Let us discuss next the compliance of these aggregates with $\prec_R$.

Proposition 4.4 (Compatibility of ADI). *If L is of dimension $\dim(X)$ in each $\ell \in L$ then $A \prec_R A' \Rightarrow ADI(A,L) < ADI(A',L)$.*

Proof. As $d(\_,A)$ and $d(\_,A')$ are continuous[2], also $d(\_,A) - d(\_,A')$ is continuous. From $A \prec_R A$ it follows that $\exists \ell_0 : 0 < d(\ell_0,A') - d(\ell_0,A)$. All of the above implies $\exists \varepsilon > 0$ such that $\forall \ell \in B_\varepsilon(\ell_0) : 0 < p \leq d(\ell,A') - d(\ell,A)$ for some positive number p. Thus $\int_{\ell \in B_\varepsilon}(d(\ell,A') - d(\ell,A))d\ell \geq \theta$ for some $\theta > 0$. Moreover, because of the non-negativity of $d(\_,A') - d(\_,A)$ on $L \setminus B_\varepsilon(\ell_0)$, it holds that $\int_{\ell \in L \setminus B_\varepsilon}(d(\ell,A') - d(\ell,A))d\ell \geq 0$. Hence $ADI(A',L) - ADI(A,L) = \int_{\ell \in B_\varepsilon}(d(\ell,A') - d(\ell,A))d\ell \geq \theta > 0$. □

Remark 4.9. If the dimension of points in L is less than the dimension of L the inclusion of these points will not have an effect on ADI. For an illustration of the problem see Figure 4.3 (left). Thus, $A \prec_R A' \Rightarrow ADI(L,A) < ADI(L,A')$ does not hold for general L.

Also MDI is a problematic aggregate with respect to $\prec_R$, as the following proposition shows.

Proposition 4.5 (Non-compatibility of MDI with $\prec_R$). *In general it is not the case that $A \prec_R A' \Rightarrow MDI(L,A) < MDI(L,A')$.*

Proof. It is easy to find a counterexample. We may, for instance, add feasible points on the right hand side of the unit disc in example in Figure 4.3 (right) which will lead to a dominating solution with respect to $\prec_R$, but not improve the maximal distance indicator (MDI). □

Proposition 4.6. $A \prec_R A' \Rightarrow (MDI(L,A) \leq MDI(L,A'))$

Proof. By $A \prec_R A'$ we know that $\forall \ell \in L, d(l,A) \leq d(l,A')$. Thus be definition of $MDI(L,A')$ we get $\forall \ell \in L, d(l,A) \leq d(l,A') \leq MDI(L,A')$ and therefore $MDI(L,A) \leq MDI(L,A')$. □

These two propositions show that even if A is strictly better than A' the indicator MDI might indicate equal quality of A and A', though it will not indicate that A' is better than A.

Next we summarize the compatibility of ADI and MDI with $\prec_C$ with the following propositions.

Proposition 4.7. *For level sets that are of dimension $\dim(L)$ it holds that: $A \prec_C A' \Rightarrow ADI(L,A) < ADI(L,A')$.*

Proof. Let us first state a lemma from probability theory:

---

[2] A proof will be given later, see Lemma 4.3.

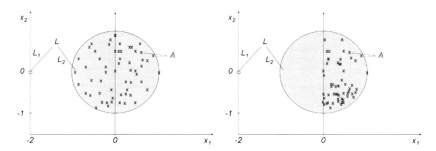

Fig. 4.3 Measuring representativeness: When the average distance (ADI) of a point in the level set $L = L_1 \cup L_2$ is taken, the small isolated region L_1 (because of its small area) contributes little to the average and thus might be neglected. When using the maximal distance indicator (MDI) in the example on the right hand side, as long as we do not cover L_1 it does not improve the maximal distance when moving some points from the right hand side to the left hand side of L_2.

Lemma 4.1. *For a probability distribution $F(x)$ of a random variable ξ it is the case that*

$$E(\xi) = \int_{x=0}^{\infty} (1 - F(x))dx - \int_{-\infty}^{0} F(x)dx.$$

Proof. For a derivation see, e.g., [10]. $\qquad\qquad\square$

As said before we can interpret $C(A,x)$ and $C(A',x)$ as a probability distribution of the distance distribution, say ξ_d, over L. Moreover $\mathrm{ADI}(L,A)$ is the mean value of this distribution and due to Lemma 4.1 it can be written as

$$\mathrm{ADI}(L,A) = \int_0^{\infty} (1 - C(A,x))dx \cdot \lambda(L).$$

If $A \prec_C A'$, that is, $C(A,x)$ dominates $C(A',x)$, then $\mathrm{ADI}(L,A) < \mathrm{ADI}(L,A')$ follows immediately. $\qquad\qquad\square$

Proposition 4.8 (Relationship between $C(A, \_)$ and MDI). *For level sets that are of dimension $\dim(L)$ in every point $\ell \in L$, it holds that:*

$$MDI(L,A) = \inf\{x \in \mathbb{R} \,|\, C(A,x) = 1\}.$$

Proof. Let $V = \{x \in \mathbb{R} \,|\, C(A,x) = 1\}$. We will show that $\inf(V) \leq MDI(L,A)$ and $MDI(L,A) \leq \inf(V)$. First we show that $\inf(V) \leq MDI(L,A)$: From the continuity of $d(\_,A)$ and the compactness of L we know that $d(\_,A)$ attains a maximum. More precisely $\exists l_{\max} \in L$ such that $\forall \ell \in L, d(\ell,A) \leq d(l_{\max},A)$. Clearly $C(A,d(l_{\max},A)) = 1$ (and $MDI(L,A) = d(l_{\max},A)$). Thus, $d(l_{\max},A) \in V$ and therefore $\inf(V) \leq d(l_{\max},A)$. Next we show that $MDI(L,A) \leq \inf(V)$. In other words, we will show that $d(l_{\max},A) \leq \inf(V)$. We will assume to the contrary that $\inf(V) < d(l_{\max},A)$ and derive a contradiction. The assumption $\inf(V) < d(l_{\max},A)$ entails the

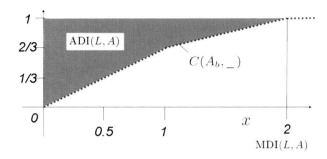

Fig. 4.4 Computation of ADI(L,A) and MDI(L,A) based on the cumulated distances curve $C(A,\_)$

existence of $x_0 \in V$ such that $\inf(V) < x_0 < d(l_{\max},A)$. Since $x_0 \in V$, we can conclude that $\{\ell \in L \mid d(l,A) = x_0\}$ is measurable and of measure $\lambda(L) > 0$. Therefore, it is non-empty. This entails that $\exists l_0 \in L$ such that $d(l_0,A) = x_0$. Since $d(\_,A)$ is continuous, there exists an ε-ball with center $l_{\max}$ ($B_\varepsilon(l_{\max})$) such that $\forall \ell \in B_\varepsilon(l_{\max})$: $d(l_0,A) < \frac{d(l_0,A)+d(l_{\max},A)}{2} < d(l,A)$. This entails that L contains two measurable, disjoint subsets $S = \{l \in L \mid d(l,A) = x_0\}$ and $B = B_\varepsilon(l_{\max})$. The first has measure $\lambda(L)$ and the second has a strict positive measure. Therefore, $\lambda(L) < \lambda(S) + \lambda(B)$. On the other hand, $\lambda(L) \geq \lambda(S) + \lambda(B)$, because λ is a measure. This is a contradiction. In other words, MDI(L,A) $\leq \inf(V)$.   □

Example 4.3. Figure 4.4 shows an example for the computation of the ADI and MDI indicators based on a known cumulative distance distribution curve. The level set is $L = [0,3]$ and the approximation set $A_b = \{1,2\}$. The ADI is given by the area of the shaded area (ADI(L,A_b) = 5/6) and the MDI by the point where the curve reaches 1, that is MDI(L,A_b) = 2.

Proposition 4.9. *In general it does* not *hold that* $A \prec_C A'$ *implies* $MDI(L,A) < MDI(L,A')$, *but for level sets with dimension* $\dim(L)$ *in every point* $\ell \in L$ *it holds that*

$$A \prec_C A' \Rightarrow \neg(MDI(A') < MDI(A)).$$

In summary, neither MDI nor ADI is fully compatible with $\prec_R$. While MDI is also capable to capture improvements in zero measure points it often will not indicate improvements. On the contrary, ADI is not sensitive to adding points in regions of L with zero measure.

4.3.4   A Preference Order for Feasibility

Next let us consider also the feasibility property or P2. For the sake of simplicity the discussion will be restricted to approximation sets of equal size.

Definition 4.6 (Lorenz Order for Feasibility $\prec_F$). Let $|A| = |A'| = n$ and $n > 0$. Then, let $d_{1:n}, i = 1, \ldots, n$ denote the distances $d(a, L), a \in A$ sorted in ascending order. Likewise, let $d'_{1:n}, i = 1, \ldots, n$ denote the distances $d(a, L), a \in A'$ sorted in ascending order. Then, $A \prec_F A'$, if and only if

$$\forall i \in \{1, \ldots, n\} : d_{i:n} \leq d'_{i:n} \text{ and}$$

$$\exists j \in \{1, \ldots, n\} : d_{j:n} < d'_{j:n}.$$

This order rewards if some infeasible points get closer to the feasible set and all other points remain unchanged. Like for the representativeness goal, we may also construct aggregate objective functions for the feasibility goal and then investigate to what extent they comply with $\prec_F$ or are refinements of it.

Definition 4.7 (Feasibility Indicators). For the comparison of finite sets of equal size, the average feasibility indicator (AFI) and the minimal feasibility indicator (MFI) are defined as follows:

$$\text{AFI}(L, A) = \frac{1}{|A|} \sum_{a \in A} d(a, L),$$

and

$$\text{MFI}(L, A) = \max_{a \in A} \{d(a, L)\}.$$

We observe some simple properties of these in relation to $\prec_F$:

Proposition 4.10 (Properties of AFI and MFI in Relation to $\prec_F$). *Let A and A' denote two approximation sets with $|A| = |A'|$, then $A \prec_F A' \Rightarrow AFI(L, A) < AFI(L, A')$, but $\neg(A \prec_F A' \Rightarrow MFI(L, A) < MFI(L, A'))$, and yet $(A \prec_F A' \Rightarrow MFI(L, A) \leq MFI(L, A'))$. Moreover, $AFI(L, A) < AFI(L, A') \Rightarrow \neg(A' \prec_F A)$, and $MFI(L, A) < MFI(L, A') \Rightarrow \neg(A' \prec_F A)$.*

Again only the average distance is compatible with the order, while the maximum distance is not. As opposed to the previously discussed ADI, the problem with isolated points does not occur, as we compute the average by a sum over a finite number of points and every point will have equal weight in this sum.

It could be debated whether $\prec_F$ is the only meaningful choice of an expedient ordering. For instance, it does not always give preference to sets that have more feasible solutions. However, as Proposition 4.10 shows, the set which has more feasible solutions is at most incomparable (if it has higher distance values in some components) but never strictly inferior.

4.3.5   Combining Representativeness and Feasibility

A question that remains is how to combine $\prec_F$ and $\prec_R$ and design indicators that comply with such combined orders. In this context, it is interesting to observe that, to a certain extent, solutions that are preferred with respect to $\prec_R$ are also preferable with respect to $\prec_F$.

In fact, if we restrict ourselves to sets of size μ, the non-dominated solutions of $\prec_F$ are sets that are completely contained in L and thus are also non-dominated solutions of $\prec_R$. The converse, however, does not hold in general. For instance if we consider the set of all singletons of $X = \{(x,y) \,|\, x^2 + y^2 \leq 1\}$ and $L = \{(x,y) \,|\, x^2 + y^2 = 1\}$ (unit circle), then the center of the circle $(0,0)$ is non-dominated with respect to $\prec_R$ but not with respect to $\prec_F$.

Moreover, we may observe conflicting pairs, i.e., sets A and A' with $A \prec_F A'$ and $A' \prec_R A$, or vice versa. For instance, let $L = \{(x,y) \in \mathbb{R}^2 \,|\, x^2 + y^2 \leq 1 \vee (x-4)^2 + y^2 \leq 1\}$ and consider $A = \{(2,0),(1,0)\}$ and A'$=\{(\frac{3}{2},0),(1,0)\}$. In this case $A \prec_R A'$ and $A' \prec_F A$.

As we would like both, feasibility and representativeness, to be improved we can introduce an order that combines these two goals:

Definition 4.8 (Combined Order for Feasibility and Representativeness). Given two approximation sets A and A' with $|A| = |A'|$ and a level set L we define $A \prec_{RF} A'$, if and only if $A \prec_R A'$ and $A' \preceq_F A$ or $A \preceq_R A'$ and $A' \prec_F A$, and $A \prec_{CF} A'$, if and only if $A \prec_C A'$ and $A' \preceq_F A$ or $A \preceq_C A'$ and $A' \prec_F A$.

Remark 4.10. Among all approximation sets of size μ, the set of non-dominated sets with respect to $\prec_{RF}$ is given by $M_\mu = \{A \subseteq L \,|\, |A| = \mu\}$. If, instead, the refined order $\prec_{CF}$ is used, then this is not the case, because sets in M_μ can become comparable.

Again we would like to find indicators that are compatible with $\prec_{CF}$ or at least with $\prec_{RF}$. Let us first study indicators based on averaging or taking the maximum by defining

$$\text{MHD}(L,A) = \max\{\text{MDI},\text{MFI}\} \tag{4.1}$$

and

$$\text{AHD}(L,A) = \frac{1}{2}(\text{ADI} + \text{AFI}). \tag{4.2}$$

Remark 4.11. The definition of $\text{MHD}(A,L)$ corresponds with the standard Hausdorff distance between the sets A and L. Moreover, $\text{AHD}(A,L)$ corresponds to the averaged Hausdorff distance for $p = 1$ as defined by Schütze et al. [13, 14] in the context of Pareto front approximation.

The compatibility of MHD and AHD with $\prec_{RF}$ and $\prec_{CF}$ is summarized in the following two propositions.

Proposition 4.11. *For finite sets A and A' with $|A| = |A'|$ it holds that $A \prec_{RF} A' \Rightarrow \neg(MHD(A',L) < MHD(A,L))$, but $\neg(A \prec_{RF} A' \Rightarrow MHD(A,L) < MHD(A',L))$.*

Proposition 4.12. *Assume all $\ell \in L$ have dimension $\dim(L)$. Then, for finite sets A and A' with $|A| = |A'|$, it holds that $A \prec_{CF} A' \Rightarrow \neg(MHD(A',L) < MHD(A,L))$, but $\neg(A \prec_{CF} A' \Rightarrow MHD(A,L) < MHD(A',L))$.*

Proposition 4.13. *Assume all $\ell \in L$ have dimension* $\dim(L)$. *Then,* $A \prec_{RF} A' \Rightarrow$ $AHD(A,L) < AHD(A',L)$ *and* $A \prec_{CF} A' \Rightarrow AHD(A,L) < AHD(A',L)$.

Proof. The proposition follows directly from the earlier propositions (Proposition 4.4, Proposition 4.7, and Proposition 4.10) on the compatibility of ADI and AFI with $\prec_{RF}$, $\prec_{CF}$ and, respectively $\prec_F$.   □

In conclusion, it turns out that the averaged Hausdorff distance is an indicator that complies well with the combined preference relations, while the Hausdorff distance, when used as a quality indicator is often not sensitive to improvements in terms of $\prec_{RF}$ or $\prec_{CF}$. Note that also in the context of Pareto front approximation the averaged Hausdorff distance was favored over the standard Hausdorff distance as it better complies with the Pareto dominance order.

4.3.6   *Diversity versus Representativeness*

Diversity-oriented search for solutions in level sets, as aimed for in [18], is not in the first place guided by the aim to represent L but rather by the dictum: *Find a set $A \subset L$ of bounded size that has maximal diversity.*

To a certain extent this goal is complementary to the representativeness goal. However, there are subtle differences that will be highlighted in this section. A natural way to define diversity would be to see it as an indicator of dissimilarity between points in A. In that respect, we may consider a set that has higher inter-point distances (or gaps between points) as being more diverse. For sets of the same size, this could be defined as the monotonicity in diversity property [19, 18] or by the following order:

Definition 4.9 (Preference Based on Diversity). Let $|A| = |A'| \geq 2$. We say that $A \prec_D A'$, if and only if there is a one-to-one function ϕ of A onto A' such that

$$\forall x \in A, \forall y \in A, x \neq y : d(x,y) \geq d(\phi(x),\phi(y)) \text{ and}$$

$$\exists x \in A, \exists y \in A, x \neq y : d(x,y) > d(\phi(x),\phi(y)).$$

Remark 4.12. Diversity and representativeness can be in conflict with each other as the following example of a conflicting pair shows: Consider the sets $A_g = \{0,3\}$ and $A_b = \{1,2\}$ for $L = [0,3]$ (as in Example 4.2), then $A_g \prec_D A_b$ and $A_b \prec_C A_g$.

Another example, would be to find a maximally diverse set on a unit ball. Diversity oriented search would tend to place points on the boundary in order to maximize the inter-point distances, while for representativeness one would like to place points for instance near the center to decrease distances of points in the ball to the approximation set.

It depends on the application domain whether diversity or representativeness should be aimed for. For instance, when searching for alternative designs in engineering, diversity is likely a better guideline. When the goal is to understand the geometry of an implicitly defined shape it would seem more natural to aim for a representative set. As diversity indicators are often more efficiently computed and to a certain extent also promote good representativeness, we will in the following discussion of quality indicators also consider indicators for diversity.

4.4   Selected Quality Indicators and Their Properties

Often the quality of a level set needs to be measured in absence of knowledge of the resulting level set, which forbids the direct application of indicators such as MHD and AHD. In particular, this is the case if the quality indicator is used for guiding a search algorithm.

In this section we will review different types of quality indicators that could be used in the context of level set approximation without a priori knowledge of the level set L. Simple and more advanced *diversity* indicators will be surveyed first. Then, an indicator that is directly motivated by representativeness and feasibility is proposed and studied.

4.4.1   Simple Spread Indicators

A simple way to maximize diversity is to maximize the minimal, or average gap of a point and the remaining points of a set. That is, we maximize one of the following sparsity indicators. We consider here, again, indicators for sets of feasible points and assume that L is a possibly non-countable set that needs to be approximated by a finite set A of feasible points.

- $IS_N(A) = \min_{x \in A} d(x, A \setminus \{x\})$ (Minimal gap),

- $IS_\Sigma(A) = \frac{1}{|A|} \sum_{x \in A} d(x, A \setminus \{x\})$ (Arithmetic mean gap),

- $IS_\Pi(A) = (\prod_{x \in A} d(x, A \setminus \{x\}))^{\frac{1}{|A|}}$ (Geometric mean gap) .

The IS_N indicator is computable in $\mathscr{O}(n \log n)$ time and updated in $\mathscr{O}(n)$ time [20]. All these indicators can be computed at most in $\mathscr{O}(n^2)$ and updated at most in time $\mathscr{O}(n)$. The minimal gap is known to be efficiently computable, but has the disadvantage that it only senses improvements in the critical pair(s). The geometric mean has been introduced next to the arithmetic mean, as it is expected to better promote solution sets with evenly sized gaps. Moreover, if multisets would be considered, then it would yield the worst indicator score (that is zero) for solution sets with duplicates. Despite their fast computation time, these indicators do not reward (or not well enough) adding points to a set. Thus they are mainly useful for comparing approximation sets of the same size.

4.4.2   Diversity Indicators

In this section we study indicators that reward the inclusion of additional points, while at the same time promote spreading points if sets are of the same size.

4.4.2.1   Weitzman Diversity

An attempt to formulate a diversity indicator on a more extensive set of desirable properties was made by Weitzman [19]. He introduced the following recursive definition for a diversity indicator of a set of points:

Definition 4.10 (Weitzman Diversity). Let A denote a set, for instance an approximation of a level set. Furthermore, let d denote a dissimilarity measure. Then the Weitzman diversity D_w is defined as

$$D_w(A) = \max_{x \in A}\{D_w(A \setminus \{x\}) + d(x, A \setminus \{x\})\}. \tag{4.3}$$

It is interesting because it has several properties that make it a nice indicator for diversity. Some of these are also interesting when approximating level sets. Others are more interesting for applications such as biodiversity assessment. The following list of properties was devised by Weitzman – we revert to writing D referring to a diversity indicator which is not necessarily the Weitzman diversity indicator. The Weitzman indicator is denoted by D_w.

Definition 4.11 (Monotonicity in Species Property). If x is added to collection A, then

$$D(A \cup \{x\}) \geq D(A) + d(x, A).$$

This property is essential also when designing algorithms that fill a set. It complies with property $P1$ stated in Section 11.1.

Definition 4.12 (Link Property). For all A, $|A| \geq 2$, there exists at least one species $x \in A$, called the *link species*, that satisfies $D(A) = D(A \setminus \{x\}) + d(x, A \setminus \{x\})$.

The link property is essential in taxonomy where genetic links are considered.

Definition 4.13 (Twin Property). Suppose that some species y outside of A is identical to some species x belonging to A, such that $d(x, y) = 0$, and $\forall z \in A, d(x, z) = d(y, z)$. Then, if y is added to A, there is no change in diversity: $D(A \cup y) = D(A)$.

If a metric space is considered, the twin property is redundant, because of the identity in zero axiom of a metric, that is, $\forall x \in A, y \in A : d(x, y) = 0 \Leftrightarrow x = y$. Recall that for dissimilarity functions only $\forall x \in A, y \in A : x = y \Rightarrow d(x, y) = 0$ is required.

Definition 4.14 (Continuity in Distances Property). Let A be an approximation set, then D is continuous at A if $\forall \varepsilon > 0, \exists \delta > 0$ such that for all approximation sets A' and all bijections ϕ from A onto A' with the following property: $\sum_{x \in A, y \in A} |d(x, y) - d(\phi(x), \phi(y))| < \delta$, it holds that $|D(A) - D(A')| < \varepsilon$.

This property is reasonable, because sets which have a similar distance structure should have also a similar diversity.

Definition 4.15 (Monotonicity in Distances Property). Let $|A| = |A'| \geq 2$. Let $\phi : A \rightarrow A'$ be a one-to-one function A onto A'. Suppose that

$$d(\phi(x), \phi(y)) \geq d(x, y), \forall x \in A, \forall y \in A, x \neq y,$$

then $D(A') \geq D(A)$.

According to this definition, if two sets have equal size, then the wider spread set should have a bigger diversity.

Definition 4.16 (Maximum Diversity that Can Be Added by a Species Property). If species y is added to collection A, then

$$D(A \cup y) \leq D(A) + \max_{x \in A}\{d(x, y)\}.$$

This property relates the distance quantitatively to the diversity.

The *computational effort* of a straightforward implementation of this indicator has a time complexity of $\mathcal{O}(|X|!)$. Weitzman proposed a faster algorithm with time complexity $\mathcal{O}(2^{|X|})$ and applied it to data with up to 35 points. For a more detailed discussion of Weitzman's diversity and its properties we refer to the original paper [19].

4.4.2.2   Solow and Polasky Indicator

Another interesting indicator is the Solow Polasky indicator (SPI) [16]. This indicator has most of the desirable properties.

Definition 4.17. Let us for a given approximation set $A = \{x_1, \ldots, x_\mu\}$ define the matrix M with $m_{i,j} = \exp(-\theta d_{x_i, x_j}), \forall i = 1, \ldots, n, \forall j = 1, \ldots, n$ for some positive parameter θ. Furthermore, in case M is non-singular, define $C = M^{-1}$ and denote the entries of C by c_{ij}, then the Solow Polasky indicator, denoted by D_{SP}, is defined as follows:

$$D_{SP}(A) = \sum_{i=1}^{n} \sum_{j=1}^{n} c_{i,j}.$$

The SPI results in a value between 1 and μ. This makes it interesting for measuring bio-diversity, as its value can be interpreted as the number of species. The extremal value of 1 is the limit with respect to the elements of the approximation set A tending to the same element. The other extreme is be obtained if pairwise distances between solutions tend to infinity. Of course, these properties need to be formulated as limit properties, but for the sake of simplicity we omit formal definitions here. The computational complexity of computing the SPI is $\mathcal{O}(n^3)$. This makes it more attractive than the Weitzman's diversity indicator as an indicator for large sets. Incremental updates of this indicator have a cheaper cost (see [18] for details). Due to its favorable properties it has been proposed for diversity-oriented search in [18, 17].

4.4.3   Indicators Based on Distances between Sets

Let us recall the Hausdorff distance ([6], page 293) as a classical indicator for computing the distance between two sets:

Definition 4.18 (Hausdorff Distance). The Hausdorff distance between a set A and a set B is defined as

$$d_H(A,B) = \max\{\sup_{a \in A}\{d(a,B)\}, \sup_{b \in B}\{d(b,A)\}\}.$$

In an earlier work by Pompeiu [9], instead of the maximum the sum of the two components was used. We may replace the supremum operator by an averaging operator as in Schütze et al. [13], yielding the:

Definition 4.19 (Averaged Hausdorff Distance). The average Hausdorff distance between a set A and a set B is defined as

$$d_{\text{avg}}(A,B) = \text{avg}\{\text{avg}_{a \in A}\{d(a,B)\}, \text{avg}_{b \in B}\{d(b,A)\}\}.$$

As opposed to d_H, the triangle inequality does not hold for d_{avg}, see [14].

Finding an approximation set A^* that minimizes the Hausdorff or average distance to L is an interesting design principle for a quality indicator, and in Section 4.3.5 it was seen that the average Hausdorff Distance has many desirable properties when it comes to the representation of level sets. Recall, that the MHD and AHD as defined in Section 4.3.5 correspond to these definitions in the context of level set approximation.

4.4.3.1   Average Distance to a Reference Set

Although we can not use knowledge about L in the more general setting of black-box level set approximation, we may know a compact set R that contains L, which, for instance, could be the search space X itself or a subset of it. For this case we propose the *average distance to a reference set* (ADI$_R$), defined as follows:

Definition 4.20 (Average Distance Indicator ADI$_R$).
Define $\text{ADI}_R(A) = 1/\text{vol}(R) \int_{x \in R} d(x,A) dx$. The normalization can be omitted for constant R.

For an illustration of ADI$_R$, see Figure 4.6. In the context of the averaged Hausdorff distance, the strategy in the absence of a complete knowledge of L is to minimize the averaged Hausdorff distance to a reference set by restricting ourselves to solutions contained in L only. As L is contained by R this will also yield a good representation of L. Here we exploit the fact that we can decide whether or not a point is in L using the function f.

In order to study the properties of the minima of the ADI, we state first some useful lemmas.

Lemma 4.2. *Let S be a metric space with distance function $d(\_,\_)$ and let $A \subseteq S$ be such that $A \neq \emptyset$. Then, $\forall s_0, s \in S, |d(s_0,A) - d(s,A)| \leq d(s_0,s)$.*

Proof. Clearly, $d(s,A)$ is defined for any $s \in S$ and any $A \subseteq S$. We want to show two inequalities: $\forall s, s_0 \in S : d(s_0,A) - d(s_0,s) \leq d(s,A)$ and $d(s,A) \leq d(s_0,A) + d(s_0,s)$. The first inequality can be rewritten as $d(s_0,A) \leq d(s_0,s) + d(s,A)$. We now proceed to exhibit the rewritten first inequality: $\forall a \in A : d(s_0,a) \leq d(s_0,s) + d(s,a)$ (the triangle inequality). From this we get $\forall a \in A : \inf_{a \in A}\{d(s_0,a)\} \leq d(s_0,a) \leq d(s_0,s) + d(s,a)$ and subsequently: $\inf_{a \in A}\{d(s_0,a)\} \leq d(s_0,s) + \inf_{a \in A}\{d(s,a)\}$. The second inequality we get by interchanging the roles of s_0 and s. $\qquad\square$

We now can easily show that the function $d(\_,A) : S \to \mathbb{R}$ is a continuous function (S is a metric space and $A \subseteq S$).

Lemma 4.3 (Continuity of $d(\_,A)$). *Let S be a metric space and A a non-empty subset of S. Then, $d(\_,A) : S \to \mathbb{R}$ is a continuous function.*

Proof. Let $s_0 \in S$ be arbitrary. We will show that $d(\_,A)$ is continuous in s_0. Let $\varepsilon > 0$. We choose $\delta = \varepsilon$. Clearly, $d(B_\delta(s_0),A) \subset B_\varepsilon(d(s_0,A))$ for by the previous lemma, if $s \in S$ and $d(s_0,s) < \delta$ (i.e., $s \in B_\delta(s_0)$), then $|d(s_0,A) - d(s,A)| \leq d(s_0,s) < \delta$. In other words, $|d(s_0,A) - d(s,A)| < \varepsilon$ and the lemma obtains. $\qquad\square$

We proceed to study subsets of L for which the ADI_R is minimal. That is, we try to see what the relationship is of such minima to the set L. We first impose rather stringent conditions on the sets R and L and in turn on the set of minima M and the reference set R. We assume that R is a metric space (possibly having the first kind of countability: any open cover has a countable subcover):

1. R is a metric space of first countability or R is compact.
2. A is compact.
3. $L \subseteq R$ and L is closed in R (in case R is a compact metric space this is equivalent to saying that L is compact).
4. Let $cl_L(A)$ denote the closure of A in L. As $d(x,A) = d(x, cl_L(A))$, we may assume for the computation of the ADI_R that A is closed in L.

Likewise, we can state the following result:

Lemma 4.4. *Assume L is a set of dimension n or lower, and R denotes a set which has dimension n in each point. Moreover, assume L is contained in R. Then,*

$$M \in \mathcal{M} = \arg\min_{A \subseteq L} \int_{x \in R} d(x,A)dx \Leftrightarrow cl_R(M) = cl_R(L)(=L),$$

where $cl_R(S)$ denotes the closure of a set S in R.

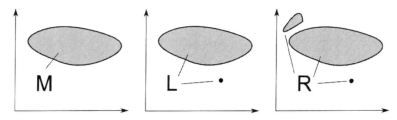

Fig. 4.5 Sets M, L and R. M is a minimizer of ADI_R, but its Hausdorff distance to L is not zero

Proof. The direction '$\Leftarrow$' rests on the facts that (1) L belongs to $\mathcal{M}$ and (2) the distance of a point to a set is equal to the distance of the same point to the closure of the set, in symbols: $A \subset R$, $d(x,A) = d(x,\text{cl}_R(A))$. We will concentrate next on the direction '$\Rightarrow$' and proof by contradiction: Given a set $M \in \arg \mathcal{M}$ such that $\text{cl}_R(M) \subset \text{cl}_R(L)$, then there exists a point x_0 in $\text{cl}_R(L) \setminus \text{cl}_R(M)$. The distance of x_0 to L is equal to 0. Moreover, the distance of x_0 to M is not equal to zero, because if the distance x_0 to M is equal to zero, then either x_0 would be an accumulation point of M or a point of M which is both not possible (as the $\text{cl}_R(M)$ is by definition M plus its accumulation points). Let $d(x_0,M) = \alpha$. Now, for x_0, an ε-ball B_0 exists such that $x_0 \in B_\varepsilon$, $\varepsilon > 0$ and in $\forall x \in B_\varepsilon : d(x,M) - d(x,L) > \alpha - \frac{1}{2}\alpha > 0$, because of continuity of $d(\_,A)$ (for any subset A of R) and we assume that R is locally $\mathbb{R}^n$. This entails that in the computation of the ADI_R of M and L on a whole neighborhood of x_0, the ADI_R-integration is strictly bigger for M compared to L. Outside of this neighborhood, the ADI_R-integration is bigger or equal for M compared to L, for $\forall x \in R, d(x,L) \leq d(x,M)$. Hence, for the whole of R the ADI_R-integration is strictly bigger for M as compared to L. This is a contradiction. Hence, $\text{cl}_R(M) = \text{cl}_R(L)$. $\square$

From Lemma 4.4 follows a corollary on an important property on the relation between the Hausdorff distance and the ADI:

Lemma 4.5. *Assume L is a set of dimension n or lower and R denotes a set which has dimension n in each point. Moreover, assume L is contained in R. Then,*

$$M \in \arg\min_{A \subseteq L} \int_{x \in R} d(x,A)dx \Rightarrow d_H(L,M) = 0.$$

Remark 4.13. To drop the condition that R has to be of dimension n in each point is problematic as the following example illustrates: In Figure 4.5 the set M is a maximizer of ADI_R and $R \supset L$. However, the Hausdorff distance between L and M is non-zero.

Lemma 4.5 leaves the question open whether or not there are other sets than L in $\arg\min_{A \subseteq L} \int_{x \in R} d(x,A)dx$. The following lemma makes it more precise how the other sets are related to L:

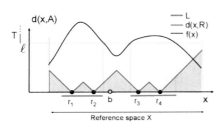

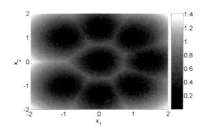

Fig. 4.6 ADI$_R$ for $R \subset \mathbb{R}$, $T = \{y \in \mathbb{R} | y > \ell\}$

Fig. 4.7 The integrand $d(x,A)$ for ADI$_R$ as described in the 2-D example

Lemma 4.6. *Let us define for an arbitrary $x \in L$ and $\varepsilon > 0$ the ε-ball in L: $B_\varepsilon(x) = \{y \in L | d(x,y) < \varepsilon\}$. Then, $ADI_R(L) < ADI_R(L \setminus B_\varepsilon(y))$ for any $y \in L$.*

Loosely speaking, it is not allowed to remove an ε-Ball from a minimal approximating set without deteriorating its optimality.

4.4.3.2 Computation of ADI

The size of the ADI integral for the uniform gap solution on the interval $[a,b]$ can be measured by elementary geometry, as it is composed of similar right triangles (see Figure 4.6).

In 1-D the integration of ADI$_R(A)$ for R being an interval $[a,b] \supset A$ can be computed in $\mathcal{O}(n \log n)$ time with

$$\text{ADI}_R(A) = \frac{1}{b-a}(0.5((x_{1:n} - a)^2 + (b - x_{n:n})^2) + \sum_{i=1}^{n-1} ((x_{i+1:n} - x_{i:n})/2)^2),$$

where $x_{i:n}$ denotes the i-th smallest point in A and $n = |A|$.

For higher dimensional X, Voronoi cells need to be used as integration areas. In Figure 4.7, an example for $f(x) = -x_1^2 - x_2^2 + 2\sqrt{x_1^2 + x_2^2}$, $T = \{2\}$ is plotted, where $A = \{(0,0), (-1,1), (0,\sqrt{2}), (1,1), (\sqrt{2},0), (1,-1), (0,-\sqrt{2}), (-1,-1)\}$, and $R = [-2,2]^2$.

Next an $\mathcal{O}(n \log(n))$ algorithm for the computation of the average distance in the 2-D plane will be derived.

Assume a 2-D box $R = [(l_1,l_2)', (u_1,u_2)']$ and a set of points within this box $A = \{\mathbf{x}^{(1)}, \ldots, \mathbf{x}^{(\mu)}\} \subset R$. The aim is to compute ADI$_R(A)$, that is the average distance of points in R to their nearest neighbor. One way to do this is to partition R into triangles such that all interior points of a given triangle share the same nearest neighbor. This nearest neighbor will be one of the corners of the triangles. For each triangle, the contribution to $\mathrm{ADI}_R(A)$ can be computed easily and the summation over the contributions of all these triangles yields $\mathrm{ADI}_R(A)$. Hence, there are two technical difficulties to be solved: 1) the partitioning into triangles and 2) the integration over a single triangle.

The Voronoi diagram is a cell complex that assigns to each point $\mathbf{x}_i \in R$ the set of points C_i for which it is the nearest Euclidean neighbor. The sets C_i form the cells of the Voronoi diagram. They can be infinite in size, though we are here only interested in $C_i' = C_i \cap R$. The cells C_i' are connected regions in R. Moreover, the boundaries of cells with non-zero measure are convex polygons. One important practical detail is that for the scenario considered here, it is required to include the bounds of the search space R as edges of the Voronoi diagram. This is to ensure that the contributions of the triangles are computed over R. In order to achieve this, one can mirror the set of points A along the boundaries of R. In 2-D, this produces four mirror sets $A_{\mathrm{left}}, A_{\mathrm{right}}, A_{\mathrm{top}}, A_{\mathrm{bottom}}$ and the Voronoi diagram of the combined sets $\{A \cup A_{\mathrm{left}} \cup A_{\mathrm{right}} \cup A_{\mathrm{top}} \cup A_{\mathrm{bottom}}\}$ contains the boundaries R. For the inner cells of this Voronoi diagram (i.e., the cells that lie within R) a triangulation can be computed. See Figure 4.8 for an example.

To triangulate R, each inner Voronoi cell can be decomposed into triangular subcells. To compute the triangles, we take the endpoints of each edge of the Voronoi cell, say $\mathbf{a}$, $\mathbf{b}$, and the center point of that cell, say $\mathbf{c}$, to form the triangle $\Delta_{\mathbf{a},\mathbf{b},\mathbf{c}}$. For each such triangle we compute the contribution to the average distance. That is, given the triangle $\Delta(\mathbf{a},\mathbf{b},\mathbf{c})$, the average distance of an interior point of this triangle to c can be obtained by integration as

$$\mathrm{ADI}_{\Delta(\mathbf{a},\mathbf{b},\mathbf{c})}(\{\mathbf{c}\}) =$$

$$\frac{d(\mathbf{a},\mathbf{b})}{6}\left(u(1+v^2) + (1/2)\cdot(1-u^2)\cdot(1-v^2)\cdot\log\frac{(u-1)}{(u+1)}\right),$$

where $u = (d(\mathbf{a},\mathbf{c}) + d(\mathbf{b},\mathbf{c}))/d(\mathbf{a},\mathbf{b})$ and $v = (d(\mathbf{a},\mathbf{c}) - d(\mathbf{b},\mathbf{c}))/d(\mathbf{a},\mathbf{b})$ [3].

Finally, to obtain the average distance the area-weighted sum over all contributions to the average distance are computed and divided by the area of the box this yields $\mathrm{ADI}_R(A)$. Let VT denote the triangulation of the area R based on the Voronoi diagram of A, then

$$\mathrm{ADI}_R(A) = \frac{1}{\mathrm{Area}(R)} \sum_{\Delta(\mathbf{a},\mathbf{b},\mathbf{c})\in VT} \mathrm{Area}(\Delta(\mathbf{a},\mathbf{b},\mathbf{c}))\mathrm{ADI}_{\Delta(\mathbf{a},\mathbf{b},\mathbf{c})}(\{\mathbf{c}\}),$$

where $\mathrm{Area}(\Delta(\mathbf{a},\mathbf{b},\mathbf{c})) = 0.5\det(\mathbf{a}-\mathbf{c}, \mathbf{b}-\mathbf{c})$ and $\mathrm{Area} = \prod_{i=1}^{2}(u_i - l_i)$.

In 2-D, the mirroring increases the number of points linearly by a factor of 2^2. The complexity of computing the Voronoi diagram is $\mathcal{O}(n\log n)$ [4, p. 160]. Finding the midpoints for all edges of the Voronoi diagram can also be done in $\mathcal{O}(n\log n)$ (note

---

[3] A derivation of this formula can be found at:

http://www.mathpages.com/home/kmath283/kmath283.htm

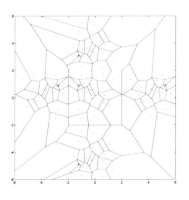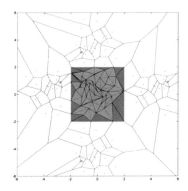

Fig. 4.8 Voronoi diagram of the reference set $[-2,2]^2$ using mirroring (left). Triangulation of the Voronoi Cells (right).

that the number of edges of the Voronoi diagram scales linearly with the number of nodes $\leq 3n - 6$ [4, p. 150]). Hence, the total complexity for computing ADI_R for a rectangular reference set in 2-D is $\mathcal{O}(n \log n)$.

4.4.3.3   Augmented Average Distance

Next we consider a quality indicator that also takes into account feasibility.

Definition 4.21 (Augmented Average Distance). Consider a level set problem with target set T and reference set R. For a set $A \subseteq R$ the Augmented Average Distance is defined as

$$\text{ADI}_R^+(A) = \text{ADI}_R(A \cap L) + \sum_{x \in A \setminus L} d(f(x), T).$$

Remark 4.14. We observe three simple properties:

E1: Sets that contain only feasible solution can be strictly worse than sets that contain also infeasible solutions.

E2: Let $A \subseteq R$, $A' \subseteq L$, and $\text{ADI}_R^+(A) \leq \text{ADI}_R^+(A')$. Then, $\text{ADI}_R(A \cap L) \leq \text{ADI}_R(A')$.

E3: Whenever an infeasible solution in an approximation set A is replaced by a feasible solution (or a feasible point is added to A without removing elements from A), then the augmented indicator is improved. More formally: $\forall x \in L$ and $\forall y \in (X \setminus L) \Rightarrow \text{ADI}_R^+((A \setminus \{y\}) \cup \{x\}) < \text{ADI}_R^+(A)$.

From this we can conclude that by improving the augmented average distance, we do not at the same time improve the feasible subset of it. However, if we proceed by single replacements of points this is indeed the case. This makes the augmented average distance an interesting quality indicator for archivers that update by adding one point at a time. In general, optimizing the augmented average distance will obtain sets that are also optimal in the average distance, as the following lemma states:

Lemma 4.7. *Assume* $L \neq \emptyset$ *and* T *is a closed set. Then* $arg\ min_{A \subseteq R} ADI_R^+(A) = arg\ min_{A \subseteq R \cap L} ADI_R(A)$.

Proof. If $A \subseteq L$, then $ADI_R(A) = ADI_R^+(A)$. Thus, if we restrict ourselves to feasible sets, the minimum is the same. Now, we show that any set that contains infeasible points is dominated (in the ADI_R^+) by a set that has feasible points, and thus is also not a candidate for a minimum. For this, we can show $A \nsubseteq L \Rightarrow ADI_R^+(A \cap L) < ADI_R^+(A)$. This is the case, because the penalty term (RHS in Definition 4.21) is strictly positive.   □

4.5   Numerical Results

Next we propose the evolutionary level set approximation algorithm (ELSA) that uses a ranking of solutions based on the following contribution:

Algorithm 4.1. Indicator-based Evolutionary Level Set Approximation (ELSA).

1: $P_0 \leftarrow$ init()
2: $t \leftarrow 0$
3: **while** not terminate **do**
4:     $q \leftarrow$ generate(P_t)
5:     $P_t' \leftarrow P_t \cup \{q\}$
6:     $r \leftarrow argmax_{p \in P_t'}\{\Delta_{QI}(p, P_t')\}$
7:     (In case of SPI, IS$_N$, IS$_\Pi$, and IS$_\Sigma$ replace argmax by argmin)
8:     $P_{t+1} \leftarrow P_t' \setminus \{r\}$
9:     $t \leftarrow t + 1$
10: **end while**
11: **return** P_t

Definition 4.22 (Quality Indicator Contribution). Given a set of solutions P, we define the individual quality indicator contribution $\Delta_{QI}(p, P)$ of a solution $p \in P$ with respect to P as

$$\Delta_{QI}(p, P) \leftarrow QI(P) - QI(P \setminus \{p\}). \qquad (4.4)$$

The ELSA algorithm is given in Algorithm 4.1. It can be considered as a simple method for approximating level sets. In the main loop, one offspring q is generated from the population P_t and added to the population P'_t. With a probability $p = 0.5$, the offspring is uniform randomly drawn from the search space or the offspring is a mutated copy of the parent. Thereafter, for each individual of the population $p \in P'_t$, the contribution $\Delta_{QI}(p, P)$ with respect to the quality indicator is computed using the quality indicator contribution. The individual with the worst contribution is deleted from the population. Note that Definition 4.22 assumes minimization of the quality indicator. Quality indicators that require maximization should be converted to minimization. The evolution loop is continued until some termination criterion is met, after which the final population P_t is returned as the level set approximation. The algorithm uses a steady-state selection scheme, as it is common in indicator-based (multiobjective) optimization. As opposed to NOAH [18] it strictly prioritizes feasibility to spread. The design of ELSA is preferable when points that are not in L should not be used to represent L.

4.5.1   Experimental Study

We perform one run of the ELSA algorithm using various augmented indicators, namely: $IS_N^+, IS_\Pi^+, IS_\Sigma^+$, Solow Polasky ($SP^+$) for $\theta = 1$, $\theta = 10$, $\theta = 1000$, ADI_R^+ with small (S), medium (M), and large (L) reference set. Here the small reference set is $[-1, 1]^2$, the medium reference set is $[-2, 2]^2$ and the large reference set is $[-8, 8]^2$. For each ELSA instance an evaluation budget of $10,000$ function evaluations is used. Each ELSA instance uses a population size of $\mu = 30$ and for mutation a Gaussian random perturbation $\sigma \mathcal{N}(0, 1)$, with $\sigma = 0.2$ is added to a copied offspring. Reflection is used for handling the box-constraints.

The simple spread indicators of Section 4.4.1 are transformed for minimization using multiplication by -1. Furthermore, for augmentation, the number of infeasible solutions is multiplied with the diameter of the search space and the cumulated distances of infeasible points to T are added to it. The SPI is also transformed for minimization using multiplication by -1 and for augmentation, the cumulated distances of infeasible points to T are added to it. For the comparisons based on average distances infeasible solution are taken care of by using the augmented average distance ADI^+. For the SPI we use the same augmentation approach as for the average distance. As for the average distance indicator, an exchange of an infeasible solution by a feasible solution always improves the indicator. The reason for this is that the SPI has the monotonicity in species properties, that is, it is always rewarded if a new point $x \in L$ is added to a given level set approximation. Obviously, the sum of distances of infeasible points to the target set T will also decrease, when an infeasible point is removed.

The test problems are the sphere function, with $L = \{(x_1, x_2)^T \in [-2, 2]^2 \mid x_1^2 + x_2^2 \leq 1\}$ and a problem based on a multimodal function, with $L = \{(x_1, x_2)^T \in [-2, 2]^2 \mid f_B(x_1, x_2) \leq 0.6\}$ and f_B is the Branke's multipeak function [3], defined as

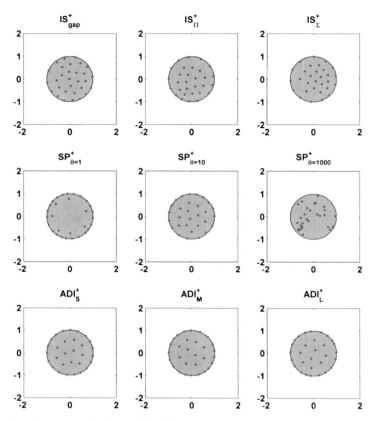

Fig. 4.9 Level set approximations of the ELSA algorithm after 10,000 function evaluations using the different set indicators on the sphere problem. Top row: $IS_N^+, IS_{\Pi}^+, IS_{\Sigma}^+$. Middle row: Solow Polasky (SP^+) with $\theta = 1$, $\theta = 10$, $\theta = 1000$. Bottom row: ADI_R^+ with small (S), medium (M), and large (L) reference sets.

$$f_B(x_1,x_2) = \frac{1}{2}((1.3 - g(x_1)) + (1.3 - g(x_2))), \qquad (4.5)$$

$$g(x_i) = \begin{cases} -(x_i+1)^2 + 1 & \text{if } -2 \le x_i < 0 \\ 1.3 \cdot 2^{-8|x_i-1|} & \text{if } 0 \le x_i \le 2 \\ 0 & \text{otherwise} \end{cases} \qquad (4.6)$$

Results are shown in Figure 4.9 and Figure 4.10. Some observations:

- The simple spread indicators are useful indicators to distribute points over the level set. It seems that P_{Π} promotes an evenly spaced approximation set, while P_{Σ} has a tendency to cluster solutions in the middle.

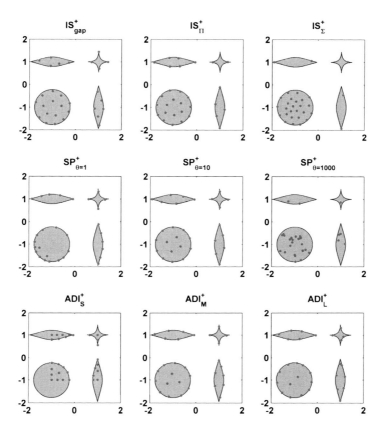

Fig. 4.10 Level set approximations of the ELSA algorithm after $10,000$ function evaluations using the different set indicators on Branke's multipeak problem. Top row: $IS_N^+, IS_{\Pi}^+, IS_{\Sigma}^+$. Middle row: Solow Polasky (SP^+) with $\theta = 1$, $\theta = 10$, $\theta = 1000$. Bottom row: ADI_R^+ with small (S), medium (M), and large (L) reference sets.

- The SPI finds evenly spaced distributions when the parameter θ is set appropriately. However, when it is too low, it pushes solutions to the boundary while when it is too high it tends to cluster.
- For the ADI indicator with an appropriate reference set, the results look good. Clearly, a too small choice of the reference set yields approximations that overlook parts of the search space, for obvious reasons. The effect of setting the reference set too large is marginal, but it introduces a tendency for the points to be placed at the boundary.

An additional result on the approximation of a thin shape using the ADI_R is given in Example 4.1. Here, $10,000$ function evaluations were used and $R = [-3, 3]^2$.

4.6 Summary and Outlook

The generalized level set approximation problem is reformulated as a multiobjective problem with respect to representativeness **P1** and feasibility **P2**. It is discovered that properties **P1** and **P2** each gives rise to a domination structure among finite approximation sets which are Lorenz orders. Aggregations of distance objectives are obtained which respect these Lorenz orders. Interestingly, **P1** and **P2**, mathematically give rise to the two components which already Pompeiu [9] and Hausdorff [6] used to define the notion of distance between two sets, and more recently the variations of this notion by Schütze et al. [14]. It is pointed out that **P1** is only to a certain extent complementary to finding maximally diverse sets of solutions. The averaged Hausdorff distance (AHD) turns out to be a favorable choice in cases where L is of the same dimension in all of its points. For cases where knowledge of the level set is absent, the ADI_R for a reference set $R \supseteq L$ is proposed. We show that in two dimensions it is computable in $\mathcal{O}(n \log n)$ time using Voronoi tessellation, though its computation will be inefficient for high-dimensional problems. In this case fast computable diversity indicators, such as IS_Π and the SPI, can be used instead. In our experiments, SPI produces slightly better results than the IS_Π indicator. However, the SPI might fail for a too large or too low choice of θ and carries a higher computational effort.

In the future, fast computation schemes for ADI in more than two dimensions need to be developed. A major area of research will be the design and analysis of (indicator-based) level set approximation algorithms. Most of the discussion can probably be generalized to metric spaces. Finally, note that support material is available at: http://natcomp.liacs.nl/code.

Acknowledgements. Michael Emmerich acknowledges financial support by NWO (The Netherlands) for the Computational Life Science grant for the BETNET project. We thank Edgar Reehuis and the reviewers for their comments.

References

1. Allgower, E.L., Georg, K.: Introduction to Numerical Continuation Methods. SIAM (2003)
2. Atkinson, A.B.: On the Measurement of Inequality. Journal of Economy 2, 244–263 (1970)
3. Branke, J.: Reducing the Sampling Variance when Searching for Robust Solutions. In: GECCO, pp. 235–242. Morgan Kaufmann (2001)
4. de Berg, M., van Kreveld, M., Overmars, M., Schwarzkopf, O.: Computational Geometry: Algorithms and Applications. Springer (2000)
5. Harada, K., Sakuma, J., Kobayashi, S., Ono, I.: Uniform Sampling of Local Pareto-Optimal Solution Curves by Pareto Path Following and its Applications in Multi-Objective GA. In: Lipson, H. (ed.) GECCO, pp. 813–820. ACM (2007)
6. Hausdorff, F.: Grundzüge der Mengenlehre, 1st edn., Berlin (1914)

 7. Gielis, J.: A Generic Transformation that Unifies a Wide Range of Natural and Abstract Shapes. American Journal of Botany 90(3), 333–338 (2003)
 8. Hillermeier, C.: Generalized Homotopy Approach to Multiobjective Optimization. JOTA 110(3), 557–583 (2001)
 9. Pompeiu, D.: Sur la Continuité des Fonctions de Variables Complexes. Annales de la Faculté des Sciences de Toulouse Sér. 2 7(3), 265–315 (1905)
10. Rényi, A.: Wahrscheinlichkeitsrechnung. VEB Verlag (1971)
11. Schütze, O., Dell'Aere, A., Dellnitz, M.: On Continuation Methods for the Numerical Treatment of Multi-Objective Optimization Problems. In: Branke, et al. (eds.) Practical Approaches to Multi-Objective Optimization. Dagstuhl Seminar Proc. 04461. IBFI, Germany (2005)
12. Schütze, O., Coello Coello, C.A., Mostaghim, S., Talbi, E.-G., Dellnitz, M.: Hybridizing Evolutionary Strategies with Continuation Methods for Solving Multi-Objective Problems. Engineering Optimization 40(5), 383–402 (2008)
13. Schütze, O., Esquivel, X., Lara, A., Coello Coello, C.A.: Some Comments on GD and IGD and Relations to the Hausdorff Distance. In: Pelikan, M., Branke, J. (eds.) GECCO (Companion), pp. 1971–1974. ACM (2010)
14. Schütze, O., Esquivel, X., Lara, A., Coello Coello, C.A.: Using the Averaged Hausdorff Distance as a Performance Measure in Evolutionary Multi-Objective Optimization. IEEE, Transact. EC (2010) (to appear)
15. Shir, O.M., Preuss, M., Naujoks, B., Emmerich, M.: Enhancing Decision Space Diversity in Evolutionary Multiobjective Algorithms. In: Ehrgott, M., Fonseca, C.M., Gandibleux, X., Hao, J.-K., Sevaux, M. (eds.) EMO 2009. LNCS, vol. 5467, pp. 95–109. Springer, Heidelberg (2009)
16. Solow, A., Polasky, S.: Measuring Biological Diversity. Environmental and Ecological Statistics 1, 95–103 (1994)
17. Ulrich, T., Bader, J., Thiele, L.: Defining and Optimizing Indicator-Based Diversity Measures in Multiobjective Search. In: Schaefer, R., Cotta, C., Kołodziej, J., Rudolph, G. (eds.) PPSN XI. LNCS, vol. 6238, pp. 707–717. Springer, Heidelberg (2010)
18. Ulrich, T., Thiele, L.: Maximizing Population Diversity in Single-Objective Optimization. In: Krasnogor, N. (ed.) GECCO, pp. 641–648. ACM (2011)
19. Weitzman, M.L.: On Diversity. The Quarterly Journal of Economics 107(2), 363–405 (1992)
20. Yuval, G.: Finding Nearest Neighbors. In: Proc. of Inf. Process. Lett., pp. 63–65 (1976)

Chapter 5
Set Oriented Methods for the Numerical Treatment of Multiobjective Optimization Problems

Oliver Schütze, Katrin Witting, Sina Ober-Blöbaum, and Michael Dellnitz

Abstract. In many applications, it is required to optimize several conflicting objectives concurrently leading to a multobjective optimization problem (MOP). The solution set of a MOP, the Pareto set, typically forms a $(k-1)$-dimensional object, where k is the number of objectives involved in the optimization problem. The purpose of this chapter is to give an overview of recently developed set oriented techniques – subdivision and continuation methods – for the computation of Pareto sets $\mathscr{P}$ of a given MOP. All these methods have in common that they create sequences of box collections which aim for a tight covering of $\mathscr{P}$. Further, we present a class of multiobjective optimal control problems which can be efficiently handled by the set oriented continuation methods using a transformation into high-dimensional MOPs. We illustrate all the methods on both academic and real world examples.

Keywords: multiobjective optimization, multiobjective optimal control, set oriented methods, subdivision, continuation.

5.1 Introduction

In a variety of applications one is faced with the problem that several objectives have to be optimized concurrently leading to a *multiobjective optimization problem* (MOP). Typically, the solution set of a MOP – the *Pareto set* – is not given by a single point as in scalar optimization but forms a $(k-1)$-dimensional object, where

Oliver Schütze
CINVESTAV-IPN, Computer Science Department, Av. IPN 2508, C. P. 07360,
Col. San Pedro Zacatenco Mexico City, Mexico
e-mail: schuetze@cs.cinvestav.mx

Katrin Witting · Sina Ober-Blöbaum · Michael Dellnitz
University of Paderborn, Chair of Applied Mathematics, Warburger Str. 100,
D-33098 Paderborn, Germany
e-mail: {witting,sinaob,dellnitz}@math.uni-paderborn.de

E. Tantar et al. (Eds.): EVOLVE- A Bridge between Probability, SCI 447, pp. 187–219.
springerlink.com © Springer-Verlag Berlin Heidelberg 2013

k is the number of objectives involved in the MOP. In case k is low (i. e., two to four), it makes sense to compute the entire solution set since this is the set of 'optimal compromises' and hence of particular interest for the decision making process.

In the literature, a huge variety of different methods for the computation of the Pareto set can be found. There exist, for instance, many scalarization methods which transform the MOP into a 'classical' scalar optimization problem (SOP). By choosing a clever sequence of SOPs a suitable finite size approximation of the entire Pareto set can be obtained (see [7, 40, 18, 32, 17, 16] and references therein). Further, there exist continuation methods that, starting from one or several solutions, perform a search along the Pareto set which is possible due to the geometry of the solution set (e. g., [26, 52, 23, 54]). Another way – and probably the most prominent one – is to use metaheuristics such as evolutionary algorithms (see [8, 20, 6] and references therein). The underlying idea is to evolve an entire set of solutions (population) during the optimization process. By this, an approximation of the entire Pareto set can be obtained by one single run of the algorithm.

The methods we consider here differ in the sense that in each iteration step a set of boxes is created with the aim to tightly cover $\mathscr{P}$. This can be done by subdivision techniques or by using certain continuation methods (so-called *recover techniques*). In the former, a sequence of nested box collections is generated that converges (ideally) to $\mathscr{P}$, and in the latter a given collection $\mathscr{C}$ is extended by a local search which is performed around promising elements (boxes) of $\mathscr{C}$. Subdivision techniques are due to their global approach highly competitive in particular if the dimension of the parameter space is moderate (say, $n < 50$), and the number of objectives is low ($k < 5$). Continuation methods are of local nature (i. e., restricted to the connected component of the solution set in which the given solution is contained), but in turn applicable to higher dimensional problems ($n \gg 1000$). Set oriented methods have been successfully applied to, for example, space mission design problems ([60, 13]), the design of electromagnetic shielding materials ([59]), the optimization of several subsystems of a rail-bound vehicle ([51, 37, 21, 22, 62, 56, 34]), an energy management problem of a tram ([33]), and the design of electrical circuits ([4]).

Next to the computation of the Pareto set of a given MOP we address the relatively young field of the numerical treatment of multiobjective optimal control problems. Whereas in multiobjective optimization one searches for Pareto optimal parameters, in optimal control one searches for optimal trajectories, which are solutions of a dynamical system given by a differential equation. Common approaches, such as *direct methods* (for an overview of different methods we refer to [3]), are based on a discretization of the trajectories and the differential equation such that in the end one is faced with a high-dimensional constrained (multiobjective) optimization problem. One of first works combining methods of direct optimal control and multiobjective optimization is e. g. [38]. In this contribution it is described how the set oriented continuation methods are combined with the recently developed direct optimal control method DMOC (Discrete Mechanics and Optimal Control [48]) which is in particular suitable for Lagrangian systems, e. g. systems in space mission design ([30, 42]), or constrained multi-body dynamics ([36, 49]). Additionally, the special case of differentially flat systems is addressed. We demonstrate on two

examples that the resulting high-dimensional MOPs can be handled by the set oriented continuation methods.

The remainder of this chapter is organized as follows: In Section 5.2, we present
the required background in multobjective optimization. In Section 5.3, we describe
a subdivision technique for the computation of relative global attractors of a given
dynamical system. In Section 5.4, we present four basic algorithms for the computation of Pareto sets, two subdivision algorithms and two continuation methods. In
Section 5.5, we present two particular approaches for the treatment of multiobjective
optimal control problems. And finally, in Section 10.6, we state some concluding remarks.

5.2   Multiobjective Optimization

In the following we consider MOPs which can be stated as follows:

$$\min_{x \in Q}\{F(x)\}, \quad Q = \{x \in \mathbb{R}^n : h(x) = 0, \, g(x) \le 0\}, \tag{5.1}$$

where F is defined as the vector of the objective functions, i. e.

$$F : Q \to \mathbb{R}^k, \quad F(x) = (f_1(x), \ldots f_k(x)), \tag{5.2}$$

with $f_1, \ldots, f_k : Q \to \mathbb{R}$, $h : Q \to \mathbb{R}^m$, $m \le n$, and $g : Q \to \mathbb{R}^q$. Though the methods
presented in the following are in principle applicable to general restriction sets Q,
we will primarily consider unconstrained problems (i. e., $Q = \mathbb{R}^n$) or domains that
result from box constraints, i. e.,

$$Q := \{x \in \mathbb{R}^n \, : \, l_i \le x_i \le u_i, \, i = 1, \ldots, n\}, \tag{5.3}$$

where $l \in \mathbb{R}^n$ and $u \in \mathbb{R}^n$ define the lower and upper bounds, respectively.

In the next definition we state the classical concept of optimality for MOPs.

Definition 5.1. (a) Let $v, w \in \mathbb{R}^k$. Then the vector v is *less than* w ($v <_p w$), if
$v_i < w_i$ for all $i \in \{1, \ldots, k\}$. The relation $\le_p$ is defined analogously.
(b) A vector $y \in \mathbb{R}^n$ is *dominated* by a vector $x \in \mathbb{R}^n$ (in short: $x \prec y$) with respect
to (9.1) if $F(x) \le_p F(y)$ and $F(x) \ne F(y)$.
(c) A point $x \in Q$ is called *Pareto optimal* or a *Pareto point* if there is no $y \in Q$
which dominates x.

In the following, we denote by P_Q the set of Pareto points (or Pareto set). The image
$F(P_Q)$ of the Pareto set is called the *Pareto front*.

In case all the objectives are differentiable, the theorem of Kuhn and Tucker
([35]) states a necessary condition for optimality. We state the result in the following
for the unconstrained case. For a more general formulation of the theorem we refer
e. g. to [40].

Theorem 5.1 ([35]). *Let x^* be a Pareto point of (9.1). Then there exist vectors $\alpha \in \mathbb{R}^k$ with $\alpha_i \geq 0, i = 1, \ldots, k$, and $\sum_{i=1}^{k} \alpha_i = 1$ such that*

$$\sum_{i=1}^{k} \alpha_i \nabla f_i(x^*) = 0. \tag{5.4}$$

Points x^* that satisfy Equation (5.4) are called Karush-Kuhn Tucker[1] (KKT) points or substationary points. The above theorem can be used to give a qualitative description of P_Q (which has first been observed in [26]). Denote by $\tilde{F} : \mathbb{R}^{n+m+k} \to \mathbb{R}^{n+m+1}$ the following map:

$$\tilde{F}(x, \alpha) = \begin{pmatrix} \sum\limits_{i=1}^{k} \alpha_i \nabla f_i(x) \\ \sum\limits_{i=1}^{k} \alpha_i - 1 \end{pmatrix}. \tag{5.5}$$

By Theorem 5.1 it follows that for every KKT point $x^* \in \mathbb{R}^n$ there exists a vector $\alpha^* \in \mathbb{R}^k$ such that

$$\tilde{F}(x^*, \alpha^*) = 0. \tag{5.6}$$

Hence, one expects – as a result of the Implicit Function Theorem – that the set of KKT-points defines a $(k-1)$-dimensional manifold. This is indeed the case under certain smoothness assumptions, see [26] for a thorough discussion of this topic.

5.3 A Subdivision Algorithm for the Computation of Relative Global Attractors

The relative global attractor of a dynamical system contains all invariant sets and is hence (among other examples, see [11, 12]) interesting for the detection of substationary points of a given MOP. In the following we present the object of interest, the framework of a subdivision technique for the computation of such objects, and describe further on a numerical realization.

5.3.1 The Relative Global Attractor

Here we define the object of interest, the relative global attractor of a dynamical system. For a more detailled discussion we refer e. g. to [11, 12].

We consider discrete dynamical systems

$$\delta \int_0^{t_f} L(q(t), \dot{q}(t)) \, dt + \int_0^{t_f} f(q(t), \dot{q}(t), u(t))) \, \delta q \, dt = 0 \tag{5.7}$$

---

[1] Named after the works of Karush [31] and Kuhn & Tucker [35] for scalar–valued optimization problems.

where $f : \mathbb{R}^n \to \mathbb{R}^n$. A subset $A \subset \mathbb{R}^n$ is called invariant if $f(A) = A$. We say an invariant set A is an attracting set if there exist a neighborhood U of A such that for every open set $V \supset A$ there is a $N \in \mathbb{N}$ such that $f^j(U) \subset V$ for all $j \geq N$. Note that for every invariant set also its closure is invariant. Hence, we can restrict ourselves to closed invariant sets A, and in this case we obtain

$$A = \bigcap_{j \in \mathbb{N}} f^j(U). \tag{5.8}$$

Hence, we can say that all the points in U are attracted by A (under iteration of f), and U is called the basin of attraction of A. If $U = \mathbb{R}^n$, then A is called the global attractor. The knowledge of the global attractor is in general beneficial since it contains all the potential interesting dynamics ([12]). For numerical aproximations, however, we have to restrict ourselves to a compact subset of the $\mathbb{R}^n$ as domain which leads directly to the notion of the relative global attractor.

Definition 5.2. Let $Q \subset \mathbb{R}^n$ be a compact set. The global attractor relative to Q is defined by

$$A_Q := \bigcap_{j \geq 0} f^j(Q). \tag{5.9}$$

Example 5.1. Consider the one-dimensional dynamical system

$$f(x) = \alpha x, \tag{5.10}$$

where $\alpha \in \mathbb{R}$ is a constant, and let $Q = [a, b]$, where $a < 0$ and $b > 0$.

(a) Let $\alpha \in (-1, 1)$. Since $|x_{j+1}| = |\alpha| |x_j|$ the relative global attractor is given by $A_Q = \{0\}$.
(b) Let $|\alpha| \geq 1$. Since for all $j \in \mathbb{N}$ it is $f^j(Q) \supset Q$ and $f^0(Q) = Q$, it is $A_Q = Q$.

As an example related to optimization consider the application of the steepest descent method ([45]) to a scalar optimization problem

$$\min_x g(x), \tag{5.11}$$

where $g : \mathbb{R}^n \to \mathbb{R}$ is a smooth function. This leads to the dynamical system

$$x_{j+1} = f(x_j) = x_j - t \nabla g(x_j), \quad j = 0, 1, 2, \ldots, \tag{5.12}$$

where $t \in \mathbb{R}_+$ is a (fixed) step size. It is important to note that the relative global attractor contains all invariant sets $A \subset Q$ and is hence interesting in the present context:

Let $x^* \in \mathbb{R}^n$ be a substationary point (i. e., $\nabla g(x^*) = 0$), then it is

$$x^* = f(x^*) \tag{5.13}$$

i. e., x^* is a fixed point of f (and in particular invariant). Note that this statement holds regardless of the choice of the step size t.

5.3.2 The Algorithm

Here we describe a subdivision technique that creates in each iteration step j a collection of sets Q_j such that each Q_j is an outer approximation of A_Q and that the sequence of Q_j's converges to A_Q in the Hausdorff sense.

Let $\mathscr{B}_0$ be an initial collection of finitely many subsets of the compact set Q such that $\cup_{B \in \mathscr{B}_0} B = Q$. Then $\mathscr{B}_j$ is inductively obtained from $\mathscr{B}_{j-1}$ in two steps:

(i) Subdivision   Construct from $\mathscr{B}_{j-1}$ a new system $\hat{\mathscr{B}}_j$ of subsets such that

$$\bigcup_{B \in \hat{\mathscr{B}}_j} B = \bigcup_{B \in \mathscr{B}_{j-1}} B \tag{5.14}$$

and

$$\mathrm{diam}(\hat{\mathscr{B}}_j) = \theta_j \, \mathrm{diam}(\mathscr{B}_{j-1}), \tag{5.15}$$

where $0 < \theta_{min} \le \theta_j \le \theta_{max} < 1$.

(ii) Selection   Define the new collection $\mathscr{B}_j$ by

$$\mathscr{B}_j = \left\{ B \in \hat{\mathscr{B}}_j : \text{there exists } \hat{B} \in \hat{\mathscr{B}}_j \text{ such that } f^{-1}(B) \cap \hat{B} \ne \emptyset \right\}. \tag{5.16}$$

Denote by Q_j the collection of compact subsets obtained after j subdivision steps, i. e.,

$$Q_j := \bigcup_{B \in \mathscr{B}_j} B \tag{5.17}$$

One can show that the limit of the Q_j's converges to the relative global attractor.

Proposition 5.1 ([12]). *Let A_Q be a global attractor relative to the compact set Q, f be a diffeomorphism, and let $\mathscr{B}_0$ be a finite collection of closed subsets with $Q_0 := \cup_{B \in \mathscr{B}_0} B = Q$. Then*

$$A_Q = \bigcap_{j=0}^{\infty} Q_j. \tag{5.18}$$

The above result can alternatively be stated as

$$\lim_{j \to \infty} d_H(A_Q, Q_j) = 0, \tag{5.19}$$

where $d_H(\cdot, \cdot)$ denotes the Hausdorff distance between two sets.

Note that the above result holds for the usage of *one* dynamical system throughout the entire iteration process. In the context of optimization, however, this might be too restrictive. As a general example, consider the dynamical system (5.12).

Instead of using a fixed step size t, one is typically interested in using several step sizes which formally leads to an entire family of dynamical systems. In the case of steepest descent this would be

$$x_{j+1} = f_i(x_j) = x_j + t_i \nabla g(x_j), \quad i \in \mathcal{I} \tag{5.20}$$

For an adaption of the subdivision technique to that context we refer to [15].

5.3.3   Realization of the Algorithm

Here we describe a possible realization of the subdivision technique.

Subdivision

For the representation of the collections $\mathcal{B}_j$ we use boxes: Let us assume that every parameter is restricted to a certain range, i. e., $a_i \leq x_i \leq b_i$, $i = 1, \ldots, n$. The search space thus is given by

$$Q = [a_1, b_1] \times \ldots \times [a_n, b_n] \subset \mathbb{R}^n. \tag{5.21}$$

Every box $B \subset \mathbb{R}^n$ can be represented by a center $c \in \mathbb{R}^n$ and a radius $r \in \mathbb{R}^n_+$ such that

$$B = B^{(c,r)} = \{x \in \mathbb{R}^n : |x_i - c_i| \leq r_i \, \forall i = 1, \ldots, n\}. \tag{5.22}$$

The box B can be subdivided with respect to the j-th coordinate. This division leads to two boxes $B_-^{(c^-, \hat{r})}$ and $B_+^{(c^+, \hat{r})}$, where

$$\hat{r}_i = \begin{cases} r_i & \text{for } i \neq j \\ r_i/2 & \text{for } i = j \end{cases}, \quad c_i^{\pm} = \begin{cases} c_i & \text{for } i \neq j \\ c_i \pm r_i/2 & \text{for } i = j \end{cases}.$$

Let $P(Q,0) := Q$, that is, $P(Q,0) = B^{(c^0, r^0)}$, where

$$c_i^0 = \frac{a_i + b_i}{2}, \quad r_i^0 = \frac{b_i - a_i}{2}, \quad i = 1, \ldots, n.$$

Denote by $\mathcal{P}(Q,d), d \in \mathbb{N}$, the set of boxes obtained after d subdivision steps starting with $B^{(c^0, r^0)}$, where in each step $i = 1, \ldots, d$ the boxes are subdivided with respect to the j_i-th coordinate, where j_i is varied cyclically. That is, $j_i = ((i-1) \bmod n) + 1$. Note that for every point $y \in Q \backslash \partial Q$ and every subdivision step d there exists exactly one box $B = B(y,d) \in \mathcal{P}(Q,d)$ with center c and radius r such that $c_i - r_i \leq y_i < c_i + r_i$, $\forall i = 1, \ldots, n$. Thus, every set of solutions $\mathcal{S}_B$ leads to a set of box collections $\mathcal{B}_d$. These collections can easily be stored in a binary tree with depth d. In Figure 5.1 a representation of five boxes with subdivision step three and three dimensions (i. e., $n = 3$) together with the corresponding set $\mathscr{B}_3$ is shown. Note that each $\mathscr{B}_d$ is completely determined by the tree structure and the initial box $B^{(c^0, r^0)}$. Using this scheme, the memory requirements grow only linearly in the dimension n of the problem.

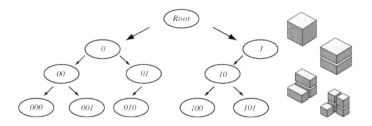

Fig. 5.1 The data structure used for the representation of the solution set

Selection

A box B is removed from the collection in the above algorithm if

$$\forall \hat{B} \in \hat{\mathscr{B}}_k : \ f^{-1}(B) \cap \hat{B} = 0 \tag{5.23}$$

Apparently, this is hard to decide apart for trivial problems. As a remedy, the following heuristic can be chosen which has shown its efficiency in numerous examples: One can discretize each box of the collection by selecting a finite set of test points (e. g. , grid points in low dimensions n of the parameter space or Monte Carlo points in higher dimensions). Then, one can replace removal strategy (5.23) by the following one:

$$f(x) \notin B \quad \text{for all test points } x \in \hat{\mathscr{B}}_k. \tag{5.24}$$

Similar strategies can be found in cell-mapping techniques (e. g., [28, 29]).

5.4   Basic Algorithms for Multiobjective Optimization

In the following, we present four different algorithms for the numerical treatment of MOPs, two subdivision algorithms and two continuation strategies. In all cases, we emphasize on the general idea, for details or comparisons to other methods we refer to the original works.

5.4.1   Subdivision Techniques

5.4.1.1   DS-Subdivision

The first algorithm we present here is in principle constructed as the one presented in (5.12), albeit tailored to the context of multiobjective optimization ([14]). Assume the MOP is unconstrained and all objectives are continuously differentiable. The following result gives a way to compute a descent direction – i. e., a direction $v \in \mathbb{R}^n$ where all objectives can be improved simultaneously – at every non-optimal point $x \in \mathbb{R}^n$.

Theorem 5.2 ([55]). *Let (MOP) be given and* $q : \mathbb{R}^n \to \mathbb{R}^n$ *be defined by*

$$q(x) = \sum_{i=1}^{k} \hat{\alpha}_i \nabla f_i(x), \tag{5.25}$$

where $\hat{\alpha}$ *is a solution of*

$$\min_{\alpha \in \mathbb{R}^k} \left\{ \left\| \sum_{i=1}^{k} \alpha_i \nabla f_i(x) \right\|_2^2 ; \alpha_i \geq 0, i = 1, \ldots, k, \sum_{i=1}^{k} \alpha_i = 1 \right\}. \tag{5.26}$$

Then either $q(x) = 0$ *or* $-q(x)$ *is a descent direction for all objective functions* $f_1, \ldots, f_k$ *in* x.

Note that since each x with $q(x) = 0$ is a substationary point, the computation of the descent direction includes a test for Pareto optimality.

Having the descent direction q, a possible dynamical system that 'pushes' the iterates toward the set of interest, the Pareto set, is now at hand: Analog to the line search method in (5.12) we can define

$$x_{j+1} = f(x_j) = x_j - tq(x_j), \quad j = 0, 1, 2, \ldots, \tag{5.27}$$

where $t \in \mathbb{R}_+$ is a chosen step size. DS-Subdivision is the subdivision technique described in Section 5.3.2 using (5.27) as dynamical system.

Example 5.2. Consider the following bi-objective problem

$$\begin{aligned} &f_1, f_2 : \mathbb{R}^2 \to \mathbb{R} \\ &f_1(x) = (x_1 - 1)^4 + (x_2 - 1)^2, \\ &f_2(x) = (x_1 + 1)^2 + (x_2 + 1)^2 \end{aligned} \tag{5.28}$$

The Pareto set of MOP (5.28) is a curve connecting the points $(-1,-1)^T$ and $(1,1)^T$. Figure 5.2 shows the result of an application of the subdivision scheme where (5.27) has been used as dynamical system. After several iteration steps a tight covering of the Pareto set can be obtained. For the evaluation of a box we have chosen the four corners as test points.

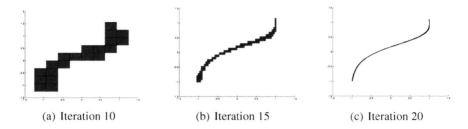

| (a) Iteration 10 | (b) Iteration 15 | (c) Iteration 20 |

Fig. 5.2 Box collections generated by DS-Subdivision applied on MOP (5.28) after 10, 15, and 20 iteration steps. Here, we have chosen $Q = [-5,5]^2$ as domain.

Indeed one can show convergence to the set of interest if it is connected.

Proposition 5.2 ([14]). *Suppose that the set $\mathscr{S}$ of substationary points is bounded and connected. Let Q be a compact neighborhood of $\mathscr{S}$. Then, an application of the subdivision algorithm to Q with respect to the iteration scheme (5.27) leads to a sequence of coverings which converges to the entire set $\mathscr{S}$; that is,*

$$d_H(\mathscr{S}, Q_j) \to 0, \quad for\ j \to \infty. \tag{5.29}$$

If e. g. the problem is convex, then it is known that $\mathscr{S}$ is equal to P_Q which is furthermore connected. Unfortunately, analog results cannot be obtained for the case where the set $\mathscr{S}$ is disconnected. The reason for this is that the relative global attractor is always connected. The following example demonstrates this in the present context.

Example 5.3. Consider the bi-objective problem as shown in Figure 5.3. This problem is constructed such that $\mathscr{S} = [0,1] \cup [1.5,2]$, where the interval $[0,1]$ contains only locally optimal solutions and the interval $[1.5,2]$ is equal to the Pareto set. An application of DS-Subdivision to $Q = [-1,3]$ will converge to the relative global attractor $A_Q = [0,2]$. To see this, one has to consider the neighborhood around the number 1: a box B that contains 1 as well as points that are bigger than one has always a nonzero intersection with its image under iteration (5.27). Further, the image of this box has also a nonzero intersection with its right neighbor B_r. Proceeding with B_r, we see that all the boxes between 1 and 1.5 have preimages in other boxes in each step of the subdivision process. Hence, the interval $(1,1.5)$ is never removed in the selection step.

However, it has to be noted that this holds for an ideal application of the algorithm. In case the removal strategy (5.24) is used in the selection strategy, it is most likely to observe convergence toward $\mathscr{S}$.

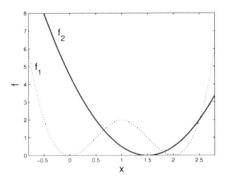

Fig. 5.3 Example of a bi-objective optimization problem where the set $\mathscr{S}$ of locally optimal solutions is disconnected

Remark 5.1. We have utilized in our studies the descent method presented in [55]. However, we have to note that there are other ways to compute descent directions (e. g., [18, 5]) which might lead to similar results.

5.4.1.2   Sampling Algorithm

Note that the algorithm described above suffers several potential drawbacks, namely:

(a) The objectives' gradients have to be at hand or have to be approximated.
(b) The set $\mathscr{S}$ of substationary points is typically a strict superset of the Pareto set, and points that are only locally optimal are typically not of interest (compare to Example 5.3).
(c) The algorithm is in principle capable of finding local Pareto points on the boundary of the domain Q. However, empirical tests have shown that in many cases a significant fraction of the boundary ∂Q is locally but not globally optimal wrt the given MOP.

The following algorithm, the *Sampling Algorithm*, tries to avoid all these potential problems. This is done by merely considering the objective values of the set of test points in each iteration. To be more precise, given a box collection $\mathscr{B}_{j-1}$, the collecion $\mathscr{B}_j$ is obtained as follows:

(i) Subdivision   This is as in Section 5.3.2.
(ii) Selection   For all $B \in \hat{\mathscr{B}}_j$, choose a set of test points $X_B \subset B$.
   $N_j :=$ nondominated points of $\bigcup_{B \in \hat{\mathscr{B}}_j} X_B$
   $\mathscr{B}_j := \{B \in \hat{\mathscr{B}}_j \ : \ \exists y \in X_B \cap N_j\}$

Note that this approach has some analogies to branch and bound strategies used for scalar optimization problems (e. g., [27]), but omits any bounding strategy. This is due to the fact that the larger the number k of objectives is, the more robust the selection strategy gets (note that for $k = 1$, N_j will typically consist of one element, which is normally not the case for $k > 1$).

Example 5.4. Consider the following bi-objective problem taken from [55]:

$$f_1, f_2 : \mathbb{R}^n \to \mathbb{R},$$

$$f_1(x) = \sum_{j=1}^{n} x_j,$$

$$f_2(x) = 1 - \prod_{j=1}^{n} (1 - w_j(x_j)),$$

(5.30)

where

$$w_j(z) = \begin{cases} 0.01 \cdot \exp(-(\frac{z}{20})^{2.5}) & \text{for} \quad j = 1, 2 \\ 0.01 \cdot \exp(-\frac{z}{15}) & \text{for} \quad 3 \le j \le n \end{cases}$$

Figure 5.4 shows a numerical result obtained by the Sampling Algorithm. Here, we have taken 10 randomly chosen test points per box. When choosing $Q = [0, 40]^3$ the set $\mathscr{S}$ contains the two faces of Q with $x_i = 0$, $i = 1, 2$. Hence, an application of DS-Subdivision leads to a tremendous effort since both faces will be kept in the box collections (see also [57]). This is avoided by the sampling approach.

5.4.2   Recover Techniques in Parameter Space

In the course of the two algorithms described above it can happen that boxes are lost that contain a part of the set of interest (e. g., due to a discretization error in the removal strategy (5.24)). The following algorithms are intended to 'heal' (or recover) the box collection. The underlying idea is that the set of interest (Pareto set or front) forms locally a manifold. That is, in the neighborhood of a 'good' box (i. e., a box with nonzero intersection with the set of interest) it is likely that there are other 'good' boxes due to the geometry of the problem. Hence, given a box collection $\mathscr{B}_j$, it makes sense to perform a local search around each box of the collection (once), and to see if neighboring boxes should be added to $\mathscr{B}_j$. It has to be noted that this approach is restricted to the connected components of the set of interest that have nonzero intersection with the given collection $\mathscr{B}_j$. On the other hand, it has turned out that the usage of the data structure is well suited to maintain a 'global' view on the part of the solution set which is already computed, and is in particular interesting for the efficient treatment of high-dimensional problems.

The idea of the recover techniques in parameter space is to recover the box collection in order to maintain a perfect covering of the Pareto set ([14]). The following

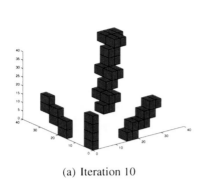

(a) Iteration 10

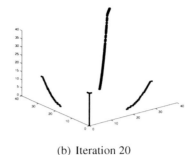

(b) Iteration 20

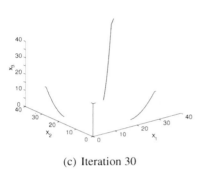

(c) Iteration 30

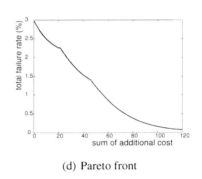

(d) Pareto front

Fig. 5.4 Numerical results for MOP (5.30) using the Sampling Algorithm.

pseudo-code gives the framework of the Recover Algorithm to extend an existing collection $\mathcal{B}_j$ (see also Figure 5.5).

(i) Step 1   Mark all boxes $B \in \mathcal{B}_j$

(ii) Step 2   (i) For all marked $B \in \mathcal{B}_k$: unmark the box and choose starting points $(s_i)_{i=1,\ldots,l}$ near B

(ii) For each s_i, $i = 1,\ldots,l$, compute a substationary point p_i starting from s_i.

(iii) For all p_i, $i = 1,\ldots,l$, if $B(y,j) \notin \mathcal{B}_k$, add $B(y,j)$ to the collection $\mathcal{B}_k$ and mark the box.

(iv) Repeat Step 2 while new boxes are added to $\mathcal{B}_k$ or until a prescribed number of steps is reached.

Note that the Recover Algorithm is similar in spirit to predictor corrector (PC) methods used for numerical (multiobjective) continuation ([53, 1, 26, 24, 52, 23]). Crucial are certainly the proper choices of the starting points s_i and the performance of the local searcher. In low dimensions it might be sufficient to choose the starting points in coordinate directions from the center of a box (as seen in Figure 5.5) together with an application of the map (5.27), i. e., to take $p_i = f^p(s_i)$, for a power $p \in \mathbb{N}$. In higher dimensions, however, this is not advisable. Instead, it makes sense to lean elements from existing PC methods applied on the map (9.11). This has been done in [58]. For an application of the recover techniques for high-dimensional problems ($n \gg 1000$) we refer to [54].

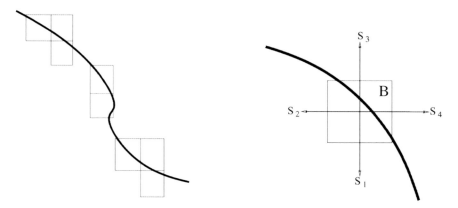

Fig. 5.5 Recover algorithm: uncomplete covering of the Pareto set (left) and possible choice of test points for a given box B (right)

Example 5.5. We consider the following MOP ([58]):

$$\min F(x) := \begin{pmatrix} (x_1 - 1)^4 + (x_2 - 1)^2 + (x_3 - 1)^2 \\ (x_1 + 1)^2 + (x_2 + 1)^4 + (x_3 + 1)^2 \\ (x_1 - 1)^2 + (x_2 + 1)^2 + (x_3 - 1)^4 \end{pmatrix} \tag{5.31}$$
$$\text{s.t. } h(x) = 1 - x_3^2 - (\sqrt{x_1^2 + x_2^2} - 4)^2 = 0$$

The MOP is given by three convex objectives which are constrained to a torus. Figure 5.6 shows a numerical result of the Recover Algorithm where the initial box collection consists of one single solution of the MOP.

We stress that the Recover Algorithm can more generally be used as a particular continuation method for the numerical solution of

$$H(x) = 0, \tag{5.32}$$

where $H : \mathbb{R}^{N+K} \to \mathbb{R}^N$ is a map (see [57, 58]). One interesting application in the present context is the numerical treatment of parameter dependent MOPs which can be expressed as follows:

$$\min_x F_\lambda : \mathbb{R}^n \to \mathbb{R}^k, \qquad \lambda \in \mathbb{R}^d \tag{5.33}$$

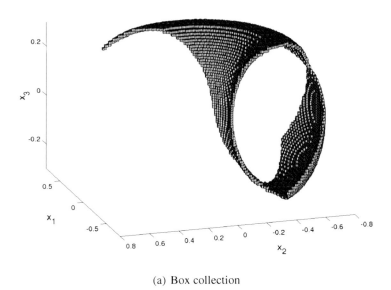

(a) Box collection

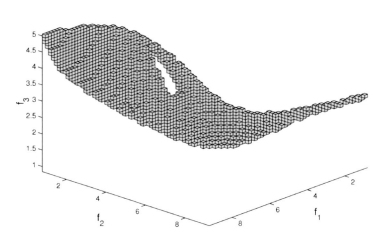

(b) Pareto front

Fig. 5.6 Numerical result of the Recover Algorithm for MOP (5.31) starting with one known Pareto optimal solution of the problem

This particular kind of problem e. g. occurs when λ is given data for the underlying system which is modelled by F and can change during the optimization process (see e. g. [51]). In case λ changes quickly it is not advisable to compute the entire Pareto set for every value of λ but it may be more efficient to approximate the set $\tilde{F}^{-1}(0)$, where

$$\tilde{F} : \mathbb{R}^{n+d+k} \to \mathbb{R}^{n+1}$$

$$\tilde{F}(x,\lambda,\alpha) := \begin{pmatrix} \sum_{i=1}^{k} \alpha_i \frac{\partial f_i}{\partial x}(x,\lambda) \\ \sum_{i=1}^{k} \alpha_i - 1 \end{pmatrix}. \tag{5.34}$$

When the auxiliary system is computed, the set of substationary points for every value $\bar{\lambda}$ is given by the projection $\tilde{F}^{-1}(0)|_{\lambda=\bar{\lambda}}$, which can easily be identified in the corresponding box collection.

Example 5.6. We consider the following parameter dependent MOP:

$$F_\lambda(x) := (1 - \lambda)F_1(x) + \lambda F_2(x), \tag{5.35}$$

where

$$F_1, F_2 : \mathbb{R}^2 \to \mathbb{R}^2$$

$$F_1(x_1, x_2) = \begin{pmatrix} (x_1 - 1)^4 + (x_2 - 1)^2 \\ (x_1 + 1)^2 + (x_2 + 1)^2 \end{pmatrix},$$

$$\tag{5.36}$$

$$F_2(x_1, x_2) = \begin{pmatrix} (x_1 - 1)^2 + (x_2 - 1)^2 \\ (x_1 + 1)^2 + (x_2 + 1)^2 \end{pmatrix}.$$

Figure 5.7 shows the set $\tilde{F}^{-1}(0)$ for problem (5.35). Two 'classical' Pareto sets for particular values of λ – using the according parts of the box collection for the auxiliary system – can be seen in Figure 5.8.

5.4.3   *Image-Set Oriented Recover Techniques*

In case the dimension of the parameter space is high and only a few objectives are beeing considered (i. e., two or three), one can alternatively generate box collections in image space ([9, 10]): For a given initial Pareto optimal solution a box on the Pareto front is generated around the image of this solution. Step by step, all neighboring boxes are inserted that contain points on the Pareto front. The insertion of boxes is based on the idea to create vectors of desired values for the objectives,

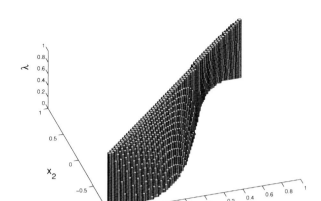

Fig. 5.7 Family of Pareto sets, see (5.35)

(a) $\lambda = 0$ (b) $\lambda = 1$

Fig. 5.8 Pareto sets for two values of λ of MOP (5.35)

so-called *targets* T, in the neighborhood of the given boxes. Then the following distance minimization problem is solved:

$$\min_{x} \|F(x) - T\|_2. \tag{5.37}$$

Using this procedure, the entire Pareto set can be covered for unconstrained multi-objective optimization problems with convex objective functions. In the nonconvex case the connected components of the Pareto front that correspond to the initial boxes can be approximated.

More precisely, the image-set oriented recover algorithm works as follows: Assume that we would like to compute Pareto optimal points within the region

$$Q_l = [f_1^{\min}, f_1^{\max}] \times \dots \times [f_k^{\min}, f_k^{\max}] \subset \mathbb{R}^k,$$

where $f_i^{\min}, f_i^{\max} \in \mathbb{R}, i = 1, \ldots, k$, are given restrictions for the objective values. The box Q_I is subdivided a set of boxes of depth d, $\mathscr{P}(Q_I, d)$, as described in Section 5.3. Then, each point $y \in Q_I$ can be assigned to a box $B(y, d)$.

The image-set oriented recover algorithm starts with a box collection $\mathscr{B}_0 \subset \mathscr{P}(Q_I, d)$. Let x_B denote a corresponding Pareto optimal solution to the box B, i. e. $F(x_B) \in B$ for $B \in \mathscr{B}_0$.

Step 1:
> Mark all $B \in \mathscr{B}_0$

Step 2:
> **for** $j = 0, \ldots,$ maximum number of steps:
> > set $\hat{\mathscr{B}}_j = \mathscr{B}_j$
> > > **for all** $B \in \mathscr{B}_j$ with B marked
> > > > **choose** target vectors $\{T_i\}_{i=1,\ldots,l}$ near B
> > > > > with $T_i \leq_p F(x_B)$
> > > > > compute $x_i^\star = \arg\min_{x \in \mathbb{R}^n} \|F(x) - T_i\|_2$ for $i = 1, \ldots, l$,
> > > > > set $F_i^\star = F(x_i^\star), i = 1, \ldots, l$,
> > > > > unmark box B
> > > > > **for all** $i = 1, \ldots, l$:
> > > > > > **if** $B(F_i^\star, d) \notin \mathscr{B}_j$
> > > > > > > set $\tilde{\mathscr{B}} = B(F_i^\star, d)$, $x_{\tilde{B}} = x_i^\star$, $F_{\tilde{B}} = F_i^\star$
> > > > > > > mark $\tilde{B}$
> > > > > > > $\hat{\mathscr{B}}_j = \hat{\mathscr{B}}_j \cup \tilde{B}$
> > **if** $\hat{\mathscr{B}}_j == \mathscr{B}_j$ **STOP**
> > $\mathscr{B}_{j+1} = \hat{\mathscr{B}}_j$

So far, it has not been explained how suitable targets T_i can be generated. Efficient strategies for the computation of target vectors can be defined by making use of local information on the Pareto set. There are different possibilities how to generate good targets. One idea presented in [10] is to generate targets along the shifted tangent space on the image of a known Pareto optimal solution. More precisely, we have to assume that $x^\star$ is Pareto optimal, $F(x^\star) = y^\star$ and $T^\star$ is the target which leads to the computation of $x^\star$. Additionally, it is required that the image of the Pareto set is smooth and forms a $(k-1)$-dimensional manifold in a neighborhood of $y^\star$. Then, new targets can be generated in two steps:

(i) Compute the normal vector to the Pareto front in the point $y^\star$. As $x^\star$ is a solution of the distance minimization problem (5.37) this normal vector is given by $n = \frac{T^\star - x^\star}{\|T^\star - x^\star\|}$. Construct a $((k\text{-}1)$-dimensional) orthonormal basis $V = \{b_1, \ldots, b_{k-1}\}$ of the tangent space at the point $y^\star$ which is orthogonal to n e. g. by computing a QR factorization of n.

(ii) Specify l targets

$$t_i = y_i^\star + \sum_{j=1}^{k-1} \alpha_{i,j} b_j + \lambda_i n, \ \ i = 1, \ldots, l.$$

The coefficients $\alpha_{i,j}$ are chosen in such a way that the points $p_i = \sum_{j=1}^{k-1} \alpha_{i,j} b_j$ are located inside neighboring boxes of the box containing $y^\star$. The value of λ_i is determined by an adaptive concept which guarantees that the targets lie below the Pareto front (but also are not too far away).

The distance minimization problem (5.37) is solved using standard optimization algorithms such as SQP which is implemented in the NAG library [46]. In Figure 5.9, a schematic representation of the image set-oriented recover algorithm is given.

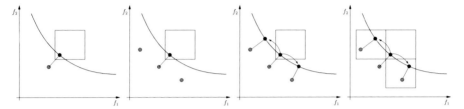

Fig. 5.9 Principal functioning of the image set-oriented recover algorithm (the black curve is the unknown Pareto front and the grey dots are the targets).

Example 5.7. Consider the bi-objective problem

$$f_1, f_2 : \mathbb{R}^{100} \to \mathbb{R},$$

$$f_1(x) = \sum_{i=1}^{100} (x_i - 1)^2$$

$$f_2(x) = \sum_{i=1}^{100} (x_i + 1)^2. \tag{5.38}$$

We restrict the optimization to a box with center $(0,0)^T$ and radius 2 in each spatial direction in image space. To demonstrate the application of the image set oriented recover algorithm this box is subdivided into boxes of depth 12 in a first study, and depth 18 in a second study. As a start, we consider the box of depth 12 or 18, respectively, containing the point $(0.25, 0.25)^T$ which lies on the Pareto front (this point can for example be computed by minimizing the weighted sum $\frac{1}{2}f_1(x) + \frac{1}{2}f_2(x)$). In Figure 5.10 the results from the application of the image-set oriented recover algorithm to this example is given for the two different box depths mentioned above. One can observe that the entire Pareto front is covered by boxes of the respective depth.

Example 5.8. The image-set oriented recover algorithm has been applied to an energy management problem of a tram which is supplied by an overhead line (cf. [9, 33]). This tram posesses an additional onboard storage system with an energy storage of high capacity which is able to store energy generated from breaking, for

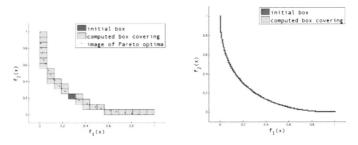

Fig. 5.10 Results of the image-set oriented recover algorithm applied to the objective functions given in Example 5.7: Box covering of the Pareto front for depth $d = 12$ (left) and $d = 18$ (right)

example. The aim is to reduce both overhead line peak power and energy consumption simultaneously during a realistic drive cycle of the tram. To compute reasonable solutions the drive cycle under consideration is divided into 1241 track sections. The energy management system has to assign a reference value to each of these sections. Thus, a multiobjective optimization problem with two objectives and 1241 optimization parameters has to be solved. Figure 5.11 shows the results. Here, both objectives are normed in such a way that they each equal one if no energy storage system would be used. Note that in the figure not the resulting boxes in image space but the solutions within these boxes are plotted.

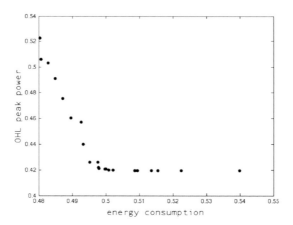

Fig. 5.11 Approximation of the Pareto front for the energy management problem of a tram described in Example 5.8

5.5   Multiobjective Optimal Control Problems

In this section we consider multiobjective optimal control problems of the form

$$\min_{x,u} J(x,u) \tag{5.39}$$

$$\text{s. t. } \dot{x}(t) = g(x(t),u(t)),$$

where $x : [0,t_f] \to \mathbb{R}^n$ is the state trajectory, $\dot{x} : [0,t_f] \to \mathbb{R}^n$ its derivative w. r. t. the time parameter $t \in [0,t_f]$, and $u : [0,t_f] \to \mathbb{R}^m$ the control trajectory. J is a vector of objective functionals,

$$J(x,u) = (J_1(x,u),\ldots,J_k(x,u))^T$$

with $J_i(x,u) = \int_0^{t_f} C_i(x(t),u(t))\,dt$, $i = 1,\ldots,k$. In contrast to the problems we considered before, the objective function depends on functions $x(t)$ and $u(t)$ rather than on single parameters. Our principal approach to solve such a trajectory optimization problem is to transform it into a nonlinear multiobjective optimization problem and solve this problem numerically using the image-set oriented recover algorithm. Such a transformation typically bases on a discretization in time such that the time-dependent functions are represented by a sequence of discrete state and control parameters that are approximations to the trajectories. (For an overview of different discretization techniques for single objective optimal control methods we refer e. g. to [2].) Thus, the mulitiobjective optimal control problem is transformed into a multiobjective optimization problem with many parameters, which consist of all time-discrete states and controls. To handle such high-dimensional multiobjective optimization problems, the presented image-set oriented recover algorithm is appropriate since it works in the low-dimensional image space rather than in the high-dimensional parameter space.

There are different possibilities how to transform the multiobjective optimal control problem into the multiobjective optimization problem, which highly depend on the system under consideration. In the following we will focus on differentially flat systems on the one hand and Lagrangian systems in general on the other hand. We will describe how the multiobjective optimal control problem can be transformed and show how these procedures can be applied to special mechatronical and mechanical systems.

5.5.1   Differentially Flat Systems

Differentially flat systems have the property that the inputs and states can be represented as a function of the flat outputs and a finite number of their derivatives wrt time (cf. [19]):

Definition 5.3 (Differential flatness [19, 43]). A system

$$\dot{x}(t) = g(x(t), u(t))$$

with states $x(t) \in \mathbb{R}^n$ and controls $u(t) \in \mathbb{R}^m$ is called *differentially flat* if there exists a fictitious output $y(t) \in \mathbb{R}^m$ with

$$y = h(x, u, \dot{u}, \ddot{u}, \ldots, u^{(p)}) \text{ such that}$$
$$x = \alpha(y, \dot{y}, \ddot{y}, \ldots, y^{(q)}) \text{ and } u = \beta(y, \dot{y}, \ddot{y}, \ldots, y^{(q)}). \tag{5.40}$$

Here, h, α and β denote real-analytic functions and $p, q \in \mathbb{N}$. y is called a *flat output*.

Differentially flat system can especially be utilized in trajectory optimization (cf. [61]). The big advantage is that in this case the trajectories can be optimized in the space of the outputs y and afterwards, the corresponding inputs and states can be computed.

Thus, a single objective optimal control problem of the form

$$\min_{x,u} j(x, u) \tag{5.41}$$
$$\text{s. t. } \dot{x}(t) = g(x(t), u(t))$$

with a differentially flat system $\dot{x}(t) = g(x(t), u(t))$ with $x(t) \in \mathbb{R}^n$ and $u(t) \in \mathbb{R}^m$ can be transformed into an optimization problem of the form

$$\min_{y} f(y), \tag{5.42}$$

where $y(t) \in \mathbb{R}^m$ denotes the flat output (cf. e. g. [61, 50] and [41]).

This concept can be easily extended to the case of a vector-valued objective functional. In this case, the optimal control problem is transformed into a conventional multiobjective optimization problem of the form

$$\min_{y} F(y), \tag{5.43}$$

where F maps from $\mathbb{R}^m$ to $\mathbb{R}^k$ and $y(t) \in \mathbb{R}^m$ denotes the flat output.

Example 5.9 (Multiobjective optimization of the guidance of a rail-bound vehicle). In the following we consider the trajectory optimization of the *guidance module* of a rail-bound vehicle (cf. [21, 22]). More precisely, this guidance module is contained in the *RailCab* vehicle, which is a linear-motor driven railway system developed by the project RailCab ("Neue Bahntechnik Paderborn", [44]) at the University of Paderborn, Germany. Figure 5.12 displays the test vehicle. It belongs to a test facility with a track length of about 530 m. The track includes a novel passive switch which allows the processing of closely following vehicles. The vehicle itself consists of a superstructure that carries the load and two undercarriages. Among many other modules the RailCab is equipped with a guidance module which is based on one

Fig. 5.12 Photograph of the RailCab test vehicle

single wheel set. It enables a driving with low attrition and allows to use the novel concept for a passive switch (cf. [25]). The guidance module allows to actively control the lateral displacement of the RailCab vehicle in the rails. Within a given clearance, the RailCab can be moved freely. This is very important, because track laying does not result in ideally straight rails and flange strikes, i. e. bumpings of the rail-heads against the flanges, have to be avoided because they cause noise and wear on the wheels and rails. Figure 5.13 shows a typical rail and a sketch of the clearance, which is the maximum distance between the flanges and the rail-heads. We assume that the measured position of the rails (the track centerline) is known a priori.

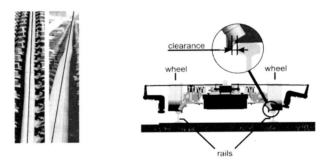

Fig. 5.13 Photograph of a rail (on the left) and sketch of the clearance (on the right) [21]

Within the clearance, the RailCab can be steered along arbitrary reference trajectories. The challenge was to compute Pareto optimal trajectories that meet several aims:

1. minimize the deviation of the vehicle from the track centerline, i. e. maximize "safety",
2. maximize the passenger comfort,
3. minimize the average energy consumption of the hydraulic actuators.

Based on a linear model of fourth order for the lateral dynamics of the RailCab vehicle (see [21] for more details), a multiobjective optimal control problem is formulated. In this model, the controlled outputs are flat. The desired reference trajectories
of length s_h for both the front and the rear axle are approximated by cubic splines.
For the computation of Pareto optimal trajectories, the image set-oriented recover
algorithm has been used. The fact that the RailCab has to stay within the clearance
is included as a constraint. Figure 5.14 shows an approximation of the Pareto front
for an exemplary track section with a length of 8 m computed by the image-set oriented recover algorithm. To each point within this Pareto front corresponds a Pareto
optimal trajectory on which the RailCab vehicle can be steered.

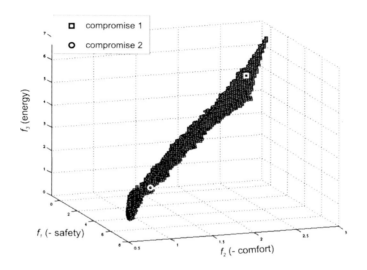

Fig. 5.14 Approximation of the Pareto front for the guidance module [21]

Two points within the Pareto front have been chosen (marked by a circle and a
square) to demonstrate the results. The circle is an example for a more safe trajectory and the square for a more comfortable one. In Figure 5.15 the corresponding
trajectories and the trajectories which stem from single objective optimizations for
each of the three objectives are given. (Here, only the optimized interpolation values at the knot points, connected with lines, are plotted.) The black line is the track
centerline and the gray lines describe the clearance around it.

One can observe that the trajectory which is more safe (line with circles) stays
close to the centerline whereas the more comfortable trajectory (line with squares)

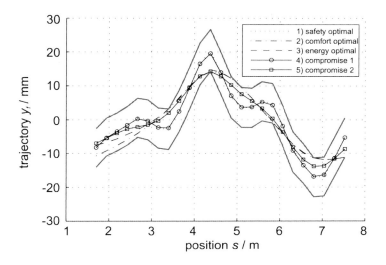

Fig. 5.15 Examples of Pareto optimal trajectories for the RailCab vehicle [21]

"cuts the corners" and is smoother. As expected, the energy optimal (dashed) and comfort optimal (dash-dot) trajectories lie close together.

5.5.2   Lagrangian Systems

We consider special kinds of dynamical systems $\dot{x}(t) = g(x(t), u(t))$, namely those systems that can be derived from a variational principle. In particular, we are interested in Lagrangian systems which comprise e. g. mechanical, but also electrical or mechatronical systems. In order to solve optimal control problems for those systems, we use DMOC (Discrete Mechanics and Optimal Control [48]), a technique that relies on a direct discretization of the variational formulation of the dynamics of the system. Based on the discretization the problem is transformed into a finite dimensional constrained optimization problem. The principal approach can be extended to the case of optimal control problems with multiple objectives. For convenience, we briefly summarize the basic idea.

Let M be an n-dimensional configuration manifold with tangent bundle TM and cotangent bundle T^*M. Consider a mechanical system with time-dependent configuration vector $q(t) \in M$ and velocity vector $\dot{q}(t) \in T_{q(t)}M$, $t \in [0, t_f]$, whose dynamical behavior is described by the Lagrangian $L : TM \to \mathbb{R}$. Typically, the Lagrangian L consists of the difference of the kinetic and potential energy. In addition, a force $f : TM \times U \to T^*M$ depending on a time-dependent control parameter $u(t) \in U \subseteq \mathbb{R}^m$ influences the system's motion. The aim is to move the mechanical system on a curve $q(t) \in M$, $t \in [0, t_f]$, from an initial state $(q^0, \dot{q}^0)$ to a final state $(q^{t_f}, \dot{q}^{t_f})$ under the influence of $f(q, \dot{q}, u)$ such that the curves q and u minimize a given objective functional

$$J(q, \dot{q}, u) = \int_0^{t_f} C(q(t), \dot{q}(t), u(t)) \, dt \qquad (5.44)$$

with $C : TM \times U \to \mathbb{R}^k$. Note, that the objective functional J involves k single objective functionals given as $J(q, \dot{q}, u) = (J_1(q, \dot{q}, u), \ldots, J_k(q, \dot{q}, u))^T$ according to (5.39) with

$$J_i(q, \dot{q}, u) = \int_0^{t_f} C_i(q(t), \dot{q}(t), u(t)) \, dt, \quad i = 1, \ldots, k,$$

and $C(q, \dot{q}, u) = (C_1(q, \dot{q}, u), \ldots, C_k(q, \dot{q}, u))^T$. At the same time, the motion $q(t)$ has to satisfy the *Lagrange-d'Alembert principle*, which requires that

$$\delta \int_0^{t_f} L(q(t), \dot{q}(t)) \, dt + \int_0^{t_f} f(q(t), \dot{q}(t), u(t))) \, \delta q \, dt = 0 \qquad (5.45)$$

for all variations δq with $\delta q(0) = \delta q(t_f) = 0$. The principle (5.45) is equivalent to the forced Euler-Lagrange equations

$$\frac{d}{dt} \frac{\partial}{\partial \dot{q}} L(q, \dot{q}) - \frac{\partial}{\partial q} L(q, \dot{q}) = f(q, \dot{q}, u), \qquad (5.46)$$

which provide as system of differential equations the equations of motion that can be summarized in the general form $\dot{x} = g(x, u)$ with $x = (q, \dot{q})$.

The optimal control problem consisting of minimizing (5.44) subject to (5.46) is numerically solved using a direct discretization approach [39, 48]. The state space TM is replaced by $M \times M$ and a path $q : [0, t_f] \to M$ by a *discrete path* $q_d : \{0, h, 2h, \ldots, Nh = t_f\} \to M$, with time step h and N a positive integer such that $q_k = q_d(kh)$ is an approximation to $q(kh)$. Similar, the control function $u : [0, t_f] \to U$ is replaced by a discrete control function $u_d : \{0, h, 2h, \ldots, Nh = t_f\} \to U$, approximating the control on each interval $[kh, (k+1)h]$ by a discrete control u_k (writing $u_k = u_d((k + \frac{1}{2})h))$.

Via an approximation of the action integral in (5.45) by a *discrete Lagrangian* $L_d : M \times M \to \mathbb{R}$,

$$L_d(q_k, q_{k+1}) \approx \int_{kh}^{(k+1)h} L(q(t), \dot{q}(t)) \, dt, \qquad (5.47)$$

and *discrete forces*

$$f_k^- \cdot \delta q_k + f_k^+ \cdot \delta q_{k-1} \approx \int_{kh}^{(k+1)h} f(q(t), \dot{q}(t), u(t)) \cdot \delta q(t) \, dt, \qquad (5.48)$$

where the left and discrete forces $f_k^{\pm}$ now depend on (q_k, q_{k+1}, u_k) we obtain the *discrete Lagrange-d'Alembert principle* (5.49). This requires to find discrete paths $\{q_k\}_{k=0}^N$ such that for all variations $\{\delta q_k\}_{k=0}^N$ with $\delta q_0 = \delta q_N = 0$, one has

$$\delta \sum_{k=0}^{N-1} L_d(q_k, q_{k+1}) + \sum_{k=0}^{N-1} f_k^- \cdot \delta q_k + f_k^+ \cdot \delta q_{k+1} = 0, \tag{5.49}$$

which is equivalent to the *forced discrete Euler-Lagrange equations*

$$D_2 L_d(q_{k-1}, q_k) + D_1 L_d(q_k, q_{k+1}) + f_{k-1}^+ + f_k^- = 0, \quad k = 1, \dots, N-1, \tag{5.50}$$

where D_i denotes the derivative w. r. t. the i-th argument. In the same manner we obtain via an approximation of the objective functional (5.44) the *discrete objective function $J_d(q_d, u_d)$*, such that we can formulate the *Discrete Constrained Multiobjective Optimization Problem* as

$$\min_{q_d, u_d} J_d(q_d, u_d) = \sum_{k=0}^{N-1} C_d(q_k, q_{k+1}, u_k) \tag{5.51}$$

subject to the discretized boundary constraints and the forced discrete Euler-Lagrange equations (5.50). Here, it holds again $J_d = (J_{d,1}, \dots, J_{d,k})^T$ and $C_d = (C_{d,1}, \dots, C_{d,k})^T$, where $J_{d,i}$ and $C_{d,i}$ are approximations to J_i and C_i, respectively, with $i = 1, \dots, k$. The number of the optimization parameters $q_d = (q_0, \dots, q_N)$ and $u_d = (u_0, \dots, u_{N-1})$ as well as the number of the equality constraints (the forced discrete Euler-Lagrange equations) of this nonlinear multiobjective optimization problem depend on the discrete grid that is used for the approximation. To meet accuracy requirements of the approximated trajectories (for a detailed convergence analysis dependent on the quadrature rules used in (5.47) and (5.48) we refer to [48]), typically a fine grid which corresponds to a small time step h is chosen, which leads to a high number of optimization parameters and equality constraints, whereas the number of objective functions J_d is independent of the time step. Thus, the image-set oriented methods described before are suitable to numerically solve this high-dimensional multiobjective optimization problem.

Example 5.10 (Underwater glider). As an application for multiobjective optimal control we consider a class of Autonomous Underwater Vehicles (AUVs) known as gliders. In order to keep the gliders autonomously operational for the greatest amount of time, it is important to minimize the amount of energy the gliders use for transport and - at the same time - minimize the time of operation when specific maneuvers are performed. The problem considered here is to find an optimal trajectory of a glider that needs to move from one location to another within a prescribed current (cf. [47]). The glider is assumed to be actuated by a gyroscopic force which implies that the relative forward speed of the glider is constant. However, the orientation of the glider cannot change instantly and the control force induces the change in the orientation of the glider. In addition to the minimization of the amount of control effort, the goal is to identify trajectories that are also time-optimal, such that the glider needs as little time as possible to reach the final destination. Thus, we have to consider a multiobjective optimization problem with the two objectives *minimize control effort* and *minimize duration time of the maneuver*. As in [63] the glider is modeled as a pointmass (with normalized mass equal to 1) in $\mathbb{R}^2$ and actuated by a gyroscopic force acting orthogonal to the relative velocity between fluid and body. Let $q(t) = (x(t), y(t))$ be the glider position, $\dot{q}(t) = (\dot{x}(t), \dot{y}(t))$ the absolute glider velocity, $V(t) = (V_x(t), V_y(t))$ the current velocity field, and $u(t) \in \mathbb{R}$ the control input representing the change in the orientation. By introducing $\dot{q}_{rel}(t) = (\dot{q}(t) - V(t))$ as relative velocity, the Lagrangian $L(q_{rel}(t), \dot{q}_{rel}(t)) = \frac{1}{2} \|\dot{q}_{rel}(t)\|^2$ in the body fixed frame is the kinetic energy of the relative motion of the glider. The gyroscopic force acting on the system is given by $f(q_{rel}(t), \dot{q}_{rel}(t), u(t)) = \left(-u(t)\dot{q}_{rel,y}(t), u(t)\dot{q}_{rel,x}(t) \right)^T$. The resulting Euler-Lagrange equations read as

$$\ddot{x}(t) = -u(t)(\dot{y}(t) - V_y(t)) + \dot{V}_x(t),$$
$$\ddot{y}(t) = u(t)(\dot{x}(t) - V_x(t)) + \dot{V}_y(t).$$

The glider has to be steered within the time span $[0, t_f]$ with free final time t_f from an initial configuration $q(0) = q^0$ to a final one $q(t_f) = q^{t_f}$, optimally with respect to the vector-valued objective functional

$$J = \begin{pmatrix} \int_0^{t_f} \|f(q_{rel}(t), \dot{q}_{rel}(t), u(t))\|^2 \, dt \\ \int_0^{t_f} 1 \, dt \end{pmatrix}.$$

In the discrete setting we model the free final time by a variable step size h that acts as an additional optimization variable bounded as $0 < h \le h_{max}$ to ensure positive step size and solutions of desired accuracy. For a fixed number of discretization points the final time is then given by $t_f = (N - 1)h$. As initial constraint we assume a prescribed initial configuration $q_{rel}(0) = (10, 0)$ and an initial relative velocity as $\dot{q}_{rel}(0) = (-10, -10)$. The final configuration is given by $q_{rel}(t_f) = (15, 2)$, while the final relative velocity is free with same magnitude as the initial one, as the control force only influences the orientation, rather than the magnitude of the relative velocity. The current velocity is assumed to be configuration-dependent in x- and zero in y-direction given as $V = (x, 0)$. The discretization of the glider model leads to a constrained nonlinear multiobjective optimization problem. It is solved making use of the image-set oriented recover algorithm. Here, the distance of the objective values to the targets is optimized subject to the constraints stemming from the system dynamics. Figure 5.16 shows the results: the approximated Pareto front (left) and some corresponding trajectories (right). As expected, the control effort increases for decreasing maneuver time. Comparing different trajectories corresponding to different Pareto points, all trajectories show the same qualitative behavior: As the initial velocity is directed away from the destination, the gyroscopic control force enforces the glider performing a circular motion starting in direction of the initial velocity. Due to the fluid velocity in x-direction, the glider moves along a loop to reach the desired final location as depicted on the right in Figure 5.16. For trajectories with shorter time duration the loop becomes smaller and the corresponding control effort becomes higher since a big change of orientation in short time requires a high force applied to the system.

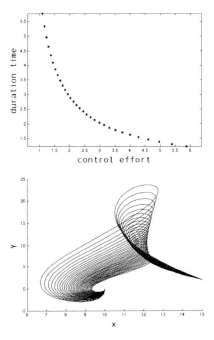

Fig. 5.16 Pareto front for the underwater glider computed with the image-set oriented recover algorithm (left), and the corresponding trajectories in configuration space (right)

5.6   Concluding Remarks

In this chapter, we have given an overview of recently developed set oriented methods for the numerical treatment of MOPs. The characteristic of these methods is that they generate box collections that aim for tight coverings of the Pareto set (or front) of a given MOP. The methods are divided into subdivision techniques and a particular kind of continuation methods (recover techniques). Subdivision techniques generate a sequence of nested box collections that converges (ideally) to the Pareto set, and the idea of the recover techniques is to extend a given collection $\mathscr{C}$ by a local search which is performed around promising elements (boxes) of $\mathscr{C}$. Subdivision techniques are of global nature and highly competitive to other state-of-the-art methods in particular if the dimension of the parameter space is moderate (say, $n < 50$), and the number of objectives is low ($k < 5$). Continuation methods are of local nature, but in turn applicable to higher dimensional problems ($n \gg 1000$). The latter has been demonstrated on several multiobjective optimal control problems which were transformed into high-dimensional MOPs.

Acknowledgements. The first author acknowledges support from CONACyT project no. 128554. This work was partly developed in the course of the Collaborative Research Center 614 – Self-Optimizing Concepts and Structures in Mechanical Engineering – University of Paderborn, and was partly funded by the Deutsche Forschungsgemeinschaft. The authors would like to thank everyone who supported them in the design and analysis of the above mentioned algorithms. Very special thanks go to Alesandro Dell'Aere, Stefan Sertl, Maik Ringkamp and Albert Seifried. Additionally we would like to thank the Chair of Control Engineering and Mechatronics (Prof. Dr.-Ing. A. Trächtler) and the Chair of Power Electronics and Electrical Drives (Prof. Dr.-Ing. J. Böcker), University of Paderborn, Germany who provided the technical applications presented in this work. In particular, we thank Jens Geisler and Tobias Knoke.

References

1. Allgower, E.L., Georg, K.: Numerical Continuation Methods. Springer (1990)
2. Betts, J.T.: Survey of numerical methods for trajectory optimization. AIAA J. Guidance, Control, and Dynamics 21(2), 193–207 (1998)
3. Binder, T., Blank, L., Bock, H.G., Bulirsch, R., Dahmen, W., Diehl, M., Kronseder, T., Marquardt, W., Schlöder, J.P., von Stryk, O.: Introduction to model based optimization of chemical processes on moving horizons. In: Grötschel, M., Krumke, S.O., Rambau, J. (eds.) Online Optimization of Large Scale Systems: State of the Art, pp. 295–340. Springer (2001)
4. Blesken, M., Rückert, U., Steenken, D., Witting, K., Dellnitz, M.: Multiobjective Optimization for Transistor Sizing of CMOS Logic Standard Cells Using Set-Oriented Numerical Techniques. In: 27th Norchip Conference (2009)
5. Bosman, P.A.N., de Jong, E.D.: Exploiting gradient information in numerical multi-objective evolutionary optimization. In: Genetic and Evolutionary Computation Conference - GECCO 2005. ACM (2005)
6. Coello Coello, C.A., Lamont, G.B., Van Veldhuizen, D.A.: Evolutionary Algorithms for Solving Multi-Objective Problems, 2nd edn. Springer, New York (2007) ISBN 978-0-387-33254-3
7. Das, I., Dennis, J.: Normal-boundary intersection: A new method for generating the Pareto surface in nonlinear multicriteria optimization problems. SIAM Journal of Optimization 8, 631–657 (1998)
8. Deb, K.: Multi-Objective Optimization using Evolutionary Algorithms. John Wiley & Sons, Chichester (2001) ISBN 0-471-87339-X
9. Dell'Aere, A.: Multi-objective optimization in self-optimizing systems. In: Proceedings of the IEEE 32nd Annual Conference on Industrial Electronics (IECON), pp. 4755–4760 (2006)
10. Dell'Aere, A.: Numerical methods for the solution of bi-level multi-objective optimization problems. PhD thesis. University of Paderborn, Germany (2008)
11. Dellnitz, M., Hohmann, A.: The computation of unstable manifolds using subdivision and continuation. In: Broer, H.W., van Gils, S.A., Hoveijn, I., Takens, F. (eds.) Nonlinear Dynamical Systems and Chaos, vol. 19, pp. 449–459. PNLDE, Birkhäuser (1996)
12. Dellnitz, M., Hohmann, A.: A subdivision algorithm for the computation of unstable manifolds and global attractors. Numerische Mathematik 75, 293–317 (1997)

13. Dellnitz, M., Ober-Blöbaum, S., Post, M., Schütze, O., Thiere, B.: A multi-objective approach to the design of low thrust space trajectories using optimal control. Celestial Mechanics and Dynamical Astronomy 105(1), 33–59 (2009)
14. Dellnitz, M., Schütze, O., Hestermeyer, T.: Covering Pareto sets by multilevel subdivision techniques. Journal of Optimization Theory and Applications 124, 113–155 (2005)
15. Dellnitz, M., Schütze, O., Sertl, S.: Finding zeros by multilevel subdivision techniques. IMA Journal of Numerical Analysis 22(2), 167–185 (2002)
16. Eichfelder, G.: Adaptive Scalarization Methods in Multiobjective Optimization. Springer, Heidelberg (2008) ISBN 978-3-540-79157-7
17. Fliege, J.: Gap-free computation of Pareto-points by quadratic scalarizations. Mathematical Methods of Operations Research 59, 69–89 (2004)
18. Fliege, J., Fux Svaiter, B.: Steepest descent methods for multicriteria optimization. Mathematical Methods of Operations Research 51(3), 479–494 (2000)
19. Fliess, M., Levine, J., Martin, P., Rouchon, P.: Flatness and defect of non-linear systems: Introductory theory and examples. International Journal of Control 61(6), 1327–1361 (1995)
20. Gandibleux, X.: Metaheuristics for Multiobjective Optimisation. Lecture notes in economics and mathematical systems. Springer (2004)
21. Geisler, J., Witting, K., Trächtler, A., Dellnitz, M.: Multiobjective optimization of control trajectories for the guidance of a rail-bound vehicle. In: 17th IFAC World Congress, Seoul, Korea, July 6-11 (2008)
22. Geisler, J., Witting, K., Trächtler, A., Dellnitz, M.: Self-optimization of the guidance module of a rail-bound vehicle. In: Gausemeier, J., Rammig, F., Schäfer, W. (eds.) 7th International Heinz Nixdorf Symposium 'Self-optimizing Mechatronic Systems: Design the Future', February 20-21, pp. 85–100. HNI-Verlagsschriftenreihe, Paderborn (2008)
23. Harada, K., Sakuma, J., Kobayashi, S., Ono, I.: Uniform sampling of local Pareto-optimal solution curves by pareto path following and its applications in multi-objective GA. In: GECCO, pp. 813–820 (2007)
24. Henderson, M.E.: Multiple parameter continuation: Computing implicitly defined k-manifolds. International Journal of Bifurcation and Chaos 12, 451–476 (2002)
25. Hestermeyer, T., Schlautmann, P., Ettingshausen, C.: Active suspension system for railway vehicles – system design and kinematics. In: 2nd IFAC - Conference on Mechatronic Systems, Berkeley, California, USA (2002)
26. Hillermeier, C.: Nonlinear Multiobjective Optimization - A Generalized Homotopy Approach. Birkhäuser (2001)
27. Horst, R., Tuy, H.: Global Optimization: Deterministic Approaches. Springer (1993)
28. Hsu, C.S.: Cell-to-cell mapping: a method of global analysis for nonlinear systems. Applied mathematical Sciences. Springer (1987)
29. Hsu, H.C.: Global analysis by cell mapping. International Journal of Bifurcation and Chaos 2, 727–771 (1992)
30. Junge, O., Marsden, J.E., Ober-Blöbaum, S.: Optimal Reconfiguration of Formation Flying Spacecraft - a decentralized approach. In: IEEE Conference on Decision and Control, San Diego, CA, USA, pp. 5210–5215 (2006)
31. Karush, W.E.: Minima of functions of several variables with inequalities as side conditions. PhD thesis, University of Chicago (1939)
32. Klamroth, K., Tind, J., Wiecek, M.: Unbiased approximation in multicriteria optimization. Mathematical Methods of Operations Research 56, 413–437 (2002)

33. Knoke, T., Romaus, C., Böcker, J., Dell'Aere, A., Witting, K.: Energy management for an onboard storage system based on multiobjective optimization. In: Proceedings of the IEEE 32nd Annual Conference on Industrial Electronics (IECON), pp. 4677–4682 (2006)
34. Krüger, M., Witting, K., Trächtler, A., Dellnitz, M.: Parametric model order reduction in hierarchical multiobjective optimization of mechatronic systems. In: 18th IFAC World Congress, Milano, Italy, August 28-September 2 (2010)
35. Kuhn, H., Tucker, A.: Nonlinear programming. In: Neumann, J. (ed.) Proceeding of the 2nd Berkeley Symposium on Mathematical Statistics and Probability (1951)
36. Leyendecker, S., Ober-Blöbaum, S., Marsden, J.E., Ortiz, M.: Discrete mechanics and optimal control for constrained systems. Optimal Control, Applications and Methods 31(6), 505–528 (2010)
37. Li, R., Pottharst, A., Witting, K., Znamenshchykov, O., Böcker, J., Fröhleke, N., Feldmann, R., Dellnitz, M.: Design and implementation of a hybrid energy supply system for railway vehicles. In: Proc. of APEC2005, IEEE Applied Power Electronics Conference 2005, Austin, Texas, USA, pp. 474–480 (2005)
38. Logist, F., Houska, B., Diehl, M., Van Impe, J.: Fast pareto set generation for nonlinear optimal control problems with multiple objectives. Structural and Multidisciplinary Optimization 42, 591–603 (2010), doi:10.1007/s00158-010-0506-x
39. Marsden, J.E., West, M.: Discrete mechanics and variational integrators. Acta Numerica 10, 357–514 (2001)
40. Miettinen, K.: Nonlinear Multiobjective Optimization. Kluwer Academic Publishers, Boston (1999)
41. Milam, M., Mushambi, K., Murray, R.: A new computational approach to real-time trajectory generation for constrained mechanical systems. In: Proceedings of the 39th IEEE Conference on Decision and Control, vol. 1, pp. 845–551 (2000)
42. Moore, A., Ober-Blöbaum, S., Marsden, J.E.: Optimization of spacecraft trajectories: a method combining invariant manifold techniques and discrete mechanics and optimal. In: 19th AAS/AIAA Space Flight Mechanics Meeting, Februar 8-12 (2009)
43. Murray, R., Rathinam, M., Sluis, W.: Differential flatness of mechanical control systems: A catalog of prototype systems. In: Proceedings of the 1995 ASME International Congress and Exposition, San Francisco (1995)
44. Neue Bahntechnik Paderborn – Project. Web-Page, http://www.railcab.de
45. Nocedal, J., Wright, S.: Numerical Optimization. Springer Series in Operations Research and Financial Engineering. Springer (2006)
46. Numerical Algorithms Group (C-Library). Web-Page, http://www.nag.co.uk/numeric/CL/CLdescription.asp
47. Ober-Blöbaum, S.: Discrete mechanics and optimal control. PhD thesis. University of Paderborn (2008)
48. Ober-Blöbaum, S., Junge, O., Marsden, J.E.: Discrete mechanics and optimal control: an analysis. Control, Optimisation and Calculus of Variations 17(2), 322–352 (2011)
49. Ober-Blöbaum, S., Timmermann, J.: Optimal control for a pitcher's motion modeled as constrained mechanical system. In: 7th International Confenrence on Multibody Systems, Nonlinear Dynamics, and Control, ASME International Design Engineering Technical Conferences, San Diego, CA, USA (2009)
50. Petit, N., Milam, M., Murray, R.: Inversion based constrained trajectory optimization. In: 5th IFAC Symposium on Nonlinear Control Systems (2001)

51. Pottharst, A., Baptist, K., Schütze, O., Böcker, J., Fröhlecke, N., Dellnitz, M.: Operating point assignment of a linear motor driven vehicle using multiobjective optimization methods. In: Proceedings of the 11th International Conference EPE-PEMC 2004, Riga, Latvia (2004)
52. Recchioni, M.C.: A path following method for box-constrained multiobjective optimization with applications to goal programming problems. Mathematical Methods of Operations Research 58, 69–85 (2003)
53. Rheinboldt, W.: On the computation of multi-dimensional solution manifolds of parametrized equations. Numerische Mathematik 53, 165–181 (1988)
54. Ringkamp, M., Ober-Blöbaum, S., Dellnitz, M., Schütze, O.: Handling high dimensional problems with multi-objective continuation methods via successive approximation of the tangent space. To appear in Engineering Optimization (2011)
55. Schäffler, S., Schultz, R., Weinzierl, K.: A stochastic method for the solution of unconstrained vector optimization problems. Journal of Optimization Theory and Applications 114(1), 209–222 (2002)
56. Schneider, T., Schulz, B., Henke, C., Witting, K., Steenken, D., Böcker, J.: Energy transfer via linear doubly-fed motor in different operating modes. In: Proceedings of the International Electric Machines and Drives Conference, Miami, Florida, USA, May 3-6 (2009)
57. Schütze, O.: Set Oriented Methods for Global Optimization. PhD thesis, University of Paderborn (2004),
    http://ubdata.uni-paderborn.de/ediss/17/2004/schuetze/
58. Schütze, O., Dell'Aere, A., Dellnitz, M.: On continuation methods for the numerical treatment of multi-objective optimization problems. In: Branke, J., Deb, K., Miettinen, K., Steuer, R.E. (eds.) Practical Approaches to Multi-Objective Optimization. Dagstuhl Seminar Proceedings, vol. 04461. Internationales Begegnungs- und Forschungszentrum (IBFI), Schloss Dagstuhl (2005),
    http://drops.dagstuhl.de/opus/volltexte/2005/349
59. Schütze, O., Jourdan, L., Legrand, T., Talbi, E.-G., Wojkiewicz, J.L.: New analysis of the optimization of electromagnetic shielding properties using conducting polymers and a multi-objective approach. Polymers for Advanced Technologies 19, 762–769 (2008)
60. Schütze, O., Vasile, M., Junge, O., Dellnitz, M., Izzo, D.: Designing optimal low thrust gravity assist trajectories using space pruning and a multi-objective approach. Engineering Optimization 41(2), 155–181 (2009)
61. Van Nieuwstadt, M.J., Murray, R.M.: Real-time trajectory generation for differentially flat systems. International Journal of Robust and Nonlinear Control 8(11), 995–1020 (1998)
62. Witting, K., Schulz, B., Dellnitz, M., Böcker, J., Fröhleke, N.: A new approach for online multiobjective optimization of mechatronic systems. International Journal on Software Tools for Technology Transfer STTT 10(3), 223–231 (2008)
63. Zhang, W., Inanc, T., Ober-Blöbaum, S., Marsden, J.E.: Optimal trajectory generation for a dynamic glider in ocean flows modeled by 3D B-Spline functions. In: IEEE International Conference on Robotics and Automation (ICRA), Pasadena, CA, USA (2008)

Part III
Landscape, Coevolution and Cooperation

Chapter 6
A Complex-Networks View of Hard Combinatorial Search Spaces

Marco Tomassini and Fabio Daolio

6.1 Hard Problems, Search Spaces, and Fitness Landscapes

According to worst-case complexity analysis, difficult combinatorial problems are those for which no polynomial-time algorithms are known (see, for instance, [15]). Thus, according to this point of view, large enough instances of these problems cannot be solved in reasonable time. The mathematical analysis is primarily based on decision problems, i.e. those that require a yes/no answer [7, 15], but the theory can readily be extended to optimization problems [16], roughly speaking, those in which we seek a solution with an associated minimum or maximum cost, which are the ones that will be dealt with here.

In spite of the above apparently negative results, there exist several approaches that all tend to provide alternative views with the purpose of approximatively solving hard problems in reasonable time. Approximation algorithms and linear programming relaxation, among others, are of this type [16]. However, if we are ready to accept to reduce our expectations even further about the quality of the solutions found, trading some rigor against more flexibility, we can include *heuristics* in our toolbox. Heuristics are approximation algorithms that seek good solutions at reasonable cost, without any guarantee about optimality or even approximability, but they usually work well in practice. Many heuristics are based on stochastic local search which, in turn, benefits from a knowledge of the structure of the given problem's solution space. Thus, the study of the structure of the search spaces of hard combinatorial problems is very relevant to understand and to improve the performance of optimization heuristics. In what follows we shall assume a discrete and finite search

Marco Tomassini · Fabio Daolio
Faculty of Business and Economics, Information System Department,
University of Lausanne, Switzerland
e-mail: {Marco.Tomassini,Fabio.Daolio}@unil.ch

E. Tantar et al. (Eds.): EVOLVE- A Bridge between Probability, SCI 447, pp. 223–245.
springerlink.com   © Springer-Verlag Berlin Heidelberg 2013

space, and a *black-box* optimization scenario that is, no information on the problem will be required except that we know, or can compute, the cost for any admissible solution of the problem.

6.1.1   Fitness Landscapes

In order to characterize finite-size combinatorial search spaces, let us begin with a few definitions. First, let S be the set of admissible solutions of the problem. One can then define a set $\mathcal{N}$ which gives the *neighborhood* of any given solution x, i.e. the solutions that can be reached from x by applying a simple move operator $op()$:

$$\mathcal{N}(x) = \{y \in S \mid y = op(x)\}.$$

This is a generalization of the neighborhood structure that would be induced by a distance function in a metric space, nonetheless one can also write:

$$\mathcal{N}(x) = \{y \in S \mid dist(y,x) = 1\},$$

which means that the neighbors y of x are those solutions that are at distance one from x, i.e. one move operation apart. We shall assume that the move operator $op()$ is a local one in the sense that it only modifies slightly the configuration x to which it is applied. Some common non-unary operators such as crossover in evolutionary algorithms are not smooth in this sense, but we shall not deal with them here.

The couple $(S, \mathcal{N})$ defines the configuration space. The third element is a function f variously called a fitness, cost, or objective function, which provides a real-valued fitness for any valid configuration in the search space:

$$f : S \to \mathbb{R},$$

In an optimization problem function f must be maximized (or minimized), i.e. one must find $x^* \in S$ such that:

$$f(x^*) \geq f(x), \ \forall x \neq x^*$$

The triple $\Lambda = (S, \mathcal{N}, f)$ defines a search space, also called a *fitness landscape* [18]. Actually, Λ defines the set of search spaces of all the instances of a given problem P, or a generic finite search space. Thus, the particular fitness landscape of a given instance $\pi \in P$ should be written as $(S_\pi, \mathcal{N}_\pi, f_\pi)$ but we shall omit the indices for the sake of simplicity as the meaning should be obvious from the context.

For continuous problems S is a real multi-dimensional space $\mathbb{R}^n$, and the function f to be optimized is $f : \mathbb{R}^n \to \mathbb{R}$. In this case a suitable neighborhood of x would be a hyper-sphere (a "ball") of radius ε centered at x. However, as stated above, in this chapter we shall restrict ourselves to discrete optimization problems only.

As an example fitness landscape one could consider S to be the binary hypercube of dimension k. $\mathcal{N}(x)$ could then be, for instance, the set of binary strings y of length

k that can be generated from a given string x by flipping a single bit uniformly at random:

$$\mathcal{N}(x) = \{y \in S \mid dist_H(y,x) = 1\},$$

where $dist_H$ is the Hamming distance. Different configurations can be evaluated according to a given fitness function $f(.)$ for the problem. For instance, the "maxone" problem calls for maximizing the number of 1s in the string and thus $f(x)$ is simply the number of ones in x in this case.

Another example comes from combinatorial problems such as the TSP in which S is the set of Hamiltonian cycles of a graph G. Using the 2-*opt* move, the neighbors of a given tour x are the tours y that can be obtained from x by removing two edges and inserting two new edges such that the result is still a tour. The fitness $f(x)$ of a tour x is its length. Many other examples can be found in the specialized literature [16, 20].

Since every solution $x \in S$ must be reachable from any other solution y, it is obvious that the fitness landscape can be viewed as a connected graph $G(V,E)$ in which the set of vertices V is identical to the elements of S, and the set of edges E represents all possible transitions between solutions according to a given move operator. In fact, the whole Λ can be viewed as a connected labeled graph in which the label of a vertex (a solution) is the solution's fitness. We assume for simplicity that the graph is undirected, i.e. if solution s_i can be reached from solution s_j, then the opposite is also true, which is very often the case. However, for some definitions of move operators the graph might be a directed one instead.

The concept of a fitness landscape turns out to be extremely useful when it comes to local search heuristics for the optimization of difficult problems. Indeed, several statistical properties of landscapes that can be readily computed by sampling the search space may provide information about the difficulty of the search and suggest effective search algorithms for such spaces. For a detailed description of suitable statistics on landscapes and how to obtain them, we refer the reader to [20] and to the specialized literature such as [10] and references therein. However, to make the chapter as self-contained as possible, we shall briefly present the more important concepts.

Local Optima

A local optimum, which is taken to be a maximum here, is a solution s_l such that

$$\exists \mathcal{N}(s_l) \mid \forall s \in \mathcal{N}(s_l), \quad f(s) < f(s_l).$$

Basins of Attraction

The basin of attraction of a local optimum $s_l \in S$ is defined here as the set $b(s_l) = \{s \in S \mid HillClimbing(s) = s_l\}$. The size of the basin of attraction of a local optimum s_l is the cardinality of $b(s_l)$. Hillclimbing means that, in each time step, we choose among all the neighbors the solution that provides the best improvement of the fitness function. Obviously, such a search must end at a local maximum. The basin definition is dependent on the local search technique used. For example, with *First improvement* instead of best improvement, the basins found will in general be different. Moreover, following Jones [9], since the fitness landscape for the same problem may appear different depending on the move operators used, the above definition only makes sense once such an operator has been defined. For instance, the TSP configuration space and its set of local optima for a given problem instance is not the same when using a 2-*opt* or a 3-*opt* neighborhood move operator, although the set of admissible solutions is identical.

Walking and Sampling a Fitness Landscape

Several landscapes features have been used to empirically characterize it from a statistical point of view. Some of them are: number and density of optimal solutions, distribution of optima, and correlation measures [20]. For example, fitness-distance correlation has the objective of establishing a relationship between the solution quality and its distance from the known global optimum using a sample of fitness values and the corresponding distances from the optimum (when it is known). From the absence or presence of a such a correlation one may conjecture some conclusions about the difficulty of the corresponding landscape. However, the analysis may deliver wrong predictions because of sampling inaccuracies and because of particular features of the landscape that render uniform random sampling inappropriate.

Random walks on the landscape are also useful. Consider a random walk Γ on the fitness landscape graph $\Gamma = \{x_0, x_1, \ldots\}$ and the corresponding sequence of fitness function values $\{f(x_0), f(x_1), \ldots\}$ seen along the walk. Autocorrelation functions of the fitness values sequence are useful to estimate the "ruggedness" of the landscape, i.e. the variability of fitness along the walk [23]. A high value of the autocorrelation function means that neighbors have similar fitnesses, indicating that the landscape tends to be smooth, whereas a low value indicates that variations in fitness between neighboring solutions are uncorrelated, giving rise to more rugged landscapes. Ruggedness is related to the difficulty of searching a landscape: in general, the more rugged the landscape, the harder it is to search for the global optimum since more rugged landscapes have more local optima in which the search may get stuck. Moreover, in more uncorrelated landscapes, a given position carries less information about the neighboring ones, making it harder to direct the search. Note, however, that the above sampling method only gives statistically reliable answers for isotropic fitness landscapes and, even in this case, its conclusions depend on the neighborhood relationship, i.e. on the move operator.

For reasons of space we end our brief description of empirical landscape analysis here. The interested reader will find many more details in [10, 20] and references therein.

6.1.2   Local Optima Networks

Another, more compressed way, of defining a fitness landscape is by focusing only on the local optima. In other words we shall assume that all the solutions that are local optima have been found somehow, for example by running a hill-climbing search starting from all the points of the fitness landscape. Of course, such an exhaustive enumeration is only feasible for small problem instances, but we shall assume for the moment that it can always be performed in principle. This approach has recently been proposed in [14, 21].

We now define a new graph G', the vertices of which constitute the set S^* of the local optima of the landscape. The edges of the graph G' will stand for possible transitions between the above optima. We consider that two local optima i and j are connected when there is at least a transition between a solution belonging to the basin b_i of i and another solution belonging to basin b_j. Of course there can be more than two solutions with this property. We therefore define the *weight* of an edge that connects two basins in the fitness landscape to account for this fact.

For each pair of solutions s and s', $p(s \rightarrow s')$ is the probability to go from s to s' with the given neighborhood structure. In the case of binary strings of size N, and the neighborhood defined by the single bit-flip operation, there are N neighbors for each solution, therefore:

$$p(s \rightarrow s') = \begin{cases} \frac{1}{N}, & \text{if } s' \in \mathcal{N}(s) \\ 0, & \text{if } s' \notin \mathcal{N}(s) \end{cases}$$

The probability to go from a solution $s \in S$ to a solution belonging to the basin b_j, is defined as:

$$p(s \rightarrow b_j) = \sum_{s' \in b_j} p(s \rightarrow s').$$

Notice that $p(s \rightarrow b_j) \leq 1$.

Thus, the total probability of going from basin b_i to basin b_j is the average over all $s \in b_i$ of the transition probabilities to solutions $s' \in b_j$:

$$p(b_i \rightarrow b_j) = \frac{1}{|b_i|} \sum_{s \in b_i} p(s \rightarrow b_j),$$

where $|b_i|$ is the size of the basin b_i.

Now we can define a *weighted local optima network* (LON) $G' = (S^*, E)$ as being the graph where the nodes are the local optima, and there is an edge $e_{ij} \in E$ with weight $w_{ij} = p(b_i \rightarrow b_j)$ between two nodes i and j if $p(b_i \rightarrow b_j) > 0$. Notice that since each maximum has its associated basin, G' also describes the interconnection of basins.

According to our definition of edge weights, $w_{ij} = p(b_i \rightarrow b_j)$ may be different from $w_{ji} = p(b_j \rightarrow b_i)$. Thus, two weights are needed in general, and we have an oriented transition graph.

The previous definitions for the inter-basin transition probabilities only depend on the basin connectivity and try to convey topological information without referring to a particular local search method. In this respect, those probabilities make sense from a sampling point of view. In fact, for metaheuristic such as Tabu or Simulated Annealing, fitness difference information would obviously be required.

Moreover, the above definitions hold for non-neutral landscapes, i.e. those landscapes in which neighbor solutions do not have the same fitness values. When many solutions have the same fitness as their neighbors in sizable parts of the landscape, we say that the landscape has a high degree of neutrality. The previous descriptions can be extended to this important case as shown in [22] but, for the sake of simplicity, here we shall deal with the non-neutral case only.

6.1.3 Some Definitions for Weighted Complex Networks

To make the chapter as self-contained as possible, in this section we give an account of a number of statistics that are useful in dealing with weighted complex networks. The treatment must necessarily be brief; readers will find a fuller exposition in the original paper [1] and in [13].

The standard clustering coefficient [13] does not consider weighted edges. We thus use the *weighted clustering* $c^w(i)$ measure proposed by [1], which combines the topological information with the weight distribution of the network:

$$c^w(i) = \frac{1}{s_i(k_i - 1)} \sum_{j,h} \frac{w_{ij} + w_{ih}}{2} a_{ij} a_{jh} a_{hi}.$$

In the previous expression a_{ij} is an element of the graph's *adjacency matrix A*, defined as $a_{ij} = 1$ if $w_{ij} > 0$, $a_{ij} = 0$ if $w_{ij} = 0$. Finally, $k_i = \sum_{j \neq i} a_{ij}$ is the degree of node i, whereas $s_i = \sum_{j \neq i} w_{ij}$ is a generalization of a node's degree for weighted networks called the node's *strength* (see also below). For each triple formed in the neighborhood of the vertex i, $c^w(i)$ counts the weight of the two participating edges of the vertex i. C^w is defined as the weighted clustering coefficient averaged over all vertices of the network.

The standard topological characterization of networks is obtained by the analysis of the probability distribution $p(k)$ that a randomly chosen vertex has degree k. For our weighted networks, a characterization of weights is obtained by the *weight distributions* $p_{in}(w)$ and $p_{out}(w)$ that a given edge has, respectively, incoming or outgoing weight w.

In the case of LONs, for each node i, the sum of outgoing edge weights is equal to 1 as they represent transition probabilities. So, an important measure is the weight w_{ii} of self-connecting edges (i.e. the probability of remaining in the same node). We have the relation: $w_{ii} + s_i = 1$. The vertex *strength*, s_i, is defined as $s_i = \sum_{j \in \mathcal{V}(i) - \{i\}} w_{ij}$, where the sum is over the set $\mathcal{V}(i) \setminus \{i\}$ of neighbors of i [1]. The strength of a node is a generalization of the node's connectivity giving information about the number and importance of the edges.

Another useful weighted network measure is *disparity* [1] $Y_2(i)$, which measures how heterogeneous are the contributions of the edges of node i to the total weight (strength):

$$Y_2(i) = \sum_{j \neq i} \left(\frac{w_{ij}}{s_i} \right)^2$$

The disparity could be averaged over the node with the same degree k. If all weights are nearby of s_i/k, the disparity for nodes of degree k is nearby $1/k$.

Finally, in order to compute the average shortest path between two nodes on the optima network of a given landscape, we consider the expected number move operations to pass from one basin to the other. This expected number can be computed by considering the inverse of the transition probabilities between basins. In other words, if we attach to the edges the inverse of the transition probabilities, this value would represent the average number of random mutations to pass from one basin to the other. More formally, the distance (e.g. expected number of bit-flip mutations) between two nodes is defined as $d_{ij} = 1/w_{ij}$, where $w_{ij} = p(b_i \rightarrow b_j)$. Now we can define the length of a path between two nodes as being the sum of these distances along the edges that connect the corresponding basins. Since the graphs are weighted and the weights are always positive, Dijkstra's algorithm is used to compute all the shortest paths.

6.2   Local Optima Networks of NK Landscapes

In this section we shall give an example of the construction and the properties of LONs for a standard family of fitness landscapes, the NK landscapes. The NK family of landscapes [11] is a model for constructing multimodal landscapes that can gradually be tuned from smooth to rugged. The model is defined on the binary hypercube of dimension N, $\mathbb{B}^N = \{0,1\}^N$, with N referring to the number of (binary) genes in the genotype (i.e. the string length), whereas K is the number of genes that influence a particular gene (the epistatic interactions). By increasing the value of K from 0 to $N-1$, the ruggedness of the NK landscapes increases.

The fitness function of a NK-landscape $f_{NK} : \{0,1\}^N \rightarrow [0,1)$ is defined on binary strings with N bits. An 'atom' with fixed epistasis level is represented by a fitness component $f_i : \{0,1\}^{K+1} \rightarrow [0,1)$ associated to each bit i. Its value depends on the allele at bit i and also on the alleles at the K other epistatic positions (K must fall between 0 and $N-1$). The fitness $f_{NK}(s)$ of $s \in \{0,1\}^N$ is the average of the values of the N fitness components f_i:

$$f_{NK}(s) = \frac{1}{N} \sum_{i=1}^{N} f_i(s_i, s_{i_1}, \ldots, s_{i_K})$$

where $\{i_1,\ldots,i_K\} \subset \{1,\ldots,i-1,i+1,\ldots,N\}$. Several ways have been proposed to choose the K other bits from N bits in the bit string. Two possibilities are mainly used: adjacent and random neighborhoods. With an adjacent neighborhood, the K bits nearest to the bit i are chosen (the genotype is taken to have periodic boundaries). With a random neighborhood, the K bits are chosen randomly on the bit string. Each fitness component f_i is specified by extension, *i.e.* a number $y^i_{s_i,s_{i_1},\ldots,s_{i_K}}$ from $[0,1)$ is associated with each element $(s_i,s_{i_1},\ldots,s_{i_K})$ from $\{0,1\}^{K+1}$. Those numbers are uniformly distributed in the range $[0,1)$.

For $K = 0$ all contributions can be optimized independently which makes f_{NK} a simple additive function with a single maximum. At the other extreme when $K = N - 1$ the landscape becomes completely random, the probability of any given configuration of being a local optimum is $1/(N+1)$, and the expected number of local optima is $2^N/(N+1)$. Intermediate values of K interpolate between these two cases and have a variable degree of "epistasis", i.e. of gene interaction [11].

Table 6.1 NK landscapes network properties for $N = 18$. Values are averages over 30 random instances, standard deviations are shown as subscripts. n_v and n_e represent the number of vertices and edges (rounded to the next integer), $\bar{C}^w$ is the mean weighted clustering coefficient. $\bar{Y}$ represents the mean disparity coefficient, $\bar{d}$ the mean path length (see text for definitions).

		$N = 18$			
K	$\bar{n}_v$	$\bar{n}_e$	$\bar{C}^w$	$\bar{Y}$	$\bar{d}$
2	50_{25}	1579_{1854}	$0.95_{0.0291}$	$0.307_{0.0630}$	73_{15}
4	330_{72}	26266_{7056}	$0.92_{0.0137}$	$0.127_{0.0081}$	174_9
6	994_{73}	146441_{18685}	$0.78_{0.0155}$	$0.076_{0.0044}$	237_5
8	$2,093_{70}$	354009_{18722}	$0.64_{0.0097}$	$0.056_{0.0012}$	273_2
10	$3,619_{61}$	620521_{20318}	$0.52_{0.0071}$	$0.044_{0.0007}$	292_1
12	$5,657_{59}$	899742_{14011}	$0.43_{0.0037}$	$0.038_{0.0003}$	297_1
14	$8,352_{60}$	1163640_{11935}	$0.36_{0.0023}$	$0.034_{0.0002}$	293_1
16	$11,797_{63}$	1406870_{6622}	$0.32_{0.0012}$	$0.032_{0.0001}$	283_1
17	$13,795_{77}$	1524730_{4818}	$0.30_{0.0009}$	$0.032_{0.0001}$	277_1

In Table 6.1 we report some basic network statistics for the LONs of NK landscapes with $N = 18$ and for K up to $K = 17$ [21]. $N = 18$ is still a size that allows an exhaustive enumeration of all the maxima in the search spaces. Some of these network statistics are related to stochastic local search difficulty on the underlying fitness landscapes and the discussion below can be taken as an example of how these relationships may be uncovered. For further details the reader is referred to the original work [21].

Clustering Coefficients

The fourth column of Table 6.1 lists the average values of the weighted clustering coefficients for $N = 18$ and all K. It is apparent that the clustering coefficients decrease regularly with increasing K. For the standard unweighed clustering, this would mean that the larger K is, the less likely that two maxima which are connected to a third one are themselves connected. Taking weights, i.e. transition probabilities into account, this means that either there are less transitions between neighboring basins for high K, and/or the transitions are less likely to occur. This confirms from a network point of view the common knowledge that search difficulty increases with K.

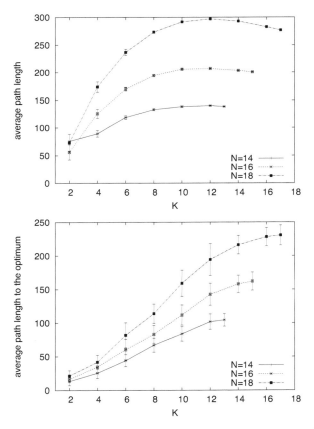

Fig. 6.1 Average distance (shortest path) between nodes (top), and average path length to the optimum from all the other basins (bottom). Error bars are shown.

Shortest Path to the Global Optimum

The average shortest path lengths $\bar{d}$ are listed in the sixth column of Table 6.1.
Fig. 6.1 (top) is a graphical illustration of the average shortest path length between
optima for NK landscapes with $N = 14, 16, 18$. Notice that the shortest path increases with N; this is to be expected since the number of optima increases exponentially with N. More interestingly, for a given N the shortest path increases with
K, up to $K = 10$, and then it stagnates and even decreases slightly for the $N = 18$.
This correlates quite well with the known fact that the search difficulty in NK landscapes increases with K. However, some paths are more relevant from the point of
view of a stochastic local search algorithm following a trajectory over the maxima
network. In order to better illustrate the relationship of this network property with
the search difficulty by heuristic local search algorithms, Fig. 6.1 (bottom) shows
the shortest path length to the global optimum from all the other optima in the landscape. The trend is clear, the path lengths to the optimum increase steadily with
increasing K.

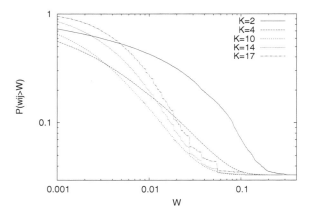

Fig. 6.2 Cumulative probability distribution of the network weights w_{ij} for outgoing edges
with $j \neq i$ in log-log scale, for $N = 18$. Averages of 30 instances for each K are reported.

Outgoing Weight Distribution

Here we report on the outgoing weight distributions $p_{out}(w)$ of the maxima network
edges. Fig. 6.2 shows the empirical cumulative probability distribution functions
for the $N = 18$ on log-log scale. One can see that the weights, i.e. the transition
probabilities to neighboring basins are small. The distributions are far from uniform
or Poissonian, empirically judging by visual inspection of the results. They are not
close to power-laws either for in this case they should appear as straight lines on

the plot at least before the degree cutoff. And indeed, we couldn't find a simple fit to the curves such as stretched exponentials or exponentially truncated power laws; however, it is apparent that the low K have longer tails. For high K the decay is faster. This seems to indicate that, on average, the transition probabilities are higher for low K. A possible explanation is that basins are shrinking with increasing K (see section 6.2.1).

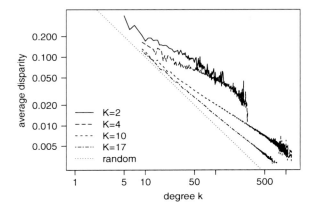

Fig. 6.3 Average disparity, Y_2, of nodes with a given degree k, for $N = 18$. Average of 30 independent instances for each K are reported. The curve $1/k$ is also reported to compare to the random case.

Disparity

Figure 6.3 depicts the disparity coefficient as defined in the previous section for $N = 18$. An interesting observation is that the disparity (i.e. inhomogeneity) in the weights of a node's out-coming links tends to decrease steadily with increasing K. This reflects that for high K the transitions to other basins tend to become equally likely, which is another indication that the landscape, and thus its representative maxima network, becomes more random and difficult to search.

When K increases, the number of edges increases and the number of edges with a weight over a certain threshold increases too (see Fig. 6.2). Therefore, for small K, each node is connected with a small number of nodes, through transitions with a relative high probability. On the other hand, for large K, the weights become more homogeneous in the neighborhood, that is, for each node, all the neighboring basins are at similar distance.

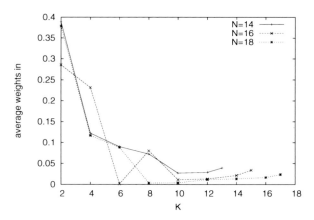

Fig. 6.4 Average value of the weights of incoming transitions into maxima nodes for $N = \{14, 16, 18\}$ and for the whole K interval.

Incoming Weights Distribution

It is also of interest to study the distribution of the weights of edges impinging into a given node $p_{in}(w)$. Instead of the histograms of $p_{in}(w)$, which do not indicate a clear dependence on K, we show in Fig. 6.4 the average w of incoming transitions, evaluated over 30 independent landscapes for $N = \{14, 16, 18\}$ as a function of K. The general trend for all values of N is that the average weight of the incoming transitions into a node quickly decreases with increasing K. This means that it is more difficult to make a transition to a given local maximum or to reach a randomly chosen one when K is large. This agrees with the fact that the basins' size is a rapidly decreasing function of K. In fact, there is a strong positive correlation between the basins' size and the weights of the transitions into the corresponding maximum, i.e. as the basin becomes larger, the number of transitions into it increases too.

6.2.1 *Basins of Attraction*

As explained in Sect. 6.1.2, the methodology used to build the optima networks also yields the associated basins of the fitness landscapes. Since the size and number of basins play an important role in search algorithms, it is useful to study their properties. Furthermore, some characteristics of the basins can be related to the optima network features. The following discussion highlights several interesting basins' features.

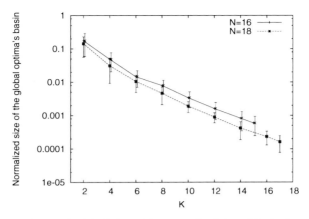

Fig. 6.5 Average with standard deviation of the normalized size of the basin corresponding to the global maximum for each K over 30 independent landscapes. The normalization is done with respect to the total size of the search space.

Basins Shape

In order to complement the previous discussion about transitions between basins, we must shed some light on the shape of these basins and of their borders. Surprisingly, over all the observed instances, except for those with $N = 14$ and $K = 2$, the average size of the basin interior is always less than 1% of the size of the basin itself. In other words, the majority of solutions sit on the basin frontier and neighboring basins are richly interconnected [21].

Global Optimum Basin Size Versus K

In Figure 6.5 we plot the average size of the basin corresponding to the global maximum for $N = 16$ and $N = 18$, and all values of K studied. The trend is clear: the basin shrinks very quickly with increasing K. This confirms that the higher the K value, the more difficult for a stochastic search algorithm to locate the basin of attraction of the global optimum.

Number of Basins of a Given Size

Figure 6.6 shows the empirical distribution of the number of basins of a given size, cumulated over all the analyzed instances with $N = 18$ for some representative values of K. It turns out that the distribution decays exponentially or faster for the lower K and it is closer to exponential for the higher K [21]. This observation is relevant to theoretical studies that estimate the size of attraction basins (see for example [8]).

These studies often assume that the basin sizes are uniformly distributed, which is not the case for the NK landscapes studied here. High values of K give rise to steeper distributions. This indicates that there are less basins of large size for large values of K. In consequence, basins are broader for low values of K, which is consistent with the fact that those landscapes are smoother.

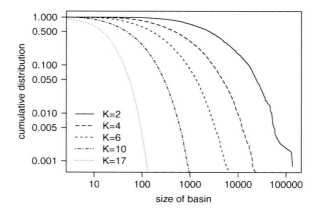

Fig. 6.6 Cumulative distribution of the number of basins of a given size. The empirical distribution is computed on 30 instances with $N = 18$ for each K value in the legend. The scaling is log-log.

Fitness of Local Optima Versus Their Basin Sizes

The scatter-plot in Fig. 6.7 illustrates the correlation between the basin sizes of local maxima (in logarithmic scale) and their fitness values. Notice that there is a clear positive correlation between the fitness values of maxima and their basins' sizes. An instance for $N = 18$ and $K = 8$ is shown but the trend is similar for all K: notably, the average Spearman correlation coefficient is above 0.8 for all K. In other words, the higher the peak the wider tends to be its basin of attraction. Therefore, on average, with a stochastic local search algorithm using the 1-bit flip operator, the global optimum would be easier to find than any other local optimum. This may seem surprising. But we have to keep in mind that as the number of local optima quickly increases with increasing K (see [11] and Table 6.1), the global optimum basin is more difficult to reach by a stochastic local search algorithm (see Fig. 6.5). This observation offers a mental picture of NK landscapes: we can consider the landscape as composed of a large number of mountains (each corresponding to a basin of attraction), and those mountains are wider the taller the hilltops. Moreover, the size of a mountain basin grows exponentially with its hight.

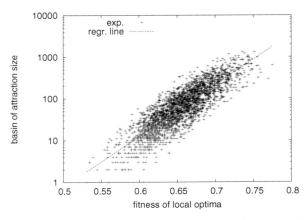

Fig. 6.7 Correlation between the fitness of local optima and their corresponding basin sizes, for a representative instance with $N = 18$ and $K = 8$

6.3   LONs for the QAP Fitness Landscapes

The family of NK landscapes are artificial spaces that are characterized by variable epistasis, randomness, and isotropy. But they are related to naturally-occurring problems such as spin glasses, a condensed-matter physics model, and more precisely to p-spin models [5], where p plays a role similar to K. In spin glasses the function analogous to f_{NK} is the energy H and the stable states are the minima of the energy hyper-surface. The *Quadratic Assignment Problem* (QAP), on the other hand, is a member of the class of computationally hard problems [7], and directly occurs in practice in industrial layout applications, even if we shall use here simplified problem generators to build instances, rather than real data.

The QAP deals with the relative location of units that interact with one another in some manner. The objective is to minimize the total cost of interactions. The problem can be stated in this way: there are n units or facilities to be assigned to n predefined locations, where each location can accommodate any one unit; location i and location j are separated by a distance a_{ij}, generically representing the per unit cost of interaction between the two locations; a flow of value b_{ij} has to go from unit i to unit j; the objective is to find an assignment, i.e. a bijection from the set of facilities onto the set of locations, which minimizes the sum of products flow $\times$ distance.

Mathematically it can be formulated as:

$$\min_{\pi \in P(n)} C(\pi) = \sum_{i=1}^{n} \sum_{j=1}^{n} a_{ij} b_{\pi_i \pi_j}$$

where $A = \{a_{ij}\}$ and $B = \{b_{ij}\}$ are the two $n \times n$ distance and flow matrixes, π_i gives the location of facility i in permutation $\pi \in P(n)$, and $P(n)$ is the set of all permutations of $\{1, 2, ..., n\}$, i.e. the QAP search space. The structure of the distance and flow matrices characterizes the class of instances of the QAP problem.

In order to perform a statistical analysis, several problem instances of at least two different problem classes have to be considered. To this purpose, the two instance generators proposed by Knowles and Corne [12] for the multi-objective QAP have been adapted and used here for the single-objective QAP. The first generator produces uniformly random instances where all flows and distances are integers sampled from uniform distributions. This leads to the kind of problem known in literature as *Tai nna*, nn being the problem dimension [19]. Distance matrix entries are, in both cases, the Euclidean distances between points in the plane. The second generator permits to obtain flow entries that are non-uniform random values. This procedure, detailed in [12], produces random instances of type *Tai nnb* that have the so called "real-like" structure since they resemble to the structure of QAP problems found in practical applications. For a general network analysis, many random uniform and random real-like instances have been generated for each problem dimension in $\{5, ..., 10\}$. Problem size 11 is the largest one for which an exhaustive sample of the configuration space is computationally feasible. Beyond that, sampling must be used. However, here we prefer to stick with exact results in order to give as accurate as possible answers.

6.3.1 General Network Features

The complete results of the statistical analysis of the above QAP landscapes appear in [4], to which the reader is referred for further information. Here we shall first comment in detail only about inter-basin transition probabilities and shortest paths and then we attempt a brief comparison with the NK case.

Figure 6.8 (top) reports for each problem dimension the average weight w_{ii} of self-loop edges. These values represent the one-step probability of remaining in the same basin after a random move. The higher values observed for real-like instances are related to their fewer optima but bigger basins of attraction. However, the trend is generally decreasing with the problem dimension. This is another confirmation that basins are shrinking with respect to their relative size.

A similar behavior characterizes the average weight w_{ij} of the outgoing links from each vertex i, as figure 6.8 (bottom) reports. Since $j \neq i$, these weights represent the probability of reaching the basin of attraction of one of the neighboring local optima. These probabilities decrease with the problem dimension. The difference between the two classes of QAP could be explained here by their different LON size.

A clear difference in magnitude between w_{ii} and w_{ij} can be observed. This means that, after a move operation, it is more likely to remain in the same basin than to reach another basin. Moreover, the decreasing trend with the problem dimension is

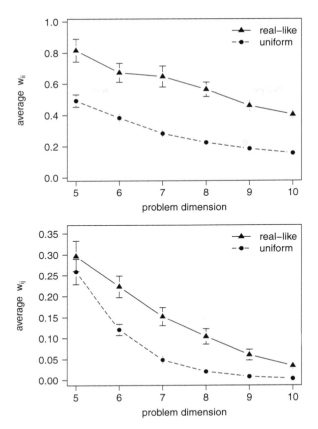

Fig. 6.8 Average transition weights w_{ii} for self-loops (top) and w_{ij} for out-going links (bottom). For each problem dimension, 30 independent and randomly generated instances are considered for each problem class (see legend). The mean value is estimated with a 95% confidence level from a one-sample t-test; error bars show the corresponding confidence intervals.

stronger for w_{ij} than for w_{ii}, especially for the uniform QAP (whose LON grows faster with the problem dimension). Therefore, even for these small instances, the probability of reaching a particular neighboring basin becomes rapidly smaller than the probability of staying in the same basin, by an order of magnitude.

Based on the weighted edges, the mean path length can be calculated as the average of all the shortest paths between any two nodes (see figure 6.9 (top)). Figure 6.9 (bottom), reports a related measure, namely the mean shortest distance from each node to the global optimum. This metric can be more interesting from the point of view of a stochastic local search heuristic trying to solve the considered QAP instance. The clear trend is that this path, as any other, increases with the problem size. Values are remarkably higher for the uniform instances, which have a larger number of local optima than the real-like instance for the same problem dimension. The figures confirm that the search difficulty increases with the domain size and the ruggedness of the fitness landscape (*i.e.* the number of local optima).

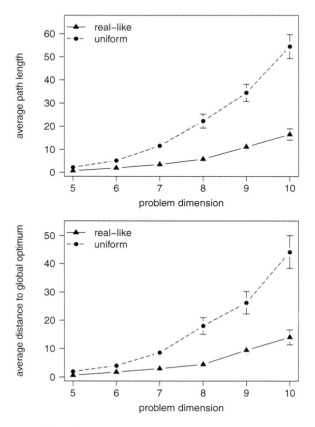

Fig. 6.9 Average path length (top) and average length of the shortest path to the global optimum (bottom). For each problem dimension, 30 independent and randomly generated instances are considered for each problem class (see legend). The mean value is estimated with a 95% confidence level from a one-sample t-test; error bars show the corresponding confidence intervals.

Now we briefly compare the nature of the *NK* and QAP problem classes LONs. Because there are no "real-like" instances of *NK*-landscapes, a comparison is made with uniform instances of QAP leading to a similar problem space size. We consider the uniform QAP with $n = 9$ and the *NK*-landscape with $N = 18$; in this case, the QAP space is of size $9! \approx 2^{18.47}$, whereas the *NK* size is 2^{18}. To have a comparable number of local optima, we take an epistasis value of $K = 3$: an $NK_{18,3}$ has on average 118.8 local optima and $118.8 \times 9!/2^{18} \approx 164.5$ is close to 137.3, the average number of local optima for uniform QAP_9.

The average sizes of the global optimum basins with respect to the total size of the search space are, respectively, 0.0719 (standard deviation 0.0384) for $NK_{18,3}$, and 0.0725 (0.0308) for QAP_9. It is also worth noting that the distribution of basin sizes with respect to the number of local optima is similar as well. Therefore, the probabilities of reaching the global optimum from a random solution are nearly equal for QAP_9 and $NK_{18,3}$, thus one may conjecture that the difficulty for searching methods based on Hill-Climbing are similar. Indeed, even if average path lengths are different in general, the average lengths of paths to the global optimum are similar as well (cf. Figs 6.1 and 6.9).

The main difference lies in connectivity: whereas the LON is nearly a complete graph for QAP_9, for $NK_{18,3}$ landscapes it is not. The average number of edges to the squared number of nodes is in fact 0.91 for QAP_9 and 0.66 for $NK_{18,3}$. A related and valuable statistics is the average probability of staying in the same basin, w_{ii}, which is 0.19 and 0.37 respectively for QAP_9 and $NK_{18,3}$. In other words, by random moves it is easier to explore the local optima of the QAP instances than those of NK. Admittedly, the pair exchange operator defines a neighborhood in the QAP permutation space that is larger than the neighborhood defined by the one bit-flip mutation in the NK binary space. This reflects on the LON connectivity and suggests that the most efficient local searcher (based on those moves) should be different: the tradeoff between exploration and exploitation should not be the same.

In spite of this, given the aforementioned similarity between problem classes as different as the NK family and the QAP ones, we might tentatively suggest that there could be some general patterns in the structure of hard combinatorial landscapes.

6.3.2   Optima Distribution and Clustering

Optima in search spaces may be distributed uniformly, as some theoretical analyses of fitness landscapes seem to assume for mathematical simplicity [8], or they may be clustered in some non-homogeneous way. Due to their randomness, LONs derived from NK landscapes have little cluster structure; in fact, they appear isotropic [23] also from the point of view of basins interconnectivity. In QAP, on the other hand, the situation could be different. There exist statistics that can be gathered on fitness landscapes that give indications about the distribution of solutions and optima (see, for instance, the book [20]). However, the purely topological approach advocated here may offer several advantages since only information about the vertices of the graph and their connections is needed. In complex network theory language, this corresponds to the detection of *communities* in the relevant LON networks. Community detection is a difficult task, but today several good approximate algorithms are available [6].

As remarked at the end of the previous section, the LONs of both the uniform and real-like instances up to size 10 are complete or almost complete since

$|E| = O(|V|^2)$. This is inconvenient for community analysis as it is difficult for any cluster detection algorithm to split-up the networks into separate communities when the graphs are so dense. However, exploiting the fact that edges with small weights represent transitions with low probability, we can filter out the network by deleting all edges whose weights are below a given threshold. This seems reasonable as most local stochastic search heuristics, such as simulated annealing, ultimately rely on highly probable transitions, although worsening moves can be accepted at the beginning. The details of the filtering procedure can be found in the original work [3]. To summarize, the dense weighted directed graph is transformed into a sparser weighted undirected one by systematically eliminating edges with small weights.

Communities or clusters in networks can be loosely defined as being groups of nodes that are strongly connected between them and poorly connected with the rest of the graph. Several detection heuristics have been proposed [6]; after a preliminary analysis, we chose two of them: Clauset et al's. method based on greedy modularity optimization [2], and Reichardt's and Bornholdt's algorithm based on a spin-glass model and simulated annealing [17]. Both methods gave consistent results on our networks and, in addition, work with undirected weighted networks, which was required in our case.

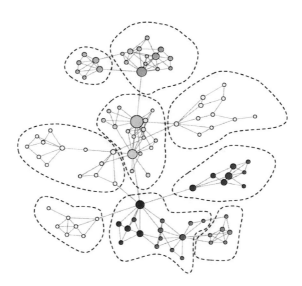

Fig. 6.10 Community structure of the LON of a real-like instance of size 11, i.e. from *Tai11b* class. The cluster partition found is highlighted. Node sizes are proportional to the corresponding basin size. Darker colors mean better fitness (lower).

Fig. 6.11 Cluster structure of the LON of a random uniform instance of size 9, i.e. from *Tai9a* class. Clusters are less well separated (see text) and cannot be clearly highlighted. Node sizes are proportional to the corresponding basin size. Darker colors mean better fitness.

The results are interesting in that random and real-like QAP instances give rise to very different LON structures from the point of view of minima distribution. While real-like instances do present a statistically significant cluster structure, the class of random uniform instances has poor community structure. This can be visually appreciated on two particular but typical cases shown in Figs. 6.10 and 6.11.

In [4] it was found that there is a positive correlation between fitness values and the corresponding basin size, especially for the random uniform problem instances. This effect is qualitatively easy to spot on the figures. The results of this community study, together with [4], shed light on an open problem in the structure of difficult combinatorial landscapes. The basin sizes of these problems have been often taken either constant or uniformly distributed at random for mathematical reasons of simplicity [8]. However, this is far from being the case for the QAP problem [4] as well as for the NK landscapes. While this conclusion cannot be generalized, it could also hold for other families of difficult combinatorial problems based, as the QAP, on permutation neighborhood such as the Traveling Salesman Problem (TSP).

6.4   Conclusions and Prospects

In this chapter we have presented a novel and potentially useful way of representing the fitness landscapes of combinatorial search problems. The representation is based on a graph structure in which vertices are local optima of the search space and oriented edges represent transition probabilities between the optima. We have shown how this information can be relevant by using two well-known discrete problems: the NK family of fitness landscapes and two classes of instances of the QAP problem. The analysis has shown that topological features such as node degree, clustering coefficient, edge weight disparity, and mean path lengths, among others, can be qualitatively related to problem difficulty in a straightforward way and could suggest structural explanations for the search difficulty. In addition, the methodology also yields the basins associated to the local optima. An analysis of the basins has unveiled the fact that, far from being uniformly distributed, basin size distribution is right-skewed. Besides, optimum fitness and basin size has been found to be positively correlated in all cases. Concerning the clustering of optima, it has been found by community detection methods that NK landscapes and random uniform instances of the QAP problem do not possess a clear cluster structure. On the contrary, optima of the real-like QAP instances cluster in a statistically significant manner. This conclusion may have interesting consequences for the design of efficient local search heuristics for those spaces.

Nevertheless, it is worth repeating that the work presented here relies on small instances of the problems. Indeed, only for those an exhaustive analysis of the fitness landscapes is feasible. However, larger instances would be more typical of real problems and should be studied as well. In order to cope with the computational complexity one has to resort to sampling the relevant networks. In fact, we are designing and analyzing efficient sampling methodologies that should allow us to tackle larger problems.

Although the mere analysis of hard combinatorial landscapes is interesting *per se*, the ultimate aim of this research is to be able to exploit LON knowledge in order to design or guide local search heuristics with the objective of making them more efficient. This activity is part of our ongoing and future research.

Acknowledgements. The authors are grateful to the Swiss National Foundation for financial support of this project under grant number 200021-124578. The work presented here is the result of our collaboration with G. Ochoa and S. Vérel: we acknowledge stimulating and useful discussions with our coworkers.

References

1. Barthélemy, M., Barrat, A., Pastor-Satorras, R., Vespignani, A.: Characterization and modeling of weighted networks. Physica A 346, 34–43 (2005)
2. Clauset, A., Newman, M.E.J., Moore, C.: Finding community structure in very large networks. Physical Review E 70(6), 66111 (2004)
3. Daolio, F., Tomassini, M., Vérel, S., Ochoa, G.: Communities of minima in local optima networks of combinatorial spaces. Physica A 390, 1684–1694 (2011)

4. Daolio, F., Vérel, S., Ochoa, G., Tomassini, M.: Local optima networks of the quadratic assignment problem. In: IEEE Congress on Evolutionary Computation, CEC 2010, pp. 3145–3152. IEEE Press (2010)
5. Derrida, B.: Random energy model: an exactly solvable model of disordered systems. Phys. Rev. B 24, 2613 (1981)
6. Fortunato, S.: Community detection in graphs. Physics Reports 486, 75–174 (2010)
7. Garey, M.R., Johnson, D.S.: Computers and Intractability. Freeman, San Francisco (1979)
8. Garnier, J., Kallel, L.: Efficiency of local search with multiple local optima. SIAM Journal on Discrete Mathematics 15(1), 122–141 (2001)
9. Jones, T.: Evolutionary algorithms, fitness landscapes and search. PhD thesis. The University of New Mexico (1995)
10. Kallel, L., Naudts, B., Rogers, A. (eds.): Theoretical Aspects of Evolutionary Computing. Springer, Heidelberg (2001)
11. Kauffman, S.A.: The Origins of Order. Oxford University Press, New York (1993)
12. Knowles, J.D., Corne, D.W.: Instance Generators and Test Suites for the Multiobjective Quadratic Assignment Problem. In: Fonseca, C.M., Fleming, P.J., Zitzler, E., Deb, K., Thiele, L. (eds.) EMO 2003. LNCS, vol. 2632, pp. 295–310. Springer, Heidelberg (2003)
13. Newman, M.E.J.: Networks: An Introduction. Oxford University Press, Oxford (2010)
14. Ochoa, G., Tomassini, M., Vérel, S., Darabos, C.: A study of NK landscapes' basins and local optima networks. In: Genetic and Evolutionary Computation Conference, GECCO 2008, pp. 555–562. ACM (2008)
15. Papadimitriou, C.H.: Computational Complexity. Addison-Wesley, Reading (1994)
16. Papadimitriou, C.H., Steiglitz, K.: Combinatorial Optimization: Algorithms and Complexity. Prentice-Hall, Englewood Cliffs (1982)
17. Reichardt, J., Bornholdt, S.: Statistical mechanics of community detection. Physical Review E 74(1), 16110 (2006)
18. Reidys, C.M., Stadler, P.F.: Combinatorial landscapes. SIAM Review 44(1), 3–54 (2002)
19. Taillard, E.D.: Comparison of iterative searches for the quadratic assignment problem. Location Science 3(2), 87–105 (1995)
20. Talbi, E.-G.: Metaheuristics: From Design to Implementation. Wiley, Hoboken (2009)
21. Tomassini, M., Vérel, S., Ochoa, G.: Complex-network analysis of combinatorial spaces: The NK landscape case. Phys. Rev. E 78(6), 066114 (2008)
22. Vérel, S., Ochoa, G., Tomassini, M.: Local optima networks of NK landscapes with neutrality. IEEE Trans. Evol. Comp. 15(6), 783–797 (2011)
23. Weinberger, E.D.: Correlated and uncorrelated fitness landscapes and how to tell the difference. Biological Cybernetics 63, 325–336 (1990)

Chapter 7
Cooperative Coevolution for Agrifood Process Modeling

Olivier Barrière, Evelyne Lutton, Pierre-Henri Wuillemin, Cédric Baudrit, Mariette Sicard, and Nathalie Perrot

Abstract. On the contrary to classical schemes of evolutionary optimisations algorithms, single population Cooperative Co-evolution techniques (CCEAs, also called "Parisian" approaches) make it possible to represent the evolved solution as an aggregation of several individuals (or even as a whole population). In other words, each individual represents only a part of the solution. This scheme allows simulating the principles of Darwinian evolution in a more economic way, which results in gain in robustness and efficiency. The counterpart however is a more complex design phase. In this chapter, we detail the design of efficient CCEAs schemes on two applications related to the modeling of an industrial agri-food process. The experiments correspond to complex optimisations encountered in the modeling of a Camembert-cheese ripening process. Two problems are considered:

- A deterministic modeling problem, phase prediction, for which a search for a closed form tree expression is performed using genetic programming (GP).
- A Bayesian network structure estimation problem. The novelty of the proposed approach is based on the use of a two step process based on an intermediate representation called *independence model*. The search for an independence model is formulated as a complex optimisation problem, for which the CCEA scheme is particularly well suited. A Bayesian network is finally deduced using a deterministic algorithm, as a representative of the equivalence class figured by the independence model.

Olivier Barrière · Evelyne Lutton
INRIA Saclay - Ile-de-France, Parc Orsay Université, 4, rue Jacques Monod,
91893 Orsay Cedex, France
e-mail: {olivier.barriere,evelyne.lutton}@inria.fr

Pierre-Henri Wuillemin
LIP6-CNRS UMR7606, 75016 Paris, France

Cédric Baudrit · Mariette Sicard · Nathalie Perrot
UMR782 Génie et Microbiologie des Procédés Alimentaires. AgroParisTech,
INRA, F-78850 Thiverval-Grignon, France

E. Tantar et al. (Eds.): EVOLVE- A Bridge between Probability, SCI 447, pp. 247–287.
springerlink.com   © Springer-Verlag Berlin Heidelberg 2013

7.1   Introduction

Cooperative Co-evolution strategies rely on a formulation of the problem as a cooperative task, where individuals collaborate in order to build a solution.

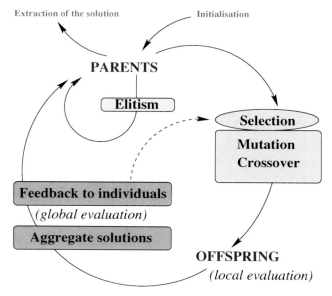

Fig. 7.1 A Parisian EA: a single population cooperative co-evolution

The large majority of these approaches deals with a co-evolution process that happens between a fixed number of separated populations. The idea is to co-evolve various species that only interact via the evaluation process. [30, 29] were the first to propose this technique, to co-evolve job-shop schedules using a parallel distributed algorithm. [53] then popularize the idea of cooperative co-evolution as an optimisation tool. It is applicable as soon as a decomposition of the problem into subcomponents can be identified. Each component then corresponds to a subpopulation that evolves simultaneously but in isolation to the other subpopulations. Individuals of a subpopulation are evaluated by aggregation with individuals of other subpopulations. Multi-species cooperative co-evolution has been applied to various problems [43, 55, 54, 22, 36, 66], including learning problems [8], and some theoretical analyses have been recently proposed, see [48, 10, 52], or [65] for an analysis considering a relationship between cooperative co-evolution and evolutionary game theory.

In this work, a different implementation of cooperative co-evolution, the so-called Parisian approach [17, 47] is used. It is derived from the classifier systems model proposed by [28]. Shown on Figure 7.1, this approach uses cooperation mechanisms within a *single* population. On the contrary to the previous model, interactions between sub-species are not limited to the evaluation step, but can also happen via genetic operators. An individual of a Parisian population, that represents only a part of the solution to the problem, can be evaluated at two levels:

- locally, using an independent evaluation (the "local" fitness), if some criteria can be designed to evaluate partial solutions (for instance, validity conditions),
- globally at each generation, via an aggregation process that builds a solution to the problem to be solved. Individuals are then rewarded via a bonus distribution.

In this way, the co-evolution of the whole population (or a major part of it) is favoured instead of the emergence of a single best individual, as in classical evolutionary schemes. The motivation is to make a more efficient use of the genetic search process within a population, and reduce the computational expense. Successful applications of such a scheme usually rely on a lower cost evaluation of the partial solutions (the individuals of the population), while computing the full evaluation only once at each generation.

The single population approach allows more interaction between subproblems, but in order to avoid trivial solutions (all individuals are the same), diversity preservation becomes a very important mechanism, to favour the evolution of subspecies, that progressively become independent from each other. At least in its early stage, a Parisian approach relies more on "exploration" mechanisms than "exploitation". Experimental tuning have proven that these two components are balanced in a different manner in classical and Parisian approaches, and that fitness sharing is an important component of Parisian scheme, that ensures an efficient convergence behaviour.

Additionally, we will see in the examples developed in this chapter, that Parisian schemes necessitate a more complex design phase. We actually need to split a problem into interdependent subproblems involving components of the same nature, which is not always possible. Questions regarding the relative efficience of different CCEA approaches, including for instance the single versus multiple population issue are very important, but still open, see for instance [62] for a first attempt in this direction.

This chapter is focussed on the design step, and presents how Parisian approaches have been developed on two examples provided by the agri-food community. The chapter is organised as follows. Section 7.2 describes the industrial process under study, cheese ripening, and the problems related to expertise modeling in this context. The two examples are then developed:

- section 7.3 deals with phase estimation using Genetic Programming : for comparison purpose, a classical GP approach is first developed, then a Parisian approach,
- section 7.4 addresses the problem of evolving the structure of a Bayesian network, with an encoding based on independence models.

Conclusions and future work are given in section 7.5.

7.2 Modeling Agri-Food Industrial Processes

This study is part of the French INCALIN research project[1]. The goal of this research project was to model agri-food industrial processes. In such food industries, manufacturing processes consist of successive operations whose underlying mechanisms are still unknown, such as the cheese ripening process. INCALIN was concerned with the understanding of the causal relationships between ingredients and physico-chemical or microbiological characteristics and on the other hand, sensory and nutritional properties. The intriguing question is how micro level properties determine or influence those on the macro level. The project aimed to explain the global behaviour of such systems.

Various macroscopic models have embedded expert knowledge, including expert systems [32, 33, 31], neural networks [35, 46], mechanistic models [1, 56], and dynamic Bayesian networks [6].

The major problem common to these techniques is related to the sparseness of available data: collecting experimental data is a long and difficult process, and resulting data sets are often not accurate or even erroneous. For example, a complete cheese ripening process lasts 40 days, and some tests are destructive, that is to say that a cheese sample is consumed during each analysis. Other measurements require the growing of bacterias in Petri dishes and then counting the number of colonies, which is very time consuming. Therefore the precision of the resulting model is often limited by the small number of valid experimental data. Also, parameter estimation procedures have to deal with incomplete, sparse and uncertain data.

7.2.1 The Camembert-Cheese Ripening Process

"Model cheeses" are produced in laboratories using pasteurized milk inoculated with *Kluyveromyces marxianus* (*Km*), *Geotrichum candidum* (*Gc*), *Penicillium camemberti* (*Pc*) and *Brevibacterium auriantiacum* (*Ba*) under aseptic conditions.

- *K. marxianus* is one of the key flora of Camembert cheese. One of its principal activity is the fermentation of lactose (noted *lo*) [14, 15] (curd de-acidification by lactose consumption). Three dynamics are apparent in the timeline of *K. marxianus* growth [38, 39]. Firstly, there is an exponential growth during about five days that corresponds to a decrease of lactose concentration. Secondly, the concentration of *K. marxianus* remains constant for about fifteen days and then decreases slowly.

- *G. candidum* plays a key role in ripening because it contributes to the development of flavour, taste and aroma of cheeses [2, 9, 40]. One of its principal activities is the consumption of lactate (noted *la*). Three dynamics are apparent

[1] "Cognitive and Viability methods for food quality control" (translation from french), supported by the French ANR-PNRA fund.

in the timeline of *G. candidum* growth [38, 39]. First, there is a latency period of about three days. Second, there is an exponential growth that corresponds to a decrease of lactate concentration and thus an increase of pH. Third, the concentration of *G. candidum* remains constant to the end of ripening.

During ripening, soft-mould cheese behave like an ecosystem (a bio-reactor), which is extremely complex to model as a whole. In such a process, human experts operators have a decisive role. Relationships between microbiological and physicochemical changes depend on environmental conditions (temperature, relative humidity ...) [39] and influence the quality of ripened cheeses [27, 38]. A ripening expert is capable of estimating the current state of some complex reactions at a macroscopic level through its perceptions (for example, sight, touch, smell and taste). Control decisions are then generally based on subjective but robust expert measurements. An important factor of parameter regulation is the subjective estimation of the current state of the ripening process. This process is split into four phases:

- Phase 1 is characterized by the surface humidity evolution of cheese (drying process). At the beginning, the surface of cheese is very wet and evolves until it is rather dry. The cheese is white with an odor of fresh cheese.

- Phase 2 begins with the apparition of a *P. camemberti*-coat (the white-coat at the surface of cheese). It is characterised by a first change of color and a "mushroom" odor development.

- Phase 3 is characterized by the thickening of the creamy under-rind. *P. camemberti* cover all the surface of cheeses and the color is light brown.

- Phase 4 is defined by strong ammoniac odor perception and the dark brown aspect of the rind of cheese.

These four phases are representative of cheese ripening. The expert's knowledge is obviously not limited to these four phases, but a correct identification of phases helps to evaluate the dynamics of ripening and to detect drift from the standard evolution.

7.2.2 Modeling Expertise on Cheese Ripening

A major problem, which was addressed in the INCALIN project, is the search for automatic procedures that mimic the way a human aggregates data through his senses, to estimate and regulate the ripening of the cheese.

Stochastic optimisation techniques, like evolutionary techniques, have already been proven successful on several agri-food problems. The interest of evolutionary optimisation methods for the resolution of complex problems related to agri-food is demonstrated by various recent publications. For example, [4] used genetic algorithms to identify the smallest discriminant set of variables to be used in certification process for an Italian cheese (validation of origin labels). [21] used GP to select the most significant wavenumbers produced by a Fourier transform infrared spectroscopy measurement device, in order to build a rapid detector of bacterial spoilage of beef. A recent overview on optimisation tools in food industries [61] discusses works based on evolutionary approaches.

We investigate here the use of cooperative co-evolution schemes (CCEAs) in the context of cheese ripening, for the modeling of expert knowledge. The next part (section 7.3) of this chapter deals with a first problem, which is phase estimation using Genetic Programming, under the form of a simple deterministic model (closed formula). Experimental as well as expert analysis made evident a simple relationship between four derivatives and the phase. A simple scheme they use in practice is based on a multilinear regression model. We will see below that a classical GP approach, that optimises a closed formula, i.e. a non-linear dependency, already improves the recognition rates, and that a Parisian scheme provides similar regognition rates with simpler structures, while keeping good recognition rates when the learning set is small.

The second part of this chapter (section 7.4) deals with a more sophisticated stochastic model of dependencies: Bayesian Network. The difficult point is now to address the problem of structure learning for a Bayesian Network (BN). Classical approaches of evolutionary computation are usually blocked by the problem of finding an efficient representation of a whole Bayesian Network. We will see that the Parisian scheme allows addressing this issue in an elegant way. In order to validate the method and compare it to the best approaches of the domain, we used classical BN benchmarks before testing it on the cheese ripening data, for which no "ground truth" model exist.

7.3 Phase Estimation Using GP

In previous work on cheese ripening modeling [6, 51], a dynamic Bayesian network (Figure 7.2) has been built, using human expert knowledge, to represent the macroscopic dynamic of each variable. The phase of the network at time t plays a determinant role for the prediction of the variables at time $t + 1$. Moreover, four relevant variables have been identified by biologists, the derivative of pH, la (lactate), Km (Kluyveromyces marxianus) and Ba (Brevibacterium auriantiacum) at time t, allowing phase prediction at time $t + 1$. This relates to a way in which experts aggregate information from their senses.

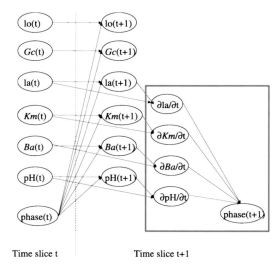

Fig. 7.2 Dynamic Bayesian Network representing dynamic variables based on the observation of ripening phases. The static Bayesian network used for comparison is in the right hand side box

7.3.1   Phase Estimation Using a Classical GP

A Genetic Programming (GP[2]) approach is used to search for a convenient formula that links the four derivatives of micro-organisms proportions to the phase at each time step t (static model), without *a priori* knowledge of the phase at $t - 1$.

When available, a functional representation of dependencies between variables is interesting (for prediction purpose for example). This problem is a symbolic regression one, however the small number of samples and their irregular distribution makes it difficult. In such a case, probabilistic dependencies (like Bayesian networks) seems usually to be more adapted, but are facing the same difficulty (robust estimation when data are sparse). A first question that could be adressed is thus to know which type of representation is more robust when data are sparse.

Results of GP estimation are compared in the sequel with the performances of a static Bayesian network, extracted from the DBN of [6], (the part within the box in Figure 7.2), and with a simple learning algorithm (multilinear prediction, see section 7.3.2.6), that was used by biologists in a first approach.

---

[2] GP is a type of EA where each individual figures a function, represented as a tree structure. Every tree node is an operator function $(+, -, /, *, \ldots)$ and every terminal node is an operand (a constant or a variable).

7.3.1.1   Overview of the Classical GP Algorithm

The classical GP algorithm consists first of an initialisation step where an initial population is randomly generated and then of a main loop where the reproduction (mutation and crossovers) and selection mechanism (ranking) are applied. The pseudo code of such an algorithm is given as follows:

Algorithm 7.1. Classical GP algorithm

Input: Maximum number of evaluations
Output: Single best individual
Creation of a random initial population
while *Maximum number of evaluations not reached* **do**
   | Create a temporary population *tmppop* using
   | selection, mutations and crossover
   | Compute the fitness of the new temporary
   | population *tmppop*
   | Select the best individuals of the current
   | population *pop+tmppop*
end
Select the best individual of the final population

7.3.1.2   Search Space

The derivatives of four variables will be considered, namely the derivative of *pH* (acidity), *la* (lactose proportion), *Km* and *Ba* (lactic acid bacteria proportions, see section 7.2.1), for the estimation of the phase (static problem). The GP will search for a phase estimator $\widehat{Phase(t)}$. That is, a function defined as follows (equation 7.1):

$$\widehat{Phase(t)} = f\left(\frac{\partial pH}{\partial t}, \frac{\partial la}{\partial t}, \frac{\partial Km}{\partial t}, \frac{\partial Ba}{\partial t}\right) \tag{7.1}$$

The function set is made of arithmetic operators: $\{+, -, *, /, \char`\^, log\}$, with protected / and *log*, and logical operators $\{if, >, <, =, and, or, xor, not\}$.

The terminal set is made of the four partial derivatives plus real constants. The constant's values are not limited and randomly initialised using one of the following laws $\mathscr{U}[0,1]$, $-\mathscr{U}[0,1]$, $\mathscr{N}(0,1)$, randomly chosen. ($\mathscr{U}$ is the uniform law, and $\mathscr{N}$ the normal law).

7.3.1.3   Fitness Function

Available data are separated in two sets: learning set and test set. Each is randomly chosen within the available data set for each run. The 16 available experiments are randomly split between learning and test sets. The size of the learning set varies from 10 to 15 experiments, while the size of the corresponding test set vary from 6 to 1 experiments (see section 7.3.2.6).

The fitness function (equation 7.2), *to be minimised*, is made of a factor that measures the quality of the fitting on the learning set, plus a "parsimony" penalisation factor in order to minimize the size, measured as the number of nodes (#*Nodes* in equation 7.2), of the evolved structures. The aim of this factor is to avoid bloat. It is divided by the number of variables (#*Variables* in equation 7.2) involved in the evaluated tree in order to favour structures that embed all four variables of the problem. Human experts use four classes to quantify the behaviour of the ripening process, and industrial processes are organised accordingly. Another type of classification (i.e. more or less classes) would have a strong impact on industrial devices. We choose to remain consistent with this expert approach. This is important in future developments where interfaces with human experts will be built. Experiments also show that recognition results are better with this constraint.

$$fitness = \frac{\sum\limits_{learning\_set} \left| f\left(\frac{\partial pH}{\partial t}, \frac{\partial la}{\partial t}, \frac{\partial Km}{\partial t}, \frac{\partial Ba}{\partial t} \right) - Phase(t) \right| + W\#Nodes}{\#Variables + 1} \tag{7.2}$$

The parameter W has been experimentally tuned. A large number of combinations were tested and it turned out that $W = 1$ is the optimal value in terms of algorithmic performance which favours evolution of structures with roughly 30 to 40 nodes. Bigger structures are so penalised that they are excluded from the population during the selection process.

7.3.1.4   Genetic Operators

A classical tree crossover (exchange of subtrees from a randomly chosen node) has been used with probability p_c (defined per tree), as a means of evolving the structure of the tree. Two types of mutations have been used:

- **Subtree mutation** (mutation of the structure), that randomly rebuilds a new subtree from a randomly chosen node, applied with probability p_{sm} (defined per tree),
- **Point mutation** (mutation of nodes content), applied with probability p_{cm} (also defined per tree) that does not modify the structure, but randomly changes the content of each node of the tree within the set of compatible functions or terminals. The probabilities (defined per node) are detailed in Table *I*. Real values are considered separately and undergo a real mutation with probability p_{rm} as a

multiplicative perturbation according to a χ^2 law of parameter N: $x' = x \frac{\sum_{i=1}^{N} \mathcal{N}(0,1)^2}{N}$ p_{rm} and N vary linearly according to generations, from 0.1 (first generation) to 0.5 (last generation) for p_{rm}, and from 1 to 1000 for N. This allows starting with rather infrequent large radius mutations and finish with more frequent mutations with smaller radius.

Table 7.1 Probabilities of point mutation operators

From	to	probability
operator	operator	0.1
variable	variable	0.1
variable	constant	0.05
constant	variable	0.05
constant	constant	p_{rm}: 0.1 to 0.5
		N: 1 to 1000

Crossover, subtree and point mutation probabilities vary along evolution according to the adapting scheme [19] available in the GPLAB toolbox [59]. p_c, p_{sm} and p_{cm} are initially fixed to $\frac{1}{3}$, and are updated according statistics of success of the various operators computed on a tuneable window of past generations.

7.3.2   *Phase Estimation Using a Parisian GP*

Instead of searching for a phase estimator as a single monolithic function, phase estimation can be split into four combined (and simpler) phase detection trees as shown in Figure 7.3. The structures searched are binary output functions (or binarised functions) that characterise one of the four phases. The individuals are split into four classes such that individuals of class k are good at characterising phase k. Finally, a global solution is made of the 5% best (at least one) individuals of each class, in order to be able to classify the sample into one of the four previous phases via a voting scheme (detailed at the end of this section).

7.3.2.1   Overview of the Parisian GP Algorithm

Unlike the classical GP algorithm, the output of Parisian GP algorithm is not a single individual but a part of the population. The main loop of the Parisian GP algorithm consists in first applying reproduction and selection mechanism, and then aggregating the current individuals in order to build a potential solution to the problem. The following pseudo code illustrates the principles of a Parisian GP:

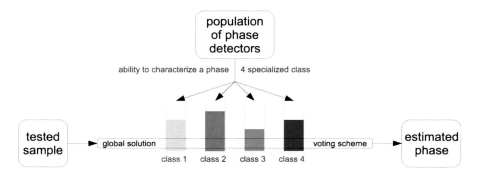

Fig. 7.3 Phase estimation using a Parisian GP. Four classes of phase detectors are defined: individuals of class k are good at characterising phase k.

7.3.2.2   Search Space

We now search for formulas of type: $I\left(\frac{\partial pH}{\partial t}, \frac{\partial la}{\partial t}, \frac{\partial Km}{\partial t}, \frac{\partial Ba}{\partial t}\right)$ with real outputs mapped to binary outputs, via a sign filtering: $(I() > 0) \rightarrow 1$ and $(I() \leq 0) \rightarrow 0$. The functions (except logical ones) and terminal sets, as well as the genetic operators, are the same as in the global approach above.

Using the available samples of the learning set, four real values can be computed, in order to measure the capability of an individual I to characterise one phase (equation 7.3):

$$k \in \{1,2,3,4\} \quad F_k(I) = 3 \sum_{i,phase=k} \frac{I(sample(i))}{\#Samples_{phase=k}} - \sum_{i,phase \neq k} \frac{I(sample(i))}{\#Samples_{phase \neq k}} \qquad (7.3)$$

in other words, if I is good for representing phase k, then $F_k(I) > 0$ and $F_{\neq k}(I) < 0$.

7.3.2.3   Local Fitness

The local adjusted fitness value, *to be maximised*, starts with a combination of three factors (equation 7.4):

$$\max\{F_1, F_2, F_3, F_4\} \times \frac{\#Ind}{\#IndPhaseMax} \times \left.\frac{NbMaxNodes}{NbNodes}\right|_{\text{if } NbNodes > NbMaxNodes} \qquad (7.4)$$

The first factor is aimed at characterising if individual I is able to distinguish one of the four phases. The second factor tends to balance the individuals between the four phases. The parameter *#IndPhaseMax* is the number of individuals representing the phase corresponding to the *argmax* of the first factor. The parameter *#Ind* is

---

Algorithm 7.2. Parisian GP algorithm

Input: Maximum number of evaluations
Output: Aggregation of individuals
Creation of a random initial population
while *Maximum number of evaluations not reached* **do**
> Create a temporary population *tmppop* using selection,
> mutations and crossover
> Compute the localfitness of the new temporary
> population *tmppop*
> Compute the adjustedfitness of the current population
> *pop+tmppop* via sharing
> Select the best individuals of the current population
> *pop+tmppop*
> Compute the globalfitness of the selected population by
> aggregating the best individuals

end

---

the total number of different individuals in the population. The third factor is a parsimony factor for avoiding large structures. *NbMaxNodes* has been experimentally tuned, and is currently fixed to 15, so that evolved structures got enough nodes to characterise the problem, but not too many, to avoid bloat effect. 15 represented a good tradeoff between accuracy and performance.

However, this is not the final formula of the local adjusted fitness. The two following subsection add two more factors, a penalising factor (μ) for individuals with too many neighbours (diversity preservation via a sharing scheme) and a bonus factor *bonus*$^\alpha$ for the best individuals.

7.3.2.4   Sharing Distance

The set of measurements $\{F_1, F_2, F_3, F_4\}$, that measures the ability of an individual to characterise each phase, provides a simplified representation of the search space in $\mathbb{R}^4$. As the aim of a Parisian evolution is to evolve distinct sub-populations, each being adapted to one of the four subtasks (to characterise one of the four phases), it is natural to use an Euclidean distance in this four dimensional phenotype space, as a basis of a simple fitness sharing scheme as stated in [20].

7.3.2.5   Aggregation of Partial Solutions and Global Fitness

At each generation, the population is shared in four classes corresponding to the phase each individual characterises the best (the argmax of $\max\{F_1, F_2, F_3, F_4\}$ for each individual). In other words, the population is split into four sub-populations (one for each class) within the population. The 5% best of each class are used via a voting scheme to decide the phase of each tested sample (see Figure 7.3). The global fitness measures the proportion of correctly classified samples on the learning set (equation 7.5):

$$GlobalFit = \frac{\sum\limits_{learning\_set} CorrectEstimations}{\#Samples} \tag{7.5}$$

The global fitness is then distributed to individuals who participated in the vote according to the following formula: $LocalFit' = LocalFit \times (GlobalFit + 0.5)^{\alpha}$.

As $GlobalFit \in [0,1]$, multiplying by $(GlobalFit + 0.5) > 1$ corresponds to a bonus. The parameter α varies along generations, for the first generations (a third of the total number of generations) $\alpha = 0$ (no bonus), and then α linearly increases from 0.1 to 1, in order to help the population to focus on the four peaks of the search space.

Several fitness measures are used to rate individuals. Namely

- the raw fitness $rawfitness$, which is the set of four values $\{F_1, F_2, F_3, F_4\}$, that measure the ability of the individual to characterise each phase,
- the local fitness $localfitness = \max(rawfitness)$ which represents the best characterised phase,
- and the adjusted fitness $adjfitness = \frac{localfitness}{\mu} \times \frac{\#IndPhaseMax}{\#Ind} \times \frac{\#NodesMax}{\#Nodes} \times bonus^{\alpha}$, which includes sharing, balance, parsimony and global fitness bonus terms.

Two sets of indicators are computed at each generation (see Figure 7.5):

- The sizes of each class, that show if each phase is equally characterised by the individuals of the population.
- The discrimination capability for each phase, computed on the 5% best individuals of each class as the minimum of: $\Delta = \max_{i \in [1,2,3,4]}\{F_i\} - \frac{\sum_{k \neq argmax\{F_i\}}\{F_k\}}{3}$

The higher the value of Δ, the better the phase is characterised.

7.3.2.6   Experimental Analysis

Available data were collected from 16 experiments during 40 days for each experiment, yielding 575 valid measurements. The data samples are relatively balanced except for phase 3, which has a longer duration, thus a larger number of samples: we have 57 representatives of phase 1, 78 of phase 2, 247 of phase 3 and 93 of phase 4. The derivatives of pH, la, Km and Ba were averaged and interpolated (spline interpolation) for some "missing" days. Indeed, due to difficulty to collect experimental data, a few values were missing. Finally, logarithms of these quantities are considered.

The parameters of both GP methods are detailed in Table II. The code has been developed in Matlab, using the GPLAB toolbox [59]. Comparative results of the four considered methods (multilinear regression, Bayesian network, GP and Parisian GP) are displayed in Figure 7.4, and a typical GP run is analysed in Figure 7.5.

Table 7.2 Parameters of the GP methods

	GP	Parisian GP
Population size	1000	1000
Number of generations	100	50
Function set	arithmetic and logical functions	arithmetic functions only
Sharing	no sharing	$\sigma_{share} = 1$ at the beginning, then linear decrease from 1 to 0.1 $\alpha_{share} = 1$ (constant)

The multilinear regression algorithm used for comparison works as follows: the data are modeled as a linear combination of the four variables:

$$\widehat{Phase}(t) = \beta_1 + \beta_2 \frac{\partial pH}{\partial t} + \beta_3 \frac{\partial la}{\partial t} + \beta_4 \frac{\partial Km}{\partial t} + \beta_5 \frac{\partial Ba}{\partial t}$$

The 5 coefficients $\{\beta_1, \ldots, \beta_5\}$ are estimated using a simple least square scheme. This model was included in the comparison because it was the model previously used by the biologists in the INCALIN project.

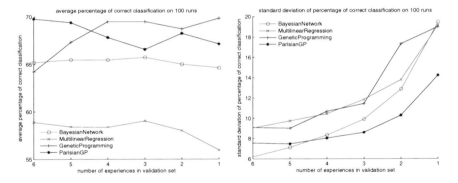

Fig. 7.4 Average (left) and standard-deviation (right) of recognition percentage on 100 runs for the 4 tested methods, the abscissa represent the size of the test-set

Experiments show that GP outperforms both multilinear regression and Bayesian network approaches in terms of recognition rates. Additionally the analysis of a typical Parisian GP run shows that it evolves much simpler structures than the classical GP. The average size of evolved structures is around 30 nodes for the classical GP approach and between 10 and 15 for the Parisian GP.

It has also to be noted in Figure 7.5 that co-evolution is balanced between the four phases. The third phase is the most difficult to characterise. This is in accordance

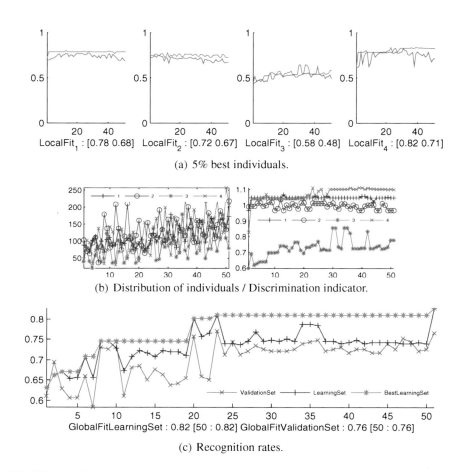

(a) 5% best individuals.

(b) Distribution of individuals / Discrimination indicator.

(c) Recognition rates.

Fig. 7.5 A typical run of the Parisian GP:
- (a): the evolution with respect to generation number of the 5% best individuals for each phase: the upper curve of each of the four graphs is for the best individual, the lower curve is for the "worst of 5% best" individuals.
- (b) left: the distribution of individuals for each phase: the curves are very irregular but numbers of representatives of each phases are balanced.
- (b) right: discrimination indicator Δ, which shows that the third phase is the most difficult to characterise.
- (c): evolution of the recognition rates of learning and test set. The best-so-far recognition rate on learning set is tagged with a star.

with human experts' judgement, for which this phase is also the most ambiguous to characterise.

The development of a cooperative co-evolution GP scheme (Parisian evolution) seems very attractive, as it allows the evolution of simpler structure in less generations, and yield results that are easier to interpret. Moreover, the computation time is almost equivalent to both presented methods (100 generations for a classical GP against 50 generations for a Parisian one), as one "Parisian" generation necessitates more complex operations, all in all). One can expect a more favourable behaviour for the Parisian scheme on more complex issues than the phase prediction problem, as the benefit of splitting the global solutions into smaller components may be higher and may yield computational shortcuts (see for example [17]).

7.4   Bayesian Network Structure Learning Using CCEAs

Bayesian networks structure learning is a NP-Hard problem [13], which has applications in many domains, as soon as one tries to analyse a large set of samples in terms of statistical dependence or causal relationship. In agri-food industries for example, the analysis of experimental data using Bayesian networks helps to gather technical expert knowledge and know-how on complex processes [6].

Evolutionary techniques were used to solve the Bayesian network structure learning problem, and were facing crucial problems like:

- Bayesian network representation (an individual being a whole structure like in [37], or a sub-structures like in [45]),
- Fitness function choice like in [45].

Various strategies were used, based on evolutionary programming [3], immune algorithms [34], multi-objective strategies [58], lamarkian evolution [64] or hybrid evolution [67].

We propose here to use an alternate representation, independence models, in order to solve the Bayesian network structure learning in two steps. Independence model learning is still a combinatorial problem, but it is easier to embed within an evolutionary algorithm. Furthermore, it is suited to a cooperative co-evolution scheme, which allows obtaining computationally efficient algorithms.

7.4.1   Recall of Some Probability Notions

The joint distribution of X and Y is the distribution of the intersection of the random variables X and Y, that is, of both random variables X and Y occurring together. The *joint probability* of X and Y is written $P(X,Y)$. The *conditional probability* is the probability of some random variable X, given the occurrence of some other random variable Y and is written $P(X|Y)$.

To say that two random variables are *statistically independent* intuitively means that the occurrence of one random variable makes it neither more nor less probable that the other occurs. If two random variables X and Y are independent, then the conditional probability of X given Y is the same as the unconditional probability of X, that is $P(X) = P(X|Y)$.

Two random variables X and Y are said to be *conditionally independent* given a third random variable Z if knowing Y gives no more information about X once one knows Z. Specifically, $P(X|Z) = P(X|Y,Z)$. In such a case we say that X and Y are conditionally independent given Z and write it $X \perp\!\!\!\perp Y \mid Z$.

7.4.2 Bayesian Networks

A Bayesian Network (BN) is a "graph-based model of a joint multivariate probability distribution that captures properties of conditional independence between random variables" as defined by [25]. On the one hand, it is a graphical representation of the joint probability distribution and on the other hand, it encodes probabilistic independences between variables. For example, a Bayesian network could represent the probabilistic relationships between diseases and symptoms. Given symptoms (resp. diseases), the network can be used to compute the probabilities of the presence of various diseases (resp. symptoms). These computations are called probabilistc inference.

Formally, a Bayesian network is represented by a directed acyclic graph (DAG) whose nodes are random variables, and whose missing edges encode conditional independences between the variables.

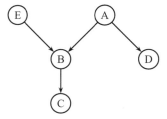

Fig. 7.6 Directed Acyclic Graph

The set of parent nodes of a node X_i is denoted by $pa(X_i)$. In a Bayesian network, the joint probability distribution of the random variables can be written using the graph structure as the product of the conditional probability distributions of each node given its parents:

$$P(X_1, X_2, \ldots, X_n) = \prod_{i=1}^{n} P(X_i | pa(X_i))$$

For instance, the joint distribution represented as a Bayesian network in Figure 7.6 can be written : $P(A,B,C,D,E) = P(A) \cdot P(B|E,A) \cdot P(C|B) \cdot P(D|A) \cdot P(E)$.

The very graph is called the structure of the Bayesian network and the values of conditional probabilities (e.g. $P(A = 0)$) for each node are called the parameters of the network.

7.4.2.1   Uses of Bayesian Networks

Using a Bayesian network can save considerable amounts of memory, if the dependencies in the joint distribution are sparse. For example, a naive way of storing the conditional probabilities of 10 binary variables as a table requires storage space for $2^{10} = 1024$ values. If the local distributions of no variable depends on more than 3 parent variables, the Bayesian network representation only needs to store at most $10 * 2^3 = 80$ values. One advantage of Bayesian networks is that it is intuitively easier for a human to understand (a sparse set of) direct dependencies and local distributions than complete joint distribution. The graph structure of a Bayesian network also allows to dramatically speed up the probabilistic inference in Bayesian network (i.e. the computation of $P(X_i|X_j)$).

Lastly, more than just a computing tool, Bayesian networks can be used to represent causal relationships and appear to be powerful graphical models of causality.

7.4.2.2   Parameter and Structure Learning

The Bayesian network learning problem has two branches: the *parameter* learning problem (in other words, how to find the probability tables of each node) and the *structure* learning problem (in other words, how to find the graph representing the Bayesian network), following the decomposition of the two constitutive parts of a Bayesian network: its structure and its parameters.

There already exists algorithms specially suited to the parameter learning problem, like expectation-maximisation (EM) that is used for finding maximum likelihood estimates of parameters.

Learning the structure is a more challenging problem because the number of possible Bayesian network structures (NS) grows superexponentially with the number of nodes [57]. For example, $NS(5) = 29281$ and $NS(10) = 4.2 \times 10^{18}$. A direct approach is intractable for more than 7 or 8 nodes, it is thus necessary to use heuristics in the search space.

In a comparative study by [23], authors identified some currently used structure learning algorithms, namely *PC* [60] or *IC/IC** [50] (causality search using statistical tests to evaluate conditional independence), *BN Power Constructor (BNPC)* [11] (also uses conditional independence tests) and other methods based on scoring criterion, such as *Minimal weight spanning tree (MWST)* [16] (intelligent weighting of the edges and application of the well-known algorithms for the problem of the minimal weight tree), *K2* [18] (maximisation of $P(G|D)$ using Bayes and a topological order on the nodes), *Greedy search* [12] (finding the best neighbour and iterate) or *SEM* [24] (extension of the EM meta-algorithm to the structure learning problem). However that may be, the problem of learning an optimal Bayesian network from a given dataset is NP-hard [13].

7.4.2.3   The PC Algorithm

PC, the reference causal discovery algorithm, was introduced by [60]. A similar algorithm, IC, was proposed simultaneously by [50]. It is based on chi-square tests to evaluate the conditional independence between two nodes. It is then possible to rebuild the structure of the network from the set of discovered conditional independences. PC algorithm starts from a fully connected network and every time a conditional independence is detected, the corresponding edge is removed. Here are the first detailed steps of this algorithm:

- Step 0: Start with a complete undirected graph G
- Step 1: Test all conditional independences of order 0 (i.e $x \perp\!\!\!\perp y \mid \emptyset$ where x and y are two distinct nodes of G). If $x \perp\!\!\!\perp y$ then remove the edge $x - y$.
- Step 2: Test all conditional independences of order 1 (i.e $x \perp\!\!\!\perp y \mid z$ where x, y, and z are three distinct nodes of G). If $x \perp\!\!\!\perp y \mid z$ then remove the edge $x - y$.
- step 3: Test all conditional independences of order 2 (i.e $x \perp\!\!\!\perp y \mid \{z_1, z_2\}$ where x, y, z_1 and z_2 are four distinct nodes of G). If $x \perp\!\!\!\perp y \mid \{z_1, z_2\}$ then remove the edge $x - y$.
- ...
- Step k: Test all conditional independences of order k (i.e $x \perp\!\!\!\perp y \mid \{z_1, z_2, \ldots, z_k\}$ where $x, y, z_1, z_2, \ldots, z_k$ are $k + 2$ distinct nodes of G). If $x \perp\!\!\!\perp y \mid \{z_1, z_2, \ldots, z_k\}$ then remove the edge between $x - y$.
- Next steps take particular care to detect some structures called *V-structures* (see section 7.4.2.4) and recursively detect orientation of the remaining edges.

The first stage is learning associations between variables for constructing an undirected structure. This requires a number of conditional independence test growing exponentially with the number of nodes. This complexity is reduced to polynomial complexity by fixing the maximal number of parents a node can have. It is of the order of N^k, where N is the size of the network and k is the upper bound on the fan-in. This implies that the value of k must remain small when dealing with big networks. In practice, k is often limited to 3. This value will be used in the sequel.

7.4.2.4   Independence Models

In this work, we do not work directly on Bayesian networks but on a more general model called *Independence Model* (IM), which can be seen as the underlying model of Bayesian networks and defined as follows:

- Let N be a non-empty set of variables, then $T(N)$ denotes the collection of all triplets $\langle X, Y | Z \rangle$ of disjoint subsets of N, $X \neq \emptyset$ and $Y \neq \emptyset$. The class of elementary triplets $E(N)$ consists of $\langle x, y | Z \rangle \in T(N)$, where $x, y \in N$ are distinct and $Z \subset N \setminus \{x, y\}$.
- Let P be a joint probability distribution over N and $\langle X, Y | Z \rangle \in T(N)$. $\langle X, Y | Z \rangle$ is called an *independence statement* (IS) if X is conditionally independent of Y given Z with respect to P (i.e., $X \perp\!\!\!\perp Y \mid Z$)
- An independence model (IM) is a subset of $T(N)$: each probability distribution P defines an IM, namely, the model $\{\langle X, Y | Z \rangle \in T(N) \; ; \; X \perp\!\!\!\perp Y \mid Z\}$, called the *independence model*[3] induced by P.

As we have seen, a Bayesian network represents a factorisation of a joint probability distribution, but there can be many possible structures that represents the same probability distribution.

For instance, the tree structures in Figure 7.7 encode the same independence statement $A \perp\!\!\!\perp B \mid C$. However, the structure in Figure 7.8, called *V-structure* (or *collider*), is not Markov equivalent to the three first ones.

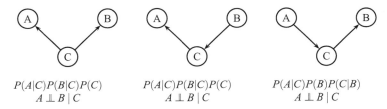

$$P(A|C)P(B|C)P(C) \qquad P(A|C)P(B|C)P(C) \qquad P(A|C)P(B)P(C|B)$$
$$A \perp\!\!\!\perp B \mid C \qquad\qquad A \perp\!\!\!\perp B \mid C \qquad\qquad A \perp\!\!\!\perp B \mid C$$

Fig. 7.7 Markov equivalent structures

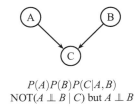

$$P(A)P(B)P(C|A,B)$$
$$\text{NOT}(A \perp\!\!\!\perp B \mid C) \text{ but } A \perp\!\!\!\perp B$$

Fig. 7.8 V-structure

Two structures are said to be *Markov equivalent* if they represent the same Independence Model. Particularly, an algorithm to learn the structure of a Bayesian network can not choose between two markov-equivalent structures.

---

[3] For more details about Independence Models and their properties, see [49].

To summarize, an independence model is the set of all the independence statements, that is the set of all $\langle X, Y|Z \rangle$ satisfied by P, and different Markov-equivalent Bayesian networks induce the same independence model. By following the paths in a Bayesian network, it is possible (even though it can be combinatorial) to find a part of its independence model using algorithms based on directional separation (d-separation) or moralization criteria. Reciprocally, an independence model is a guide to produce the structure of a Bayesian network.

Consequently, as the problem of finding an independence model can be turned to an optimisation problem, we investigate here the use of an evolutionary algorithm. More precisely, we build an algorithm that let a population of triplets $\langle X, Y|Z \rangle$ evolve until the whole population comes near to the independence model, which corresponds to a cooperative co-evolution scheme.

7.4.3   Evolution of an Independence Model

As in section 7.3, our algorithm (Independence Model Parisian Evolutionary Algorithm - IMPEA) is a *Parisian* cooperative co-evolution. However, in a pure Parisian scheme (Figure 7.1), a multi-individuals evaluation (global fitness computation) is done at each generation and redistributed as a bonus to the individuals who participated in the aggregation. Here, IMPEA only computes the global evaluation at the end of the evolution, and thus do not use any feedback mechanism. This approach, which is an extreme case of the Parisian CCEA, has already been used with success for example in real-time evolutionary algorithms, such as the *flies* algorithm [41].

IMPEA is a two steps algorithm. First, it generates a subset of the independence model of a Bayesian network from data by evolving elementary triplets $\langle x, y|Z \rangle$, where x and y are two distinct nodes and Z is a subset of the other ones, possibly empty. Then, it uses the independence statements that it found at the first step to build the structure of a representative network.

7.4.3.1   Search Space and Local Fitness

Individuals are elementary triplets $\langle x, y|Z \rangle$. Each individual is evaluated through a chi-square test of independence which tests the null hypothesis H_0: "The nodes x and y are independent given Z". The chi-square statistic χ^2 is calculated by finding the difference between each observed O_i and theoretical E_i frequencies for each of the n possible outcomes, squaring them, dividing each by the theoretical frequency, and taking the sum of the results: $\chi^2 = \sum_{i=1}^{n} \frac{(O_i - E_i)^2}{E_i}$. The chi-square statistic can then be used to calculate a *p-value* p by comparing the value of the statistic χ^2 to a chi-square distribution with $n - 1$ degrees of freedom, as represented on Figure 7.9.

p represents the probability to make a mistake if the null hypothesis is not accepted. It is then compared to a significance level α (0.05 is often chosen as a cut-off for significance) and finally the independence is rejected if $p < \alpha$. The reader

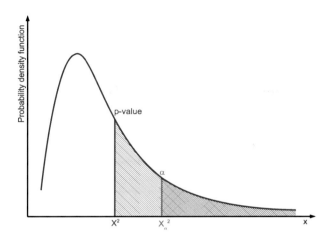

Fig. 7.9 Chi-square test of independence

has to keep in mind that rejecting H_0 allows one to conclude that the two variable are dependent, but not rejecting H_0 means that one cannot conclude that these two variable are dependent (which is not exactly the same as claiming that they are independent). Given that the higher the p-value, the stronger the independence, p seems to be a good candidate to represent the local fitness (which measures the quality of individuals). Nevertheless, this fitness suffers from two drawbacks:

- When dealing with small datasets, individuals with long constraining set Z tends to have good p-values only because dataset is too small to get enough samples to test efficiently the statement $x \perp\!\!\!\perp y \mid Z$.
- Due to the exponential behaviour of the chi-square distribution, its tails vanishes so quickly that individuals with poor p-values are often rounded to 0, making then indistinguishable.

First, p has to be adjusted in order to promote independence statements with small Z. This is achieved by setting up a parsimony term as a positive multiplicative malus $parcim(\#Z)$ which decrease with $\#Z$, the number of nodes in Z. Then, when $p < \alpha$ we replace the exponential tail with something that tends to zero slower. This modification of the fitness landscape allows avoiding *plateaus* which would prevent the genetic algorithm to travel all over the search space. Here is the adjusted local fitness[4]:

$$AdjLocalFitness = \begin{cases} p \times parcim(\#Z) & if\ p \geq \alpha \\ \alpha \times parcim(\#Z) \times \frac{X_\alpha^2}{X^2} & if\ p < \alpha \end{cases}$$

---

[4] *Note:* This can be viewed as an "Ockham's Razor" argument.

7.4.3.2 Genetic Operators

The genome of an individual, being $\langle x, y | Z \rangle$ where x and y are simple nodes and Z is a set of nodes is straightforward: It consists in an array of three cells (see Figure 7.10), the first one containing the index of the node x, the second cell containing the index of y and the last one is the array of the indexes of the nodes in Z.

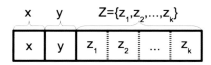

Fig. 7.10 Representation of $\langle x, y | Z \rangle$

This coding implies specific genetic operators because of the constraints resting upon a chromosome: there must not be doubles appearing when doing mutations or crossovers. A quick-and-dirty solution would have been to first apply classical genetic operators and then apply a *repair operator* a posteriori. Instead, we propose wise operators (which do not create doubles), namely two types of mutations and an robust crossover.

- Genome content mutation
  This mutation operator involves a probability p_{mG} that an arbitrary node will be changed from its original state. In order to avoid the creation of doubles, this node can be muted into any nodes in N except the other nodes of the individual, but including itself (see Figure 7.11).

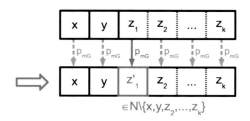

Fig. 7.11 Genome content mutation

- Add/remove mutation
  The previous mutation randomly modifies the content of the individuals, but does not modify the length of the constraining set Z. We introduce a new mutation operator called *add/remove mutation*, represented on Figure 7.12, that allows randomly adding or removing nodes in Z. If this type of mutation is selected,

with probability P_{mAR}, then new random nodes are either added with a probability P_{mAdd} or removed with $1 - P_{mAdd}$. These probabilities can vary along generations. Moreover, the minimal and the maximal number of nodes allowed in Z can evolve as well along generations, for tuning the growth of Z.

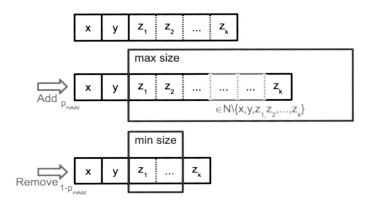

Fig. 7.12 Add/remove mutation

- Crossover
  The crossover consists in a simple swapping mechanism between x, y and Z. Two individuals $\langle x, y | Z \rangle$ and $\langle x', y' | Z' \rangle$ can exchange x or y with probability p_{cXY} and Z with probability p_{cZ} (see Figure 7.13). When a crossover occurs, only one swapping among $x \leftrightarrow x'$, $y \leftrightarrow y'$, $x \leftrightarrow y'$, $y \leftrightarrow x'$ and $Z \leftrightarrow Z'$ is selected via a wheel mechanism which implies that $4p_{cXY} + p_{cZ} = 1$. If the exchange is impossible, then the problematic nodes are automatically muted in order to keep clear of doubles.

7.4.4   Sharing

So as not to converge to a single optimum, but enable the genetic algorithm to identify multiple optima, we use a sharing mechanism that maintains diversity within the population by creating *ecological niches*. The complete scheme is described in [20] and is based on the fact that fitness is considered as a shared resource, that is to say that individuals having too many neighbours are penalised. Thus we need a way to compute the distance between individuals so that we can count the number of neighbours of a given individual. A simple Hamming distance was chosen: two elementary triplets $\langle x, y | Z \rangle$ and $\langle x', y' | Z' \rangle$ are said to be neighbours if they test the same two nodes (i.e., $\{x, y\} = \{x', y'\}$), whatever Z. Finally, dividing the fitness of each individual by the number of its neighbours would result in sharing the

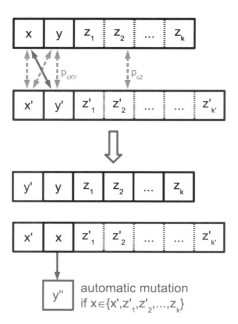

Fig. 7.13 Robust crossover

population into sub-populations whose size is proportional to the height of the peak they are colonising [26]. Instead, we take into account the relative importance of an individual with respect to its neighbourhood, and the fitness of each individual is divided by the sum of the fitnesses of its neighbours [42]. This scheme allows one to equilibrate the sub-populations within peaks, whatever their height.

7.4.5 *Immortal Archive and Embossing Points*

Recall that the aim of IMPEA is to construct a subset of the independence model, and thus the more independence statements we get, the better. Using a classical Parisian Evolutionary Algorithm scheme would allow evolving a number of independence statements equal to the population size. In order to be able to evolve larger independence statements sets, IMPEA implements an *immortal archive* that gather the best individuals found so far. An individual $\langle x, y | Z \rangle$ can become immortal if any of the following rules applies:

- Its p-value is equal to 1 (or numerically greater than $1 - \varepsilon$, where ε is the precision of the computer)
- Its p-value is greater than the significance level and $Z = \emptyset$
- Its p-value is greater than the significance level and $\langle x, y | \emptyset \rangle$ is already immortal

This archive serves two purposes: the most obvious one is that at the end of the generations, not only we get all the individuals of the current population but also all the immortal individuals, which can make a huge difference. But this archive also plays a very important role as *embossing points*: when computing the sharing coefficient, immortal individuals that are not in the current population are added to the neighbours counting. Therefore a region of the search space that has already been explored but that has disappeared from the current population is *marked as explored* since immortals individuals count as neighbours and thus penalise this region, encouraging the exploration of other zones.

7.4.5.1   Clustering and Partial Restart

Despite the sharing mechanism, we experimentally observed that some individuals became over-represented within the population. Therefore we add a mechanism to reduce this undesirable effect: if an individual has too many redundant representatives then the surplus is eliminated and new random individuals are generated to replace the old ones.

7.4.6   Description of the Main Parameters

The Table 7.3 describes the main parameters of IMPEA and their typical values or range of values, in order of appearance in the text above. Some of these parameters are scalars, like the number of individuals, and are constant along the whole evolution process. Others parameters, like the minimum or maximum number of nodes in Z, are arrays indexed by the number of generations, allowing these parameter to follow a profile of evolution.

7.4.7   Bayesian Network Structure Estimation

The last step of IMPEA consist in reconstructing the structure of the Bayesian network. This is achieved by aggregating all the immortal individuals and only the *good ones* of the final population. An individual $\langle x, y | Z \rangle$ is said to be *good* if its p-value does not allow rejecting the null hypothesis $x \perp\!\!\!\perp y \mid Z$. There are two strategies in IMPEA: a pure one, called *P-IMPEA*, which consists in strictly enforcing independence statements and a constrained one, called C-IMPEA, which adds a constraint on the number of desired edges.

Table 7.3 Parameters of IMPEA. Values are chosen within their typical range depending on the size of the network and the desired computation time

Name	Description	Typical value
MaxGens	Number of generations	$50\ldots200$
Ninds	Number of individuals	$50\ldots500$
Alpha	Significance level of the χ^2 test	$0.01\ldots0.25$
Parcim (#Z)	Array of parsimony coefficient (decreases with the length of Z)	$0.5\ldots1$
PmG	Probability of genome content mutation	$0.1/(2+\text{\#}Z)$
PmAR	Probability of adding or removing nodes in Z	$0.2\ldots0.5$
PmAdd (#Gen)	Array of probability of adding nodes in Z along generations	$0.25\ldots0.75$
MinNodes (#Gen)	Array of minimal number of nodes in Z along generations	$0\ldots2$
MaxNodes (#Gen)	Array of maximal number of nodes in Z along generations	$0\ldots6$
Pc	Probability of crossover	0.7
PcXY	Probability of swapping x and y	$1/6$
PcZ	Probability of swapping Z	$1/3$
Epsilon	Numerical precision	10^{-5}
MaxRedundant	Maximal number of redundant individuals in the population	$1\ldots5$

7.4.7.1 Pure Conditional Independence

Then, as in PC, P-IMPEA starts from a fully connected graph, and for each individual of the aggregated population, applies the rule "$x \perp\!\!\!\perp y \mid Z \Rightarrow$ *no edge between x and y*" to remove edges whose nodes belong to an independence statement. Finally, the remaining edges (which have not been eliminated) constitute the undirected structure of the network.

7.4.7.2 Constrained Edges Estimation

C-IMPEA needs an additional parameter which is the desired number of edges in the final structure. It proceeds by accumulation: it starts from an empty adjacency matrix and for each $\langle x, y \mid Z \rangle$ individual of the aggregated population, it adds its fitness to the entry (x, y). An example of a matrix obtained this way is shown on Figure 7.14.

At the end of this process, if an entry (at the intersection of a row and a column) is still equal to zero, then it means that there was no independence statement with this pair of nodes in the aggregated population. Thus these entries exactly correspond to the strict application of the conditional independences. If an entry has a

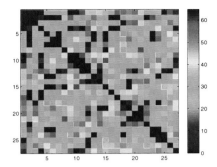

Fig. 7.14 Accumulated adjacency matrix of a network with 27 nodes (from Insurance network)

low sum, then it is an entry for which IMPEA found only a few independence statements (and/or independence statements with low fitness) and thus there is a high expectancy of having an edge between its nodes. Therefore to add more edges in the final structure (up to the desired number of edges), we just have to select edges with the lowest values and construct the corresponding network.

This approach seems to be more robust since it allows some "errors" in the chi-square tests, but strictly speaking, if an independence statement is discovered, there cannot be any edge between the two nodes.

7.4.8   Experiments and Results

Prior to a test on the the cheese-ripening data, the experimental analysis has been first performed on simulated data, where the true BN structure is known. A first experiment has been done on a toy-problem (section 7.4.8.1) in order to analyse the behaviour of IMPEA on a case where the complexity of the dependencies is controlled (i.e. where there is one independence statement that involves a long conditional set Z). A second test has been made on a classical benchmark of the domain, the insurance network (section 7.4.8.2), where input data are generated from a real-world BN. The test on cheese ripening data is detailed in section 7.4.8.3.

7.4.8.1   Test Case: Comb Network

To evaluate the efficiency of IMPEA, we forge a test-network which looks like a *comb*. A n-comb network has $n+2$ nodes: x, y, and $z_1, z_2, \ldots, z_n$, as one can see on Figure 7.15. The Conditional Probability Tables (CPT) are filled in with a uniform law. It can be seen as a kind of classifier: given the input $z_1, z_2, \ldots, z_n$, it classifies the output as x or y. For example, it could be a classifier that accepts a person's salary details, age, marital status, home address and credit history and classifies the person as acceptable/unacceptable to receive a new credit card or loan.

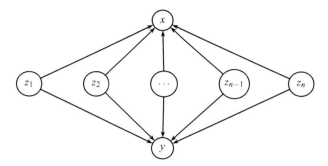

Fig. 7.15 A n-comb network

The interest of such a network is that its independence model can be generated (using semi-graphoid rules) from the following independence statements:

$$\forall i, j \text{ such as } i \neq j, z_i \perp\!\!\!\perp z_j$$
$$x \perp\!\!\!\perp y \mid \{z_1, z_2, \dots, z_n\}$$

Thus it has only one complex independence statement and a lot of simple (short) ones. In particular, the only way to remove the edge between x and y using statistical chi-square tests is to test the triplet $\langle x, y \mid \{z_1, z_2, \dots, z_n\}\rangle$. This cannot be achieved by the PC algorithm as soon as $k < n$ (recall that k is limited to 3 due to combinatorial complexity).

Typical run: We choose to test P-IMPEA with a simple 6-comb network. It has been implemented using an open source toolbox, the *Bayes Net Toolbox for Matlab* [44] available at http://bnt.sourceforge.net/. We draw our inspiration from PC and initialise the population with individuals with an empty constraining set and let it grow along generations up to 6 nodes, in order to find the independence statement $x \perp\!\!\!\perp y \mid \{z_1, \dots, z_6\}$. As shown on Figure 7.16, the minimal number of nodes allowed in Z is always 0, and the maximal number is increasing on the first two third of the generations and is kept constant to 6 on the last ones. The average number of nodes in the current population is also slowly rising up but remains rather small since in this example, there are a lot of small *easy to find* independence statements and only a single big one.

The correct structure (Figure 7.17) is found after 40 (out of 50) generations.

The Figure 7.18 represents the evolution of the number of errors along generations. The current evolved structure is compared with the actual structure: an *added* edge is an edge present in the evolved structure but not in the actual comb network, and a *deleted* edge is an edge that has been wrongly removed. The total number of

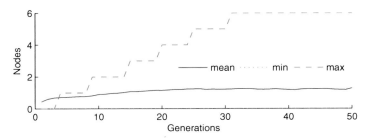

Fig. 7.16 Evolution of Minimal, Maximal and Average number of nodes in Z along generations

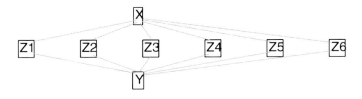

Fig. 7.17 Final evolved structure for the comb network

errors is the sum of added and deleted edges. Note that even if the number of errors of the discovered edges is extracted at each generation, it is by no means used by IMPEA or reinjected in the population because this information is only relevant in that particular test-case where the Bayesian network that generated the dataset is known.

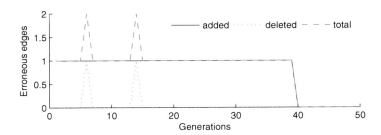

Fig. 7.18 Evolution of the number of erroneous edges of the structure along generations

Statistical results: The previous example gives an idea of the behaviour of P-IMPEA, but to compare it fairly with PC we must compare them not only over multiple runs but also with respect to the size of the dataset. So we set up the following experimental protocol:

- A 4-comb network is created and we use the same Bayesian network (structure and CPT) throughout the whole experiment.
- We chose representative sizes for the dataset: $\{500, 1000, 2000, 5000, 10000\}$, and for each size, we generate the corresponding number of cases from the comb network.
- We run 100 times both PC and P-IMPEA, and extract relevant information (see Tables 7.4 and 7.5):

  – How many edges were found? Among these, how many were erroneous? (added or deleted)
  – What is the percentage of runs in which the $x - y$ edge is removed?

- PC is tuned with a fan-in k limited to 3 (a larger fan-in is not used as PC is performing a full combinatorial research) and P-IMPEA is tuned with 50 generation of 50 individuals in order to take the same computational time as PC. 50 generation are more than enough to converge to a solution due to the small size of the problem. Both algorithms share the same significance level α.

The actual network contains 8 edges and 6 nodes. Therefore the number of possible alternative is $2^6 = 64$ and if we roughly want to have 30 samples per possibility, we would need approximatively $64 * 30 \approx 2000$ samples. That explains why performances of the chi-square test are very poor with only 500 and 1000 cases in the dataset. Indeed, when the size of the dataset is too small, PC removes the $x - y$ edge (see the last column of Table 7.4) while it does not even test $\langle x, y \mid \{z_1, z_2, z_3, z_4\}\rangle$ because it is limited by k to 3 nodes in Z.

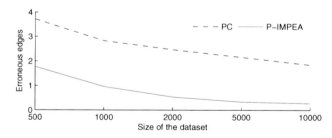

Fig. 7.19 Number of erroneous edges (added+deleted) for PC and P-IMPEA, depending on the size of the dataset

Regarding the global performance, the Figure 7.19 puts up the average number of erroneous nodes (either *added* or *deleted*) of both algorithms. As one can expect, the number of errors decreases with the size of the dataset, and it is clear that P-IMPEA clearly outperforms PC in every case.

Finally, if one has a look to the average number of discovered edges, it is almost equal to 8 (which is the actual number of edges in the 4-comb structure) for P-IMPEA (Table 7.5) whereas it is greater than 9 for the PC algorithm since it can't remove the $x - y$ edge (Table 7.4).

Table 7.4 Averaged results of PC algorithm after 100 runs

Cases	Edges	Added	Removed	Errors	x-y?
500	5.04 ± 0.85	0.38 ± 0.50	3.34 ± 0.78	3.72 ± 1.01	97%
1000	6.50 ± 1.24	0.66 ± 0.71	2.16 ± 1.01	2.82 ± 1.23	83%
2000	8.09 ± 1.18	1.27 ± 0.80	1.18 ± 0.68	2.45 ± 0.91	39%
5000	9.71 ± 0.74	1.93 ± 0.57	0.22 ± 0.46	2.15 ± 0.73	0%
10000	9.84 ± 0.58	1.84 ± 0.58	0 ± 0	1.84 ± 0.58	0%

Table 7.5 Averaged results of P-IMPEA algorithm after 100 runs

Cases	Edges	Added	Removed	Errors	x-y?
500	6.64 ± 0.79	0.05 ± 0.21	1.73 ± 1.90	1.78 ± 1.94	100%
1000	7.32 ± 0.91	0.18 ± 0.50	0.78 ± 1.01	0.96 ± 1.24	100%
2000	8.87 ± 1.04	0.24 ± 0.51	0.29 ± 0.60	0.53 ± 0.82	97%
5000	8.29 ± 0.32	0.30 ± 0.59	0.03 ± 0.17	0.33 ± 0.63	90%
10000	8.27 ± 0.31	0.27 ± 0.54	0 ± 0	0.27 ± 0.54	89%

7.4.8.2   Classical Benchmark: The Insurance Bayesian Network

Insurance is a network for evaluating car insurance risks developped by [7]. The Insurance Bayesian network contains 27 variables and 52 arcs. It is a large instance. A database of 50000 cases generated from the network has been used for the experiments below.

Once again, we start from a population with small Z and let it increase up to 4 nodes. The Figure 7.20 illustrates this growth: the average size of the number of nodes in Z of the current population follows the orders given by the minimum and the maximum values.

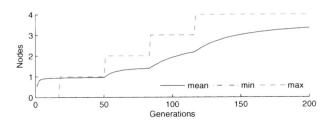

Fig. 7.20 Evolution of Minimal, Maximal and Average number of nodes in Z along generations

Concerning the evolution of the number of erroneous edges represented on Figure 7.21, it quickly decreases during the first half of the generation (the completely connected graph has more than 700 edges) and then stagnates. At the end, P-IMPEA finds 39 edges out of 52 among which there is no added edge, but 13 which are wrongly removed. It is slightly better than *PC* which also wrongly removes 13 edges, but which adds one superfluous one.

The best results are obtained with C-IMPEA and a desired number of edges equal to 47. Then, only 9 errors are made (see Table 7.6). When asking for 52 edges, the actual number of edges in the Insurance network, it makes 14 errors (7 additions and 7 deletions).

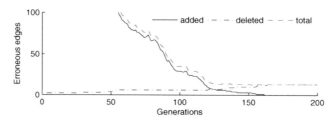

Fig. 7.21 Evolution of the number of erroneous edges of the structure along generations

Table 7.6 Number of detected edges for all algorithms

Algorithm	Edges	Added	Removed	Errors
PC	40	1	13	14
P-IMPEA	39	0	13	13
C-IMPEA	47	2	7	9
C-IMPEA	52	7	7	14

7.4.8.3 Real Dataset: Cheese Ripening Data from the INCALIN Project

The last step is to test our algorithm on real data. Our aim is to compare the result of IMPEA with a part of the dynamic Baysian network, already described at section 7.3, built with human expertise in the scope of the INCALIN project. We are interested in the part of the network that predicts the current phase knowing the derivatives of some bacteria proportions. We used the same data as in the first part of the report (see section 7.3.2.6), made of the derivatives of pH, la, Km and Ba and estimation of the current phase done by an expert.

After 10 generations of 25 individuals each, P-IMPEA converges to a network whose structure is almost the same as the one proposed by expert. As one can see on the right of Figure 7.22, no extra edge is added, but one edge is missing, between the derivative of la and the phase.

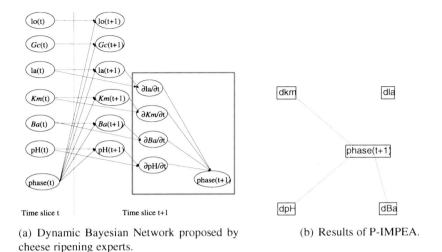

(a) Dynamic Bayesian Network proposed by   (b) Results of P-IMPEA.
cheese ripening experts.

Fig. 7.22 Comparison between the model proposed by experts and the network found by IMPEA on a real dataset from the INCALIN project

Table 7.7 Features of the two Parisian schemes

	Parisian Phase Prediction	**IMPEA**
Individuals	Phase predictors	IS : Independence Statements $\langle x, y \vert Z\rangle$
Population/groups	Classifier	IM : Independence Model
Nb of cooperating components	4	variable
Aggregation	clustering + selection of the 5% best	at the end of the evolution only
Local fitness	capability to characterise a phase $\max\{F_1, F_2, F_3, F_4\}$ × pressure toward simple structure	adjusted *p-value* × pressure toward small conditional parts
Global fitness	voting scheme + evaluation on the learning set	none
Sharing	Euclidean distance on $\{F_1, F_2, F_3, F_4\}$	Hamming distance on $\{x, y\}$
Specific features	variable population size inflation / deflation	archive embossing points

7.4.9   *Analysis*

We compared performances on the basis of undirected graphs produced by both algorithms. The edge directions estimation has not been yet programmed in IMPEA, this will be done in future developments, using a low combinatorial strategy similar to PC. Comparisons between both algorithms do not depend on this step.

The two experiments of section 7.4.8 prove that IMPEA favourably compares to PC, actually, besides the fact that IMPEA relies on a convenient problem encoding,

PC performs a deterministic and systematic search while IMPEA uses evolutionary mechanisms to prune computational efforts and to concentrate on promising parts of the search space. The limitation of PC according to problem size is obvious in the first test (Comb network): PC is unable to capture a complex dependency, even on a small network. Additionally it is to be noticed that IMPEA better resists to a current problem of real life data, that is the insufficient number of available samples.

7.5   Conclusion

Parisian CCEAs and cooperative co-evolution in general, when applicable, yield efficient and robust algorithms. As we have seen in this chapter, the main concern is the design of adequate representations for cooperative co-evolution schemes, in other words, representations that allow a collective evolution mechanism. One has to build an evolution mechanism that uses pieces of solutions instead of complete solutions as individual. It is also needed to evaluate the pieces of solutions (local fitness) before being able to select the best pieces that can be considered as components of a global solution.

In the example of section 7.3, we first designed a classical GP, where the phase estimator was searched as a single best "monolithic" function. Although it already outperforms the previous other methods, we obtained additional improvements by splitting the phase estimation into four combined (and simpler) "phase detectors". We actually used additional a priori informations about the problem. The structures evolved here were binary output functions that characterised one of the four phases. Their aggregation was made via a robust voting scheme. The resulting phase detector has almost the same recognition rate as the classical GP but with a lower variance, evolves simpler structure during less generations, and yield results that are easier to interpret.

In section 7.4, the cooperative coevolution algorithm IMPEA has allowed overcoming a known drawback of the classical approach, that is to find an efficient representation of a direct acyclic graph. We have shown that the cooperative scheme is particularly adapted to an alternate representation of Bayesian Networks: Independence Models (IM). IM represent data dependencies via a set of Independence Statements (IS) and IS can directly be considered as individuals of a CCEA.

The major difficulty, which is to build a Bayesian Network representative at each generation has been overcome for the moment by a scheme that only built a global solution at the end of the evolution (second step of IMPEA). Future work on this topic will be focused on an improvement of the global fitness management within IMPEA. The major improvement of IMPEA is that it only performs difficult combinatorial computations when local mechanisms have pushed the population toward "interesting" area of the search space, thus avoiding to make complex global computations on obviously "bad" solutions. In this sense, CCEAs take into account a priori information to avoid computational waste, in other words, complex computations in unfavourable areas of the search space.

Table 7.7 gives an overview of the features of the two CCEAs schemes presented in this chapter. There are some major differences between the two approaches.

- With respect to the nature of the cooperation within the population: the Parisian phase prediction is relying on components that are structured in 4 clusters (each individual only votes for the phase it characterises the best), while IMPEA collects the best individuals of its population in an archive to build the global independence model.
- With respect to the synchonisation of the global fitness calculation: the Parisian phase prediction computes a global fitness at each generation and use a bonus distribution mechanism, while IMPEA only relies on local calculations at each generation. The global calculation is made only once at the end of the evolution.

It is interesting to note that IMPEA may be considered as an incomplete Parisian scheme, as it does not use any global calculation. Future work on this algorithm will be aimed at evaluating if a global calculation may accelerate its convergence and robustness. Note however that for instance the fly algorithm [63, 41] does not use any global fitness either, but is able to provide extremely rapid results: the cooperation mechanisms may operate in some cases without global fitness.

The common characteristics of these two examples is that the cooperative scheme has allowed representing in an indirect way some complex structures (classification rules in the first example and Bayesian Networks in the second one). This way of exploiting the artificial evolution scheme is versatile enough to facilitate the integration of constraints and the development of various strategies (archive and embossing points as in section 7.4, or variable population size as stated in [5] for instance). The experiments described in this chapter join previous studies on "Parisian evolution", that experimentally proved that very efficient algorithms can be built on this cooperation-coevolution basis, in terms of rapidity [41], or in terms of size and complexity of the problems [17, 63].

Acknowledgements

 This work has been funded by the French ANR (National Research Agency), via a PNRA fund (Agri-food research program, 2007-2009).

References

1. Aldarf, M., Fourcade, F., Amrane, A., Prigent, Y.: Substrate and metabolite diffusion within model medium for soft cheese in relation to growth of penicillium camembertii. J. Ind. Microbiol. Biotechnol. 33, 685–692 (2006)
2. Arfi, K., Amrita, F., Spinnler, H.E., Bonnarme, P.: Catabolism of volatile sulfur compounds precursors by brevibacterium linens and geotrichum candidum, two microorganisms of the cheese ecosystem. J. Biotechnol. 105(3), 245–253 (2003)

3. Tucker, A., Liu, X.: Extending evolutionary programming methods to the learning of dynamic bayesian networks. In: Proceedings of Genetic and Evolutionary Computation Conference (GECCO 1999), Orlando, Florida, USA. Morgan Kaufmann (July 1999)

4. Barile, D., Coisson, J.D., Arlorio, M., Rinaldi, M.: Identification of production area of ossolano italian cheese with chemometric complex aproach. Food Control 17(3), 197–206 (2006)

5. Barrière, O., Lutton, E.: Experimental analysis of a variable size mono-population cooperative-coevolution strategy. In: Proceedings of International Workshop on Nature Inspired Cooperative Strategies for Optimization (NICSO 2008), Puerto de La Cruz, Tenerife (November 2008)

6. Baudrit, C., Wuillemin, P.-H., Sicard, M., Perrot, N.: A Dynamic Bayesian Network to Represent a Ripening Process of a Soft Mould Cheese. In: Lovrek, I., Howlett, R.J., Jain, L.C. (eds.) KES 2008, Part II. LNCS (LNAI), vol. 5178, pp. 265–272. Springer, Heidelberg (2008)

7. Binder, J., Koller, D., Russell, S., Kanazawa, K.: Adaptive probabilistic networks with hidden variables. Machine Learning 29, 213–244 (1997)

8. Bongard, J., Lipson, H.: Active coevolutionary learning od deterministic finite automata. Journal of Machine Learning Research 6, 1651–1678 (2005)

9. Boutrou, R., Guguen, M.: Interests in geotrichum candidum for cheese technology. Int. J. Food Microbiol. 102, 1–20 (2005)

10. Bucci, A., Pollack, J.B.: On identifying global optima in cooperative coevolution. In: GECCO 2005: Proceedings of the 2005 Conference on Genetic and Evolutionary Computation, pp. 539–544. ACM, New York (2005)

11. Cheng, J., Bell, D.A., Liu, W.: Learning belief networks from data: An information theory based approach. In: Proceedings of Sixth ACM International Conference on Information and Knowledge Management, Las Vegas, Nevada, USA, pp. 325–331 (November 1997)

12. Chickering, D.M., Boutilier, C.: Learning equivalence classes of bayesian-network structures. Journal of Machine Learning Research, 150–157 (1996)

13. Chickering, D.M., Heckerman, D., Meek, C.: Large-sample learning of bayesian networks is np-hard. Journal of Machine Learning Research 5, 1287–1330 (2004)

14. Choisy, C., Desmazeaud, M.J., Gripon, J.C., Lamberet, G., Lenoir, J.: La biochimie de l'affinage. In: Le Fromage, pp. 86–105. Lavoisier, Paris (1997)

15. Choisy, C., Desmazeaud, M.J., Gueguen, M., Lenoir, J., Schmidt, J.L., Tourneur, C.: Les phénomènes microbiens. In: Le Fromage, pp. 86–105. Lavoisier, Paris (1997)

16. Chow, C., Liu, C.: Approximating discrete probability distributions with dependence trees. IEEE Transactions on Information Theory 14(3), 462–467 (1968)

17. Collet, P., Lutton, E., Raynal, F., Schoenauer, M.: Polar ifs + parisian genetic programming = efficient ifs inverse problem solving. Genetic Programming and Evolvable Machines Journal 1(4), 339–361 (2000)

18. Cooper, G.F., Herskovits, E.: A bayesian method for the induction of probabilistic networks from data. Machine Learning 9(4), 309–347 (1992)

19. Davis, L.: Adapting operators probabilities in genetic algorithms. In: Proceedings of the Third Conference on Genetic Algorithms, pp. 61–69. Morgan-Kaufmann (June 1989)

20. Deb, K., Goldberg, D.E.: An investigation of niche and species formation in genetic function optimization. In: Proceedings of the Third Conference on Genetic Algorithms, pp. 42–50 (June 1989)

21. Ellis, D.I., Broadhurst, D., Goodacre, R.: Rapid and quantitative detection of the microbial spoilage of beef by fourier transform infrared spectroscopy and machine learning. Analytica Chimica Acta 514(2), 193–201 (2004)

22. Eriksson, R., Olsson, B.: Cooperative coevolution in inventory control optimisation. In: Proceedings of the Third International Conference on Artificial Neural Networks and Genetic Algorithms, East Lansing, MI, USA. Springer (July 1997)
23. Francois, O., Leray, P.: Etude comparative d'algorithmes d'apprentissage de structure dans les réseaux bayésiens. Technical report. Rencontres des Jeunes Chercheurs en IA (2003)
24. Friedman, N.: Learning belief networks in the presence of missing values and hidden variables. In: Proceedings of 14th International Conference on Machine Learning, Nashville, Tenessee, USA, pp. 125–133. Morgan Kaufmann (July 1997)
25. Friedman, N., Linial, M., Nachman, I., Pe'er, D.: Using bayesian network to analyze expression data. J. Computational Biology 7, 601–620 (2000)
26. Goldberg, D.E., Richardson, J.: Genetic algorithms with sharing for multimodal function optimization. In: Proceedings of Second International Conference on Genetic Algorithms and Their Application, Cambridge, MA, USA, pp. 41–49. Lawrence Erlbaum Associates, Inc. (July 1987)
27. Gripon, A.: Mould-ripened cheeses. In: Cheese: Chemistry, Physics and Microbiology, pp. 111–136. Chapman and Hall, London (1993)
28. Holland, J.H., Reitman, J.S.: Cognitive systems based on adaptive algorithms. SIGART Bull. 63, 49–49 (1977)
29. Husbands, P.: Distributed coevolutionary genetic algorithms for multi-criteria and multi-constraint optimisation. In: Selected Papers from AISB Workshop on Evolutionary Computing, London, UK, pp. 150–165. Springer (April 1994)
30. Husbands, P., Mill, F.: Simulated co-evolution as the mechanism for emergent planning and scheduling. In: Proceedings of the Fourth International Conference on Genetic Algorithms, San Diego, CA, USA, pp. 264–270. Morgan Kaufman (July 1991)
31. Ioannou, I., Mauris, G., Trystram, G., Perrot, N.: Back-propagation of imprecision in a cheese ripening fuzzy model based on human sensory evaluations. Fuzzy Sets And Systems 157, 1179–1187 (2006)
32. Ioannou, I., Perrot, N., Curt, C., Mauris, G., Trystram, G.: Development of a control system using the fuzzy set theory applied to a browning process - a fuzzy symbolic approach for the measurement of product browning: development of a diagnosis model - part i. Journal of Food Engineering 64, 497–506 (2004)
33. Ioannou, I., Perrot, N., Mauris, G., Trystram, G.: Development of a control system using the fuzzy set theory applied to a browning process - towards a control system of the browning process combining a diagnosis model and a decision model - part ii. Journal of Food Engineering 64, 507–514 (2004)
34. Jia, H., Liu, D., Yu, P.: Learning dynamic bayesian network with immune evolutionary algorithm. In: Proceedings of Fourth International Conference on Machine Learning and Cybernetics, Guangzhou, China (August 2005)
35. Jimenez-Marquez, S.A., Thibault, J., Lacroix, C.: Prediction of moisture in cheese of commercial production using neural networks. Int. Dairy J. 15, 1156–1174 (2005)
36. De Jong, E.D., Stanley, K.O., Wiegand, R.P.: Introductory tutorial on coevolution. In: GECCO 2007: Proceedings of the 2007 GECCO Conference Companion on Genetic and Evolutionary Computation, London, UK (July 2007)
37. Larranaga, P., Poza, M.: Structure learning of bayesian networks by genetic algorithms: A performance analysis of control parameters. IEEE Journal on Pattern Analysis and Machine Intelligence 18(9), 912–926 (1996)
38. Leclercq-Perlat, M.N., Buono, F., Lambert, D., Latrille, E., Spinnler, H.E., Corrieu, G.: Controlled production of camembert-type cheeses. part i: Microbiological and physico-chemical evolutions. J. Dairy Res. 71, 346–354 (2004)

39. Leclercq-Perlat, M.N., Picque, D., Riahi, H., Corrieu, G.: Microbiological and biochemical aspects of camembert-type cheeses depend on atmospheric composition in the ripening chamber. J. Dairy Sci. 89, 3260–3273 (2006)
40. Lenoir, J.: The surface flora and its role in the ripening of cheese. Int. Dairy Fed. Bull. 171, 3–20 (1984)
41. Louchet, J., Guyon, M., Lesot, M.J., Boumaza, A.M.: Dynamic flies: a new pattern recognition tool applied to stereo sequence processing. Pattern Recognition Letters 23, 335–345 (2002)
42. Lutton, E., Martinez, P.: A genetic algorithm with sharing for the detection of 2d geometric primitives in images. In: AE 1995: Selected Papers from the European Conference on Artificial Evolution, Lille, France, pp. 287–303 (October 1995)
43. Moriarty, D.E., Miikkulainen, R.: Forming neural networks through efficient and adaptive coevolution. Evolutionary Computation 5, 373–399 (1998)
44. Murphy, K.: The bayes net toolbox for matlab. Computing Science and Statistics 33(2), 1024–1034 (2001)
45. Myers, J.W., Laskey, K.B., DeJong, K.A.: Learning bayesian networks from incomplete data using evolutionary algorithms. In: Proceedings of Genetic and Evolutionary Computation Conference (GECCO 1999), Orlando, Florida, USA, vol. 1, pp. 458–465. Morgan Kaufmann (July 1999)
46. Ni, H.X., Gunasekaran, S.: Food quality prediction with neural networks. Food Technology 52, 60–65 (1998)
47. Ochoa, G., Lutton, E., Burke, E.: Cooperative royal road functions. In: Evolution Artificielle, Tours, France, pp. 29–31 (October 2007)
48. Panait, L., Luke, S., Harrison, J.F.: Archive-based cooperative coevolutionary algorithms. In: GECCO 2006: Proceedings of the 8th Annual Conference on Genetic and Evolutionary Computation, Seattle, Washington, USA (July 2006)
49. Pearl, J.: Probabilistic Reasoning in Intelligent Systems: Networks of Plausible Inference. Morgan Kaufmann (1988)
50. Pearl, J., Verma, T.: A theory of inferred causation. In: Proceedings of Second International Conference on the Principles of Knowledge Representation and Reasoning, Cambridge, MA, USA (April 1991)
51. Pinaud, B., Baudrit, C., Sicard, M., Wuillemin, P.-H., Perrot, N.: Validation et enrichissement interactifs d'un apprentissage automatique des paramètres d'un réseau bayésien dynamique appliqué aux procédés alimentaires. In: Journées Francophone sur les Réseaux Bayésiens, Lyon, France (May 2008)
52. Popovici, E., De Jong, K.: The effects of interaction frequency on the optimization performance of cooperative coevolution. In: Proceedings of Genetic and Evolutionary Computation Conference, GECCO 2006, Seattle, Washington, USA (July 2006)
53. Potter, M.A., De Jong, K.: A Cooperative Coevolutionary Approach to Function Optimization. In: Davidor, Y., Männer, R., Schwefel, H.-P. (eds.) PPSN 1994. LNCS, vol. 866, pp. 249–257. Springer, Heidelberg (1994)
54. Potter, M.A., De Jong, K.: Cooperative coevolution: An architecture for evolving coadapted subcomponents. Evolutionary Computation 8(1), 1–29 (2000)
55. Potter, M.A., De Jong, K.: The Coevolution of Antibodies for Concept Learning. In: Eiben, A.E., Bäck, T., Schoenauer, M., Schwefel, H.-P. (eds.) PPSN 1998. LNCS, vol. 1498, pp. 530–539. Springer, Heidelberg (1998)
56. Riahi, M.H., Trelea, I.C., Leclercq-Perlat, M.N., Picque, D., Corrieu, G.: Model for changes in weight and dry matter during the ripening of a smear soft cheese under controlled temperature and relative humidity. International Dairy Journal 17, 946–953 (2000)

57. Robinson, R.W.: Counting unlabeled acyclic digraphs. In: Combinatorial Mathematics V: Proceedings of the Fifth Australian Conference, Melbourne, Australia, pp. 28–43. Springer (2000)
58. Ross, J., Zuviria, E.: Evolving dynamic bayesian networks with multi-objective genetic algorithms. Applied Intelligence 26(1), 13–23 (2007)
59. Silva, S.: GPLAB A Genetic Programming Toolbox for MATLAB (2008), http://gplab.sourceforge.net/
60. Spirtes, P., Glymour, C., Scheines, R.: Causation, Prediction, and Search, 2nd edn. The MIT Press (2001)
61. Tarantilis, C.D., Kiranoudis, C.T.: Operational research and food logistics. Journal of Food Engineering 70(3), 253–255 (2005)
62. Tonda, A., Lutton, E., Squillero, G.: Lamps: A test problem for cooperative coevolution. In: NICSO 2011, the 5th International Workshop on Nature Inspired Cooperative Strategies for Optimization, Cluj Napoca, Romania, October 20-22 (2011)
63. Vidal, F.P., Lazaro-Ponthus, D., Legoupil, S., Louchet, J., Lutton, É., Rocchisani, J.-M.: Artificial Evolution for 3D PET Reconstruction. In: Collet, P., Monmarché, N., Legrand, P., Schoenauer, M., Lutton, E. (eds.) EA 2009. LNCS, vol. 5975, pp. 37–48. Springer, Heidelberg (2010)
64. Wang, S.-C., Li, S.-P.: Learning Bayesian Networks by Lamarckian Genetic Algorithm and Its Application to Yeast Cell-Cycle Gene Network Reconstruction from Time-Series Microarray Data. In: Ijspeert, A.J., Murata, M., Wakamiya, N. (eds.) BioADIT 2004. LNCS, vol. 3141, pp. 49–62. Springer, Heidelberg (2004)
65. Wiegand, R.P., Liles, W., De Jong, K.: Analyzing cooperative coevolution with evolutionary game theory. In: Proceedings of the 2002 Congress on Evolutionary Computation CEC 2002, Honolulu, Hawaii, pp. 1600–1605 (May 2000)
66. Wiegand, R.P., Potter, M.A.: Robustness in cooperative coevolution. In: GECCO 2006: Proceedings of the 8th Annual Conference on Genetic and Evolutionary Computation, Seattle, Washington, USA (July 2006)
67. Wong, M.L., Leung, K.S.: An efficient data mining method for learning bayesian networks using an evolutionary algorithm-based hybrid approach. IEEE Transactions on Evolutionary Computation 8, 378–404 (2004)

Chapter 8
Hybridizing cGAs with PSO-like Mutation

E. Alba and A. Villagra

Abstract. Over the last years, interest in hybrid metaheuristics has risen considerably in the field of optimization. Combinations of operators and metaheuristics have provided very powerful search techniques. In this chapter we incorporate active components of Particle Swarm Optimization (PSO) into the Cellular Genetic Algorithm (cGA). We replace the mutation operator by a mutation based on concepts of PSO. We present two hybrid algorithms and analyze their performance using a set of different problems. The results obtained are quite satisfactory in efficacy and efficiency, outperforming in most cases existing algorithms for a set of problems.

8.1   Introduction

Evolutionary Algorithms (EAs) applied to optimization problems represent a very intense line of research during the last decade [1]. Well-accepted subclasses of EAs are Genetic Algorithms (GA), Genetic Programming (GP), Evolutionary Programming (EP), and Evolution Strategies (ES).

These EAs algorithms work over a set (population) of potential solutions (individuals) which undergoes stochastic variations in order to search for better solutions. Most EAs work on a single population (panmictic) of individuals, applying operators to the population as a whole. However, working on a structured population in which individuals interact in neighborhoods is also possible and successful compared to panmictic techniques.

E. Alba
Department of Computer Science
University of Málaga, Spain
e-mail: eat@lcc.uma.es

A. Villagra
Emerging Technologies Laboratory
Universidad Nacional de la Patagonia Austral, Argentine
e-mail: avillagra@uaco.unpa.edu.ar

E. Tantar et al. (Eds.): EVOLVE- A Bridge between Probability, SCI 447, pp. 289–302.
springerlink.com   © Springer-Verlag Berlin Heidelberg 2013

Among the many types of structured EAs (where the population is somehow decentralized), distributed and cellular algorithms are the most popular optimization tools [3], [4], [20], [24].

In a cellular Evolutionary Algorithm (cEA) the population is represented by a two dimensional grid of individuals. Each individual in the grid has a neighborhood that overlaps with the neighborhoods of its nearby individuals. The individuals can only interact with their neighbors in the reproductive cycle where the variation operators are applied.

This overlapping neighborhoods provide cEAs with an implicit slow diffusion mechanism. The slow dispersion of the best solutions over the population is the cause of a good balance between the exploitation and the exploration efficacy, something really sought when an efficient and accurate algorithm is designed. Nevertheless, this characteristic produces a slow convergence to the optimum and thus decreases the efficiency of the algorithm. An open research line then consists in creating new algorithmic models that try to improve the efficiency of a cEA by incorporating active components of other algorithms. This is not the only means to leverage the efficiency of a cEA, but it is an structured and novel way of approaching it that we are proposing in this chapter.

Over the last years, interest in hybrid metaheuristics has risen considerably in the field of optimization [11]. Combinations of algorithms such as several metaheuristics in a single technique have provided very powerful search procedures. In this work we intend to generate new functional and efficient hybrid algorithms in a methodological and structured way. In particular our global idea is to use as the base core technique a cGA algorithm and insert in its basic behavior some "active principles" of other metaheuristics: movement of a particle from PSO, transition probability from ACO, and Boltzmann probability from SA, among others. In this case, concepts of Particle Swarm Optimization (PSO)[14] that are capital for its good search properties are isolated and transferred to the base cGA.

PSO was originally designed and introduced by Eberhart and Kennedy [9], [14], [16] in 1995. The PSO is a population based search algorithm inspired in the social behavior of birds, bees or a shoal of fishes. Each individual within the swarm is represented by a vector in multidimensional search space. It has been shown that this simple model can deal with difficult optimization problems efficiently. Many versions and improvements to the original PSO have been proposed [5], [6], [12], [18], [21].

This chapter is organized as follows. In Section 8.2 we show basic concepts of cGA and PSO. In Section 8.3 we describe our hybrid algorithms. In Section 8.4 we show the experiments and results. Finally, in Section 8.5, we describe some conclusions and suggest future research lines.

8.2   Basic Concepts

In this section we briefly describe the metaheuristic techniques used in this work: Particle Swarm Optimization and Celullar Genetic Algorithm.

8.2.1   *Particle Swarm Optimization*

The Particle Swarm Optimization was developed by Kennedy and Eberhart in 1995 [14]. This is a population-based technique inspired by social behavior of the movement of flocks of birds or schools of fish. In PSO the potential solutions, called particles, "fly" or "move" through the problem space by following some simple rules. All of the particles have fitness values based on their position and have velocities which direct the flight of the particles. PSO is initialized with a group of random particles (solutions), and then searches for optima by updating them through generations. In every iteration, each particle is updated by following two "best" values. The first one is the best solution (according to fitness) that particle has found so far. This value is called *pbest*. Another "best" value that is tracked by the particle swarm optimizer is the best value obtained so far by any particle in the population. This best value is a global best and thus it is called *gbest*.

Every particle updates its velocity and position with the following equations:

$$v_{n+1} = \omega_i v_n + \underbrace{\varphi_1 * rand * (pbest_n - x_n)}_{cognitive} + \underbrace{\varphi_2 * rand * (gbest_n - x_n)}_{social} \qquad (8.1)$$

$$x_{n+1} = x_n + v_{n+1} \qquad (8.2)$$

ω_i is the inertia coefficient which avoid big fluctuations over time; v_n is the particle velocity; x_n is the current particle position in the search space; $pbest_n$ and $gbest_n$ are defined as the "personal" best and global best seen so far; *rand* is a random number between $(0,1)$; and φ_1, φ_2 are learning factors.

It is important to highlight in Equation 8.1 that the second term represents what the particle itself has learned, and it is sometimes referred to as the "cognitive" term. The cognitive component in the velocity equation represents the best performance of the particle so far. The third term, sometimes referred as "social term" represents the group best solution so far.

PSO was originally developed to solve real-value optimization problems. To extend the real-value version of PSO to a binary/discrete space, Kennedy and Eberhart [15] proposed a binary PSO (BPSO) method. In their model a particle will decide on "yes" or "no", "true" or "false", etc. also binary values can be a representation of a real value in binary search space. In this binary version, the particle's personal best and global best is updated as in continuous version. The velocities of the particles are defined in terms of probabilities that a bit will change to one. Using this definition, a velocity must be restricted within the range $[0,1]$. So a transformation is used to map all real valued numbers of velocity to the range $[0,1]$ [15]. The normalization function used here is a sigmoid function s:

$$s(v_n^j) = \frac{1}{1 + exp(-v_n^j)} \qquad (8.3)$$

where j represents the $j - th$ component of the velocity.

Also, the Equation 8.1 is used to update the velocity vector of the particle. And the new position of the particle is obtained using Equation 8.4:

$$x_{n+1}^j = \begin{cases} 1 & \text{if } r < s(v_n^j) \\ 0 & \text{otherwise} \end{cases} \tag{8.4}$$

where r is a random number in the range $[0, 1]$.

8.2.2   Cellular Genetic Algorithm

A cGA is a subclass of Genetic Algorithms (GA) in which the population is structured in a specified topology, so that individuals may only interact with their neighbors. These overlapped small neighborhoods help in exploring the search space because the induced slow diffusion of solutions through the population provides a kind of exploration, while exploitation takes place inside each neighborhood by genetic operations. In Algorithm 8.1 we can see a pseudocode of the cGA [2], [24]. We can view that after the generation and evaluation of the initial population, genetic operators (selection, recombination, mutation, and replacement) are applied to each individual within the environment of their neighborhoods iteratively until the termination condition is met.

Algorithm 8.1. Pseudocode of a cGA

```
 1: /* Algorithm parameters in 'cga' */
 2: Steps-Up(cga)
 3: for s ⟵ 1 to MAX_STEPS do
 4:    for x ⟵ 1 to WIDTH do
 5:       for y ⟵ 1 to HEIGHT do
 6:          nList ⟵ ComputeNeigh (cga,position(x,y));
 7:          parent1 ⟵ IndividualAt(cga,position(x,y));
 8:          parent2 ⟵ LocalSelect(nList);
 9:          /* Recombination */
10:          DPX1(cga.Pc,nList[parent1],nList[parent2],auxInd.chrom);
11:          /* Mutation */
12:          BitFlip(cga.Pm,auxInd.chrom);
13:          auxInd.fit ⟵ cga.Fit(Decode(auxInd.chrom));
14:          InsertNewInd(position(x,y),auxInd,[if_not_worse],cga,auxPop);
15:       end for
16:    end for
17:    cga.pop ⟵ auxPop;
18:    UpdateStatistics(cga)
19: end for
```

The population is structured in a two-dimensional (2-D) toroidal grid, and the neighborhood defined on it (line 6) contains five individuals: the considered one $[position(x, y)]$ plus the north, east, west and south string (called NEWS or Linear5). The considered individual itself is always selected for being one of its two parent (line 7). The second parent is selected by Tournament Selection (line 8). Genetic operators are applied to individuals in lines 10 and 12. We use in this chapter a two point crossover operator (DPX1) and traditional binary mutation operator - *bit-flip*. After applying these operators, the algorithm calculates the fitness value of the new individual (line 13) and insert it on its equivalent place in the new population (line 14) only if its value is better or equal than the old one (always adding the new individual to the next population).

After applying the above mentioned operators to the individuals we replace the old population by the new one (line 17), and calculate some statistics (line 18). The resulting behavior is a synchronous elitist cGA.

8.3   Active Components of PSO into cGA

We begin this section by describing the basic ideas that have given rise to this work. The modification of the canonical cGA is based in a global idea of capturing the main characteristics of a different metaheuristic and incorporing them into a cGA with the intention of improving its performance. In this chapter we used concepts of PSO but in further works other metaheuristics will be considered like ACO (Ant Colony Optimization), SA (Simulated Annealing), and VNS (Variable Neighborhood Search), among others. Figure 8.1 graphically shows the global schema where this work fits: in our travel in incorporating ideas of existing algorithms into a cGA we start in this chapter with PSO, and plan to go for other techniques in the near future.

To incorporate active components from PSO we maintain information about cognitive and social factors during the execution of a regular synchronous cGA with the intention of improving its performance.

In this work we propose two algorithms called hyCP-local and hyCP-global. In both algorithms we will treat each individual as a particle. We maintain its velocity, position and information about its personal (*pbest*), and social (*gbest*) knowledge to update the information (velocity and position). Then a mutation based on PSO is used and line 12 (mutation in the canonical cGA Algorithm 8.1) is replaced with the following lines:

```
1:   UpdateVelocity;
2:   UpdateIndividual (cga.Pm, auxInd.chrom);
```

The first line updates the velocity of the particle using Equation 8.1. The second line modifies the individual taking into account the mechanism with the sigmoid function

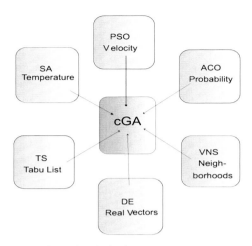

Fig. 8.1 Inserting concepts of metaheuristics into a cGA

using Equation 8.4. The pseudo-code of the algorithms proposed is described in Algorithm 8.2.

Algorithm 8.2. Pseudocode of hyCP-local and hyCP-global
___

1: /∗ Algorithm parameters ∗/
2: Steps-Up(cga)
3: **for** $s \longleftarrow$ 1 to MAX_STEPS **do**
4:    **for** $x \longleftarrow$ 1 to $WIDTH$ **do**
5:       **for** $y \longleftarrow$ 1 to $HEIGHT$ **do**
6:          nList $\longleftarrow$ ComputeNeigh (cga,position(x,y));
7:          parent1 $\longleftarrow$ IndividualAt(cga,position(x,y));
8:          parent2 $\longleftarrow$ LocalSelect(nList);
           /∗ Recombination ∗/
9:          DPX1(cga.Pc,nList[parent1],nList[parent2],auxInd.chrom);
           /∗ Mutation based on PSO ∗/
10:         **UpdateVelocity**;
11:         **UpdateIndividual (cga.Pm, auxInd.chrom)**;
12:         auxInd.fit $\longleftarrow$ cga.Fit(Decode(auxInd.chrom));
13:         InsertNewInd(position(x,y),auxInd,[if_not_worse],cga,auxPop);
14:       **end for**
15:    **end for**
16:    cga.pop $\longleftarrow$ auxPop;
17:    UpdateStatistics(cga)
18: **end for**
___

Both algorithms will apply this mutation based on PSO, with the difference that hyCP-local uses the local neighborhood (Linear5), and then selects one neighbor from there as *gbest*. For hyCP-global the global optimum of the all population is used as *gbest*.

8.4   Experiments and Analysis of Results

In this section we present the set of problems chosen for this study. Then for the hyCP-local algorithm we determine which is the best way to select a neighbor from the neighborhood. Finally, we present and analyze the results obtained when solving all problems with our hybrids (hyCP-local and hyCP-global) and cGA, always with a constant neighborhood shape (Linear5).

We have chosen a representative set of problems. The benchmarks contains many different interesting features in optimization, such as epistasis, multimodality, and deceptiveness. The problems used are Massively Multimodal Deceptive Problem (MMDP) [10], Frequency Modulation Sounds (FMS) [23], Multimodal Problem Generator (P-PEAKS) [13], COUNTSAT [8](an instance of MAXSAT [22]), Error Correcting Code Design (ECC)[19], and Maximum Cut of a Graph (MAXCUT) [17]. The problems selected for this benchmark are explained bellow:

Massively Multimodal Deceptive Problem (MMDP), made up of k deceptive subproblems of 6 bits each one, whose value depends on the number of ones (*unitation*) a binary string has. The global optimum has a value of k and it is attained when every subproblem is composed of zero or six ones. We use here a instance of $k = 40$ subproblems, and its maximum value is 40.

Frequency Modulation Sounds problem (FMS), is defined as determining six real-parameters of frequency modulated sound model. The parameters are defined in the range $[-6.4, +6.35]$, and we encode each parameter into a 32 bit substring in the individual. The optimum value for this problem is 0.0.

Multimodal Problem Generator (P-PEAKS), the idea is to generate P random $N-$bit string that represent the location of P peaks in the search space. In this chapter, we have used an instance of $P = 100$ peaks of length $N = 100$ bits each. The maximum fitness value for this problem is 1.0.

COUNTSAT problem, is an instance of MAXSAT. In this problem the solution value is the number of clauses that are satisfied by $n-$bit input string. The optimum value is having all the variables set to 1. In this work, an instance of $n = 20$ variables has been used, with the optimum value of 6860.

Maximum Cut of a Graph (MAXCUT), for coding the problem we use a binary string of length n. We have considered a graph of made up of 20 vertices. The global optimal solution for this instance is 56.740064.

Error Correcting Code Design Problem (ECC), we will consider a three-tuple (n, M, d), where n in the length of each codeword (number of bits), M is the number of codewords, and d is the minimum Hamming distance between any pair of codewords. We consider in the present chapter an instance where $M = 24$ and $n = 12$ which has an optimal fitness value of 0.0674.

The common parameterization used for all algorithms is described in Table 8.1, where L is the length of the string representing the chromosome of the individuals. One parent is always the individual itself while the other one is obtained by using *Tournament Selection* (TS). The two parents are ensured to be different.

Table 8.1 Parameterization used in our algorithms

Population Size	400 individuals
Selection of Parents	itself + Tournament Selection
Recombination	DPX1, $p_c = 1.0$
Bit Mutation	(Bit-flip for cGA), $p_m = 1/L$
Replacement	Replace If Not Worse
Inertia coefficient	w = 1
Leaning factors	$\varphi_1, \varphi_2 = 1$
Random value	$rand = UN(0, 1)$

In the recombination operator, we obtain just one offspring from two parents: the one with the largest part of the best parent. The *DPX1* recombination is applied always (probability $p_c = 1.0$). The bit mutation probability is set to $p_m = 1/L$. The exceptions are COUNTSAT, where we use $p_m = (L - 1)/L$ and the FMS problem, for which a value of $p_m = 1/(2 * L)$ is used. These two values are needed because the algorithms otherwise had a negligible solution rate with the standard $p_m = 1/L$ probability.

We will replace a given individual on each generation only if the offspring fitness is not worse than this given individual. The cost of solving a problem is analyzed by measuring the number of evaluations of the objective function made during the search. The stop condition for all algorithms is to find a solution or to achieve a maximum of one millon function evaluations. The last three rows represent the values used only for the algorithms based on PSO.

To indicate the strength of the relations between different factors and performance measures we perform here a statistical validation of results [7]. First, we should decide between non-parametric and parametric tests: when the data is non-normally distributed we should use non-parametric methods, otherwise we use parametric tests. To evaluate the normality of all results we applied a Kolmogorov-Smirnov test. Then, if we are comparing the results of two algorithms and the distribution of the results is normal we will apply t-test, whereas if not normally distributed we will apply a Wilcoxon test. When a comparison of more than two algorithms is needed and the distribution is normal we will perform a ANOVA test, in other case the test we will apply is Kruskal-Wallis.

First of all, for hyCP-local we implement two ways of selecting the social knowledge (*gbest*) from the neighborhood of the individual. The first proposal selects in a random way one neighbor from the neighborhood and uses it as social knowledge (*gbest*). The second proposal selects the best neighbor from the neighborhood and uses it as social knowledge (*gbest*).

To decide which is the best form to select this social knowledge we use both of them to solve all problems. To test whether differences between the results of the two proposals are statistically significantly we applied t-test (when the results follow a normal distribution) and Wilcoxon test (when the results do not follow a normal distribution). As regards to hyCP-global *gbest* is the best value found by any particle with in the population. Throughout the paper all best values are **bolded**.

Table 8.2 Results obtained using hyCP-local with Random Neighbor and hyCP-local with the Best Neighbor for a set of problems

Problem	Best	Random Neighbor		Best Neighbor		t-test/Wilcoxon	
		Evals	Time	Evals	Time	Evals	Time
ECC	0.07	162600	4345	**141400**	**3369**	(-)	(+)
P-PEAKS	1.00	**37800**	3413	39600	**3126**	(-)	(+)
MMDP	40	264800	9955	**182000**	**5051**	(+)	(+)
FMS	0.00	**412000**	24795	462800	**24269**	(-)	(-)
COUNTSAT	6860	**224800**	1491	348200	1848	(-)	(-)
MAXCUT	56.74	**7600**	54	**7600**	**41**	(-)	(-)

All algorithms are implemented in Java, and run on a 2.53 GHz Intel i5 processor with 4 GB RAM under Windows 7.

In Table 8.2 the median results of 30 independent runs are included. The first column (Problem) represents the name of the problem resolved, the second column (Best) the better found value and then for each algorithm (hyCP-local with a random neighbor select from the neighborhood and hyCP-local with best neighbor select from the neighborhood) we show the median value for the number of evaluations (columns Evals) needed to solve each problem and the time in *ms* consumed (columns Time). Finally, the last column (t-test|Wilcoxon) represents the *p*-values computed by performing t-test or Wilcoxon test as appropriate. We will consider a 0.05 level of significance. Statistical significant differences among the algorithms are shown with symbols "(+)", while nonsignificance is marked with "(-)".

We can observe that the use of the best neighbor in hyCP-local in general produces better results than the ones got by using a random neighbor from the neighborhood. In three problems addressed (P-PEAKS, FMS, and COUNTSAT) hyCP-local with a random neighbor required less number of evaluations to obtain the optimum than hyCP-local with the best neighbor. Regarding the time required to reach the optimum value in five of the six problems resolved hyCP-local with the best neighbor requires less time to find optimal value compared to the time required by hyCP-local with random neighbor. In addition in three of the six problems the difference are statically significant. Taking into account this results we decide to use the best neighbor in hyCP-local in the forthcoming study of this chapter.

In Table 8.3 we show the percentage of success in 150 independent runs for the three algorithms. We can see that the success rate for our hybrids is higher (or equal in some cases) than the cGA algorithm. Moreover, cGA obtained a very undesirable (0%) hit rate for the COUNTSAT problem.

In Table 8.4 the following information is shown. The first column (Problem) represents the name of the problem resolved, the second column (Best) the better found value and then for each algorithm (hyCP-local, hyCP-global, and cGa) the number of evaluations (columns Evals) needed to solve each problem and the time in ms consumed (columns Time). Finally, the last column (ANOVA|K-W) represents the *p*-values computed by performing ANOVA or Kruskal-Wallis tests as

Table 8.3 Percentage of success obtained by hyCP-local, hyCP-global, and cGA for a set of problems out of 150 independent runs

Problem	hyCP-local	hyCP-global	cGA
ECC	**100%**	**100%**	**100%**
P-PEAKS	**100%**	**100%**	**100%**
MMDP	58%	**61%**	54%
FMS	**83%**	81%	25%
COUNTSAT	**80%**	36%	0%
MAXCUT	**100%**	**100%**	**100%**

appropriate, on the time and evaluations results to assess the statistical significance of them (columns Evals and Time). We will consider a 0.05 level of significance. Statistical significant differences among the algorithms are shown with symbols "(+)", while non-significance is shown with "(-)".

We can observe that our hybrid algorithms reduce the number of evaluations required to reach the optimum value and also in two problems (FMS and COUNTSAT) these differences are statistically significant. Also, our hybrids required smaller time to obtain the optimum for four of the six problems (P-PEAKS, FMS, COUNTSAT, and MAXCUT). Nevertheless, the difference are statically significant only in two cases (P-PEAKS and COUNTSAT) so we still need to investigate more on this issue ofrunning time. This is an expected behavior since the mutation based on PSO requires some additional calculations to keep individual updated and be able to use individual and social knowledge as appropriate.

Table 8.4 Results obtained by cGA and our hybrids for a set of problems

Problem	Best	hyCP-local		hyCP-global		cGA		ANOVA/K-W	
		Evals	Time	Evals	Time	Evals	Time	Evals	Time
ECC	0.07	**141400**	3369	157600	4370	150000	**2512**	(-)	(+)
P-PEAKS	1.00	39600	**3126**	**38200**	3376	39200	3283	(-)	(+)
MMDP	40	182000	5051	211200	6457	**144000**	**2295**	(+)	(+)
FMS	0.00	462800	24269	**367400**	**22183**	646800	29326	(+)	(-)
COUNTSAT	6860	**348200**	**1848**	577200	3468	1000000	2342	(+)	(+)
MAXCUT	56.74	7600	**41**	**7000**	51	8000	49	(-)	(-)

Figure 8.2 shows the number of evaluations needed for each algorithm to reach the optimum value in each problem. We can observe the median values and how the results are distributed for the six problems. In Figures 8.2(a) and (b) the results obtained for each algorithm are very similar and the difference among the results are not statistically significant, but our hybrids seem to tend to require less number of evaluations than cGA. In Figure 8.2(c) cGA obtained the higher median value and the difference among the results are statistically significant. In Figure 8.2(d) we can observe that median value was obtained by hyCP-global and also in this case the difference among the results are statistically significant. In Figure 8.2(e) we can

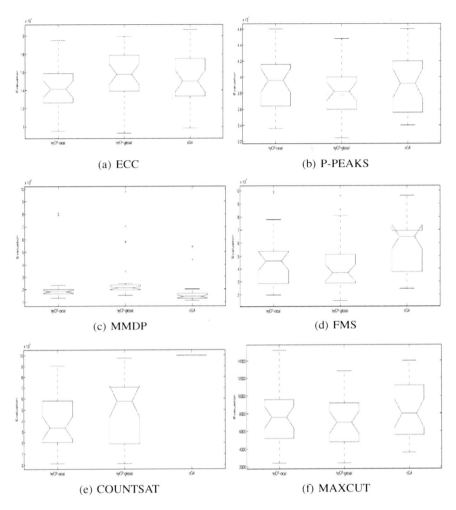

Fig. 8.2 Box-plots of the number of evaluations required for the algorithms considering: (a) ECC, (b) P-PEAKS, (c) MMDP, (d) FMS, (e) COUNTSAT, and (f) MAXCUT problems

see a marked difference in favor of hyCP-local. Recall that for this problem cGA never found the optimal value. Finally, in Figure 8.2(f) hyCP-global obtained the minimum value. Though clear best results are not always showing up, the trend and the punctual over performance tells us of a comparable or better technique as to the search effort.

Figure 8.3 shows the Time (ms) needed for each algorithm to reach the optimum value for each problem. In Figures 8.3(b), (d), (e), and (f) we can observe that our hybrids obtained the minimum median values in four of the six problems, and in two

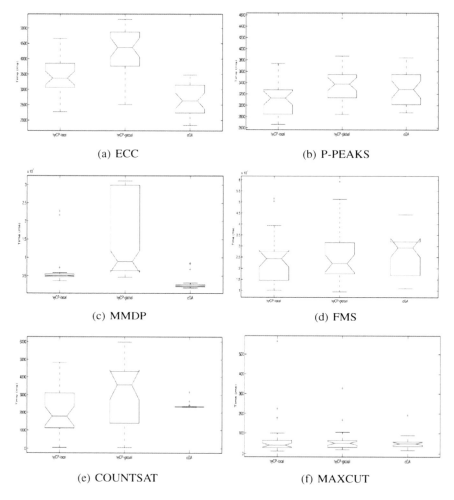

Fig. 8.3 Box-plots of the Time required for the algorithms considering: (a) ECC, (b) P-PEAKS, (c) MMDP, (d) FMS, (e) COUNTSAT, and (f)MAXCUT problems

cases the difference among the results are statistically significant. In Figures 8.3(a) and (c) we can see how cGA obtained the minimum median values, and in both cases the difference among the results are statistically significant.

8.5   Conclusions and Further Work

In this work we have presented the insertion of characteristics from a population-based technique, PSO, into a cellular genetic algorithm. The motivation for this work was to improve the performance of cGA with the addition of components that make efficient other metaheuristics, with the idea of getting even better results than those already obtained by the core technique.

In this case we introduce a mutation based on local particle swarm optimization (hyCP-local) and a mutation based on global particle swarm optimization (hyCP-global). In nearly all the analyzed problems our hybrids obtained equal or better results than the obtained without them. In five of the six problems analyzed the best performance in terms of the number of evaluations was obtained by our hybrids. Meanwhile, in three of the six problems analyzed the hybrid algorithms obtained higher percentage of success than the obtained by cGA. As regards to the time required to reach the optimum values, only in four of the six problems discussed our hybrids obtain the minimums values. This behavior is expected because the introduction of the PSO concepts into cGA also introduces more processing time and this affects the time required to reach the optimum values. In fact, getting faster running times in four out of six problems is a very interesting result because of the expected overhead in them.

These results encourage us to expand the set of problems discussed in future works and to incorporate other active components from other metaheuristics.

Acknowledgements. This research was partially funded by Spanish TIN2008-06491-C04-01, TIN2011-28194, and Andalusian P07-TIC-03044. The second author acknowledges the constant support provided by the Universidad Nacional de la Patagonia Austral.

References

1. Bäck, T. (ed.): Seventh International Conference on Genetic Algorithms. Morgan Kaufmann Publishers (1997)
2. Alba, E., Dorronsoro, B.: Cellular Genetic Algorithms. Springer (2008)
3. Alba, E., Tomassini, M.: Parallelism and evolutionary algorithms. IEEE Transactions on Evolutionary Computation 6(5), 443–462 (2002)
4. Cantú-Paz, E.: Eficient and Accurate Parallel Genetic Algorithms, 2nd edn. Book Series on Genetic Algorithms and Evolutionary Computation, vol. 1. Kluwer Academic (2000)
5. Chen, X., Li, Y.: A modified pso structure resulting in high exploration ability with convergence guaranteed. IEEE Trans. Syst., Man, Cybern. B, Cybern. 37(5), 1271–1289 (2007)
6. Chen, Y.-P., Peng, W.-C., Jian, M.-C.: Particle swarm optimization with recombination and dynamic linkage discovery. IEEE Trans. Syst., Man, Cybern. B, Cybern. 37(6), 1460–1470 (2007)
7. Derrac, J., García, S., Molina, D., Herrera, F.: A practical tutorial on the use of nonparametric statistical tests as a methodology for comparing evolutionary and swarm intelligence algorithms. Swarm and Evolutionary Computation 1(1), 3–18 (2011)
8. Droste, S., Jansen, T., Wegener, I.: A natural and simple function which is hard for all evolutionary algorithms. In: 3rd Asia-Pacific Conf. Simulated Evol. Learning, pp. 2704–2709 (2000)

9. Eberhart, R., Kennedy, J.: A new optimizer using particles swarm theory. In: Sixth International Symposium on Micro Machine and Human Science, Nagoya, Japan, pp. 39–43. IEEE Service Center, Piscataway (1995)

10. Goldberg, D., Deb, K., Horn, J.: Massive multimodality, deception, and genetic algorithms. In: Männer, R., Manderick, B. (eds.) Int. Conf. Parallel Prob. Solving from Nature II, pp. 37–46 (1992)

11. Hart, W., Krasnogor, N., Smith, J.: Recent Advances in Memetic Algorithms. Springer (2005)

12. Janson, S., Middendorf, M.: A hierarchical particle swarm optimizer and its adaptive variant. IEEE Trans. Syst., Man, Cybern. B, Cybern. 35(6), 1272–1282 (2005)

13. De Jong, K., Potter, M., Spears, W.: Using problem generators to explore the effects of epistasis. In: 7th Int. Conf. Genetic Algorithms, pp. 338–345. Morgan Kaufmann (1997)

14. Kennedy, J., Eberhart, R.: Particle Swarm Optimization. In: IEEE Int. Conf. Neural Netw., vol. 4, pp. 1942–1948 (1995)

15. Kennedy, J., Eberhart, R.C.: A discrete binary version of the particle swarm algorithm. In: IEEE International Conference on Systems, Man, and Cybernetics, Computational Cybernetics and Simulation, vol. 5, pp. 4104–4108. IEEE (1997)

16. Kennedy, J., Eberhart, R.: Swarm Intelligence. Morgan Kaufmann (2001)

17. Khuri, S., Bäck, T., Heitkötter, J.: An evolutionary approach to combinatorial optimization problems. In: 22nd Annual ACM Computer Science Conference, pp. 66–73 (1994)

18. Krohling, R.A., Coelho, L.S.: Coevolutionary particle swarm optimization using gaussian distribution for solving constrained optimization problems. IEEE Trans. Syst., Man, Cybern. B, Cybern. 36(6), 1407–1416 (2006)

19. MacWilliams, F., Sloane, N.: The Theory of Error-Correcting Codes. North-Holland (1977)

20. Manderick, B., Spiessens, P.: Fine-grained parallel genetic algorithm. In: Schaffer, J.D. (ed.) Third International Conference on Genetic Algorithms (ICGA), pp. 428–433. Morgan Kaufmann (1989)

21. Mendes, R., Kennedy, J., Neves, J.: The fully informed particle swarm: Simpler, maybe better. IEEE Trans. Evol. Comput. 8(3), 204–210 (2004)

22. Papadimitriou, C.: Computational Complexity. Adison-Wesley (1994)

23. Tsutsui, S., Fujimoto, Y.: Forking genetic algorithm with blocking and shrinking modes. In: Forrest, S. (ed.) 5th International Conference on Genetic Algorithms, pp. 206–213 (1993)

24. Whitley, D.: Cellular genetic algorithms. In: Forrest, S. (ed.) Fifth International Conference on Genetic Algorithms (ICGA), p. 658. Morgan Kaufmann (1993)

Part IV
Multi-objective Optimization, Heuristic Conversion Algorithms

Chapter 9
On Gradient-Based Local Search
to Hybridize Multi-objective Evolutionary
Algorithms

Adriana Lara, Oliver Schütze, and Carlos A. Coello Coello

Abstract. Using evolutionary algorithms when solving multi-objective optimization problems (MOPs) has shown remarkable results during the last decade. As a consolidated research area it counts with a number of guidelines and processes; even though, their efficiency is still a big issue which lets room for improvements. In this chapter we explore the use of gradient-based information to increase efficiency on evolutionary methods, when dealing with smooth real-valued MOPs. We show the main aspects to be considered when building local search operators using the objective function gradients, and when coupling them with evolutionary algorithms. We present an overview of our current methods with discussion about their convenience for particular kinds of problems.

9.1   Introduction

Over the last few decades, a huge development on theoretical and practical approaches on solving multi-objective optimization problems (MOPs) has been done; either as multi-objective mathematical programming [14], or also as stochastic heuristics like evolutionary algorithms—named in this case as multi-objective evolutionary algorithms (MOEAs) [10, 9]. MOEAs are suitable to numerically approximate solutions of MOPs for several reasons; we can specially mention that, by nature, they spring an entire set of solutions on each run—instead of just one solution point as the traditional methods do. Having a set-oriented procedure is very

Adriana Lara
Mathematics Department ESFM-IPN, Edif. 9 UPALM, 07300 Mexico City, Mexico
e-mail: adriana@esfm.ipn.mx

Carlos A. Coello Coello · Oliver Schütze
Computer Science Department, CINVESTAV-IPN, Av. IPN 2508, Col. San Pedro Zacatenco,
07360 Mexico City, Mexico
e-mail: {ccoello, schuetze}@cs.cinvestav.mx

E. Tantar et al. (Eds.): EVOLVE- A Bridge between Probability, SCI 447, pp. 305–332.
springerlink.com   © Springer-Verlag Berlin Heidelberg 2013

convenient when solving MOPs—we will clarify this idea later, using the next definitions.

Definition 9.1.1. *The multi-objective problem (MOP) is defined as minimizing*

$$F(x) := [f_1(x), f_2(x), \ldots, f_k(x)]^T \tag{9.1}$$

subject to:

$$g_i(x) \leq 0 \quad i = 1, 2, \ldots, m \tag{9.2}$$

where $x = [x_1, x_2, \ldots, x_n]^T \in \mathbb{R}^n$ is the vector of decision variables (or *decision parameters*), $f_i : \mathbb{R}^n \to \mathbb{R}$, $i = 1, \ldots, k$, are the objective functions and $g_i : \mathbb{R}^n \to \mathbb{R}$, $i = 1, \ldots, m$, are the constraint functions, which define the feasible region $\mathscr{X}$ of the problem. The problem above is also known as a *vector optimization* problem. If functions g_i are not present, we are dealing with an *unconstrained* MOP.

Solving a MOP is very different than solving a single-objective optimization problem (*i.e.* $k = 1$). Since some of the f_i are normally "in conflict" with each other,[1] the solution of a MOP is not given by a unique point; this is because normally no single solution exists that provides the best possible value for all the objectives. Consequently, solving a MOP implies finding a trade-off between all the objective functions; this requires the generation of a set of possible solutions instead of a single one—as in the single-objective optimization case. The notion of "optimality" that we just informally described, was originally proposed by Francis Ysidro Edgeworth in 1881 [13] and it was later generalized by Vilfredo Pareto, in 1896 [37]. This concept is known today as *Pareto optimality* and will be formally introduced next.

Definition 9.1.2. *Given two vectors $x, y \in \mathbb{R}^n$, we say that x **dominates** y (denoted by $x \prec y$) if $f_i(x) \leq f_i(y)$ for $i = 1, \ldots, k$, and $F(x) \neq F(y)$.*

Definition 9.1.3. *We say that a vector of decision variables $x \in \mathscr{X} \subseteq \mathbb{R}^n$ is **non-dominated** with respect to $\mathscr{X}$, if there does not exist another $x' \in \mathscr{X}$ such that $x' \prec x$.*

Definition 9.1.4. *We say that a vector of decision variables $x^* \in \mathscr{X} \subset \mathbf{R}^n$ is **Pareto optimal** if it is non-dominated with respect to $\mathscr{X}$.*

Definition 9.1.5. *a) The **Pareto set** $\mathscr{P}^*$ is defined by:*

$$\mathscr{P}^* = \{x \in \mathscr{X} \,|\, x \text{ is Pareto optimal}\}.$$

*b) The **Pareto front** $\mathscr{P}\mathscr{F}^*$ is defined by:*

$$\mathscr{P}\mathscr{F}^* = \{F(x) \in \mathbb{R}^k \,|\, x \in \mathscr{P}^*\}$$

---

[1] For example, one objective may refer to manufacture cost and another to quality of the product.

We thus wish to determine the Pareto optimal set, from the set $\mathscr{X}$ of all the decision variable vectors that satisfy the constraints of the problem. Note, however, that in practice just a finite representation of the Pareto optimal set is normally achievable.

Assuming x^* as a Pareto point of (9.1), there exist [27] a vector $\alpha \in \mathbb{R}^k$, with $0 \le \alpha_i, i = 1, \ldots, k$ and $\sum_{i=1}^k = 1$ such that

$$\sum_i^k \alpha_i \nabla f_i(x^*) = 0. \tag{9.3}$$

A point x^* that satisfies (9.3) is called a **Karush-Kuhn-Tucker (KKT) point.**

Example 9.1. Consider the following unconstrained MOP:

$$\text{minimize } F(x,y) := \left[x^2 + y^2, (x-10)^2 + y^2 \right]^T, \tag{9.4}$$

where $x, y \in \mathbb{R}$.

The Pareto set of this problem is the line segment $[(0,0),(10,0)] \subset \mathbb{R}^2$, and the Pareto front is shown, as a continuous line, in Figure 9.1.

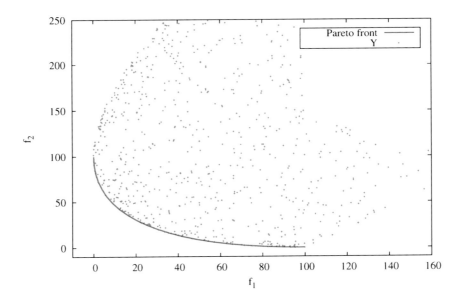

Fig. 9.1 This figure illustrates the Pareto front for the Example 9.1, the axes represent the value regarding each function—what we call the objective space. The points of Y are the images of randomly taken vectors inside the domain.

The set oriented nature of MOEAs had made them very popular among currently available methods for solving MOPs (see [8] for application examples). Even though, efficiency is the main drawback when they are used, since the evaluation of the objective functions is constantly made, each generation, for the whole population. Current research about improving MOEAs includes their hybridization with local searchers, in order to make a guided search at a particular moment of the procedure. The coupling of evolutionary algorithms with any local search procedure is also known as a *memetic algorithm* [35].

Local search operators depend on the domain, since the goal is precisely to take advantage of previous knowledge of the problem. A lot of research has been done on local searchers, and memetic MOEAs, for discrete and combinatorial optimization problems (see for example [25, 24, 26]). For multi-objective problems with continuous domains, the local search has not been deeply studied as itself; this is probably due to the fact that neighborhoods on continuous spaces have an infinite cardinality. The vicinity notion in continuous domain problems is related with open balls or manifolds, so the natural choice is to study them by the differential properties of the functions; this leads to the use of gradient information to compute better solutions.

In this chapter we focus on local searchers, for continuous MOPs, built to use (implicitly and/or explicitly) the gradient information of the objective functions. We focus mainly on unconstrained cases, but give some insights for the extension to constrained search spaces.

In the rest of this chapter we develop the main ideas from particular work [31, 28, 32, 29, 30] on local searchers for continuous memetic algorithms. We introduce in the next section the basic concepts to study the gradient geometry in the case of MOPs. In Section 9.3 we present algorithms, based on multi-objective line search, emphasizing their particular features. The applicability of these methods, when hybridizing MOEAs, is presented in Section 9.4 as well as some discussion over the main hybridization aspects. Finally some conclusions and possible extensions are included in Section 9.5.

9.2   Descent Cones and Directions

When solving an optimization problem with several objective functions, we have to deal with several gradients—one for each function. Some gradient-based MOEAs have been built to perform descent movements alternating the single gradient of each objective function [16, 21]; this could have certain applications on preferences management. However, when designing a special local searcher for continuous multi-objective problems, our focus lies on finding a suitable mechanism to improve solutions using simultaneously every function gradient. The reason is that such a good local searcher must have the feature that after its application, over a single solution x_i, it throws a new solution x_{i+1} which is better in the Pareto sense (*i.e.* x_{i+1} *dominates* x_i). This feature is very important when analyzing the global convergence of the methods.

Using gradient information to guide the simultaneous descent of several functions can not be done in a straightforward way. The main reason is that, since the objective functions are each other in conflict, their gradients typically point toward different directions. This turns the task of finding a common improving direction into another multi-objective problem. To understand the local behavior of gradient-based methods in multi-objective optimization, descent cones constitute the main tool since they were first used in the multi-objective evolutionary context by Brown and Smith [6, 7].

We state next the main concepts and introduce the notation required for further discussion.

Let $f_1, \ldots, f_k : \mathbb{R}^n \to \mathbb{R}$ be continuous and differentiable, and $\langle \cdot, \cdot \rangle$ denote the standard inner product in $\mathbb{R}^n$.

Let

$$\nabla f_i(x) = \left(\frac{\partial f_i(x)}{\partial x_1}, \ldots, \frac{\partial f_i(x)}{\partial x_n} \right)$$

be the gradient of the function f_i at x. Then, for each $x \in \mathbb{R}^n$ and each $i \in \{1, \ldots, k\}$, with $\nabla f_i(x) \neq 0$, we define:

$$H_{x,i} = \left\{ v \in \mathbb{R}^n : \langle \nabla f_i(x), v \rangle = 0 \right\},$$

$$H_{x,i}^+ = \left\{ v \in \mathbb{R}^n : \langle \nabla f_i(x), v \rangle \geq 0 \right\},$$

and

$$H_{x,i}^- = \left\{ v \in \mathbb{R}^n : \langle \nabla f_i(x), v \rangle \leq 0 \right\}.$$

Since the set $H_{x,i}$ is the orthogonal complement of the vector $\nabla f_i(x)$, it is (if $\nabla f_i(x) \neq 0$) in general a hyperplane[2] of $\mathbb{R}^n$; also, it divides the space in two n–dimensional sets: $H_{x,i}^+$ and $H_{x,i}^-$ (see Figure 9.2).

Definition 9.2.1. *We denote*

$$C_x(-, -, \ldots, -) = \bigcap_{i=1}^k H_{x,i.}^- \setminus \{0\}.$$

as the descent cone *pointed at x (see Figure 9.3). Similarly, the* ascent cone *is defined by* $C_x(+, +, \ldots, +) = \bigcap_{i=1}^k H_{x,i.}^+ \setminus \{0\}$, *and the* diversity cones *are the intersection of hyperplanes when they are not all of the form* $H_{x,i}^+$ *and neither all of them of the form* $H_{x,i}^-$.

When having k functions in a MOP, each function f_i determines a gradient ∇f_i and a hyperplane $H_{x,i}$ for a certain solution x. Summarizing, for each point x in the search

---

[2] A hyperplane of an n-dimensional space is a subspace with dimension $n1$.

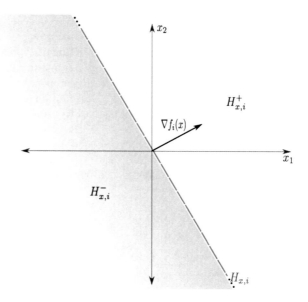

Fig. 9.2 This figure shows the division of $\mathbb{R}^2$ into the two half-spaces, $H_{x,i}^+$ and $H_{x,i}^-$, induced by the gradient of the function f_i at the point x

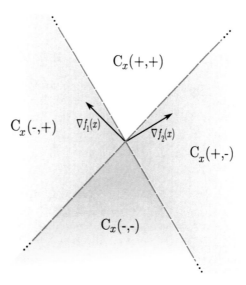

Fig. 9.3 This figure shows the ascent cone $C_x(+,+)$, the descent cone $C_x(-,-)$, and the diversity cones $C_x(+,-)$ and $C_x(-,+)$, for a certain point x, for a bi-objective problem

space, the k functions determine a division of this space into one descent cone, one ascent cone and $2^k - 2$ diversity cones—as they were previously named in [6][3].

Descent cones describe the dynamics of the search from a gradient-based geometry point of view, and this has several important implications (see for example the results presented in [42]) that are useful when building local search operators for algorithms over real numbers. One primary observation [6] is that when the point x is far away from any Pareto optimal point the descent cone is big. This happens because the gradients are almost aligned; then, the chance of randomly generating a direction/solution which simultaneously improves all the functions is high (nearly 50%). On the other hand, when the point x is near a Pareto optimal point, the gradients are almost linearly dependent, which means that the descent cone shrinks, and the possibility of randomly generating a better point is low (see Figure 9.4).

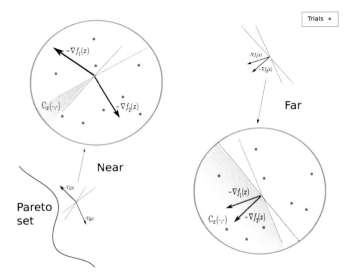

Fig. 9.4 This figure shows two cases, when the point is near and when it is far from a Pareto set; the descent cone $C_x(-,-)$ shrinks down when the Pareto set is reached

The above analysis could explain why MOEAs have good performance at the beginning of the search, and a slow convergence rate at latter stages—when points are near to the Pareto front and the chance of generating randomly better points is reduced. This observation inspires guidelines for a suitable hybridization; if it is possible to identify when the evolutionary search is no longer producing good results, this is then the time when the gradient-based local search can take part of the process—in order to certainly improve solutions in a deterministic way.

---

[3] In [6] the descent cones are defined, for illustrative purposes, by pictures of the corresponding affine hyperplanes described here. We are stating the formal definitions using no affine hyperplanes just to be consistent with our approach.

Improving solutions implies performing movements in specific search directions. In multi-objective optimization, as we have noticed, these directions should be able to throw (at least locally) better solutions regarding all the functions simultaneously; this is formally said by the next definition.

Definition 9.2.2. *A vector $v \in \mathbb{R}^n$ is called a multi-objective descent direction of the point $x \in \mathbb{R}^n$ if*

$$v \in C_x(-,-,\ldots,-).$$

In other words, a multi-objective descent direction is such that the directional derivatives with respect to v in x are non-positive, *i.e.* $\langle \nabla f_i(x), v \rangle \leq 0$ for all $i \in 1,\ldots,m$ without allowing them to be all equal to zero. This means that if we perform a small movement over v, we obtain a local improvement (decrease) simultaneously for all the objective functions. In the next section, we present different ways to calculate descent directions and to perform movements toward (and along) the Pareto set. In particular, we present a result for the construction of a descent direction in the simplest multi-objective case—two objectives.

9.3   Practical Approaches

9.3.1   *Movements toward the Optimum*

When running a MOEA, new points are generated by variation operators in order to move (evolve) the population of solutions toward an approximation of the Pareto set. Due to the stochastic nature of the process, a probability to generate non-improving points does exist. On the other hand, when using gradient information within a MOEA, the resulting hybrid algorithm is able to perform directed accurate movements toward an improved solution. This newly computed point dominates the original one when multi-objective descent search directions—together with a suitable step size control—are used. As we mentioned before, computing such directions is also a multi-objective problem [2]; since each objective provides its own (gradient-based) range of movements for descent, all of these possible directions need to be properly combined into a single one in order to efficiently guide a MOEA.

9.3.1.1   A Simple Bi-objective Descent Direction

The simplest way to combine two gradients, in order to get a common descent direction, is by adding them. This fact has been already observed (e.g. [11]) but it has not been exploited in memetic algorithms yet. The next result shows that this operation throws, in fact, a bi-objective descend direction. Unfortunately, Proposition 9.1 cannot be generalized for more than two objective functions. For that case, applying other approaches—mentioned next—is necessary to obtain the descent direction; even when this represents a higher computational cost.

Proposition 9.1. *Let $x \in \mathbb{R}^n$, and let $f_1, f_2 : \mathbb{R}^n \to \mathbb{R}$ define a bi-objective MOP. Then, the direction*

$$\overline{\nabla}_x = - \left(\frac{\nabla f_1(x)}{||\nabla f_1(x)||} + \frac{\nabla f_2(x)}{||\nabla f_2(x)||} \right), \tag{9.5}$$

where $|| \cdot || = || \cdot ||_2$, is a descent direction at x for the MOP.

Proof: Let us denote $\nabla_i := \frac{\nabla f_i(x)}{||\nabla f_i(x)||}$ for $i = \{1, 2\}$, and θ be the angle between ∇_1 and ∇_2. Then

Similarly, $\langle \overline{\nabla}_x, \nabla_2 \rangle \leq 0$; then, $\overline{\nabla}_x$ is a descent direction of the point x for the defined MOP. Note that if $\nabla f_i(x) = 0$, for $i = 1$ or $i = 2$, then x is a KKT point. □

One of the main issues when using gradient-based tools is how to validate the increase of the computational cost, after using the method, with the achieved improvements. The currently available MOEAs that use descent directions as local search engines have two sources of computational cost. The first one is associated to the fitness function evaluations required to estimate the gradients and to perform the line search. The second source is related to the computation of the descent direction itself. In this sense, and unlike previous approaches [15, 2, 20], this way (Equation 9.5) of calculating a direction has the advantage of having a zero cost for the computation of the descent direction. We claim that this procedure is the simplest way to combine the gradients of two functions, but it can not be generalized to more than two functions, since a similar arithmetic combination of them does not produce a descent direction in general, see the following example:

Example 9.2. Assuming a three-objective problem such that, for a certain x,

$$\nabla f_1(x) = (1.000, 1.000, 1.000)$$
$$\nabla f_2(x) = (-0.944, 0.970, 0.374)$$
$$\nabla f_3(x) = (0.836, -0.177, -0.334).$$

Then, computing

$$\overline{\nabla}_x = - \left(\frac{\nabla f_1(x)}{||\nabla f_1(x)||} + \frac{\nabla f_2(x)}{||\nabla f_2(x)||} + \frac{\nabla f_3(x)}{||\nabla f_3(x)||} \right)$$
$$= (-0.3826, -0.5262, -0.3730)$$

leads to

$$\langle \overline{\nabla}_x, \nabla_1 \rangle = -1.2818 < 0.$$
$$\langle \overline{\nabla}_x, \nabla_2 \rangle = 0.4423 > 0.$$
$$\langle \overline{\nabla}_x, \nabla_3 \rangle = -0.2889 < 0.$$

with $\nabla_i := \frac{\nabla f_i(x)}{||\nabla f_i(x)||}$ for $i = \{1, 2, 3\}$. Then $\overline{\nabla}_x$ is not a common descent direction.

9.3.1.2   General Approaches

The most commonly used proposals for computing multi-objective descent directions are the one from Fliege and Svaiter [15] and the one from Schäffler *et al.* [38]. Theoretical results about their efficacy and convergence are available. These methods return a common descent direction for all the objectives after solving a quadratic optimization problem—which could be derived into a linear one. These approaches have been already incorporated into multi-objective memetic strategies [44], [28], [31]. Their main drawback is that, being greedy methods used to perform descents, they can present a bias in case of unbalanced magnitudes of the gradient vectors. However, these methods present some advantages, like having an intrinsic stopping criterion and being quite effective in practice.

As an unbiased alternative, it is possible to use normalized gradients and to work with the proposals developed by Bosman and DeJong [2], and, to manage constraints, by Harada *et al.* [20]. These methods look into the entire Pareto set of descent directions and choose (randomly, at each step) one direction within it. The cost of this approach is again related to solving a system of linear equations.

It is worth noting that other approaches [7, 31], which do not explicitly compute the gradients, have also an acceptable performance using less resources. They are completely based on the information extracted from the descent cones, at a particular moment during the search. One of these methods will be presented in the next section.

9.3.2   Movements along the Pareto Set

Descent directions are not the only interesting directions to move along during the multi-objective optimization search. Sometimes it is also beneficial to perform movements along the Pareto set; or also, specifically directed movements toward a particular region. Moving in a direction along the Pareto set is also possible using gradient-based continuation methods such as those described in [22, 41, 19], or those with no estimation of the gradients required, as in [7] and [31]. We describe next an operator which is able to perform movements either toward or along the Pareto set, making an automatic switch between these two types of movements when a KKT point is almost[4] reached.

9.3.2.1   The Hill Climber with Side-step

In [31] a novel point-wise iterative search procedure, called the Hill Climber with Side-step (HCS) has been proposed. This procedure is designed to perform local search over continuous domain MOPs. Based on the descent cone size and the

---

[4] We set a tolerance parameter to consider if an approximation is 'good enough'.

Karush-Kuhn-Tucker conditions [27], the HCS is capable of moving both toward or along the set of (local) Pareto points, depending on the location of the current iterate. Since proximity and a good distribution of solutions are features for the approximations of Pareto sets, this local search operator has shown potential when combined with MOEAs.

Two variants of the HCS have been proposed, a gradient-free version (denoted as HCS1) and a version that exploits explicit gradient information (denoted as HCS2). Both of them can be used as standalone algorithms to explore parts of the Pareto set, starting with one single solution, and are able to handle constraints of the model to some extent. In the following we will explain the two approaches as standalone algorithms, complemented with the recommendations to be used within a memetic algorithm.

HCS1 is based on the observation that the objective gradients are practically aligned when the initial point is far away from the optima—and the descent cone is almost equal to the half-spaces associated with each objective. So, the procedure starts with an initial point x_0, and the next iterate x_1 is randomly chosen from a vicinity $B(x_0, r)$ with radius r. If $x_1 \prec x_0$ the movement direction for improvement is set as $v = x_1 - x_0$; if $x_0 \prec x_1$, then the direction is flipped.

When a solution is near to a Pareto point, the gradients point nearly toward opposite directions, and the probability of generating a dominated or a dominating point—like in the case above—is low (see Figure 9.4). So, when $\tilde{x}_1$ is not comparable against x_0, the point is stored, and labeled, as a point which corresponds to a specific diversity cone; then, a new trial point is generated. After N_{nd} trials obtaining mutually non-dominated solutions, the proximity with the optima is assumed and this triggers a 'sidestep' movement over the Pareto front.

To perform this sidestep movement, the stored points $\tilde{x}_1, \ldots \tilde{x}_{N_{nd}}$ are used in the following way. If $\tilde{x}_1 - x_0$ is, for example, in the cone $C(+, -)$, then $x_0 - \tilde{x}_1$ is in the opposite cone $C(-, +)$ (for the bi-objective MOPs, the general k-objective case is analogue). When the limit for unsuccessful trials is reached, a search along $C(-, +)$ is performed; Taking advantage of the accumulated information, the following direction is used:

$$v_{acc} = \frac{1}{N_{nd}} \sum_{i=1}^{N_{nd}} s_i \frac{\tilde{x}_i - x_0}{\|\tilde{x}_i - x_0\|}, \tag{9.6}$$

where

$$s_i = \begin{cases} 1 & \text{if } f_1(\tilde{x}_i) < f_1(x_0). \\ -1 & \text{else.} \end{cases} \tag{9.7}$$

By construction, v_{acc} is in $C(-, +)$, and by averaging the trial search directions we aim to obtain a direction[5] which is ideally 'perpendicular' to the (small) descent cone. Note that in this case v_{acc} is indeed a 'sidestep' to the upward movement of the hill climbing process as desired, but this search direction does not necessarily have to point along the Pareto set (see next subsection for an alternative method with better guidance properties); also, there is no guarantee that v_{acc} indeed points

---

[5] This direction has previously been proposed as a local guide for a multi-objective particle swarm algorithm in [5].

to a diversity cone, but in other case, there will be an improvement on the solution anyway. This means that, even with these two considerations, this sidestep is still a good option in practice, when working with few objectives or when coupling the operator with evolutionary methods.

Algorithm 9.1. HCS1 (without Using Gradient Information)

Require: starting point $x_0 \in Q$, radius $r \in \mathbb{R}_+^n$, number N_{nd} of trials, MOP with $k = 2$
Ensure: sequence $\{x_l\}_{l \in \mathbb{N}}$ of candidate solutions
1: $a := (0, \ldots, 0) \in \mathbb{R}^n$
2: $nondom := 0$
3: **for** $l = 1, 2, \ldots$ **do**
4:      set $x_l^1 := x_{l-1}^b$ and choose $x_l^2 \in B(x_l^1, r)$ at random
5:      choose $i_0 \in \{1, 2\}$ at random
6:      **if** $x_l^1 \prec x_l^2$ **then**
7:          $v_l := x_l^2 - x_l^1$
8:          compute $t_l \in \mathbb{R}_+$ and set $\tilde{x}_l^n := x_l^2 + t_l v_l$.
9:          choose $x_l^b \in \{\tilde{x}_l^b, x_l^1\}$ such that $f(x_l^b) = \min(f(\tilde{x}_l^n), f(x_l^1))$
10:         $nondom := 0, \quad a := (0, \ldots, 0)$
11:      **else if** $x_l^2 \prec x_l^1$ **then**
12:          proceed analogous to case "$x_l^1 \prec x_l^2$" with
13:          $v_l := x_l^1 - x_l^2$ and $\tilde{x}_l^n := x_l^1 + t_l v_l$.
14:      **else**
15:          **if** $f_{i_0}(x_l^2) < f_{i_0}(x_l^1)$ **then**
16:             $s_l := 1$
17:          **else**
18:             $s_l := -1$
19:          **end if**
20:          $a := a + \frac{s_l}{N_{nd}} \frac{x_l^2 - x_l^1}{\|x_l^2 - x_l^1\|}$
21:          $nondom := nondom + 1$
22:          **if** $nondom = N_{nd}$ **then**
23:             compute $\tilde{t}_l \in \mathbb{R}_+$ and set $\tilde{x}_l^n := x_l^1 + \tilde{t}_l a$.
24:             $nondom := 0, \quad a := (0, \ldots, 0)$
25:          **end if**
26:      **end if**
27: **end for**

Algorithm 9.1 shows the pseudocode of the HCS1 operator as a standalone process. The sidestep direction is determined by the value of i_0 (see line 5 and lines 15-20). For simplicity, the value of i_0 is chosen at random. In order to introduce an orientation to the search, the following modifications can be done in the bi-objective case: in the beginning, i_0 is fixed to 1 for the following iteration steps. When the sidestep (line 23) has been performed N_s times during the run of an algorithm, this indicates that the current iteration is already near to the (local) Pareto set, and this vector is stored in x_p. If in the following no improvements can be achieved according to f_1 within a given number N_i of sidesteps, the HCS 'jumps' back to x_p, and a similar process is started but aiming for improvements according to f_2. That is, i_0 is set to -1 for the following steps (see Figure 9.5). A possible stopping criterion, hence, could be to stop the process when no improvements can be achieved according to f_2 within another N_i sidesteps along $C(+,-)$. This is in fact used as stopping criterion. Finally, for the computation of t_l more attention has to be paid (see Section 9.3.4). In the original proposal, a strategy analog to [12] was used for the presented experiments.

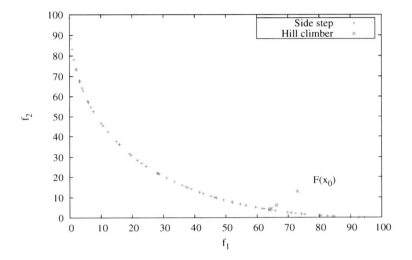

Fig. 9.5 This figure shows the performance of the HCS1 as standalone algorithm for Example 9.1; the 'anchor' picture shows the entire Pareto front, built using the steering mechanism previously described.

The second version of the HCS consists of a movement toward the Pareto front, using either a gradient-based descent direction, or a continuation-based movement in case this descent direction does not exist. A possible realization of the HCS2 is by using the descent direction presented in [38] (or the one in [15] as an alternative), which is described next.

Let a MOP be given and $q : \mathbb{R}^n \to \mathbb{R}^n$ be defined by

$$q(x) = \sum_{i=1}^{k} \hat{\alpha}_i \nabla f_i(x), \tag{9.8}$$

where $\hat{\alpha}$ is a solution of

$$\min_{\alpha \in \mathbb{R}^k} \left\{ \| \sum_{i=1}^{k} \alpha_i \nabla f_i(x) \|_2^2; \alpha_i \geq 0, i = 1, \ldots, k, \sum_{i=1}^{k} \alpha_i = 1 \right\}. \tag{9.9}$$

Then either $q(x) = 0$ or $-q(x)$ is a descent direction for all objective functions $f_1, \ldots, f_k$ in x. This states that for every point $x \in Q$ which is not a KKT–point a descent direction (i.e., a direction where all objectives' values can be improved) can be found by solving the quadratic optimization problem (9.9). In case $q(x) = 0$ the point x is a KKT–point. Thus, a test for optimality is automatically performed when computing the descent direction for a given point $x \in Q$. Given such a point x, the quadratic optimization problem (9.9) can be solved leading to the vector $\hat{\alpha}$. In case

$$\| \sum_{i=1}^{k} \hat{\alpha}_i \nabla f_i(x) \|_2^2 \geq \varepsilon_{\mathscr{P}}, \tag{9.10}$$

i.e., if the square of the norm of the weighted gradients is larger than a given threshold $\varepsilon_{\mathscr{P}} \in \mathbb{R}_+$, the candidate solution x can be considered to be 'away' from a local Pareto point, and thus, it makes sense to seek for a dominating solution. For this, the descent direction (9.8) can be taken together with a suitable step size control. If the value of the term in (9.10) is less than $\varepsilon_{\mathscr{P}}$, this indicates that x is already in the vicinity of a local Pareto point. In that case one can integrate elements from (multi-objective) continuation [22, 1] to perform a search along the Pareto set. For simplicity we assume that we are given a KKT–point $\hat{x}$ and the respective weight $\hat{\alpha}$ obtained by (9.9). Then the point $(\hat{x}, \hat{\alpha}) \in \mathbb{R}^{n+k}$ is contained in the zero set of the auxiliary function $\tilde{F} : \mathbb{R}^{n+k} \to \mathbb{R}^{n+1}$ of the given MOP which is defined as follows:

$$\tilde{F}(x, \alpha) = \begin{pmatrix} \sum\limits_{i=1}^{k} \alpha_i \nabla f_i(x) \\ \sum\limits_{i=1}^{k} \alpha_i - 1 \end{pmatrix}. \tag{9.11}$$

In [22] it has been shown that the zero set $\tilde{F}^{-1}(0)$ can be linearized around $\hat{x}$ by using a QU-factorization of $\tilde{F}'(\hat{x}, \hat{\alpha})^T$, i.e., the transposed of the Jacobian matrix of $\tilde{F}$ at $(\hat{x}, \hat{\alpha})$. To be more precise, given a factorization

$$\tilde{F}'(\hat{x}, \hat{\alpha})^T = QU \in \mathbb{R}^{(n+k)\times(n+k)}, \tag{9.12}$$

where $Q = (Q_N, Q_K) \in \mathbb{R}^{(n+k)\times(n+k)}$ is orthogonal with $Q_N \in \mathbb{R}^{(n+k)\times(n+1)}$ and $Q_K \in \mathbb{R}^{(n+k)\times(k-1)}$, the column vectors of Q_K form—under some mild regularity assumptions on $\tilde{F}^{-1}(0)$ at $(\hat{x}, \hat{\alpha})$, see [22]—an orthonormal basis of the tangent space of $\tilde{F}^{-1}(0)$. Hence, it can be expected that each column vector $q_i \in Q_K$, $i = 1, \ldots, k-1$, points (locally) along the Pareto set and is thus well suited for a sidestep direction. The step size control is explained in detail in [31].

Based on the above discussion, the HCS2 is presented in Algorithm 9.2. It is worth to remark that this is one possible realization and that there exist certainly other ways leading, however, to similar results. For instance, alternatively to the descent direction used in Algorithm 9.2, the ones proposed in [15] and [4] can be taken as well. The threshold $\varepsilon_{\mathscr{P}}$ is used for the vicinity test of a given local Pareto point. This value is certainly problem dependent, but can be made 'small' due to convergence properties of the hill climber (e.g., [15]).

Further, the movement along the Pareto set can be realized by predictor-corrector methods [22, 1] which consist, roughly speaking, of a repeated application of a predictor step obtained by a linearization of $\tilde{F}^{-1}(0)$ and a corrector step which can be done, e.g. via a Gauss-Newton method.

It is worth noting that although the HCS2 is proposed for the unconstrained case, an extension to the constrained case for the hill climber is possible (see, e.g., [15] for possible modifications); but, this is not straightforward for the movement along the Pareto set (i.e., the sidestep). Though it is possible to extend system (9.11) using equality constraints (e.g., by introducing slack variables to transform the inequality constraints into equality constraints); but according to [22], this could lead to efficiency problems in the numerical treatment.

Algorithm 9.2. HCS2 (Using Gradient Information)

Require: starting point $x_0 \in Q$
Ensure: sequence $\{x_l\}_{l \in \mathbb{N}}$ of candidate solutions
1: **for** $l = 0, 1, 2, \ldots$ **do**
2:      compute the solution $\hat{\alpha}$ of (9.9) for x_l.
3:      **if** $\| \sum_{i=1}^{k} \hat{\alpha}_i \nabla f_i(x_l) \|_2^2 \geq \varepsilon_{\mathscr{P}}$ **then**
4:          $v_l := -q(x_l)$
5:          compute $t_l \in \mathbb{R}_+$ and set $x_{l+1} := x_l + t_l v_l$
6:      **else**
7:          compute $\tilde{F}'(\hat{x}, \hat{\alpha})^T = (Q_N, Q_K)U$ as in (9.12)
8:          choose a column vector $\tilde{q} \in Q_K$ at random
9:          compute $\tilde{t}_l \in \mathbb{R}_+$ and set $x_{l+1} := x_l + \tilde{t}_l \tilde{q}$.
10:     **end if**
11: **end for**

The HCS shows large potential when used within multi-objective memetic algorithms; mainly because it performs an efficient local search which starts with one point and ends not only with an improvement of this point, but also, with two candidates for spread. Figure 9.6 ilustrates the application of the HCS as a local searcher over a population of three individuals, when the points are far, a hill climber movement (HC) is perfomed; and the hill climber sith side step (HCS) is applied when the optima is 'almost' reached. The operator can repeat the descent step—hill climber—several times until a sidestep is triggered. An analysis of the performance of these two versions, and comparisons in terms of efficiency and cost, can be found in [31].

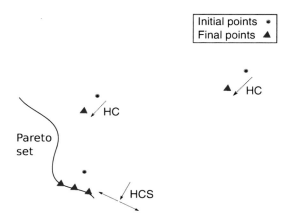

Fig. 9.6 This figure shows the way to use the HCS as local search operator, in order to be coupled with a set oriented heuristic. The local search starts with a point and ends with an improved point, or with three points, the one obtained by the gradient-based descent and the other two obtained by sidesteps.

9.3.3   Directed Movements

It is possible, furthermore, to perform directed movements not only toward and along the Pareto front, but also to any desired direction in the objective space; for this, the Directed Search Method (DS) was recently proposed [40, 39]. This method has the advantage of performing movements along determined paths in the objective function space.

9.3.3.1   The Directed Search Method

When working with MOPs, performing local search movements toward a particular region is sometimes desired. In this scope, a proposal [40] for the computation of directed search movements was recently introduced. The complexity of this operator is again linear and only first order gradient information is necessary in order to use it. This approach calculates a gradient-based descent direction using a controlled bias toward regions of interest determined in objective space; because of that, this proposal has many potential applications in the context of designing hybrid MOEAs.

Under the assumption that $x_0 \in \mathbb{R}^n$ is not a KKT point—in particular is also not a Pareto optimal point—a descent direction must exist such that all the directional derivatives must be non positive. In other words, once a vector $-\alpha \in \mathbb{R}^k$ is chosen, with $\alpha \in \mathbb{R}^k, 0 \leq \alpha, \sum_{i=1}^{k} \alpha_i = 1$, representing a desired search direction in image space. Then, a search direction $v \in \mathbb{R}^n$ in parameter space is sought such that for $y_0 = x_0 + tv$, where $t \in \mathbb{R}_+$ is the step size, it holds:

$$\lim_{t \searrow 0} \frac{f_i(y_0) - f_i(x_0)}{t} = \langle \nabla f_i(x_0), v \rangle = -\alpha_i, \quad i = 1, \ldots, k. \tag{9.13}$$

Using the Jacobian J of F,

$$J(x_0) = \begin{pmatrix} \nabla f_1(x_0)^T \\ \vdots \\ \nabla f_k(x_0)^T \end{pmatrix} \in \mathbb{R}^{k \times n}, \tag{9.14}$$

Equation (9.13) can be stated in matrix vector notation as

$$J(x_0)v = -\alpha. \tag{9.15}$$

Then, this search direction v can be computed by solving a system of linear equations. It is important to remark that system (9.15) is typically highly underdetermined—since in most cases the number of parameters is (much) higher than the number of objectives in a given MOP—which implies that its solution is not unique.

To find a solution of this system, one option is to take the greedy choice, i.e., the solution with the smallest norm, which is given by

$$v = J(x_0)^+(-\alpha), \tag{9.16}$$

where $J(x_0)^+ \in \mathbb{R}^{n \times k}$ denotes the pseudo inverse of $J(x_0)$. In case the rank of $J := J(x_0)$ is maximal (which we will assume in the following), the pseudo inverse is given by $J^+ = J^T(JJ^T)^{-1}$. Given a 'descent direction' $-\alpha \in \mathbb{R}^k$, a sequence of dominating points in 'direction $-\alpha$' can thus be found by numerically solving the following initial value problem:

$$\begin{aligned} x(0) &= x_0 \in \mathbb{R}^n \\ \dot{x}(t) &= J(x(t))^+(-\alpha), \quad t > 0. \end{aligned} \tag{IVP$_{(-\alpha)}$}$$

One observation worth noting is that even if $-\alpha$ is a descent direction, there is no guarantee that the solution curve c of (IVP$_{(-\alpha)}$) always leads to a Pareto optimal solution. When the image $F(\mathbb{R}^n)$ is bounded below, however, c leads to a boundary point x^* of the image. Since for x^* it holds $rank(J(x^*)) < k$, this can be used to trace numerically the end point of (IVP$_{(-\alpha)}$) in a certain way—the numerical integration can be stopped if the condition number $\kappa_2(J(x_i))$ exceeds a given (large) threshold. In [34], more insights about efficient computation of such end points, e.g. specialized predictor-corrector (PC) methods are presented. For the case of m active inequality constraints, the details can be found in [40, 33].

Even when setting α looks like imposing a weights vector, this method is able to also reach non-convex regions on the front; this is illustrated in Figure 9.7. For several reasons, this method would be a good choice to be used as a local engine inside a gradient-based memetic algorithm. For example, it is a good alternative to

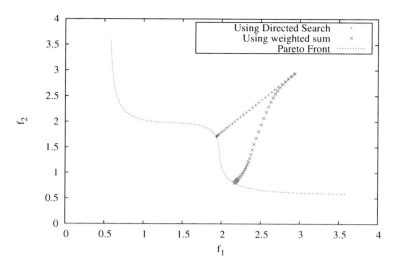

Fig. 9.7 Numerical result and comparison for the Directed Search method and the weighted sum method, starting from the same point and performing the movement with direction $\alpha = [0.55, 0.45]^T$, toward a non-convex region

greedy approaches like [15] and [38]; it can also perform a movement similar to the one proposed in [4], but in a user-preference controlled way. This means that, using a reference point $Z \in \mathbb{R}^k$ we can compute α as follows

$$\alpha(x_0, Z) := \frac{F(x_0) - Z}{||F(x_0) - Z||_1},$$

having in this way a guided route for the descent (see Figure 9.8). When applying greedy strategies, for a multi-objective gradient-based descent, for functions with unbalanced magnitudes, it is possible to get an undesired bias. This method avoids this potential problem. This method also gives us the best direction (nearest point) according to the reference point Z, by applying the following continuation strategy to perform a search along the Pareto set as follows:

Assume we are given a (local) Pareto point x and the related KKT weight α, i.e., such that

$$\sum_{i=1}^{k} \alpha_i \nabla f_i(x) = 0 \qquad (9.17)$$

and further we assume that

$$rank(J(x)) = k - 1 \qquad (9.18)$$

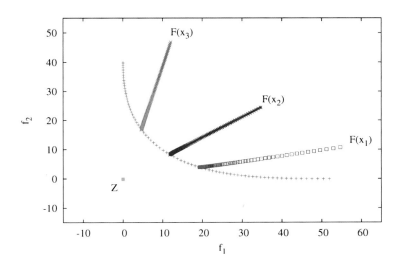

Fig. 9.8 Solution paths thrown by the Directed Search method for three different starting points, using the same target point Z

It is known (e.g., [22]) that in this case α is orthogonal to the Pareto front at $y = F(x)$, i.e.,

$$\alpha \perp T_y \partial F(\mathbb{R}^n), \tag{9.19}$$

where $\partial F(\mathbb{R}^n)$ denotes the border of the image $F(\mathbb{R}^n)$. Thus, a search orthogonal to α (again in image space) could be promising to predict a new solution near x along the Pareto set. To use (9.15), for instance a QR-factorization of α can be computed, i.e.,

$$\alpha = QR, \tag{9.20}$$

where $Q = (q_1, \dots, q_k) \in \mathbb{R}^{k \times k}$ is an orthogonal matrix and q_i, $i = 1, \dots, k$, its column vectors, and $R = (r_{11}, 0 \dots, 0)^T \in \mathbb{R}^{k \times 1}$ with $r_{11} \in \mathbb{R} \backslash \{0\}$ (for the computation of such a factorization we refer e.g. to [36]). Since by Equation (9.20) $\alpha = r_{11} q_1$, i.e., $\alpha \in span\{q_1\}$, and Q orthogonal it follows that the column vectors $q_2, \dots, q_k$ build an orthonormal basis of the hyperplane which is orthogonal to α. Thus, a promising well-spread set of search directions v_i may be the ones which satisfy

$$J(x)v_i = q_i, \quad i = 2, \dots, k. \tag{9.21}$$

In a next step, the predicted points $p_i = x + t_i v_i$, where $t_i \in \mathbb{R}$ are step sizes, can be corrected back to the Pareto set. For this, one can e.g. solve numerically $(IVP_{(-\alpha)})$ using p_i as initial value and $-\alpha$ as direction in $(IVP_{(-\alpha)})$. Continuing this procedure iteratively leads to a particular PC variant for MOPs. Note that here no Hessians of the objectives have to be computed which is indeed the case for other existing multi-objective PC methods.

To conclude, we mention that when mixing local searchers with evolutionary algorithms, having a method to steer the search—like the Directed Search method—has a lot of potential. We will show this in the next section.

9.3.4 Step-Length Computation

As a final remark for the section, we note that once the movement direction is set, choosing a suitable step size is not an easy task in multi-objective optimization. In practice, different proposals have been tested with good results [12, 43, 31]. One possibility [15] is also to adapt the well known Armijo-Goldstein rule to the multi-objective case, and accept any step length t that holds

$$F(x+tv) \leq F(x) - ct J(x) v,$$

where $F : \mathbb{R}^n \to \mathbb{R}^k$ is the multi-objective function and $J(x) : \mathbb{R}^n \to \mathbb{R}^n$ is the Jacobian matrix of F at x. The value $c \in (0,1)$ is a control parameter to decide how fine grained, numerically speaking, the descent will be. A bad choice of c can highly increase the cost related to function evaluations. With a suitable initial step length, this method is easily applicable; however, finding an efficient approach in the general multi-objective case is still an open problem.

9.4 Toward the Hybridization

9.4.1 Main Issues

We have shown, in the previous sections, that there are some options available to compute directions—based on gradient information—in the context of multi-objective optimization; but, the question of how to efficiently integrate them into a population-based context—as in the set oriented algorithms, such as MOEAs are—remains wide open. In this sense, it is also worth noting that the suitable choice of the movement direction relies also on the location of the point, and on the location of all the other population individuals (see Figure 9.9).

Once the descent direction v, for a specific point $x \in \mathbb{R}^n$, is obtained, the new line search function is defined as

$$f_v^i : \mathbb{R} \longrightarrow \mathbb{R}$$
$$t \longmapsto f_i(x+tv).$$

Now, the difficulty turns into the computation of the step size for f_v^i at x, because it is again a multi-objective problem (see Figure 9.10). Even when approximations of the optimal step size for each function are easy to estimate, the question is how

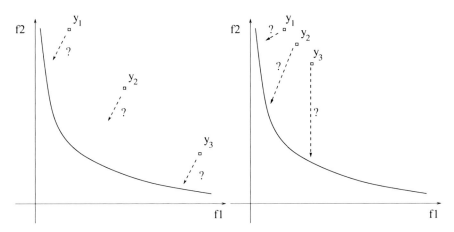

Fig. 9.9 This figure shows that, in a population context, an efficient search direction for each individual depends on several factors

to combine this information to find a common step to maximize the improvement. In this case an exact step size calculation is not possible; but, the use of inexact methods, like those described at Section 9.3.4, is a good option in practice. Step size is an important issue since it compromises the efficiency of the local search and the memetic algorithm as well. Even when the computation of the search direction and the step length are apparently independent of each other, a bad choice of the second can raise the cost of the procedure by several times.

Talking about hybridization with gradient-based local searchers, another important issue is that it is not possible to *a priori* determine a specific amount of resources, to be specifically devoted, for the local search procedure and the global one as well. In this sense, in order to produce efficient algorithms, an adaptive mechanism to control the use of local search is advised; but, this is itself a non-trivial problem. Gradient-based local search is typically a high-cost procedure; then, such a balance mechanism must be capable of determining when the gradient method outperforms the pure evolutionary search, during the solution of a specific problem—which means that the procedure is cost-effective. One possible option is to incorporate local search, as a method to refine solutions, only at the end of the search (as suggested in [18]); but this leads to a two-stage algorithm, and the precise time to start the local search is critical. In this sense, a proposal about doing the switch during running-time is presented in [29]. The main idea is to start the second stage when the evolutionary procedure is not improving the solutions anymore; for example, when all the individuals are mutually non-dominated, and the selection mechanism of the MOEA faces troubles because of that. In this case a refinement with a certain direction for improvements is desirable—precisely what is done with gradient-based methods. On the other hand, using deterministic search directions—over the stochastic technique—may accelerate the 'convergence' of the

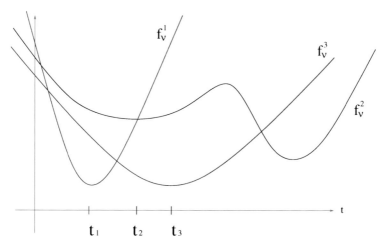

Fig. 9.10 Simultaneous line search for three functions along the direction v. Even when solving separately the m line searches getting t_1, t_2 and t_3, it is not possible to say which is the suitable step length for all of them together.

search when dealing with problems with very smooth fitness landscapes; then, in this cases the use of gradient-based procedures from the beginning of the search can be advantageous [28]. In conclusion, an adaptive switch mechanism between the two search engines (the local and the global one) during running-time is the desirable choice; but this is still an open research problem.

A final consideration comes with the fact that when dealing with population-based algorithms, it is important to keep a bounded archive of improved solutions. When using evolutionary algorithms, there are several available mechanisms to limit the size of this archive. This turns into an issue when resources have been spent using local search, since this bounding mechanisms typically delete solutions without notice if it has been previously improved by an expensive mechanism or not. When using gradient information at the end of the search, solutions are accurate in proportion to the amount of resources we want to spend in the line search. Finally, when building hybrid algorithms, saving the previously refined solutions from the truncation mechanism is mandatory; then, it is necessary to set special mechanisms to archive solutions in these cases.

9.4.2   *Early Hybrids*

Early attempts to combine MOEAs with gradient-based information use well-known MOEAs as their baseline algorithms, and simply replace the mutation operator by a line search procedure [44][45]. Other proposals have used gradient-based local search as an additional operator to be applied under certain rules during the

MOEA run [2][3]. Within this same line of thought, other mathematical programming techniques such as SQP [17] or the reference point method [49] have also been coupled with MOEAs [23][46]. It is worth noting for completeness, that in hybrid MOEAs using separately the gradients of single objectives is also possible, only when dealing with few-objective problems [21].

In [28] it is presented a two-stage algorithm (GBMES) which uses gradient-based line search in a first stage, in order to quickly reach points that are close to the Pareto set. During this stage, a MOEA with a reduced population size was combined with the gradient-based local search. The balance of resources to apply the local search was naturally given by the selection of the best individuals, from the small population. After spending a certain amount of resources, the second stage attempted to reconstruct the front (see [28] for details). This proposal has the natural drawbacks of being a two-stage procedure and has some limitations to be applied in general. Nevertheless, this work showed the potential advantages of using a descent direction of movement when dealing with problems with a high number of parameters; mostly because when the space is highly dimensional, the evolutionary techniques—avid to keep a uniform distribution of the population—can easily get lost, making very profitable to count with certain search directions.

In [31] the HCS operator is coupled with two state-of-the-art MOEAs. The effectiveness of this hybrid algorithm was assessed in conventional test problems for MOEAs. In most cases, the advantages of the hybrid method over the original MOEA were very clear. The balance of the resources for this local search operator was made through an *a priori* set probability function. We confirmed (previously stated in [24]) with this work that the part of the balance (local vs. global search) is, in general, the most important issue in terms of efficiency, for this type of algorithms.

Also, an attempt to perform a dynamic balance control (between the two operators) is presented in [29]. Here, an indicator over the improvements made by the evolutionary search is combined with a probability function, which controls the number of individuals to be modified by the local search. This approach shows promising results on traditional benchmarks.

9.5   Conclusions and New Trends

We presented, in this chapter, different gradient-based local search operators with diverse features. Apart from the cost of computing the gradients of the objective functions, we described a computational zero-cost descent direction, suitable for bi-objective problems. We also presented operators to perform movements both toward and along the Pareto set. The HCS has been presented in two versions, with and without explicit use of gradient information, and its main feature lies on the automatic switch between the two movements (hill climbing and sidestep) which makes it a powerful local searcher.

The terms *convergence* and *spread* are commonly used when talking about approximations of sets in the multi-objective context (mostly the Pareto set or its image, the Pareto front). Convergence is about the proximity of the solutions toward the set of interest, while spread relates to the minimal distance of these solutions to each other—which should be maximized in order to 'capture' as much as possible from the set of interest. The feature of our methods, to operate between a range of movements—toward, along and directed—is very promising when solving MOPs, because of the importance of the balance between convergence and spread. Even when these method are compromised, like other gradient based local searchers, when used in problems with a high number of local Pareto points, they have been found to be efficient in combination with MOEAs.

Although particular descent directions, to improve the convergence of approximation sets, have been suggested for MOPs (e.g., [15, 38]), their use within memetic strategies is not widely accepted. This is maybe due to an undesired bias of the chosen descent directions. The presented DS method goes beyond and allows the search to be steered in a particular controlled direction, which has so far not been considered. Here, the greedy direction from a given solution can be redefined according to preferences—in order to steer the search along all the regions of the front, or those that are difficult to explore by conventional MOEA mechanisms.

Regarding open research problems, we can state the adaptation of inexact methods for step-size control (such as Wolfe conditions, Armijo conditions, etc.) to the multi-objective search. Ensuring convergence, and the study of speed of convergence, for these methods are important issues to address when building an efficient interleaving between MOEAs and local search. We also mention that adaptive control of the resources allowed for the local search during runtime is one of the main issues of this hybridization. This control should automatically determine when the evolutionary operators are not producing improvements and when the introduction of gradient-based local search is cost effective. Another promising possibility is to combine several local search heuristics in a same hybrid algorithm by an adaptive control mechanism, like in [47, 48].

Finally, a very important aspect of hybrid MOEAs is the archive management. It is not desirable that our archiving strategy (like crowding or truncation) destroys the refinement previously done to certain solutions. Hence, every gradient-based algorithm should be coupled with a suitable archiving strategy. Interleaving the selection process and the gradient-based improvements with the archiving strategy of a MOEA is also a promising path for future research.

Acknowledgements. The first author acknowledges scholarship support from CONACyT as a Ph.D. student, and from IPN project no. 20121478. The second author acknowledges support from CONACyT project no. 128554. The third author acknowledges support from CONACyT project no. 103570.

References

1. Allgower, E.L., Georg, K.: Numerical Continuation Methods. Springer (1990)
2. Bosman, P.A.N., de Jong, E.D.: Exploiting Gradient Information in Numerical Multi-Objective Evolutionary Optimization. In: Beyer, H.-G., et al. (eds.) 2005 Genetic and Evolutionary Computation Conference (GECCO 2005), vol. 1, pp. 755–762. ACM Press, New York (2005)
3. Bosman, P.A.N., de Jong, E.D.: Combining Gradient Techniques for Numerical Multi-Objective Evolutionary Optimization. In: Keijzer, M., et al. (eds.) 2006 Genetic and Evolutionary Computation Conference (GECCO 2006), Seattle, Washington, USA, vol. 1, pp. 627–634. ACM Press (July 2006) ISBN 1-59593-186-4
4. Bosman, P.A.N., Thierens, D.: The Naive MIDEA: A Baseline Multi–objective EA. In: Coello Coello, C.A., Hernández Aguirre, A., Zitzler, E. (eds.) EMO 2005. LNCS, vol. 3410, pp. 428–442. Springer, Heidelberg (2005)
5. Branke, J., Mostaghim, S.: About Selecting the Personal Best in Multi-Objective Particle Swarm Optimization. In: Runarsson, T.P., Beyer, H.-G., Burke, E.K., Merelo-Guervós, J.J., Whitley, L.D., Yao, X. (eds.) PPSN 2006. LNCS, vol. 4193, pp. 523–532. Springer, Heidelberg (2006)
6. Brown, M., Smith, R.E.: Effective Use of Directional Information in Multi-objective Evolutionary Computation. In: Cantú-Paz, E., Foster, J.A., Deb, K., Davis, L., Roy, R., O'Reilly, U.-M., Beyer, H.-G., Kendall, G., Wilson, S.W., Harman, M., Wegener, J., Dasgupta, D., Potter, M.A., Schultz, A., Dowsland, K.A., Jonoska, N., Miller, J., Standish, R.K. (eds.) GECCO 2003. LNCS, vol. 2723, pp. 778–789. Springer, Heidelberg (2003)
7. Brown, M., Smith, R.E.: Directed Multi-objective Optimization. International Journal of Computers, Systems and Signals 6(1), 3–17 (2005)
8. Coello Coello, C.A., Lamont, G.B. (eds.): Applications of Multi-Objective Evolutionary Algorithms. World Scientific, Singapore (2004)
9. Coello Coello, C.A., Lamont, G.B., Van Veldhuizen, D.A. (eds.): Evolutionary Algorithms for Solving Multi-Objective Problems, 2nd edn. Springer, New York (2007) ISBN 978-0-387-33254-3
10. Deb, K.: Multi-Objective Optimization using Evolutionary Algorithms. John Wiley & Sons, Chichester (2001) ISBN 0-471-87339-X
11. Dellnitz, M., Schütze, O., Hestermeyer, T.: Covering Pareto Sets by Multilevel Subdivision Techniques. Journal of Optimization Theory and Applications 124(1), 113–136 (2005)
12. Dennis, J.E., Schnabel, R.B.: Numerical Methods for Unconstrained Optimization and Nonlinear Equations. Prentice-Hall (1983)
13. Edgeworth, F.Y.: Mathematical Physics. P. Keagan, London (1881)
14. Ehrgott, M., Wiecek, M.M.: Multiobjective Programming. In: Multiple Criteria Decision Analysis: State of the Art Surveys, vol. 78, pp. 667–722. Springer (2005)
15. Fliege, J., Svaiter, B.F.: Steepest descent methods for multicriteria optimization. Mathematical Methods of Operations Research 51(3), 479–494 (2000)
16. García, P., Fátima, M.L., Julián, C.F., Rafael, C.C., Carlos, A., Hernández-Díaz, A.G.: Hibridación de métodos exactos y heurísticos para el problema multiobjetivo. Rect@, Actas 15(1) (2007)
17. Gill, P.E., Murray, W., Saunders, M.A.: Snopt: An sqp algorithm for large-scale constrained optimization. SIAM Journal on Optimization 12(4), 979–1006 (2002)

18. Harada, K., Ikeda, K., Kobayashi, S.: Hybridizing of Genetic Algorithm and Local Search in Multiobjective Function Optimization: Recommendation of GA then LS. In: Keijzer, M., et al. (eds.) 2006 Genetic and Evolutionary Computation Conference (GECCO 2006), Seattle, Washington, USA, vol. 1, pp. 667–674. ACM Press (July 2006) ISBN 1-59593-186-4

19. Harada, K., Sakuma, J., Kobayashi, S., Ono, I.: Uniform sampling of local pareto-optimal solution curves by pareto path following and its applications in multi-objective GA. In: Proceedings of the 9th Annual Conference on Genetic and Evolutionary Computation, p. 820. ACM (2007)

20. Harada, K., Sakuma, J., Kobayashi, S.: Local Search for Multiobjective Function Optimization: Pareto Descent Method. In: GECCO 2006: Proceedings of the 8th Annual Conference on Genetic and Evolutionary Computation, pp. 659–666. ACM Press, New York (2006)

21. Hernández-Díaz, A.G., Coello Coello, C.A., Pérez, F., Caballero, R., Molina, J., Santana-Quintero, L.V.: Seeding the Initial Population of a Multi-Objective Evolutionary Algorithm using Gradient-Based Information. In: 2008 Congress on Evolutionary Computation (CEC 2008), Hong Kong, pp. 1617–1624. IEEE Service Center (June 2008)

22. Hillermeier, C.: Nonlinear Multiobjective Optimization: A Generalized Homotopy Approach. International Series of Numerical Mathematics, vol. 135. Birkhäuser (2001)

23. Hu, X., Huang, Z., Wang, Z.: Hybridization of the Multi-Objective Evolutionary Algorithms and the Gradient-based Algorithms. In: Proceedings of the 2003 Congress on Evolutionary Computation (CEC 2003), Canberra, Australia, vol. 2, pp. 870–877. IEEE Press (December 2003)

24. Ishibuchi, H., Yoshida, T., Murata, T.: Balance Between Genetic Search and Local Search in Memetic Algorithms for Multiobjective Permutation Flowshop Scheduling. IEEE Transactions on Evolutionary Computation 7(2), 204–223 (2003)

25. Jaszkiewicz, A.: Genetic local search for multiple objective combinatorial optimization. European Journal of Operational Research 137(1), 50–71 (2002)

26. Knowles, J.D.: Local-Search and Hybrid Evolutionary Algorithms for Pareto Optimization. PhD thesis, The University of Reading, Department of Computer Science, Reading, UK (January 2002)

27. Kuhn, H.W., Tucker, A.W.: Nonlinear Programming. In: Proceedings of the Second Berkeley Symposium on Mathematics Statistics and Probability, University of California Press (1951)

28. Lara, A., Coello Coello, C.A., Schütze, O.: Using Gradient-Based Information to Deal with Scalability in Multi-objective Evolutionary Algorithms. In: 2009 IEEE Congress on Evolutionary Computation (CEC 2009), Trondheim, Norway, pp. 16–23. IEEE Press (May 2009)

29. Lara, A., Coello Coello, C.A., Schütze, O.: A painless gradient-assisted multi-objective memetic mechanism for solving continuous bi-objective optimization problems. In: 2010 IEEE Congress on Evolutionary Computation (CEC), pp. 1–8. IEEE Press (2010)

30. Lara, A., Coello Coello, C.A., Schütze, O.: Using gradient information for multi-objective problems in the evolutionary context. In: Proceedings of the 12th Annual Conference Comp on Genetic and Evolutionary Computation, pp. 2011–2014. ACM (2010)

31. Lara, A., Sanchez, G., Coello Coello, C.A., Schütze, O.: HCS: A New Local Search Strategy for Memetic Multi-Objective Evolutionary Algorithms. IEEE Transactions on Evolutionary Computation 14(1), 112–132 (2010)

32. Lara, A., Schütze, O., Coello Coello, C.A.: New Challenges for Memetic Algorithms on Continuous Multi-objective Problems. In: GECCO 2010 Workshop on Theoretical Aspects of Evolutionary Multiobjective Optimization, Portland, Oregon USA, pp. 1967–1970. ACM (July 2010)

33. Mejía, E.: The directed search method for constrained multi-objective optimization problems. Master's thesis, CINVESTAV-IPN, México City (November 2010)

34. Mejía, E., Schütze, O.: A predictor corrector method for the computation of boundary points of a multi-objective optimization problem. In: 2010 7th International Conference on Electrical Engineering Computing Science and Automatic Control (CCE), pp. 395–399. IEEE (2010)

35. Moscato, P.: On evolution, search, optimization, genetic algorithms and martial arts: Towards memetic algorithms. Caltech Concurrent Computation Program, C3P Report 826 (1989)

36. Nocedal, J., Wright, S.: Numerical Optimization, Series in Operations Research and Financial Engineering. Springer, New York (2006)

37. Pareto, V.: Cours Déconomie Politique. Lausanne, F. Rouge, Paris (1896)

38. Schäffler, S., Schultz, R., Weinzierl, K.: Stochastic method for the solution of unconstrained vector optimization problems. Journal of Optimization Theory and Applications 114(1), 209–222 (2002)

39. Schütze, O., Lara, A., Coello Coello, C.A.: The directed search method for unconstrained multi-objective optimization problems. In: A Bridge Between Probability, Set Oriented Numerics and Evolutionary Computation, EVOLVE 2011 (2011)

40. Schütze, O., Lara, A., Coello Coello, C.A.: The directed search method for multi-objective optimization problems. Technical report (2009),
http://delta.cs.cinvestav.mx/~schuetze/
technical_reports/index.html

41. Schütze, O., Lara, A., Coello Coello, C.A.: Evolutionary Continuation Methods for Optimization Problems. In: 2009 Genetic and Evolutionary Computation Conference (GECCO 2009), Montreal, Canada, July 8-12, pp. 651–658. ACM Press (2009) ISBN 978-1-60558-325-9

42. Schütze, O., Lara, A., Coello Coello, C.A.: On the influence of the number of objectives on the hardness of a multiobjective optimization problem. IEEE Transactions on Evolutionary Computation 15(4), 444–455 (2011)

43. Schütze, O., Talbi, E.-G., Coello Coello, C., Santana-Quintero, L.V., Pulido, G.T.: A Memetic PSO Algorithm for Scalar Optimization Problems. In: Proceedings of the 2007 IEEE Swarm Intelligence Symposium (SIS 2007), Honolulu, Hawaii, USA, pp. 128–134. IEEE Press (April 2007)

44. Shukla, P.K.: Gradient Based Stochastic Mutation Operators in Evolutionary Multi-objective Optimization. In: Beliczynski, B., Dzielinski, A., Iwanowski, M., Ribeiro, B. (eds.) ICANNGA 2007. LNCS, vol. 4431, pp. 58–66. Springer, Heidelberg (2007)

45. Shukla, P.K.: On Gradient Based Local Search Methods in Unconstrained Evolutionary Multi-objective Optimization. In: Obayashi, S., Deb, K., Poloni, C., Hiroyasu, T., Murata, T. (eds.) EMO 2007. LNCS, vol. 4403, pp. 96–110. Springer, Heidelberg (2007)

46. Sindhya, K., Deb, K., Miettinen, K.: A Local Search Based Evolutionary Multi-objective Optimization Approach for Fast and Accurate Convergence. In: Rudolph, G., Jansen, T., Lucas, S., Poloni, C., Beume, N. (eds.) PPSN 2008. LNCS, vol. 5199, pp. 815–824. Springer, Heidelberg (2008)

47. Vasile, M., Zuiani, F.: A hybrid multiobjective optimization algorithm applied to space trajectory optimization. In: 2010 IEEE Congress on Evolutionary Computation (CEC), pp. 1–8. IEEE (2010)
48. Vasile, M., Zuiani, F.: Multi-agent collaborative search: an agent-based memetic multi-objective optimization algorithm applied to space trajectory design. Proceedings of the Institution of Mechanical Engineers, Part G: Journal of Aerospace Engineering (2011)
49. Wierzbicki, A.P.: Reference point methods in vector optimization and decision support. Working Papers ir98017. International Institute for Applied Systems Analysis (April 1998)

Chapter 10
On the Integration of Theoretical Single-Objective Scheduling Results for Multi-objective Problems

Christian Grimme, Markus Kemmerling, and Joachim Lepping

Abstract. We present a modular and flexible algorithmic framework to enable a fusion of scheduling theory and evolutionary multi-objective combinatorial optimization. For single-objective scheduling problems, that is the optimization of task assignments to sparse resources over time, a variety of optimal algorithms or heuristic rules are available. However, in the multi-objective domain it is often impossible to provide specific and theoretically well founded algorithmic solutions. In that situation, multi-objective evolutionary algorithms are commonly used. Although several standard heuristics from this domain exist, most of them hardly allow the integration of available single-objective problem knowledge without complex redesign of the algorithms structure itself. The redesign and tuned application of common evolutionary multi-objective optimizers is far beyond the scope of scheduling research. We therefore describe a framework based on a cellular and agent-based approach which allows the straightforward construction of multi-objective optimizers by compositing single-objective scheduling heuristics. In a case study, we address strongly NP-hard parallel machine scheduling problems and compose optimizers combining the known single-objective results. We eventually show that this approach can bridge between scheduling theory and evolutionary multi-objective search.

10.1 Introduction

Since almost a century, scheduling problems are subject to theoretical research. Many principles in the assignment of tasks to sparse resources over time are well

Christian Grimme · Markus Kemmerling
Robotics Research Institute, TU Dortmund, Germany
e-mail: {christian.grimme,markus.kemmerling}@udo.edu

Joachim Lepping
INRIA Rhône-Alpes, Grenoble University, France
e-mail: joachim.lepping@inria.fr

E. Tantar et al. (Eds.): EVOLVE- A Bridge between Probability, SCI 447, pp. 333–363.

understood and a variety of optimal algorithms or heuristic rules are available. For many approaches the theoreticians can give approximation factors, which provide worst-case guarantees for certain heuristics in comparison to the optimal case. If there is no approximation factor, however, theoretical foundations at least provide some deeper insights into problem properties. In the simplest case, this knowledge might be expressed in rules of thumb to address problems heuristically. A good example for the incorporation of such knowledge is the large amount of dispatching rules, see Haupt [21] for a survey, that address scheduling problems in a simple yet effective way.

However, algorithms and rules are almost exclusively available for single-objective optimization problems. In cases where multiple objectives must be optimized simultaneously, it is seldom possible to provide algorithmic solutions. Instead, research mainly focuses on the complexity of such problems. This is due to the very specific definition of multi-objective optimality: instead of a single solution, such problems usually comprise a whole set of optimal solutions in which each solution is incomparable to others—defining the so called Pareto-front. To solve multi-objective problems, scheduling theoreticians resort to general purpose approaches like the ε-Constraint or Tchebycheff method. They allow the integration of single-objective knowledge as local search methods but do not include global mechanisms for the solution set approximation.

Alternatively, there are nowadays various randomized search heuristics for multi-objective problems available. Especially, the multi-objective evolutionary algorithms have been constantly adapted to more and more challenging problems. Although such approaches yield convincing results on various problem instances, they are only accepted with reservation by theoretical scheduling researchers due to their inherent random processes. The random manipulation of solutions by an evolutionary algorithm is even sceptically considered as blind guesswork. In the single-objective case, it is recommended to support the random search by already achieved problem knowledge. This might be either realized by starting from a heuristically generated solution or by integrating problem knowledge into special tailored search operators. Consequently, one would expect to support multi-objective search by the clever combination of corresponding single-objective problem knowledge in a similar way. This is expected to improve an algorithm's problem related performance.

However, for multi-objective evolutionary algorithms the integration of single-objective problems knowledge is difficult. A problem specific adoption of these algorithms usually implies a complex redesign and requires detailed insight into the algorithm's theoretical background as well as into its implementation. The optimizer can be considered as monolithic black box where each algorithmic component has its own special purpose and the sophisticated interplay between components make the whole procedure efficiently work. A monolithic algorithm does not provide any interfaces to enable specific modifications. As such, the tuning of common evolutionary multi-objective optimizers is far beyond the scope of scheduling research.

To overcome this drawback, we provide a modular and flexible algorithmic framework to enable a fusion of scheduling theory and evolutionary combinatorial optimization especially for multi-objective problems. We define such a framework based on a cellular and agent-based approach that is motivated from the predator-prey model proposed by Laumanns and colleagues [27] in 1998. There, predators represent a single objective each and locally threaten prey individuals represent the multi-objective problem's solutions. As inherent effect of multiple predators the authors expected an adaptation of the prey to all predator influences and the emergence of trade-off solutions.

We start with a detailed analysis of single-objective scheduling heuristics and discuss their combination and applicability in multi-objective problems considering different scheduling problems. Then, we review the agent-based system on a spatial population and provide a detailed review of its properties. Based on these results, we define a general framework which allows the straightforward composition of single-objective scheduling heuristics to build multi-objective optimizers. In a case study, we finally show that this approach can bridge between scheduling theory and evolutionary multi-objective search.

10.2   Scheduling Problems and Theoretical Results

Scheduling is the assignment of jobs $J = \{1,\ldots,n\}$ to resources $M = \{1,\ldots,m\}$ while a set of constraints must be met. Each job j has a processing time p_j and often some further properties like a release date r_j, a due date d_j, and a weight w_j. The aim is to find an assignment such that one or many conflicting objectives are optimized. Typically, scheduling problems are formulated in the $\alpha|\beta|\gamma$ three-field notation [15], where α denotes the machine environment, β a set of constraints or additional restrictions, and the γ-field contains the list of objectives. We use the pure summation sign (Σ) as abbreviation for a sum over all jobs $(\Sigma_{j=1}^{n})$ for the matter of simplicity.

An example of an easy scheduling problem is the check-in counter queue problem, denoted as $1|r_j|\Sigma U_j$. Here, 1 denotes a single check-in counter (single machine) where passenger j arrives at time r_j. The objective is to ensure that only a minimum number of passengers miss their flight. That is expressed by minimizing the number of tardy jobs ΣU_j. Naturally, this requires a due date d_j for each passenger.[1] $P_m||C_{max}$ extends the environment to m counters. In this case, the objective is to check-in all passengers in shortest possible time. Each passenger j is checked at time C_j, the completion time of the job, while C_{max} denotes the time the last passenger is admitted $(\max_{j=1\ldots n}\{C_j\})$. Assuming that all passengers already wait before the counters in the beginning $(r_j = 0)$, our problem becomes a load balancing task.

$P_m|r_j,M_j|\Sigma w_j T_j$ is another example for the parallel machines environment and describes a system (e.g., a compute cluster) with m identical machines in parallel. Job j arrives at release date r_j and has to leave by the due date d_j. According to some conditions (e.g., memory, operating system) job j may be processed only on one of the machines belonging to the subset M_j. If job j is not completed in time, a penalty

---

[1] Although d_j can be regarded as restriction, we omit it from the β part as its existence is implied by the due date related objective.

$w_j T_j$ depending on the job's importance w_j and the tardiness $T_j = \max\{0, C_j - d_j\}$ is incurred.

As Pinedo [35] states, practitioners have already dealt with scheduling problems in manufacturing at the beginning of the last century. However the first publications on this topic appeared in the early fifties. In the seventies, research focused mainly on the complexity of single-objective scheduling. The eighties then were identified by Hoogeveen [23] as starting point for detailed investigations of multi-objective scheduling problems. The obtained complexity results for single and multi-objective—of which a few will be summarized in this section—took away the hope of solving many problems optimally in polynomial time. This led to the development of heuristics, which try to approximate optimal solutions in reasonable time.

10.2.1   Single-Objective

In the last sixty years, a lot of research was done in the field of single-objective scheduling. As long as not stated otherwise, the following results are taken from Pinedo [35] who gives a good overview of the developed algorithms and complexity results. Another good starting point is Brucker's and Knust's web page[2]. Since we will deal with identical parallel machines, this section is focused on the results for this environment. The single-machine environment is a special case of P_m. This implies when a problem with one machine is NP-hard, then the general case with multiple machines is also NP-hard.

The minimization of makespan for two parallel machines ($P_2||C_{max}$) is equivalent to PARTITION [35] and therefore NP-hard in the ordinary sense. The general case with m machines is NP-hard in the strong sense, which can be shown by a reduction of 3-PARTITION. A common heuristic to solve this problem is the *Longest Processing Time First* (LPT) rule that has an approximation factor $\frac{C_{max}(LPT)}{C_{max}(OPT)} \leq \frac{4}{3} - \frac{1}{3m}$. If preemptions are allowed the problem becomes easy and can be solved optimally with the *Longest Remaining Processing Time First* (LRPT) rule in polynomial time.

The objective $\sum C_j$, referring to the total completion time of the schedule, can be solved easily for the parallel machine environment. The *Shortest Processing Time First* (SPT) rule yields an optimal schedule. This is even true when preemptions are allowed. Unfortunately, the more general *Weighted Shortest Processing Time First* (WSPT) rule is not optimal for the total weighted completion time $\sum w_j C_j$. Bruno et al. [4] showed that the problem with two machines is NP-hard. However, WSPT is a good heuristic with an approximation factor $\frac{\sum w_j C_j (WSPT)}{\sum w_j C_j (OPT)} < \frac{1}{2}(1 + \sqrt{2})$, see Kawaguchi and Kyan [26]. Recently Schwiegelshohn [37] provided an alternative proof for this result. Further, Sahni [36] proposed a *Fully Polynomial Time Approximation Scheme* (FPTAS) that requires $O(n(\frac{n^2}{\varepsilon})^{m-1})$ time and calculates a solution that is not worse than $(1 + \varepsilon)$ times the optimal value.

---

[2] http://www.informatik.uni-osnabrueck.de/knust/class/

Due date related objectives for parallel machines are usually hard. The maximum lateness problem $P_m||L_{max}$—which can be solved easily in the single-machine environment $m = 1$ by the *Earliest Due Date First* (EDD) rule—is NP-hard in the ordinary sense for the general case. The special case where each job has due date $d_j = 0$ is equivalent to $P_m||C_{max}$. Introducing release dates renders the problem even for single machines $(1|r_j|L_{max})$ strongly NP-hard, which can be proved by a reduction of 3-PARTITION.

$1||\sum T_j$ is the problem of minimizing the total tardiness for single machines and NP-hard in the ordinary sense. But a pseudo-polynomial algorithm based on dynamic programming exists. This approach uses the dominance result that if $p_j \leq p_k$ and $d_j \leq d_k$, then there is an optimal schedule in which job j is processed before job k. Further, there is a FPTAS that computes in $O(\frac{n^7}{\varepsilon})$ a solution which is not worse than $(1 + \varepsilon) \sum T_j(OPT)$. This approximation scheme rescales the processing times and due dates and solves the resulting problem with the dynamic programming approach. The problem becomes strongly NP-hard when weights or release dates are introduced. The NP-completeness of $P_2||\sum T_j$ was proven by Lenstra et al. [30]. They reduced KNAPSACK to $P_2||L_{max}$ and $P_2||L_{max}$ to $P_2||\sum T_j$. Jouglet and Savourey [25] investigated the strongly NP-hard problem $P_m|r_j|\sum w_j T_j$ and introduced dominance rules since most others do not hold when release dates are considered. It may be noticed that according to Baptiste [1] the special case $P_m|p_j = p, r_j|\sum T_j$ is easy.

The number of tardy jobs $(\sum U_j)$ can be solved easily for the single-machine environment. Jobs are scheduled in increasing order of their due dates. If a job is scheduled and will be completed late, then the already scheduled job with the longest processing time is marked late and will be put to the end of the schedule. This is known as Moore's algorithm [33]. The special problem $1|d_j = d|\sum w_j U_j$ with constant due date and weights is already NP-hard in the ordinary sense and Pinedo [35] shows, that WSPT is not bounded in its approximation factor on this problem. Still, due to its equivalence to KNAPSACK there exists a pseudo-polynomial algorithm. The problem $1|r_j|\sum U_j$ with release dates, however, is strongly NP-hard. For parallel machines, Garey and Johnson [14] showed that the problem $P_m||\sum U_j$ is NP-complete. Well known heuristics based on Moore's algorithm were proposed by Ho and Chang [22] as well as by Süer, Báez, and Czajkiewjcz [39]. In this work, we refer to aspects of the SBC3 heuristic to be integrated as expert knowledge.

10.2.2   Multi-objective

In real life, scheduling problems often have multiple conflicting objectives. However, research mostly focused on problems with only one objective. Not till the eighties multiple objectives were usually aggregated and considered as single objective, see Hoogeveen [23]. According to Lei [29], the investigations on multi-objective can be divided in time before and after 1995.

Before 1995, multi-objective scheduling was tackled with implicit enumeration techniques like branch and bound and mathematical programming. Typical ways to deal with multiple objectives in this domain are the application of iterative methods like weighting (see Section 10.2.2.2), the Tchebycheff method (see Section 10.2.2.3), and the ε-Constraint method (see Section 10.2.2.4) as meta-search strategies. Nagar et al. [34] gives an overview about the literature on multi-objective scheduling till the early nineties. They stated that these "techniques have not been very successful". They point out that techniques like Genetic Algorithms (GAs) are promising but, due to their generic nature, "need to be complemented with problem specific knowledge".

Since 1995, the adoption of GAs have increased. Lei [29] gives a high-level overview about the corresponding literature. GAs and Evolutionary Algorithms (EAs) maintain a set of solution rather than a single solution, which is advantageous when we strive for a set of Pareto-optimal solutions (see Section 10.2.2.1 for a definition). Due to that, EAs are widely used to solve multi-objective optimization problems, see Deb [9] or Coello Coello et al. [5] for a detailed overview. One of the most successful multi-objective evolutionary algorithms (MOEAs) is NSGA-II (see Section 10.2.2.5).

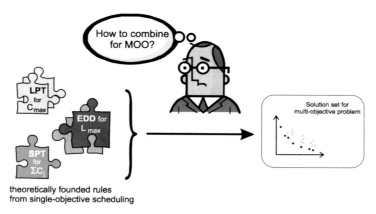

Fig. 10.1 Initial situation of the scheduling scientist: Although he/she has theoretical insights into many single-objective problems and even specific rules at hand, he/she has almost no algorithmic foundation to use this knowledge in multi-objective optimization (MOO) problems

Although MOEAs can be applied as black-box optimizers to arbitrary problems without adaptation, it is possible to improve their results by integrating problem specific knowledge. Research has provided some well performing, partially optimal, heuristics for several objectives (see Section 10.2.1). Unfortunately, it is rather complex to combine these single-objective heuristics for multi-objective problems as symbolically illustrated in Figure 10.1. We just need to consider the problem $P_m||\sum C_j, C_{max}$ to clarify this: our single-objective solutions suggest to order jobs ascending regarding processing time for $\sum C_j$ and in inverse order to get a good

C_{max} solution. The two building blocks are contradicting in such a way that an easy combination of both approaches seems impossible.

In rare cases researchers like van Wasserhove and Gelder [41] have managed to combine expertise and constructed an efficient algorithm, e.g., for $1||\sum C_j, L_{max}$. Still, also these approaches are inspired from generic methods like the ε-Constraint approach. In the following paragraphs, we review some of these approaches before we discuss a rather famous representative of evolutionary approaches for multi-objective algorithms. Eventually, we will again discuss whether these standard approaches suffice to integrate expertise.

10.2.2.1   Dominance and Pareto-Optimality

In this paper, we exclusively consider the problem of enumerating all Pareto-optima for z objectives $f_1, \ldots, f_z$, see T'kindt and Billaut [40]. To define Pareto-optimality, we first need to define a relation between solutions regarding all objectives, the dominance relation. We define S as the set of all possible solutions.

If a solution x dominates a solution y, we denote it as $x \prec y$, if and only if x is not worse than y for each objective f_i and better for at least one objective:

$$x \prec y \Leftrightarrow \forall i \in \{1, \ldots, z\} : f_i(x) \le f_i(y) \land \exists i \in \{1, \ldots, z\} : f_i(x) < f_i(y). \quad (10.1)$$

Regarding the dominance relation, there are usually incomparable solutions.

A solution x is (strongly) Pareto-optimal if and only if x is not dominated by any other solution y:

$$\neg \exists y \in S : y \prec x. \quad (10.2)$$

Note that a Pareto-optimal solution does not necessarily dominate all other possible solutions.

A solution x is weakly Pareto-optimal if and only if there does not exist another solution y which is better for all objectives:

$$\neg \exists y \in S : \forall i \in \{1, \ldots, z\} : f_i(y) < f_i(x). \quad (10.3)$$

A strongly Pareto-optimal solution is also weakly Pareto-optimal. But a weakly Pareto-optimal solution is not necessarily strongly Pareto-optimal.

10.2.2.2   The Weighting Method

A very simple and often used approach for finding specific multi-objective solutions is the weighting method. It is based on reformulating a given multi-objective problem into a single-objective problem by aggregating all z objectives with respect to a user-defined weighting. Using the weighting vector $\mathbf{w} = (w_1, \ldots, w_z)^T, 0 \le w_i \in \mathbb{R}$ each objective is assigned a specific priority and the original problem can be transfered to a single-objective problem as shown in equation (10.4).

$$\text{minimize } \sum_{i=1}^{z} w_i f_i(\mathbf{x}) \tag{10.4}$$

$$\text{with } \mathbf{x} \in S \text{ and } \sum_{i=1}^{z} w_i = 1$$

By choosing an adequate weighting vector and applying some single-objective search strategy a Pareto-optimal solution can be determined. If we repeatedly apply this search with different weighting vectors we might find multiple solutions and get an approximation of the whole Pareto-front. However, in the real-valued case this method is restricted to convex problem instances. For those problems, Das and Dennis [8], for instance, give an intuitive trigonometric proof showing that non-convex parts of a Pareto-front cannot be reached by weighting.[3] Still, in many real-world optimization approaches, this method is used due to its easy applicability.

10.2.2.3 The Zenith-Point or Tchebycheff Method

In 1976, Bowman [3] used the Tchebycheff norm as optimization objective to find multi-objective solutions. Again, this methods is based on weighting, however, the minimization process always concentrates on minimizing the largest distance between a solution and a given reference point $\mathbf{u}^*$. As this point is sometimes called zenith point the method is also known as *Zenith-Point Method*. The zenith-point represents a virtual goal in function space given by the user and represents some preferences regarding the multi-objective solution.

$$\text{minimize } \max_{i=1,\ldots,z} \left\{ w_i \cdot \| f_i(\mathbf{x}) - \mathbf{u}_i^* \| \right\} \tag{10.5}$$

$$\text{with } \mathbf{x} \in S$$

Equation (10.5) formulates the minimization problem using the weighted Tchebycheff norm with a weigthing vector $\mathbf{w}$, the zenith-point $\mathbf{u}^*$ and z objectives. In contrast to simple weighting, this method also finds solutions in non-convex parts of the Pareto-front. However, also weakly Pareto-optimal solutions are found which may necessitate a costly filtering of solutions. Furthermore, the user has to provide some information by placing a reference point in function space.

10.2.2.4 The ε-Constraint Method

As allegedly most prominent method for practical application in the scheduling community the ε-Constraint method focuses mainly on a single objective. One objective is considered to be optimized while all remaining objectives are transformed

---

[3] Although the standard definition of convexity is restricted to $\mathbb{R}$, this also holds for supported solutions in the discrete case: only supported solutions from the linear combination of two points can be generated.

into constraints by defining an upper bound for each of them. Consequently, the new problem can be formulated as shown in Equation (10.6).

$$\text{minimize } f_c(x) \qquad\qquad\qquad (10.6)$$
$$\text{subject to } f_i(x) \leq \varepsilon_i, \quad \forall i = 1,\dots,z$$
$$\text{with } i \neq c, c \in \{1,\dots,z\}, \text{ and } x \in S$$

With proper chosen bounds ε_i, the approach always reaches a Pareto-optimal point. A repeated application of this method therefore guarantees to find every Pareto-optimal solution regardless of the Pareto-front's convexity properties [32]. However, apart from the necessity to filter weakly dominated solutions, the method's performance mainly depends on the suitable selection of ε_i for all objectives to ensure feasible solutions. Further, the use of hard constraints is rarely adequate for expressing true design objectives and for many problems it becomes even impossible to determine the constraint value a priori, see also Laumanns, Thiele, and Zitzler [28].

10.2.2.5   NSGA-II

The Non-Dominated Sorting Genetic Algorithm - II (NSGA-II), which was introduced by Deb et al. (see [10] and [11]), is one of the most famous MOEAs. To deal with incomparable solutions, it applies non-dominated ranking and sorting based on crowding. All non-dominated solutions of a population are assigned to the first front. All solutions, which are only dominated by solutions in the first front, are assigned to the second front and so on. For selection, solutions are mainly compared based on their mutual domination. If two solutions are in the same front, they are judged according to their crowding distance. The crowding distance of a solution denotes the sum of normalized distances along each coordinate direction in objective space between the solution and its direct neighbors, see also Deb et al. [11]. A higher value indicates a less crowded environment of the solution and a larger diversity. Thus, selecting according to this measure as secondary criterion ensures diversity. For offspring creation, standard binary tournament selection, recombination, and mutation are used. The next generation is selected consecutively from the best non-dominated fronts of the combined population of current generation and their offspring. NSGA-II is widely used to solve scheduling problems. Recent research was done by SongFa and Ying [38], who developed a MOEA based on NSGA-II to optimize task assignment in distributed grids. Yuan and Quanfeng [42] adopted NSGA-II to schedule jobs with machine dependent processing times and fuzzy due dates on parallel machines considering the makespan and a fuzzy grade of satisfaction depending on the tardiness. Li et al. [31] optimized the makespan and the total tardiness of a parallel machine scheduling problem with NSGA-II.

10.2.3 The Gap between Single-Objective Theory and Multi-objective Approaches

Although there are several general methods to address multi-objective problems and as such also scheduling problems, their structure seldom supports the integration of expertise. Principally, it is possible to integrate some single-objective results as local search mechanisms in the discussed iterative methods like the weighting, zenith-point, or ε-Constraint method. However, all those searches can be very time-consuming. Additionally, one may have no idea which theoretical result may contribute to local search and which aspect hinders convergence.

In their original (single-objective) form, evolutionary algorithms provide a simple mechanism to integrate expert knowledge: the variation operators. Here theoretical ideas can be applied in a more or less randomized way to support convergence during offspring generation: A bias in mutation or recombination operators—based on knowledge on the search space structure or problem characteristics—can directly contribute to an accelerated convergence. The integration of direct local search heuristics is a more sophisticated way of expertise integration. However, in the multi-objective case, solution quality is evaluated regarding contradicting objectives and biased variation or local search often support only single objectives. The combination of expertise to gain a cooperative bias for all objectives is again difficult to construct. It just shifts the problem of expertise combination from the direct heuristic design shown in Figure 10.1 to the variation operator level in generic algorithms, see Figure 10.2.

Further, MOEAs usually have a monolithic structure which makes it hard to adapt them except from changing variation operators. Already for small adjustments, one has to understand first the algorithm's structure and second the chosen implementation. This alone constitutes a great barrier for a domain expert. The slightest adjustment of one part of an algorithm may affect other parts and render useless or even worsen the resulting.

A modularly composed algorithm is easier to understand because the components can be considered separately. But the most desirable optimizer for a domain expert is the one that allows to plug in knowledge at well-defined points without bothering about the system as a whole. This is the great advantage of the here discussed Predator Prey Model (PPM) (see Section 10.3). A domain expert can define agents one by one without bothering about their interplay. Even more, since he can attach arbitrary rules and objectives to each agent, he does not have to think about a complex combination of single-objective heuristics. In the following, we will discuss this approach in detail and evaluate its benefits.

10.3 The Modular Predator-Prey Model

Predator and prey interaction is a principle of natural co-evolutionary interplay between species that leads to mutual adaption and development. The here discussed predator-prey model (PPM) is certainly nature-inspired, however, no real

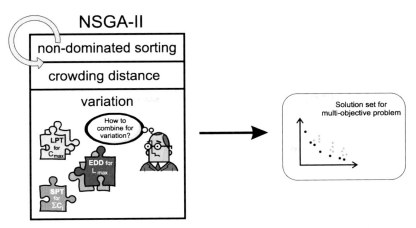

Fig. 10.2 Situation of integrating theoretical knowledge into the NSGA-II variation operator. The scheduling expert is still not knowing how to combine his basic heuristics to a multi-objective search operator. Here, the initial problem has just been shifted into the NSGA-II algorithmic framework.

co-evolutionary approach. The analogy to the predator and prey relationship was initially used by Laumanns et al. [27] as abstract point of view to describe the emergence of solutions for a multi-objective optimization problems: As much as a solution for a multi-objective problem has to meet all objectives as good as possible, a prey has to resist many predators to survive as long as possible. *Prey* are usual individuals of an evolutionary multi-objective algorithm representing possible solutions of the optimization problem. Placed immobile at vertices of a two-dimensional toroidal grid, they represent a spatially distributed population. The *predators* roam across the spatial structure according to a random walk scheme, which is retained as a uniformly distributed movement in the neighborhood of the position of a predator, see Figure 10.3 (a). After its movement, a predator chases prey only within its current spanned neighborhood. In more detail, this "hunting" consists of evaluating all prey in the direct neighborhood of the predator's position according to a *single objective*. This objective is bound to and carried by the predator. Usually, the worst prey within this neighborhood is "eaten". As soon as a vertex on the grid becomes free, a reproduction takes place that refills the empty vertex with a new prey. The offspring prey is created out of neighboring prey using variation operators as schematically depicted in Figure 10.3 (b).

Obviously, a central component of the PPM is the predator action. It is more formally summarized in Algorithm 10.1. In our realization the predator first executes its movement step, see line 2, and subsequently spans a von-Neumann selection neighborhood SN with radius v.[4] In that neighborhood, all prey are evaluated regarding the predator's single objective f_c resulting in the worst one marked for deletion.

---

[4] A von-Neumann neighborhood of radius v contains—starting from a given position—all neighboring positions that can be reached by at most v steps along the connecting edges.

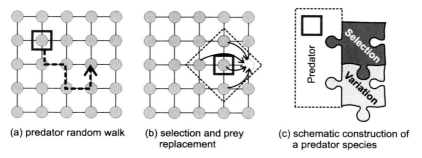

(a) predator random walk   (b) selection and prey   (c) schematic construction of
replacement   a predator species

Fig. 10.3 Depiction of the PPM working principle, see (a). and (b). Architecture of a predator individual constructed from a selection and variation building block (c).

Then, in line 5 an offspring is generated out of the non-marked individuals in *SN* using variation operator *Op*. In our realization, the final replacement approach follows an elitist philosophy: the marked worst prey is only replaced, if the offspring is better regarding the predators objective. Afterwards, the process is repeated until a termination criterion is reached. As the described action is restricted to each predator and completely self-contained, multiple predators can act in parallel and bring their action to the distributed population.

Algorithm 10.1. General predator behavior.

Preconditions: *predator* $:= (f_c, \pi, Op, v)$
  1: **while** not terminated **do**
  2:    $\pi = \text{move}(\pi)$
  3:    span selection neighborhood *SN* of size v
  4:    evaluate prey in *SN* regarding f_c and mark worst prey for removal
  5:    *offspring* $:= Op(SN \backslash worstPrey)$
  6:    evaluate(*offspring*) regarding f_c
  7:    **if** *offspring* better than worstPrey **then**
  8:        remove worstPrey and introduce *offspring*
  9:    **end if**
10: **end while**

A further key aspect of the PPM—in its modified version presented by Grimme and Lepping in 2007 [17]—is the modular nature of the predators agents and the fine-grained steering of selection and variation influences through them. As schematically depicted in Figure 10.3 (c), a predator species is constructed from a selection as well as a variation building block. The selection block determines according to which objective the predator evaluates the prey. The variation block affects the production of new offspring in a very restricted neighborhood. As such— by defining different predator species—it is quite simple to bring multiple local influences (standard mutation, recombination, or specific local search or expert knowledge) to the prey, and as such to the overall evolutionary process. Grimme and Lepping showed this first for combining different mutation and recombination influences in real-valued problems [17].

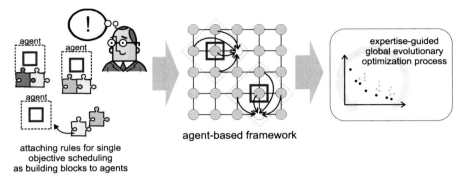

Fig. 10.4 In the PPM situation, the expert just attaches his single-objective knowledge as building blocks to the acting agents. Eventually, via their interaction on the prey, solution each heuristic participates in solution generation.

This modular framework provides an excellent starting point to solve the problem of knowledge integration. In contrast to the previously described situations, where we always lacked a strategy to combine single-objective expertise, the PPM provides a seamless way to plug in theoretical findings to predators, see Figure 10.4. If we can represent expert knowledge as biased variation operators, we are directly able to apply them locally to the population. From the PPM's original motivation we can hope that a good combination of their effects emerges from the predators cooperation over time.

For the real valued case without expertise integration, however, earlier investigations by Grimme, Lepping, and Papaspyrou [19] have shown that the PPM generally tends to lose diversity and thus solution quality. The expected almost magical emergence of trade-off solutions cannot be achieved for this case. This effect is rooted in the strict single-objective and locally restricted selection mechanism: over time, relatively stable areas of similar solutions are generated which hinder the emergence of intermediate solutions. Instead of breeding and conserving compromises, they are extincted immediately. This is, because they are inferior if only compared locally regarding a single objective. In many situations, this may lead to the development of mere extremal solutions for each objective and a complete loss of intermediate trade-offs.

Still, the situation is different in the case of expertise integration. While standard mutation in the real valued case is usually unbiased [2], expertise-guided variation operators in our case favor a certain search direction, i.e. a specific single objective selection. As such, in our building block environment with prey consisting of selection and variation modules, we are able to construct different kinds of predators. It is not only possible to attach a specific biased variation operator to the corresponding single-objective (the one it supports) but also to not favored, contradicting objectives. This leads to a kind of lexicographic order of objectives in predator action. Still, the selection is performed regarding the predator's objective and certainly conserves good solutions. However simultaneously, the application of a contradicting variation operator may lead to a downstream optimization of a secondary objective under the initial optimality conservation. This may favor cooperative solutions and can slow down the loss of diversity. This way, trade-off solutions may be generated and conserved. We will discuss this effect again in the next section, when the PPM has been adapted to scheduling problems.

10.4   Adopting the Predator-Prey Model to Scheduling Problems

In this section, we show how the predator-prey model can be adopted to the scheduling domain. We therefore describe the problem encoding scheme first and then explain a general purpose variation operator that allows to express certain problem knowledge. The evaluation in this section demonstrate the practical application of the concept and proves its effectiveness.

Since we solve scheduling problems with n jobs, we use a standard permutation encoding of length n to represent the genotype of the problem. As we consider only offline scheduling problems, an algorithm has to find the set of optimal job permutations. The schedule construction from our chosen permutation representation is quite simple for parallel machine environments: the jobs are FCFS-dispatched in permutation order to the machines, see Figure 10.5. In this example, we assume $n = 6$ jobs $j = 1, \ldots, 6$ and $m = 2$ machines while release dates r_j, processing times p_j, and due dates d_j of jobs are given as problem instance. Each due date can be principally met as the condition $r_j + p_j \leq d_j$ holds for all jobs. The individual encoding in Figure 10.5 results in a schedule where each job is assigned to the next free machine. In this case, the encoded sequence results in a schedule with unnecessary idle time that occurs for job 3 (shaded area). This job is released at time $r_3 = 5$ but is sequenced after job 5 which has a earliest start time of $r_5 = 7$. Thus, also the start of job 3 is delayed as we strictly stick to the FCFS order. With this encoding every possible schedule is representable. The resulting objectives can be computed by summing up all completion times ($\sum C_j = 35$) and all unit penalties ($\sum U_j = 1$).

10.4.1   Variation Operator Design

Investigations by Grimme and Lepping [17] show that special variation operators, triggered by autonomously acting predators, yield good approximation results especially for multi-objective problems. We adapt this methodology by integrating problem specific expert knowledge into the operator design: The variation operators

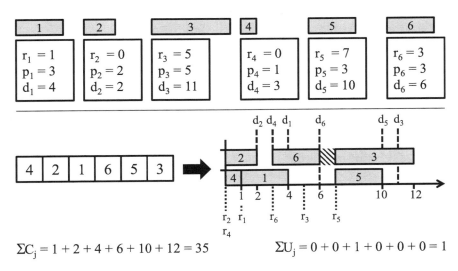

$\Sigma C_j = 1 + 2 + 4 + 6 + 10 + 12 = 35$ $\Sigma U_j = 0 + 0 + 1 + 0 + 0 + 0 = 1$

Fig. 10.5 Encoding scheme and resulting schedule obtained by FCFS-scheduling

used for the reproduction of prey apply those strategies for single-objective scheduling which are explained in Section 10.2.

For many single-objective scheduling problems, expertise on solution strategies can be expressed by optimal sorting rules. Based on the discussion in Section 10.2.1, a general variation operator is designed which allows to bring in the effects of SPT, LPT, EDD, or any other sorting schemes into the population. Figure 10.6 exemplarily depicts the application of this operator to a given sequence with processing times p_j: a position is selected randomly in the permutation representation of the genotype. Then, a subsequence of $2\delta + 1$ genes is sorted according to a certain rule. Here, we show the application of SPT sorting: A position is randomly selected while δ elements are involved upstream and downstream from this position. In this way, we have an easy mechanism for steering the mutation operator's strength varying the amount changed genes in the individual. However it is reasonable to prefer slight changes over strong perturbations as the inner structure of individuals should be principally conserved over generations. Thus, we chose the δ-window from a normal distribution with an externally adjustable step size of σ. This value is chosen as a constant at the beginning of the optimization and is not adjusted during the application. Obviously, $\delta = 0$ has no effect as only the initial gene at the current position is selected. A larger δ leads to a higher probability of a completely ordered genome which limits the Pareto-front regarding the respective objective. In this example, the SPT sorting of a subsequence improves the total completion time objective from $\Sigma C_j = 62$ to $\Sigma C_j = 59$.

Due to the modular character of the considered agent-based optimization framework, predator agents and variation operators can be coupled in various ways. Two main classes of coupling can be identified: either a predator is coupled with its

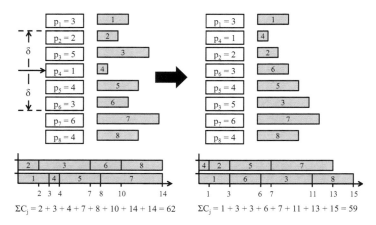

Fig. 10.6 Schematic depiction of the mutation operator concept with $\delta = 2$ and SPT-mutation

corresponding knowledge-based variation operator or coupled with an operator that supports another objective. In the first case, an operator that supports the predator's objective can be expected to favor convergence towards extremal solutions. That is, solutions rapidly converge to the objective's optimal solution and fully neglect other objectives. A similar behavior was also observed by Grimme et al. [19] and beneficially used to reach the outer perimeter of the desired Pareto-front.

In the second case, the selection/mutation coupling favors an implicit lexicographical ordering which conserves good solutions of the predator's objective. Additionally, this configuration favors a subsequent optimization of the other objectives because the sorting is different from the actual selection. This approach can provide means to maintain good solutions while simultaneously exploring the search space regarding another objective. For the evaluation, we consequently configure the PPM with all possible mutation/selection combinations to take advantage of both described effects. An analysis of these effects and a proof of concept will be given during the following evaluation.

10.4.2 Evaluation

For demonstrating and investigating the effects and benefits of knowledge integration according to our approach, we will refer to the evaluation done in a recent separate research paper of the authors [20]. There, the principle of adding order-based single-objective scheduling results was extensively tested. We repeat some of the most important results and discuss consequences.

To get a broad understanding of the PPM's performance on our considered parallel machine scheduling problems, we generated a set of 100 synthetic test problem instances. Half of the set we denote as $\mathscr{J}_1^{50} \dots \mathscr{J}_{50}^{50}$ and consists of

50 jobs each. The second half ($\mathcal{J}_1^{100} \ldots \mathcal{J}_{50}^{100}$) comprises 100 jobs in each instance. We sampled all sets with characteristics $p_j = \lfloor \mathcal{U}(1,50) \rfloor$, $\forall j = 1 \ldots n$ and $d_j = p_j + \lfloor \mathcal{U}(1,100) \rfloor$, $\forall j = 1 \ldots n$. For the parallel setup, we further consider both 8 and 12 identical machines, i. e., in all P_m problems we set $m = 8$ and $m = 12$ leading to overall 200 tested configurations. The complete sets of job instances as well as the later discussed results are available via the authors' web page.[5] Further, we instantiated the PPM by defining four fundamental building blocks constituting the runtime environment and the agents' architecture.

- The *spatial population structure* is represented by a two-dimensional toroidal grid with a size of 10×10 nodes initialized with 100 random individuals.
- The *movement* of a predator follows a uniformly distributed random walk pattern of stepping size 1, ensuring that each position is visited equally often.
- The *number of steps* for each model run is restricted to 6,000 function evaluations. This value is independent of the actual number of involved predators.
- The *selection and reproduction neighborhood* of a predator is fixed to a radius of 1, resulting in a selection set of five prey individuals.

Starting with these basic settings, the PPM environment can be extended for arbitrary objectives by just adding a new predator species. In the same way, expert knowledge can be attached as building block to each predator individually. We detail our experimental settings regarding selection and variation building blocks but also regarding a reference algorithm in the next section.

10.4.2.1   Experimental Setup

As literature states that there is no specific solution strategy to the bi-objective problem $P_m||C_{max}, \sum C_j$, see Dutot et al. [12], we are forced to apply a standard heuristic to generate reference results. As representative of often applied general-purpose optimizers we apply NSGA-II. For now, we exclusively investigate the knowledge integration into PPM and merely use NSGA-II as reference to judge the performance of our approach. Therefor, NSGA-II is treated as given and an attempt to modify the variation operators is delayed until section 10.5. To fine-tune NSGA-II, we performed several simulation studies concerning various population sizes and variation probabilities and used the recommended NSGA-II standard configuration, see Deb [9] (mutation and recombination), miscellaneous standard recombination operators like two-point crossover and order crossover, as well as the exclusive application of unbiased mutation. For comparison, we consider the best solutions achieved by NSGA-II using a population size of 100 with exclusive swap mutation. This mutation randomly swaps eight jobs in the sequence and is applied to each individual (variation probability of 1.0). To be comparable with PPM we also restrict the number of steps per run to 6,000 function evaluation.

   The PPM, however, uses two special tailored operators derived from the general mutation operator scheme: SPT-mutation derived from the SPT rule, which solves

---

[5] http://tinyurl.com/3ql3hpe

$P_m||\sum C_j$ optimally as well as LPT-mutation that approximates $P_m||C_{max}$ and takes its cue from the earlier discussed LPT rule. We create a predator species for each objective and connect them to both operators. Further, we discovered in preparatory experiments that LPT-mutation should be emphasized over the quite efficient SPT-mutation and choose therefore $\sigma = 10$ while SPT is applied with $\sigma = 5$. Regardless of the number of predators, we run the PPM for overall 6,000 function evaluations per experiment and performed 50 independent runs of each instance out of our synthetically generated problem set (i.e., all 200 configurations where tested 50 times). Figures 10.7 and 10.8 show single solutions out of all runs for later explanation purposes only.

We analyzed the results of our algorithmic runs by applying two metrics: the hypervolume metric and the ε-dominance metric. The hypervolume metric computes the normalized volume in function space which is covered by the solution front and bounded by a reference point. The larger this value becomes, the better is the approximation. This metric considers convergence behavior and diversity preservation at the same time. The used reference points are $\mathbf{R}_1 = [1000, 5000]$ for problem instances with $n = 50$ and $\mathbf{R}_2 = [1000, 20000]$ for the instances with $n = 100$.

To provide an alternative viewpoint on our solutions, we also apply the ε-dominance metric $I_\varepsilon(A, B)$. It is a binary metric [44] which determines, whether a solution set A dominates another solution set B completely. In detail, the metric value denotes by how much units in function space sets A or B have to be shifted in origin direction to completely dominate the respective other set. The metric has to be applied two-sided to be interpreted correctly. Only if $I_\varepsilon(A, B) > 0$ and the inverse comparison $I_\varepsilon(B, A) \leq 0$ hold, the set A dominates set B entirely. Otherwise, intersections of the determined solution fronts do not allow a domination statement.

10.4.2.2   Results and Discussion

For hypervolume evaluation, we find that all results of PPM enclose a larger volume than the results of NSGA-II. For significance analysis we apply a Wilcoxon rank-sum test ($p = 0.05$) on the determined hypervolume values for all 50 independent experiments of our 50 synthetically generated instances. Further, we test the variance of both algorithms to state on robustness (Fligner-Killeen test [13] as most robust test for variance analysis according to Conover et al. [6], $p = 0.05$). We find that the variance is significant smaller for the PPM than for NSGA-II and conclude that the new approach behaves more robust on problems due to the integrated expertise knowledge. Exemplary, Figure 10.7 shows both obtained Pareto-fronts (PPM and NSGA-II) and single-objective SPT as well as LPT solutions in the objective space. In that case, PPM clearly outperforms NSGA-II in both convergence and diversity although, on the first sight, one can get the impression that PPM generates a less diverse front than NSGA-II. However, for the PPM the gray shaded "overall area of non-domination", i.e., the hypervolume, is larger. As PPM converges to $\sum C_j(OPT)$ value, only the minimum C_{max} value is at that point considered for the Pareto-front.

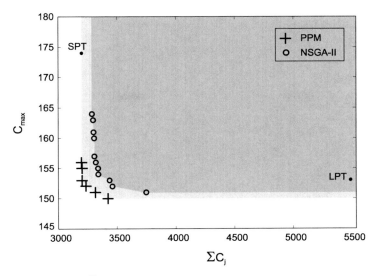

Fig. 10.7 Results for $\mathscr{J}_1^{50}$ with $m = 8$. Circles: NSGA-II; crosses: PPM.

Table 10.1 Results of the ε-Indicator evaluation, which judges on complete domination between two result sets. The columns show mean and variance of (in %) how often the results of PPM dominate the results of NSGA-II completely and vice versa over all instances and for different machine configurations. The operator $A \rhd_\varepsilon B$ denotes the percentaged domination count of A over B with respect to ε-dominance.

Job sets	m	Ref. Point	PPM $\rhd_\varepsilon$ NSGA-II		NSGA-II $\rhd_\varepsilon$ PPM
			mean (in %)	std. (in %)	%
$\mathscr{J}_1^{50} \ldots \mathscr{J}_{50}^{50}$	8	R_1	72.76	16.31	0
$\mathscr{J}_1^{50} \ldots \mathscr{J}_{50}^{50}$	12	R_1	91.66	6.56	0
$\mathscr{J}_1^{100} \ldots \mathscr{J}_{50}^{100}$	8	R_2	56.54	12.32	0
$\mathscr{J}_1^{100} \ldots \mathscr{J}_{50}^{100}$	12	R_2	100	0.02	0

Looking at total domination with the ε-Indicator, we find the results shown in Table 10.1. Usually, the PPM completely dominates the NSGA-II results, while NSGA-II never dominates the PPM. However, there are some cases in which results are marked as incomparable, distinctively observable in the 8-machine scenario. Figure 10.8 shows this effect visually: a single solution of NSGA-II dominates a solution of the PPM. Although the hypervolume is larger for the PPM solution, not all solutions of NSGA-II are dominated. Interestingly, this effect vanishes for larger setups with 12 machines. We hypothesize that this is due to the size of the solution set which may become smaller for multiple machine setups. Results from the real-valued problem domain have already proven the PPM's tendency to converge more in extremal regions of the Pareto-set [19].

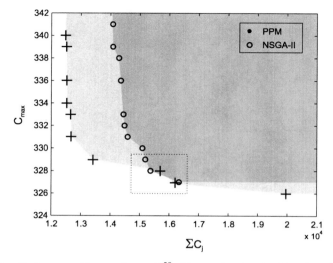

Fig. 10.8 Results for a specific case from $\mathcal{J}_{15}^{50}$ with $m = 8$, where PPM and NSGA-II results are incomparable regarding the ε-indicator. The area that hinders complete dominance of PPM is marked by the dotted box. Circles: NSGA-II; crosses: PPM.

Although diversity as described in Section 10.3 is basically preserved due to the lexicographic combination of influences via selection and specific variation operators, the general problem of diversity loss can also be seen in Figure 10.8. We therefore present in the rest of the paper a possible extension to the model that preserves diversity in the solution set without any external preservation mechanism. That means, no external archive is used which retroacts into population development. We emphasize that for the proposed diversity mechanism only genuine model components are used.

10.5   Integrating a Self-adaptive Mechanism for Diversity Preservation

The previous section already showed promising results for the knowledge integration mechanism of the PPM. However, the PPM reveals a disadvantage regarding the diversity of its results. The disadvantage of extremal convergence in the PPM may also be exploited in combination with an automatic steering mechanism to overcome preserve diversity throughout the optimization process. A steering mechanism has to be able to potentially target all locations on the Pareto-front to cover the whole solution set. Instead of an archive that retroacts into population development, we use a simple storage to remember already found good solutions. Again, we emphasize that this storage is completely independent of the PPM and serves only for result preservation.

The presented modifications were proposed in a recent study by the authors [18]. There, they extend the predators to represent different species which are able to adapt their selection behavior during the evolutionary process. In this way, they realized a sort of niching strategy which favors specialized predators on the one hand. On the other hand, the cooperation between predators is supported allowing exploration of the whole search space. Eventually, this property favors the overall coverage of the actual Pareto-front.

According to previous theoretical findings on the predator prey model's convergence behavior [19], this approach incorporates aspects of the ε-Constraint method, see Section 10.2.2.4. The extended PPM is referred to as ε-PPM. We recapitulate the details about the extension in the following section and evaluate the proposed mechanism in a *case study* for the problem $P_m|r_j|\sum C_j, \sum U_j$ including release dates for jobs.

10.5.1   Algorithmic Extension and Implementation

We combine the advantageous aspects of the PPM and compensate its drawbacks by integrating the ε-Constraint concept: as usual, each predator selects regarding a single objective f_c. In addition, each predator respects $z - 1$ bounds out of the set $\beta = \{\beta_i \mid \beta_i \in \mathbb{R}, i = 1, \ldots, z\}$ of bounds for each objective i. For objective f_c we do not consider any bound and set $\beta_c = \infty$. This is analogously to the ε-Constraint method. We can argue that the local and elitist evolutionary process is replaced by an ε-Constraint selection-guided evolutionary process.

Algorithm 10.2. General predator behavior.

Preconditions: $predator := (f_c, \pi_0, \beta, Op, v)$
 1: **while** not terminated **do**
 2:     $\pi_{i+1} = \text{move}(\pi_i)$
 3:     span selection neighborhood *SN* of size v
 4:     evaluate prey in *SN* regarding f_c and β and mark worst prey for removal
 5:     *offspring* $:= Op(SN\backslash worstPrey)$
 6:     evaluate(*offspring*) regarding f_c and β
 7:     **if** *offspring* $\leq worstPrey$ **then**
 8:         remove worstPrey and introduce *offspring*
 9:     **end if**
10: **end while**

Algorithm 10.2 summarizes the central behavior of a predator which still selects regarding one objective f_c and starts its action at position π_0 in the spatial population. Now, the bounds from β with $\beta_c = \infty$ are considered and selection as well as reproduction using operator Op are applied to prey in a neighborhood of size v around the position π_i. For all following considerations and analysis in this case study, we assume Gaussian mutation as reproduction operator. Later on, in scheduling applications we will again use the new operator defined in Section 10.4.1 that easily includes expertise in a heuristic way.

The predator moves throughout the spatial population and selects the worst prey from its own position's neighborhood, see Lines 2–5 in Algorithm 10.2. After breeding and evaluating an offspring, the worst prey is removed if the offspring's fitness is better, see Algorithm 10.3. The evaluation is done for objective f_c. Additionally, violations of all $z - 1$ remaining objectives induce exponentially weighted penalty. This penalty is computed during evaluation of prey by the predator regarding its specific objective. Each predator applies Algorithm 10.2 independently—respecting its specific objective and bounds β—to the prey population. The value for penalization is computed based on the remaining objective values which are not favored by the predator. For each objective value f_i the violation of b_i is computed normalized to the interval of maximum and minimum allowed bound value max_{feas} and min_{feas} in component i. Both values are determined during predator interaction. We discuss their construction and emergence (based on the utopia and nadir points) in Section 10.5.1.1.

Algorithm 10.3. ε-Constraint evaluation.

Preconditions: $evaluate = (prey, F_c, \beta, min_{feas}, max_{feas})$
1: $penalty = 0$
2: $f = prey.fitness$
3: **for all** $f_i \in f$ with $i \neq c$ and $\beta_i < f_i$ **do**
4:     $interval(i) = max_{feas}(i) - min_{feas}(i)$
5:     $distance(i) = (f_i - \beta_i)/interval(i)$
6:     $penalty = penalty + distance(i) \cdot max_{feas}(i)$
7: **end for**
8: **return** $f_c + \exp(\max(0, penalty)) - 1$

The right hand side of Figure 10.9 shows the expected effect of predators' actions when two predators that select regarding objective f_1 are considered. Due to the single-objective selection, f_1 is minimized until the upper bound ε_2 for objective f_2 is reached. The same holds for the second predator with another specific bound ε_2' for the secondary objective f_2. Locality of selection and penalization of bound violations lead to the known extremal behavior. This results in a certain niching behavior which generates different intermediate trade-offs along the Pareto-front.

10.5.1.1   Automatic Detection of the Feasible Area

Using bounds for the ε-Constraint method implicitly assumes a range for possible bound values. In Algorithm 10.3 we denoted this range by the difference of $max_{feas} - min_{feas}$. However, this range depends on each objective and thus on the whole multi-objective problem. A simple and often used way to detect the feasible

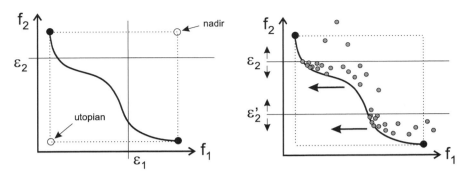

Fig. 10.9 Schematic depiction of the feasible area limited by utopian and nadir vector (left) and ε-restricted search space that prevents extremal convergence (right)

area involves interaction of the decision maker: he is asked to specify lower and upper bounds for all objectives in advance. However, as we want to construct a rather general method, the bound ranges should be determined automatically. Thus, we let predators interact among each other such that they can propagate information about already found solutions.

Generally, two special locations in the solution space can be used to define the feasible area: the *utopian point*, see Section 10.2.2.3, and the *nadir point*. Both define the lower and upper bounds of Pareto-optimality. The utopia point $\mathbf{u} = (u_1, \ldots, u_z)$ is the hypothetical location in solution space where all objectives are minimal. This point is either not part of the Pareto-optimal solution set or a unique solution for a (more or less trivial) multi-objective problem. In our approach, the predators cooperate to determine the utopia point: if two predators with the same objective meet on the spatial structure[6] they exchange information about their best discovered objective value and adopt it for further propagation to other predators. The nadir vector $\mathbf{n} = (n_1, \ldots, n_z)$ is more difficult to obtain as it defines the best upper bound of all remaining objectives assuming optimality of the considered objective. In other words, it shows for a certain objective the minimum pay-off, when another objective is optimal (see left side of Figure 10.9). In our algorithm, the nadir information is generated when the current utopian value remains unchanged and it is then propagated among predators. That means, also nadir point information is exchanged between predators when they meet on the toroidal structure. Additionally, nadir information is always forgotten by the predators, when utopian values improve during information exchange. This supports the exploration of the decision space and the renewal of information during predator interaction. As such, the nadir information always emerges downstream to utopian values and adapts to further decision space exploration. Exemplarily, the left hand side in Figure 10.9 shows the feasible area together with utopia and nadir point.

---

[6] "Meeting of predators" is both agents residing on the same position in the spacial structure.

10.5.1.2   Predator-Induced Niching Behavior

Using the information about the detected feasible area, each predator's individual search bounds are dynamically updated to induce the desired niching behavior. As depicted in Figure 10.9, dedicated areas of the Pareto-front can be reached using specific ε-bounds. Initially, all predators start without any bounds while trying to cooperatively achieve utopia and nadir point information, see previous paragraph. If utopia and nadir information is available, the predator adapts its bounds respecting the utopia point's component u_i as lower bound and the nadir point's component n_i as upper bound for objective f_i. The predator's ε-bound is then chosen randomly from the lower bound/upper bound interval.

During the evolutionary process, predators' ε-bounds are continuously updated. However, to guarantee a permanent predator influence along with an expected extremal convergence towards the boundaries, predators only update their individual ε-bounds when they meet each other in the spatial structure. Then, a normal distributed perturbation, see Equation 10.7,

$$pert = \mathcal{N}\left(0, \frac{\max(\beta_i - u_i, n_i - \beta_i)}{(n_i - u_i)}\right) \tag{10.7}$$

is added to the current bound leading to a moderate change of the predators' selection pressure. In contrast to a fully random generation of new bounds, the approach of "minor environmental change" allows the prey population to react more rapidly to a new situation as possibly only minor changes in solution structure are necessary to adapt to the new bounds. Additionally, it also allows to leave the predefined area of utopia and nadir points to still allow exploration through mutation outside the feasible area.

Technically, the perturbation is normalized using the detected feasible area and the component-wise maximum distance between the current value and the lower and upper bounds. This approach neither incorporates information from the evolutionary process nor information provided by other predators.

10.5.2   Evaluation

The extended PPM was evaluated by the authors in [18]. There, they showed that the integrated ε-Constraint approach successfully supports the PPM in preserving and finding a diverse Pareto-front even for non-convex problems. In this chapter, we again focus on the exploitation of theoretical funded single-objective knowledge and reflect on some experiments and results performed in another resent study by the authors [16]. Beside the simplicity of expertise integration into the ε-PPM we exemplarily show the performance gain through expertise integration by comparing the ε-PPM with and without integrated expertise to NSGA-II. Second, we evaluate whether a similar and intentionally straightforward application of expertise to a standard approach like NSGA-II can yield similar results. The obtained algorithm will be called *modified NSGA-II* (mNSGA-II).

10.5.2.1   Experimental Setup

To evaluate the approach, we rely again on the distribution from our previous experiments in Section 10.4 to generate scheduling problems. However, we increase the problem complexity by adding release dates to the setup. For a proof of concept, we create an exemplary instance with $n = 50$ jobs for $P_m|r_j|\sum C_j, \sum U_j$ with $n = 50$ by uniformly sampling processing times $p_j = \mathcal{U}(1, 50), \forall j = 1, \ldots, n$ and due dates $d_j = p_j + \mathcal{U}(1, 100), \forall j = 1, \ldots, n$. Release dates are generated randomly based on p_j and d_j as

$$r_j = \begin{cases} \mathcal{U}(0, d_j - p_j) , & \mathcal{U}(0, 1) < 0.9 \\ 0 & , \text{otherwise.} \end{cases}$$

For the parallel machine setup, we consider $m = 8$ identical machines. The modified ε-PPM approach for addressing scheduling problems (mε-PPM) is constructed out of 10 predator species where five species select regarding one objective. They are equipped with specific mutation operators that reflect partial expertise. For each objective there is a predator with EDD-biased mutation (EDDMut, $\delta = 2$), with SPT-biased mutation (SPTMut, $\delta = 2$), with release-date-biased mutation (RDMut, $\delta = 3$), with random swap mutation (RSwap, $\delta = 4$), and finally with a mutation inspired by SBC3 that shifts a random number of jobs to the end of a sequence (ShiftMut, $\delta = 4$). The predators move on a 10×10 torus which is populated with 100 randomly initialized individuals. In total, predators do at most $50,000$ function evaluations.

For comparison reasons, we apply standard ε-PPM with mere random mutation as well as the standard NSGA-II implementation with a population size of 100 and random swap mutation. Additionally, we modified NSGA-II (mNSGA-II) to construct a simple enhanced version of NSGA-II by integrating the expertise-biased mutation operators from mε-PPM into the selection mechanism of NSGA-II: during the application of the genetic operators, the current individual is compared to the best known fitness for each objective which leads to two different ratios. Based on these ratios, the mutation decision is biased in favor of the worse objective: EDD-biased mutation for objective $\sum U_j$ and SPT-biased mutation for objective $\sum C_j$. These are generally executed secondary to release date sorting. Additionally, a completely random mutation is added with a probability of $p = 0.1$. Again, we restrict for both algorithms the runtime to $50,000$ function evaluations.

For statistical soundness of our results, the obtained problem instance is solved 30 times with ε-PPM, mε-PPM, NSGA-II, and mNSGA-II each. To compare the four algorithms, we apply the hypervolume indicator [43] and the derived hypervolume ratio (HR) for pairwise comparison of the Pareto-fronts as performance measure. The hypervolume is normalized with reference point $[4000, 30]$ and lower bound $[3500, 6]$. On the hypervolume results, a pairwise *Wilcoxon rank-sum test with 5 %*

significance level and *Bonferroni correction* is adopted to ensure the significance of our results, see Dalgaard [7] for an explanation of this test. Further, we use the binary additive ε-Indicator to determine how often a single algorithm's run dominates another result. This indicator yields only the value 1 if a solution set dominates another set completely. Otherwise the indicator returns the value 0.

10.5.2.2   Results

The experimental results are shown in Tables 10.2 and 10.3 as well as Figure 10.10. In this paragraph, we briefly describe those results but their implications are detailed afterwards in Section 10.5.2.3.

Table 10.2   Hypervolume results for 30 experiments with ε-PPM, mε-PPM, NSGA-II, and the modified version of NSGA-II on the considered scheduling problem

Hypervolume			
Method	Mean	Median	Std.
ε-PPM	0.5628	0.5656	0.0271
mε-PPM	**0.5980**	**0.6049**	**0.026**
NSGA-II	0.5568	0.5615	0.038
mNSGA-II	0.4949	0.5050	0.050

Table 10.2 summarizes the results as mean and median hypervolume for 30 experiments. Additionally, the standard deviation is given. According to the pairwise Wilcoxon rank-sum test, the four algorithm perform significantly different. The p-value obtained by the test when comparing two algorithm is at most $1.4 \cdot 10^{-5}$. The bold written results indicate the best performing algorithm for the tested instance.

Table 10.3 focuses on the presentation of comparison data between the algorithms. In each row, two algorithms are compared regarding hypervolume and mutual domination. The first three columns give some statistical data on the pairwise hypervolume ratio, while the last column gives a pairwise domination count based on the binary and additive ε-Indicator. In total, 900 pairwise comparisons were done for each algorithm pair. Finally, Figure 10.10 exemplary shows the Pareto-fronts computed by ε-PPM, mε-PPM, NSGA-II and mNSGA-II in a typical run. A typical run is defined by Jägersküpper and Preuss [24] as the run which comes closest to median performance value. Since we repeat the experiments 30 times, no run can directly be assigned to the median performance value. Therefore, we choose a run with the performance nearest to the median randomly.

Table 10.3 Hypervolume ratio and domination results for 30 experiments with ε-PPM, mε-PPM, NSGA-II, and the modified version of NSGA-II on the considered scheduling problem. To determine domination the additive ε-Indicator was applied. The #Dom column shows the domination count of the first over the second approach, while #iDom shows the inverse domination count.

Hypervolume Ratio				ε-Ind.	
Method	Mean	Median	Std.	#Dom.	#iDom
mε-PPM/NSGA-II	1.0794	1.0687	0.094	210	1
NSGA-II/mNSGA-II	1.1376	1.1174	0.145	312	5
mε-PPM/mNSGA-II	1.2217	1.1981	1.142	690	0
mε-PPM/ε-PPM	1.0596	1.0585	0.072	224	20
ε-PPM/NSGA-II	1.0220	1.0105	0.095	71	10
ε-PPM/mNSGA-II	1.1426	1.1141	0.137	263	4

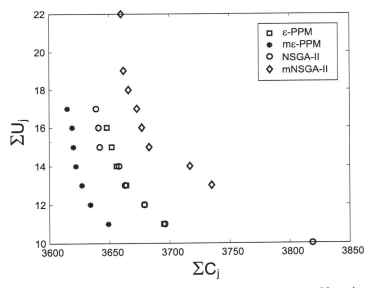

Fig. 10.10 Typical runs of all four examined algorithms on a problem instance of $P_m|r_j|\sum C_j, \sum U_j$

10.5.2.3   Discussion

Our case study shows, that the ε-PPM and its modification for scheduling problems (mε-PPM) can preserve diversity. It has further the same properties to seamlessly integrate expert knowledge as the original PPM. We can also observe the benefit of expertise integration into our approach. With this approach we can significantly outperform both ε-PPM and NSGA-II. Although this is not surprising at all, the results for mNSGA-II are all the more confirming the arguments in favor of PPM as framework for expertise integration. Throughout the work we stated that the monolithic structure of traditional approaches may hinder the seamless integration of expert knowledge. Our results show, that the straightforward application of single-objective knowledge inside the variation operators of NSGA-II leads to a deterioration compared to standard mutation. Certainly, we may be able to restructure and adapt NSGA-II such that this problem class can be approximated in a similar quality as by the PPM. However, an engineer is probably not willing to cope with algorithm design issues to finally reach his solution. Here, the PPM obviously offers a simple way to directly integrate his expertise without algorithmic modifications.

10.6   Conclusion

We proposed a modularly composed, easy to understand, and simple to adapt evolutionary framework for solving multi-objective optimization problems: the predator-prey model (PPM). The modular concept of predators composed from single objective selection and locally operating variation is the key advantage of the PPM over standard, often dominance-based multi-objective evolutionary algorithms (MOEAs) for expertise integration. The latter are usually of rather monolithic structure and therefore hard to understand and difficult to adapt. Thus, for their usage in real-life a domain expert usually has to modify the whole algorithm for his purposes. That leaves a major gap between the framework character of the general approach and its applicability.

In this work, we prove a considerably smaller gap for the PPM: the expert directly applies his available knowledge to the considered single objectives by plugging it into predators. Eventually, the common application of all these influences leads to an implicit cooperation of strategies and to the emergence of good trade-off solutions. This way, it is easy to build a problem specific, multi-objective optimizer transparent to the algorithm's details. To illustrate the seamless application of the PPM framework, we addressed NP-hard multi-objective problems. We considered two scheduling test problems: $P_m||C_{max}, \sum C_j$ and $P_m|r_j| \sum C_j, \sum U_j$. Further, we proposed an extension of the general PPM scheme using a mechanism which is inspired by the ε-Constraint method. That way, we acquire more diverse solutions. Complementing this method with an automatic detection of the feasible area through predator cooperation, allows to apply this mechanism in the PPM model directly. Thus, the advantages of the PPM and the ε-Constraint method are combined. The seamless integration of expertise is preserved and no additional parameters are needed.

References

1. Baptiste, P.: Scheduling equal-length jobs on identical parallel machines. Discrete Applied Mathematics 103(1-3), 21–32 (2000)
2. Beyer, H.-G.: The Theory of Evolution Strategies. Springer, Berlin (2001)
3. Bowman Jr., V.J.: On the Relationship of the Tchebycheff Norm and the Efficient Frontier of Multi-Criteria Objectives. In: Thiriez, H., Zionts, S. (eds.) Multiple Criteria Decision Making. Lecture Notes in Economics and Mathematical Systems, vol. 130, pp. 76–85. Springer, Berlin (1976)
4. Bruno, J., Coffman Jr., E.G., Sethi, R.: Scheduling independent tasks to reduce mean finishing time. Communications of the ACM 17(7), 382–387 (1974)
5. Coello Coello, C., Lamont, G.B., van Veldhuizen, D.A.: Evolutionary Algorithms for Solving Multi-Objective Problems, 2nd edn. Springer, New York (2007)
6. Conover, W.J., Johnson, M.E., Johnson, M.M.: A comparative study of tests for homogeneity of variances, with applications to the outer continental shelf bidding data. Technometrics 4(23), 351–361 (1981)
7. Dalgaard, P.: Introductory Statistics with R. In: Statistics and Computing, Springer, New York (2002)
8. Das, I., Dennis, J.E.: A closer look at drawbacks of minimizing weighted sums of objectives for Pareto set generation in multicriteria optimization problems. Structural and Multidisciplinary Optimization 14, 63–69 (1997)
9. Deb, K.: Multi-Objective Optimization using Evolutionary Algorithms, 1st edn. Wiley-Interscience Series in Systems and Optimization. Wiley (2001)
10. Deb, K., Agrawal, S., Pratab, A., Meyarivan, T.: A Fast Elitist Non-dominated Sorting Genetic Algorithm for Multi-objective Optimization: NSGA-II. In: Deb, K., Rudolph, G., Lutton, E., Merelo, J.J., Schoenauer, M., Schwefel, H.-P., Yao, X. (eds.) PPSN 2000. LNCS, vol. 1917, pp. 849–858. Springer, Heidelberg (2000)
11. Deb, K., Pratap, A., Agarwal, S., Meyarivan, T.: A fast and elitist multiobjective genetic algorithm: NSGA-II. IEEE Transactions on Evolutionary Computation 6(2), 182–197 (2002)
12. Dutot, P.-F., Rzadca, K., Saule, E., Trystram, D.: Multi-Objective Scheduling. In: Introduction to Scheduling, 1st edn., pp. 219–251. CRC Press (2010)
13. Fligner, M.A., Killeen, T.J.: Distribution-free two-sample tests for scale. Journal of the American Statistical Association 71(353), 210–213 (1976)
14. Garey, M.R., Johnson, D.S.: Computers and Intractability: A Guide to the Theory of NP-Completeness. Freeman, San Francisco (1979)
15. Graham, R.L., Lawer, E.L., Lenstra, J.K., Rinnooy Kan, A.H.G.: Optimization and approximation in deterministic sequencing and scheduling: A survey. Annals of Discrete Mathematics 5, 287–326 (1979)
16. Grimme, C., Kemmerling, M., Lepping, J.: An expertise-guided multi-criteria approach to scheduling problems. In: Proceedings of the International Genetic and Evolutionary Computation Conference (GECCO), pp. 47–48. ACM, New York (2011)
17. Grimme, C., Lepping, J.: Designing Multi-objective Variation Operators Using a Predator-Prey Approach. In: Obayashi, S., Deb, K., Poloni, C., Hiroyasu, T., Murata, T. (eds.) EMO 2007. LNCS, vol. 4403, pp. 21–35. Springer, Heidelberg (2007)
18. Grimme, C., Lepping, J.: Integrating niching into the predator-prey model using epsilon-constraints. In: Proceedings of the International Genetic and Evolutionary Computation Conference (GECCO), pp. 109–110. ACM, New York (2011)

19. Grimme, C., Lepping, J., Papaspyrou, A.: Exploring the Behavior of Building Blocks for Multi-Objective Variation Operator Design using Predator-Prey Dynamics. In: Thierens, D., et al. (eds.) Proceedings of the International Genetic and Evolutionary Computation Conference (GECCO), London, pp. 805–812. ACM (June 2007)

20. Grimme, C., Lepping, J., Schwiegelshohn, U.: Multi-Criteria Scheduling: An Agent-based Approach for Expert Knowledge Integration. Journal of Scheduling, 1–15 (2011)

21. Haupt, R.: A Survey of Priority Rule-Based Scheduling. OR Spectrum 11(1), 3–16 (1989)

22. Ho, J.C., Chang, Y.-L.: Minimizing the number of tardy jobs for m parallel machines. European Journal of Operational Research 84(2), 343–355 (1995)

23. Hoogeveen, H.: Multicriteria scheduling. European Journal of Operational Research 167(3), 592–623 (2005)

24. Jägersküpper, J., Preuß, M.: Empirical Investigation of Simplified Step-Size Control in Metaheuristics with a View to Theory. In: McGeoch, C.C. (ed.) WEA 2008. LNCS, vol. 5038, pp. 263–274. Springer, Heidelberg (2008)

25. Jouglet, A., Savourey, D.: Dominance rules for the parallel machine total weighted tardiness scheduling problem with release dates. Computers & Operations Research 38(9), 1259–1266 (2011)

26. Kawaguchi, T., Kyan, S.: Worst case bound of an lrf schedule for the mean weighted flow-time problem. SIAM Journal on Computing 15(4), 1119–1129 (1986)

27. Laumanns, M., Rudolph, G., Schwefel, H.-P.: A Spatial Predator-Prey Approach to Multi-objective Optimization: A Preliminary Study. In: Eiben, A.E., Bäck, T., Schoenauer, M., Schwefel, H.-P. (eds.) PPSN 1998. LNCS, vol. 1498, pp. 241–249. Springer, Heidelberg (1998)

28. Laumanns, M., Thiele, L., Zitzler, E.: An efficient, adaptive parameter variation scheme for metaheuristics based on the epsilon-constraint method. European Journal of Operational Research 169(3), 932–942 (2006)

29. Lei, D.: Multi–objective production scheduling: a survey. The International Journal of Advanced Manufacturing Technology 43(9), 926–938 (2009)

30. Lenstra, J.K., Rinnooy Kan, A.H.G., Brucker, P.: Complexity of machine scheduling problems. Annals of Discrete Mathematics 1, 343–362 (1977)

31. Li, X., Amodeo, L., Yalaoui, F., Chehade, H.: A multiobjective optimization approach to solve a parallel machines scheduling problem. Advances in Artificial Intelligence, 1–10 (2010)

32. Miettinen, K.: Nonlinear Multiobjective Optimization. Kluwer's International Series in Operations Research & Management Science. Kluwer Academic Publishers, Boston (1999)

33. Moore, J.M.: An n–job, one machine sequencing algorithm for minimizing the number of late jobs. Management Science 15(1), 102–109 (1968)

34. Nagar, A., Haddock, J., Heragu, S.: Multiple and bicriteria scheduling: A literature survey. European Journal of Operational Research 81(1), 88–104 (1995)

35. Pinedo, M.: Scheduling: Theory, Algorithms, and Systems, 3rd edn. Springer (2009)

36. Sahni, S.K.: Algorithms for scheduling independent tasks. Journal of the ACM 23(1), 116–127 (1976)

37. Schwiegelshohn, U.: An alternative proof of the Kawaguchi-Kyan bound for the Largest-Ratio-First rule. Technical Report 0111, TU Dortmund University (2011)

38. SongFa, H., Ying, Z.: NSGA-II based grid task scheduling with multi-qos constraint. In: Proceedings of the 3rd International Conference on Genetic and Evolutionary Computing (WGEC 2009), pp. 306–308. IEEE (2009)

39. Süer, G.A., Báez, E., Czajkiewicz, Z.: Minimizing the number of tardy jobs in identical machine scheduling. Computers and Industrial Engineering 25(1-4), 243–246 (1993)

40. T'kindt, V., Billaut, J.-C.: Multicriteria Scheduling. Theory, Models and Algorithms, 2nd edn. Springer, Berlin (2006)

41. van Wassenhove, L.N., Gelders, F.: Solving a Bicriterion Scheduling Problem. European Journal of Operational Research 2(4), 281–290 (1980)

42. Yuan, X., Quanfeng, L.: Bicriteria parallel machines scheduling problem with fuzzy due dates based on NSGA-II. In: Proceedings of the International Conference on Intelligent Computing and Intelligent Systems (ICIS 2010), vol. 3, pp. 520–524. IEEE (2010)

43. Zitzler, E.: Evolutionary Algorithms for Multiobjective Optimization: Methods and Applications. PhD thesis, ETH Zürich (1999)

44. Zitzler, E., Thiele, L.: Multiobjective Evolutionary Algorithms: A Comparative Case Study and the Strength Pareto Approach. IEEE Transactions on Evolutionary Computation 3(4), 257–271 (1999)

Chapter 11
Analysing the Robustness of Multiobjectivisation Approaches Applied to Large Scale Optimisation Problems

Carlos Segura, Eduardo Segredo, and Coromoto León

Abstract. Multiobjectivisation transforms a mono-objective problem into a multi-objective one. The main aim of multiobjectivisation is to avoid stagnation in local optima, by changing the landscape of the original fitness function. In this contribution, an analysis of different multiobjectivisation approaches has been performed. It has been carried out with a set of scalable mono-objective benchmark problems. The experimental evaluation has demonstrated the advantages of multiobjectivisation, both in terms of quality and saved resources. However, it has been revealed that it produces a negative effect in some cases. Some multiobjectivisation schemes require the specification of additional parameters which must be adapted for dealing with different problems. Multiobjectivisation with parameters has been proposed as a method to improve the performance of the whole optimisation scheme. Nevertheless, the parameter setting of an optimisation scheme which considers multiobjectivisation with parameters is usually more complex. In this work, a new model based on the usage of hyperheuristics to facilitate the application of multiobjectivisation with parameters has been proposed. Experimental evaluation has shown that this model has increased the robustness of the whole optimisation scheme.

11.1 Introduction

Many real world problems require the application of *Optimisation Strategies*. Several exact approaches have been designed to deal with optimisation problems. However, exact approaches are not affordable for many real world applications, so a wide variety of Approximation Algorithms has been developed with the aim of obtaining good quality solutions in a restricted time. Metaheuristics [24] are a family of

Carlos Segura · Eduardo Segredo · Coromoto León
Dpto. de Estadística, I.O. y Computación. Universidad de La Laguna
La Laguna, 38271, Santa Cruz de Tenerife, Spain
e-mail: {csegura,esegredo,cleon}@ull.es

E. Tantar et al. (Eds.): EVOLVE- A Bridge between Probability, SCI 447, pp. 365–391.
springerlink.com

Approximation Techniques that have become popular to solve optimisation problems. They are high-level strategies that guide a set of heuristic techniques in the search of an optimum. Among them, Evolutionary Algorithms (EAs) [23] are one of the most popular strategies. They are population-based algorithms inspired on the biological evolution. EAs have shown great promise for calculating solutions to large and difficult optimisation problems. To define a configuration of an EA several components or parameters, as the survivor selection mechanism, and the genetic and parent selection operators must be specified. Therefore, several flavours of EAs are seen to exist. Usually, the quality of the obtained solutions highly depends on such components and parameters. As a result, the process of making the parameter setting of an EA usually takes too much user and computational effort [17].

In many problems, EAs may have a tendency to converge towards local optima. The likelihood of this occurrence depends on the shape of the fitness landscape [11]. Several methods have been designed with the aim of dealing with local optima stagnation [24]. Some of the simplest techniques rely on restarting the approach when stagnation is detected. In other cases, a component which inserts randomness or noise in the search is used. Maintaining some memory, in order to avoid exploring the same zones several times, is also a typical approach. Finally, population-based strategies intrinsically try to maintain the diversity of a solution set. By recombining such solutions, a larger area of the decision space may be explored. In the particular case of EAs, stagnation may be alleviated by using several mechanisms, as increasing the mutation rate or using selection schemes with a lower selection pressure. Another alternative mechanism resides on the usage of *multiobjectivisation*.

The term *multiobjectivisation* was introduced in [30] to refer to the reformulation of originally mono-objective problems as multi-objective ones. Multiobjectivisation changes the fitness landscape, so it can be useful to avoid local optima [26], and consequently to make easier the resolution of the considered problem. However, it can also produce a harder problem [5]. Multiobjectivisation can be carried out following two general schemes. The first one is based on decomposing the original objective, while the second one is based on adding new objective functions. The addition of alternative functions can be performed by considering problem-dependent or problem-independent information. The alternative functions can depend on a single individual or on a full set of individuals [36]. The latter option has been addressed in the current study. In order to deal with a multiobjectivised problem, a multi-objective technique must be applied. Multi-Objective Evolutionary Algorithms (MOEAs) have been proposed with the aim of dealing with multi-objective optimisation problems. Some multiobjectivisation schemes require the specification of some additional parameters. The quality of the obtained solutions might be improved by using this kind of schemes [41]. However, the parameter setting of such schemes can be even more time-consuming. This hinders the proper usage of EAs.

Hyperheuristics are a promising approach to facilitate the application of EAs. A hyperheuristic can be viewed as a heuristic that iteratively chooses between a set of given low-level metaheuristics in order to solve an optimisation problem [7]. Hence, by using several EA configurations that combine different components or parameters, the drawbacks of parameter setting can be mitigated. Hyperheuristics operate at a higher level of abstraction than metaheuristics, because they have no knowledge about the problem domain. The underlying principle in using a hyperheuristic approach is that different metaheuristics have different strengths and weaknesses, and it makes sense to combine them in an intelligent manner. Hyperheuristics provide two main benefits. First, it might detect which is the most suitable metaheuristic for a given problem or instance. Thus, the user and computational effort associated to the parameter setting might be decreased. In addition, different values of parameters might be optimal at different stages of the optimisation process [28]. In such cases, hyperheuristics might grant more resources to the fittest low-level metaheuristic on each stage of the optimisation process. Therefore, the quality of the solutions obtained by the hyperheuristic might be higher than the quality of the solutions obtained by any of the involved low-level metaheuristics.

The aim of this contribution is twofold. First, a deep analysis of the validity of multiobjectivisation has been performed. A comparison among mono-objective EAs and the Non-dominated Sorting Genetic Algorithm II [19] (NSGA-II) applied to a set of multiobjectivised problems has been carried out. Such algorithms have been selected due to their popularity, instead of considering them because of their performance with the considered problems. Therefore, the aim of the analysis has not been to improve the best-known approaches for such problems. It has focused on analysing how multiobjectivisation can improve single-objective algorithms. Several promising multiobjectivisation methods proposed in previous works [40] have been applied. The analysis has shown the different benefits of multiobjectivisation. However, it has also shown some drawbacks in terms of its robustness. In fact, multiobjectivisation has not provided benefits in some of the tested problems. A novel multiobjectivisation which requires the specification of parameters has also been proposed. An analysis of this kind of techniques has been accomplished. This analysis has revealed that the fittest values of the parameters might depend on the considered optimisation problem, and on the stage of the optimisation process. Second, a novel optimisation scheme that merges hyperheuristics and multiobjectivisation with parameters has been proposed. The aim of this last optimisation scheme is to avoid the requirement of specifying a particular value for the multiobjectivisation parameters. Computational results have demonstrated that such an approach has increased the robustness and ease of use of multiobjectivisation. The different analyses have been carried out considering a set of well-known mono-objective scalable problems [33]. The relationship between the number of considered variables and the performance of the proposals has been analysed. Experiments with up to 5000 variables have been performed.

The rest of this Chapter is organised as follows. In Section 11.2, the applied mono-objective and multi-objective EAs are detailed. The multiobjectivisation techniques used in this work are described in Section 11.3. The notions and background of parameter setting and hyperheuristics are given in Section 11.4. In Section 11.5, the way in which hyperheuristics can increase the robustness of multiobjectivisation with parameters is explained. The particular hyperheuristic used in this contribution is also described at this point. Then, the experimental evaluation is presented in Section 11.6. Finally, conclusions are given in Section 11.7.

Algorithm 11.1. EA Pseudocode

1: Generate an initial population with N individuals
2: Evaluate all individuals in the population
3: **while** (not stopping criterion) **do**
4:     Mating selection: select parents to generate the offsprings
5:     Variation: Apply genetic operators to the mating pool to create a child population
6:     Evaluate the offsprings
7:     Select individuals for the next generation
8: **end while**

11.2   Optimisation Schemes

In order to analyse the behaviour of multiobjectivisation approaches, they must be tested with a particular optimisation scheme. Metaheuristics are high-level approaches designed to deal with optimisation problems. They can be classified into two groups: trajectory-based and population-based algorithms. Trajectory-based approaches maintain at any instant of the optimisation process a single solution. In population-based algorithms, several solutions are simultaneously taken into account. Both population-based and trajectory-based approaches have been successfully applied to a large amount of complex real-world applications [24]. In multiobjective optimisation, population-based approaches are more popular because the maintenance of diversity is particularly important, and because a single preferred direction can not be identified for trajectory-based approaches [13]. Thus, this work focuses on analysing multiobjectivisation with population-based metaheuristics.

Evolutionary computation [23] is a special brand of computing which draws its inspiration from natural evolutionary processes. In this work, a subset of evolutionary computation techniques -EAs- has been used to test the behaviour of multiobjectivisation. EAs are a set of population-based approaches inspired by biological evolution. They are based on the same generic framework (Algorithm 11.1), but in order to obtain a particular configuration of an EA several details have to be specified. Consequently, several variants of EAs are seen to exist. Among these components, the survivor selection mechanism, and the genetic and parent selection operators have been widely analysed.

In mono-objective EAs the survivor selection operator chooses which individuals will be allowed in the next generation. This decision is usually based on the fitness value, favouring those with higher quality, although the concept of age might be another possibility [23]. In this work, a set of mono-objective EAs has been applied. Each one of them implements a different survivor selection mechanism:

- **Steady-State (SS-EA):** In each generation, one offspring is generated. If it is better than any of the individuals of the population, the worst of them is replaced by this new offspring.

- **Generational with Elitism (GEN-EA):** In each generation, $N - 1$ offsprings are generated, being N the population size. All parents, except the fittest one, are discarded, and they are replaced by the generated offsprings.
- **Replace Worst (RW-EA):** In each generation, N offsprings are generated. The N fittest individuals, between parents and offsprings, are selected to survive.

The aforementioned survivor selection schemes do not consider information about the variables domain. However, multiobjectivisation usually considers such information. By this way, it can be useful to avoid premature convergence. In order to apply multiobjectivisation to a particular problem, a multi-objective approach must be used. In this work, a MOEA has been used. Several MOEAs are seen to exist. Among them, the NSGA-II is clearly the most popular one.

The NSGA-II (Algorithm 11.2) is a non-dominated sorting based MOEA. Two of the most important characteristics of this algorithm are the following. First, it uses a fast non-dominated sorting approach. Second, it applies a selection operator which combines previous populations with new generated ones, ensuring elitism in the approach. Both aforementioned features are guided by the *crowded comparison operator* $(\geq_n)$. This operator assigns two attributes to every individual i of the population: the *non-domination rank* (i_{rank}) and the *local crowding distance* ($i_{distance}$).

The non-domination rank makes use of the Pareto Dominance concept. The procedure to calculate it is as follows. First, the set of non-dominated individuals of the population are assigned to the first rank. Then the process is repeated considering only the individuals that do not have a rank assigned. The rank assigned at each step is increased by one. The process ends when all individuals in the population have its corresponding rank established.

The local crowding distance is used to estimate the density of solutions surrounding a particular individual. First, the size of the largest cuboid enclosing the individual i without including any other individual that belongs to its rank is calculated. Then, the crowding distance is calculated as the average side-length of the cuboid. It is worthy to mention that the local crowding distance of the boundary individuals of every rank is assigned to an infinite value. Finally, the partial order given by $\geq_n$ is the following:

$$i \geq_n j \quad if \quad (i_{rank} < j_{rank}) \; or \; ((i_{rank} = j_{rank}) \; and \; (i_{distance} > j_{distance})) \quad (11.1)$$

In order to complete the definition of the mono-objective and the multi-objective EAs, other components must also be specified. A direct encoding of the candidate solutions has been considered, i.e. they have been represented by a vector of D real numbers, being D the number of variables of the considered optimisation problems. The parent selection mechanism has been the well-known *Binary Tournament* [23]. The *Uniform Mutation* [23] (UM) with a probability $p_m = \frac{1}{D}$, and the *Simulated Binary Crossover* [18] (SBX) with a probability $p_c = 1$, have been applied.

Algorithm 11.2. NSGA-II Pseudocode

1: Initialisation: Generate an initial population P_0 with N individuals. Assign $t = 0$.
2: **while** (not stopping criterion) **do**
3:     Fitness assignment: Calculate fitness values of individuals in P_t. Use only the non-domination rank in the first generation, and the crowded comparison operator in other generations.
4:     Mating selection: Perform binary tournament selection on P_t in order to fill the mating pool.
5:     Variation: Apply genetic operators to the mating pool to create a child population CP.
6:     Combine P_t and CP, selecting the best individuals using the crowded comparison operator to constitute P_{t+1}.
7:     $t = t + 1$
8: **end while**

11.3   Multiobjectivisation

The term *multiobjectivisation* was introduced in [30] to refer to the reformulation of originally mono-objective problems as multi-objective ones. There are two main ways of multiobjectivising a problem. The first one is based on decomposing the original objective into several sub-objectives. The Pareto Front of the new defined problem should contain a solution with the original optimal value. The second one is based on aggregating a new alternative function as the second objective. Such a function is used in conjunction with the original fitness function. In this last case, the Pareto Front always contains a solution with the original optimal value. The main advantage of the second approach is that it can be performed by considering solely problem-independent information. Thus, by following such a scheme, general multiobjectivisation approaches useful for several optimisation problems might be designed. In this work, several alternative functions have been considered.

Multiobjectivisation usually decreases the selection pressure of the original approach. Therefore, when used in combination with EAs, some low quality individuals could survive in the population with a higher probability. However, if properly configured, in the long term these individuals could help to avoid stagnation in local optima, so higher quality solutions might be obtained.

Several options have been proposed to define the artificial objective [1, 44, 6]. Some of them are based on the Euclidean distance on the decision space [44]. It is worthy to mention that these functions are a direct measure of the diversity:

- **DCN**: The distance to the closest neighbour of the population has to be maximised.
- **ADI**: The average distance to all individuals of the population has to be maximised.
- **DBI**: The distance to the best individual of the population, i.e. the one with the lowest fitness, has to be maximised.

Other authors propose the usage of objectives which may preserve diversity without using a direct measure for diversity [1]. Among them, the following ones have been defined:

- **Random**: A random value is assigned as the second objective to be minimised. Smaller random values may be assigned to some low-quality individuals that would get a chance to survive.
- **Inversion**: In this case, the optimisation direction of the original objective function is inverted and is used as the artificial objective. This approach highly decreases the selection pressure.
- **Time stamp**: The artificial objective is calculated as a time stamp of when an individual is generated. Each individual in the initial population is stamped with a different time stamp represented by a counter which gets incremented every time a new individual is created. From the second population all new generated individuals get the same time stamp that is set to the population size plus the generation index. This time stamp must be minimised.

Note that among these options, DCN and ADI take more time to calculate than the others, because they need to look at $O(N^2)$ distances. Since in real-world applications the evaluation time is usually highly related to the number of evaluations, the time to calculate distances is assumed to be negligible compared to the time to evaluate a candidate solution. Consequently, in this work comparisons have been performed solely on the basis of function evaluations.

In [6], an analysis about the behaviour of the aforementioned artificial functions in dynamic environments was performed. It revealed the superiority of the distance-based artificial objectives. In [41], multiobjectivisation was successfully applied to a packing problem. A strategy based on the DBI multiobjectivisation achieved the best results. This strategy also took into account the penalty of low-quality candidate solutions. This work is focused on the distance-based artificial functions. In addition, a novel alternative objective that applies some of the ideas shown in [41] is proposed. Specifically, it is a problem-independent technique that tries to increase the diversity of the population. It starts from the DCN alternative objective, but it incorporates the usage of a threshold ratio ($th \in [0, 1]$) which must be specified by the user. Such a multiobjectivisation is named DCN-THR. The threshold ratio is used to avoid the survival of individuals with a very low quality. Being *bestFit* the fitness value of the best individual of the population, and *shift* a value that ensures that *bestFit* − *shift* ≥ 0 in the whole execution, the threshold value (v) is defined as:

$$v = \frac{(bestFit - shift)}{th} + shift \qquad (11.2)$$

The alternative objective of individuals whose fitness value is higher than v is assigned to 0. As a result, individuals that are not able to achieve the fixed threshold are penalized. In the special case where $th = 0$, individuals are never penalized. Thus, DCN-THR with $th = 0$ has the same behaviour than the DCN function.

Figure 11.1 shows the behaviour of the DCN function when integrated with the NSGA-II. The maximisation of the DCN function and the minimisation of the fitness

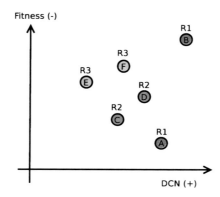

Fig. 11.1 Behaviour of the NSGA-II without threshold

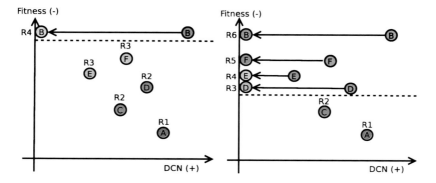

Fig. 11.2 Behaviour of the NSGA-II with two different values of threshold

function are assumed. Each candidate solution is tagged with a label that indicates its corresponding ranking assigned by the NSGA-II. By this way, the label *Ri* means that the corresponding candidate solution belongs to the rank number *i*. Figure 11.2 shows the effect of incorporating the threshold to the DCN function. The broken line represents the value of *v*. It is worthy to mention that the higher the threshold ratio the lower *v* is. Such a Figure shows that every candidate solution which does not fulfil the minimum quality level established by the threshold ratio is shifted in the objective space. Specifically, a value equal to 0 is assigned to its alternative objective. Afterwards, the NSGA-II crowded comparison operator is applied. The effect of the shift is that the corresponding candidate solution will usually belong to a worse rank. Thus, its survival probability will be usually decreased.

In cases where a low threshold ratio is applied, the objective function is similar to the original DCN function. In the opposite case, i.e. when a high threshold ratio is used, the algorithm behaves as if multiobjectivisation was not applied. Since the performance of multiobjectivisation might depend on the tackled problem, by properly tuning the threshold ratio, the approach can be adapted to the problem or instance to solve.

11.4 Parameter Setting

In order to successfully apply a metaheuristic, several components and parameters must usually be specified. For instance, it has been mentioned that in order to define a particular EA, several details must be defined. It is customary to call these details the EA parameters. Among other parameters, in EAs [23], the mutation, crossover, and selection operators, as well as several probabilities must be fixed. The quality of the solutions highly depends on such components and parameterisations [2, 34, 35]. Consequently, it is very important to perform the parameterisation in a proper way.

Finding the appropriate parameter values of an EA is a source of research which emerged several decades ago [17], and during all that time, researchers have been trying to find answers to questions like:

- Is there a generic set of parameter values applicable to all problems?
- Is there a generic set of parameter values applicable to a particular set of problems?
- Is it better to preset the parameter values before the execution or modify the parameter values during the execution of the algorithm?
- What parameter subset really affects the behaviour of an algorithm?

Usually, when a novel problem is tackled, there is no *a priori* information of which metaheuristic is the most suitable one. Therefore, in order to obtain high-quality solutions, several metaheuristics, as well as several parameterisations for each of them must be tested. As a result, the effort required to obtain high-quality solutions, both in terms of user-effort and computational-effort, might be very large. The selection of the parameters can be done in several ways [4]:

- Checking in a systematic way some ranges of the parameter values and assessing the performance of each value.
- Based on the experiences reported in the literature for similar applications.
- Performing a theoretical analysis of the behaviour of the metaheuristic to determine the optimal parameter setting.
- Using "standard" values.

In order to reduce the waste of resources of a systematic testing of the different parameter ranges, and with the aim of minimising the user effort, several studies have analysed the automation of EAs and other metaheuristics parameterisation [42, 32]. Parameter setting strategies are divided into two categories: *parameter tuning* and *parameter control*. In parameter tuning [42] the objective is to identify the best set of parameters for a given metaheuristic. Then the algorithms are executed using the same parameterisation during all the run. By contrast, in parameter control, the optimisation strategies start the execution with a set of initial values, and modify those values during the execution, depending on the behaviour of the approach in the different stages of the optimisation process. The main advantage of parameter control is that high-quality solutions might be found in a single run.

Parameter tuning is usually done by experimenting with different values and selecting the ones that produce the best results. The most frequently used tuning strategy is based on a multiple execution in which several parameter values are tested to find the appropriate combination of them for a particular problem. This process can be automated by the use of a top-level driver that chooses different parameter values in a systematic way. Metaheuristic parameters usually interact in highly non-linear ways [32]. Moreover, there is a large amount of parameterisation options, but only little knowledge about the effect of the parameters. Thus, the problem of finding the best set of parameters is very hard. In addition, the stochastic behaviour of metaheuristics also hinders the design of the tuning strategies. Several studies [22] have concluded that the usage of a static set of parameters during a metaheuristic run seems to be inappropriate. In fact, it has been empirically and theoretically demonstrated that different values of parameters might be optimal at different stages of the optimisation process [28]. Therefore, the main drawback of parameter tuning is that there is no guarantee that a fixed set of parameters leads to optimal performance because the demands of optimisation algorithms change during an optimisation run from exploration in early stages to exploitation in later stages. Since different parameter settings are needed to emphasise either exploration or exploitation, it follows that optimal parameter settings might vary over time.

Parameter control strategies allow changing the parameter values during the metaheuristic runs. Thus, the aim of parameter control is to design a control strategy that selects the parameters to use at each stage of the optimisation process. Several metaheuristics as Simulated Annealing (SA) and Evolution Strategies (ESs) provide self-adaptive parameters to deal with the requirements of each optimisation stage. Since they have been of great value in several fields, it seems promising to incorporate these ideas into other metaheuristics. Moreover, it would be of a great value to integrate these ideas in a general way, so that a large number of metaheuristics can profit from it. Several strategies to adapt the parameters of the algorithms have been designed. Most of them [46] depend on the formulation of the metaheuristic, so they can not be applied in a straightforward manner to other metaheuristics.

Hyperheuristics are popular general methods which can be applied to perform the parameter control. The main advantage of hyperheuristics is that they are independent of the metaheuristics that they control. A hyperheuristic can be viewed as a heuristic that iteratively chooses between a set of given low-level metaheuristics in order to solve an optimisation problem [7]. Hyperheuristics operate at a higher level of abstraction than heuristics, because they have no knowledge about the problem domain. The motivation behind the approach is that, ideally, once a hyperheuristic algorithm has been developed, several problem domains and instances could be tackled by only replacing the low-level metaheuristics. Thus, the aim in using a hyperheuristic is to raise the level of generality at which most current metaheuristic systems operate.

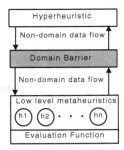

Fig. 11.3 Hyperheuristic Framework

The main motivation of hyperheuristics is to design problem-independent strategies. Thus, a hyperheuristic is not concerned with solving a given problem directly as in the case of most heuristics. In fact, the search is on a metaheuristic search space rather than a search space of potential problem solutions. The hyperheuristic solves the problem indirectly by recommending which solution method to apply at which stage of the optimisation process. Generally, the goal of raising the level of generality is achieved at the expense of reduced - but still acceptable - solution quality when compared to tailor-made approaches. A diagram of a general hyperheuristic framework [7] is shown in Figure 11.3. It shows a problem domain barrier between the low level metaheuristics and the hyperheuristic itself. The data flow received by the hyperheuristic could include the quality of achieved solutions (average, improvement, best, worst), the resources (time, processors, memory) invested to achieve such solutions, etc. Based on such information, the hyperheuristic make its decisions. The data flow coming from the hyperheuristic could include information about which heuristic must be executed, its parameters, stopping criteria, etc.

Hyperheuristics can be classified in several ways. The adaptation level refers to the historical knowledge that is considered by the hyperheuristic. Based on the adaptation level, hyperheuristics can be classified as global or local approaches [37]. In global approaches all the historical knowledge is considered, while in local ones, some of the information is discarded. Hyperheuristics can also be classified in terms of the characteristics of the low-level metaheuristics into two groups [9], the ones which operate with constructive techniques and the ones which operate with improvement techniques. Constructive techniques are used to build solutions from scratch. At each step, they determine a subpart of the solution. Improvement metaheuristics are iterative approaches which take an initial solution, and modify it with the aim of improving the objective value. Some hyperheuristics have been designed to operate specifically with one kind of low-level metaheuristics, while other ones, can use both constructive and improvement methods.

Previously to the appearance of the concept of hyperheuristic, some research was performed analysing similar ideas. Composer [25] was one of the first proposals which used a search space constituted by heuristics. In such a case, the search was performed by using a hill-climbing strategy. Other proposals consisted in hybridising genetic algorithms and heuristics [43]. Several ways of incorporating the ideas of hyperheuristics into an optimisation problem have been proposed. The hyperheuristics which deal with mono-objective optimisation problems are much more extensive. In [8] a tabu search based hyperheuristic was presented. The same hyperheuristic was used inside a SA algorithm [21]. The hyperheuristic was used to combine several neighbourhood definitions. Other metaheuristics which have inspired the creation of hyperheuristics are genetic algorithms [14] and ant colony optimisation [10, 12, 20]. The choice functions have also been used multiple times [15, 16, 29]. In such cases, a scoring function is used to assess the performance of each low-level heuristic. The resources are granted to the low-level heuristic which maximise such a function. In [45], a choice function was also used to score each method. However, the resources were assigned using a probability function which is based on the assigned score. Based on this idea, a parallel optimisation model was proposed in [31]. This model has been successfully applied to several real-world applications [39, 38].

11.5   Increasing the Robustness of Multiobjectivisation

Multiobjectivisation changes the fitness landscape, so it can be useful to avoid local optima [26], and consequently, to make easier the resolution of the problem. However, it can also produce a harder problem [5]. In the case of multiobjectivising a problem with the DCN-THR function, the value of the threshold ratio might affect the behaviour of the NSGA-II. Specifically, the operation of the parent and survivor selection mechanisms is modified. Given that the most suitable parent and survivor selection operators depend on the problem or instance to solve, a fixed threshold ratio will not probably be the most suitable for every problem and instance. Moreover, the fittest parameter values might also depend on the optimisation stage. Therefore, by using fixed values for such parameters, suboptimal results might be obtained.

The combination of DCN-THR and hyperheuristics can produce two main benefits. First, fixing a specific threshold ratio is not required when a hyperheuristic is used. Instead, the practitioner might specify a set of ratios that should be tested. This facilitates the decisions that must be taken by the practitioners, increasing the usability of the scheme. Moreover, with a fixed parameterisation of the hyperheuristic a larger amount of problems can be tackled than with a fixed parameterisation of the low-level metaheuristics. Thus, by combining hyperheuristics and multiobjectivisation the robustness of the whole optimisation scheme is increased.

The hyperheuristic presented in [45] has been applied in this work. Such a hyperheuristic (HH_imp) is based on using a scoring strategy and a selection strategy for picking up the low-level configuration that must be executed. Once a strategy is picked up, it is executed until a local stopping criterion is achieved. Then, another low-level configuration is selected, and it is executed taking as initial population the last population of the previous selected approach. This process continues until a global stopping criterion is reached. The selection of the low-level strategy that must be executed is as follows. First, the scoring strategy assigns a score to each low-level configuration. This score estimates the improvement that each low-level metaheuristic or configuration can achieve when it starts from the currently obtained solutions. In order to perform such an estimate, the previous fitness improvements achieved by each configuration are used. The improvement (imp) is defined as the difference, in terms of the fitness value, between the best achieved individual and the best initial individual. Considering a configuration $conf$, which has been executed j times, the score - $s(conf)$ - is calculated as a weighted average of its latest k improvements (Equation 11.3). In such an Equation, $imp[a][b]$ represents the improvement achieved by the configuration a in its execution number b. Depending on the value of k, the adaptation level of the hyperheuristic can be set. The weighted average assigns greater importance to the latest executions.

$$s(conf) = \frac{\sum\limits_{i=1}^{min(k,j)} (min(k,j) + 1 - i) \cdot imp[conf][j-i]}{\sum\limits_{i=1}^{min(k,j)} i} \quad (11.3)$$

The stochastic behaviour of the involved low-level metaheuristics may lead to variations in the results achieved by them. Therefore, it is appropriate to make some selections based on a random scheme. The hyperheuristic can be tuned by means of the parameter β, which represents the minimum selection probability that should be assigned to a low-level configuration. Being n_h the number of involved low-level configurations, a random selection following a uniform distribution is performed in $\beta \cdot n_h$ percentage of the cases. Therefore, the probability of selecting each configuration $conf$ is given by:

$$prob(conf) = \beta + (1 - \beta \cdot n_h) \cdot \left[\frac{s(conf)}{\sum\limits_{i=1}^{n_h} s(i)} \right] \quad (11.4)$$

11.6 Experimental Evaluation

This section shows the experimental evaluation performed with the optimisation schemes described in this Chapter. The optimisation schemes have been implemented using METCO [31] (*Metaheuristic-based Extensible Tool for Cooperative Optimisation*). Tests have been run on a Debian GNU/Linux computer with four AMD ® Opteron ™ (model number 6164 HE) at 1.7 GHz and 64 GB RAM. The compiler that has been used is GCC 4.4.5. Different experiments have been applied to the F1-F11 mono-objective benchmark problems proposed in [33]. They are a set of scalable continuous optimisation problems, which combine different properties regarding the modality, the separability, and the ease of optimisation dimension by dimension, i.e. whether the fitness can be optimised by independently adjusting each variable or not.

Since experiments have involved the use of stochastic algorithms, each execution has been repeated 30 times. Comparisons have been performed following the statistical analysis applied in [3]. First, a *Shapiro-Wilk test* is performed in order to check whether the values of the results follow a normal (Gaussian) distribution or not. If so, the *Levene test* checks for the homogeneity of the variances. If samples have equal variance, an ANOVA *test* is done. Otherwise, a *Welch test* is performed. For non-Gaussian distributions, the non-parametric *Kruskal-Wallis* test is used to compare the medians of the algorithms. A confidence level of 95% has been considered.

11.6.1   Performance of Multiobjectivisation

The first analysis has been devoted to measure the performance of multiobjectivisation. A comparison between mono-objective and multiobjectivised approaches has been performed. The mono-objective and multi-objective EAs described in Section 11.2 have been used. In the case of the mono-objective EAs, they have been executed with three different survivor selection mechanisms: SS-EA, GEN-EA, and RW-EA. In the case of the NSGA-II, it has been executed with the multiobjectivised versions of the benchmark problems. The multiobjectivisation has been performed with three different mechanisms: DCN, ADI, and DBI. The number of decision variables D of the optimisation problems has been fixed to 50 and 500. Each algorithm has been executed with a population size N of 5, 10, and 20 individuals. The stopping criterion has been fixed to a total number of $5000 \cdot D$ evaluations.

Table 11.1 shows, for each population size, the best mono-objective and multiobjectivised approaches considering $D = 50$. Comparisons have been made in terms of the median of the fitness achieved at the end of each execution. The same information is shown in Table 11.3 considering $D = 500$. In both cases, the superiority of the GEN-EA and the DCN strategies is clear. However, for some problems, they are not the best-behaved schemes.

Mono-objective and multiobjectivised techniques have been compared in terms of the achieved fitness. The median of the error achieved by the best mono-objective and multiobjectivised approaches, for each population size, is given in Table 11.2, considering $D = 50$. The error has been defined as the difference between the achieved fitness and the optimal fitness. It has been calculated considering an accuracy equal to $1 \cdot 10^{-6}$. Moreover, a statistical analysis between the best mono-objective approach and the best multiobjectivised one has been carried out for each population size. For cases in which differences have been statistically significant, data of the best of both algorithms is shown in bold. Table 11.4 shows the same information for the case of $D = 500$. For both values of D, algorithms with lower population sizes have obtained lower errors. In addition, the superiority of the multiobjectivised approaches has been clearly demonstrated. However, in some cases the best mono-objective approach has obtained statistically better results than the

Table 11.1 Best mono-objective and multi-objective approaches - $D = 50$

	F1	F2	F3	F4	F5	F6
Mono_5	GEN-EA	GEN-EA	GEN-EA	GEN-EA	GEN-EA	GEN-EA
Mono_10	GEN-EA	GEN-EA	GEN-EA	GEN-EA	GEN-EA	GEN-EA
Mono_20	GEN-EA	GEN-EA	GEN-EA	RW-EA	GEN-EA	GEN-EA
Multi_5	DCN	DCN	DCN	DCN	DCN	DCN
Multi_10	DCN	DCN	DCN	DCN	DCN	DCN
Multi_20	DCN	DCN	DCN	DCN	DCN	DCN

	F7	F8	F9	F10	F11
Mono_5	GEN-EA	GEN-EA	GEN-EA	GEN-EA	GEN-EA
Mono_10	GEN-EA	GEN-EA	GEN-EA	GEN-EA	GEN-EA
Mono_20	GEN-EA	GEN-EA	GEN-EA	GEN-EA	GEN-EA
Multi_5	DCN	ADI	DCN	DCN	DCN
Multi_10	DCN	DCN	DCN	DCN	DCN
Multi_20	DCN	DCN	DCN	DCN	DCN

Table 11.2 Median of the error for the best approaches - $D = 50$

	F1	F2	F3	F4	F5	F6
Mono_5	$< 1 \cdot 10^{-6}$	$1.76 \cdot 10^{0}$	$1.55 \cdot 10^{2}$	$< 1 \cdot 10^{-6}$	$2.25 \cdot 10^{-2}$	$3.00 \cdot 10^{-3}$
Multi_5	$< 1 \cdot 10^{-6}$	$\mathbf{5.67 \cdot 10^{-1}}$	$1.55 \cdot 10^{2}$	$< 1 \cdot 10^{-6}$	$\mathbf{1.10 \cdot 10^{-2}}$	$\mathbf{1.00 \cdot 10^{-3}}$
Mono_10	$< 1 \cdot 10^{-6}$	$1.63 \cdot 10^{0}$	$3.44 \cdot 10^{2}$	$< 1 \cdot 10^{-6}$	$1.50 \cdot 10^{-2}$	$\mathbf{2.00 \cdot 10^{-3}}$
Multi_10	$5.00 \cdot 10^{-4}$	$\mathbf{1.05 \cdot 10^{0}}$	$\mathbf{1.52 \cdot 10^{2}}$	$1.00 \cdot 10^{-3}$	$1.00 \cdot 10^{-2}$	$4.00 \cdot 10^{-3}$
Mono_20	$6.00 \cdot 10^{-3}$	$2.01 \cdot 10^{0}$	$3.38 \cdot 10^{2}$	$8.60 \cdot 10^{-2}$	$2.10 \cdot 10^{-2}$	$1.60 \cdot 10^{-2}$
Multi_20	$\mathbf{3.00 \cdot 10^{-3}}$	$\mathbf{1.64 \cdot 10^{0}}$	$1.97 \cdot 10^{2}$	$\mathbf{1.00 \cdot 10^{-3}}$	$1.90 \cdot 10^{-2}$	$\mathbf{9.00 \cdot 10^{-3}}$

	F7	F8	F9	F10	F11
Mono_5	$4.57 \cdot 10^{-3}$	$1.44 \cdot 10^{9}$	$7.24 \cdot 10^{0}$	$1.84 \cdot 10^{-4}$	$6.80 \cdot 10^{0}$
Multi_5	$\mathbf{1.37 \cdot 10^{-3}}$	$\mathbf{4.50 \cdot 10^{8}}$	$\mathbf{4.65 \cdot 10^{0}}$	$\mathbf{3.11 \cdot 10^{-5}}$	$\mathbf{4.46 \cdot 10^{0}}$
Mono_10	$2.59 \cdot 10^{-3}$	$1.02 \cdot 10^{9}$	$7.77 \cdot 10^{0}$	$\mathbf{1.16 \cdot 10^{-4}}$	$7.05 \cdot 10^{0}$
Multi_10	$2.55 \cdot 10^{-3}$	$\mathbf{6.49 \cdot 10^{8}}$	$\mathbf{5.62 \cdot 10^{0}}$	$4.69 \cdot 10^{-4}$	$\mathbf{6.12 \cdot 10^{0}}$
Mono_20	$1.82 \cdot 10^{-2}$	$1.34 \cdot 10^{9}$	$1.79 \cdot 10^{1}$	$5.40 \cdot 10^{-3}$	$1.98 \cdot 10^{1}$
Multi_20	$\mathbf{3.70 \cdot 10^{-3}}$	$8.77 \cdot 10^{8}$	$\mathbf{6.92 \cdot 10^{0}}$	$\mathbf{1.47 \cdot 10^{-3}}$	$\mathbf{8.00 \cdot 10^{0}}$

best multiobjectivised technique. From a total number of 66 statistical tests, in 41 of them the best multiobjectivised approach has been statistically superior. The best mono-objective algorithm has been statistically superior in 9 tests. Finally, 16 statistical tests have shown no significant differences between both strategies. In addition, it is important to know which kinds of problems are more adequate to be solved by multiobjectivised approaches. Multiobjectivisation has provided more benefits when it has been applied to a unimodal problem. In fact, for a fixed population size, the best multiobjectivised approach has been statistically better than the best mono-objective algorithm in a 76.2% of the cases in which unimodal problems have been

Table 11.3 Best mono-objective and multi-objective approaches - $D = 500$

	F1	F2	F3	F4	F5	F6
Mono_5	GEN-EA	RW-EA	GEN-EA	GEN-EA	GEN-EA	GEN-EA
Mono_10	GEN-EA	RW-EA	GEN-EA	GEN-EA	GEN-EA	GEN-EA
Mono_20	GEN-EA	RW-EA	GEN-EA	GEN-EA	SS-EA	GEN-EA
Multi_5	DCN	ADI	DCN	DCN	DCN	DCN
Multi_10	DCN	DCN	DCN	DCN	DCN	DCN
Multi_20	DCN	DCN	DCN	DCN	DCN	DCN

	F7	F8	F9	F10	F11
Mono_5	GEN-EA	GEN-EA	GEN-EA	GEN-EA	GEN-EA
Mono_10	GEN-EA	GEN-EA	GEN-EA	GEN-EA	GEN-EA
Mono_20	GEN-EA	GEN-EA	GEN-EA	GEN-EA	GEN-EA
Multi_5	DCN	DCN	DCN	DCN	DCN
Multi_10	DCN	DCN	DCN	DCN	DCN
Multi_20	DCN	DCN	DCN	DCN	DCN

Table 11.4 Median of the error for the best approaches - $D = 500$

	F1	F2	F3	F4	F5	F6
Mono_5	$3.00 \cdot 10^{-3}$	$1.67 \cdot 10^{1}$	$1.32 \cdot 10^{3}$	$3.00 \cdot 10^{-3}$	$8.00 \cdot 10^{-3}$	$3.00 \cdot 10^{-3}$
Multi_5	$< 1 \cdot 10^{-6}$	$\mathbf{1.14 \cdot 10^{1}}$	$1.26 \cdot 10^{3}$	$3.00 \cdot 10^{-3}$	$< 1 \cdot 10^{-6}$	$\mathbf{1.00 \cdot 10^{-3}}$
Mono_10	$\mathbf{2.00 \cdot 10^{-3}}$	$1.41 \cdot 10^{1}$	$1.47 \cdot 10^{3}$	$3.95 \cdot 10^{-2}$	$< 1 \cdot 10^{-6}$	$\mathbf{3.00 \cdot 10^{-3}}$
Multi_10	$5.00 \cdot 10^{-3}$	$1.81 \cdot 10^{1}$	$1.48 \cdot 10^{3}$	$1.00 \cdot 10^{0}$	$1.00 \cdot 10^{-3}$	$4.00 \cdot 10^{-3}$
Mono_20	$6.50 \cdot 10^{-2}$	$\mathbf{1.54 \cdot 10^{1}}$	$1.78 \cdot 10^{3}$	$9.28 \cdot 10^{-1}$	$8.00 \cdot 10^{-3}$	$1.70 \cdot 10^{-2}$
Multi_20	$\mathbf{3.70 \cdot 10^{-2}}$	$2.23 \cdot 10^{1}$	$1.87 \cdot 10^{3}$	$\mathbf{8.95 \cdot 10^{-2}}$	$\mathbf{4.50 \cdot 10^{-3}}$	$\mathbf{1.30 \cdot 10^{-2}}$

	F7	F8	F9	F10	F11
Mono_5	$4.02 \cdot 10^{-2}$	$1.93 \cdot 10^{11}$	$7.53 \cdot 10^{1}$	$1.84 \cdot 10^{-3}$	$7.40 \cdot 10^{1}$
Multi_5	$\mathbf{1.37 \cdot 10^{-2}}$	$\mathbf{1.41 \cdot 10^{11}}$	$4.49 \cdot 10^{1}$	$\mathbf{3.46 \cdot 10^{-4}}$	$4.44 \cdot 10^{1}$
Mono_10	$\mathbf{2.57 \cdot 10^{-2}}$	$1.83 \cdot 10^{11}$	$8.42 \cdot 10^{1}$	$\mathbf{1.32 \cdot 10^{-3}}$	$8.24 \cdot 10^{1}$
Multi_10	$2.99 \cdot 10^{-2}$	$\mathbf{1.58 \cdot 10^{11}}$	$6.30 \cdot 10^{1}$	$4.60 \cdot 10^{-3}$	$\mathbf{6.19 \cdot 10^{1}}$
Mono_20	$1.77 \cdot 10^{-1}$	$2.04 \cdot 10^{11}$	$2.08 \cdot 10^{2}$	$6.24 \cdot 10^{-2}$	$2.08 \cdot 10^{2}$
Multi_20	$\mathbf{3.83 \cdot 10^{-2}}$	$\mathbf{1.73 \cdot 10^{11}}$	$\mathbf{7.78 \cdot 10^{1}}$	$2.76 \cdot 10^{-2}$	$\mathbf{7.63 \cdot 10^{1}}$

considered. This ratio has decreased to a 37.5% when multimodal problems have been taken into account.

The previous analysis has demonstrated the validity of multiobjectivisation in terms of the achieved quality level. However, it is important to quantify the improvement that can be achieved by using multiobjectivisation, in terms of the invested number of evaluations. Some of the notions of Run-Length Distributions (RLD) [27] have been used. RLD show the relationship between success ratios and time. Success ratio is defined as the probability of achieving a certain quality level by a given metaheuristic configuration. The quality level has been fixed so that all executions

Table 11.5 Percentage of saved evaluations by the best multiobjectivisation

	F1	F2	F3	F4	F5	F6	F7	F8	F9	F10	F11
$D = 50$	26.1%	45.9%	40.0%	-28.9%	33.3%	32.6%	29.2%	45.5%	31.0%	23.5%	33.3%
$D = 500$	26.9%	-7.7%	25.0%	-21.2%	29.4%	28.0%	25.9%	38.7%	29.4%	35.3%	29.4%

Table 11.6 Statistical comparison among different threshold values

	F1	F2	F3	F4	F5	F6	F7	F8	F9	F10	F11
Best Threshold	0	0.8	0.2, 0.4	0.4, 0.6	0	0, 0.2	0.2	0, 0.2	0.8	0	0.8

have been able to achieve it. The number of evaluations required to achieve a 50% of success ratio have been calculated for the best-behaved mono-objective and multiobjectivised approaches, regardless of the population size. Table 11.5 shows the percentage of saved evaluations by the best multiobjectivised approach in relation to the best mono-objective one. A negative value means that the mono-objective algorithm has converged faster than the corresponding multiobjectivised strategy to the fixed quality level. Once again, the superiority of multiobjectivisation can be noted. The best-behaved multiobjectivised approach has provided benefits in 19 cases from a total number of 22. Moreover, in the majority of the cases, the percentage of saved evaluations has been large. In fact, in some cases, about a 45% of evaluations has been saved.

11.6.2   On the Usage of Multiobjectivisation with Parameters

The second experiment has focused on analysing multiobjectivisation with parameters. Specifically, the considered benchmark problems have been multiobjectivised with the DCN-THR strategy. The aim of the analysis has been to discover the relationship between the values of the threshold ratio th, and the quality of the obtained results. The multiobjectivised versions of the benchmark problems have been solved with the NSGA-II. The parameterisation has been as follows. The number of decision variables D has been fixed to 500. Since lower errors have been obtained with lower population sizes in the previous experiments, the population size has been fixed to 5 individuals. The following values have been used for the threshold ratio th: 0, 0.2, 0.4, 0.6, and 0.8. Finally, the stopping criterion has been fixed to a total number of $2 \cdot 10^7$ evaluations.

Table 11.6 shows, for each benchmark problem, the threshold ratio of the configuration which has obtained the best median of the fitness in $1.25 \cdot 10^6$ evaluations. It also shows the threshold ratios of those configurations which have not been statistically different than the best one. There has been no value which has been able to be among the best ones for every problem. Thus, in order to decide the proper value to use, *a priori* information of the problem to solve is required.

Table 11.7 Number of evaluations to achieve a fixed quality level

	0	0.2	0.4	0.6	0.8
F1	$\mathbf{1.25 \cdot 10^6}$	$2.2 \cdot 10^6$	$3.05 \cdot 10^6$	$4.7 \cdot 10^6$	$6.05 \cdot 10^6$
F2	$1.35 \cdot 10^6$	$1.35 \cdot 10^6$	$1.4 \cdot 10^6$	$1.4 \cdot 10^6$	$\mathbf{1.25 \cdot 10^6}$
F3	$1.75 \cdot 10^6$	$\mathbf{1.25 \cdot 10^6}$	$1.35 \cdot 10^6$	$1.7 \cdot 10^6$	$1.8 \cdot 10^6$
F4	$2.15 \cdot 10^6$	$1.45 \cdot 10^6$	$\mathbf{1.25 \cdot 10^6}$	$\mathbf{1.25 \cdot 10^6}$	$1.5 \cdot 10^6$
F5	$\mathbf{1.25 \cdot 10^6}$	$4 \cdot 10^6$	$5.15 \cdot 10^6$	$6.9 \cdot 10^6$	$1.74 \cdot 10^7$
F6	$1.25 \cdot 10^6$	$\mathbf{1.2 \cdot 10^6}$	$1.95 \cdot 10^6$	$2.5 \cdot 10^6$	$3.75 \cdot 10^6$
F7	$2.3 \cdot 10^6$	$\mathbf{1.25 \cdot 10^6}$	$2 \cdot 10^6$	$3.85 \cdot 10^6$	$8.6 \cdot 10^6$
F8	$\mathbf{1.25 \cdot 10^6}$	$1.3 \cdot 10^6$	$1.45 \cdot 10^6$	$1.4 \cdot 10^6$	$1.45 \cdot 10^6$
F9	$1.9 \cdot 10^6$	$1.85 \cdot 10^6$	$1.75 \cdot 10^6$	$1.65 \cdot 10^6$	$\mathbf{1.25 \cdot 10^6}$
F10	$\mathbf{1.25 \cdot 10^6}$	$1.8 \cdot 10^6$	$2.4 \cdot 10^6$	$3.25 \cdot 10^6$	$5 \cdot 10^6$
F11	$1.85 \cdot 10^6$	$1.85 \cdot 10^6$	$1.85 \cdot 10^6$	$1.6 \cdot 10^6$	$\mathbf{1.25 \cdot 10^6}$

In order to measure the impact over the performance, RLD have been used. The quality level has been fixed as the median of the fitness achieved by the best configuration in $1.25 \cdot 10^6$ evaluations. Table 11.7 shows, for each threshold ratio, the number of evaluations required to obtain a success ratio of 50%, when the aforementioned quality level has been taken into account. Considering the number of evaluations required by suboptimal approaches, the importance of correctly fixing the parameter is clear. For instance, considering the problem F5, the number of evaluations required by the best configuration approximately represents a 7.18% of the evaluations required by the worst one.

Another open research question is whether the proper value of *th* depends on the optimisation stage or not. With the aim of answering this question the following experiment has been performed. First, four different optimisation stages have been defined. To define such stages the median of the fitness values achieved by the best configuration in $2.5 \cdot 10^5$, $5 \cdot 10^5$, $7.5 \cdot 10^5$, and $1 \cdot 10^6$ evaluations have been considered. Then, for each calculated value, 30 individuals with a quality similar to such a value have been generated. In order to generate the 30 individuals, the best configuration has been executed 30 times. For each one of these 30 executions, the generated individual has been the first one that has achieved an original objective value lower than the fixed value for the corresponding stage.

After it, the NSGA-II has been tested with different values of the threshold ratio for each considered stage. Considering each one of the stages, each one of the executions of the NSGA-II has included one of the 30 individuals generated by the aforementioned method in its initial population. The remaining individuals of the population have been randomly generated. By this way, the performance of the NSGA-II with different threshold values can be tested when it starts from different quality levels. The stopping criterion has been fixed to $5 \cdot 10^5$ evaluations. The different configurations of the NSGA-II have been compared in terms of the achieved fitness. Tables 11.8, 11.9, and 11.10 show the comparison among the considered threshold values for the problems F4, F5, and F11, respectively. For each stage,

Table 11.8 Statistical comparison among different threshold values by stages - F4

	Stage 0					Stage 1				
	0	0.2	0.4	0.6	0.8	0	0.2	0.4	0.6	0.8
0	↔	↔	↔	↔	↓	↔	↔	↔	↓	↓
0.2	↔	↔	↔	↔	↓	↔	↔	↔	↓	↓
0.4	↔	↔	↔	↔	↓	↔	↔	↔	↔	↔
0.6	↔	↔	↔	↔	↔	↑	↑	↔	↔	↔
0.8	↑	↑	↑	↔	↔	↑	↑	↔	↔	↔

	Stage 2					Stage 3				
	0	0.2	0.4	0.6	0.8	0	0.2	0.4	0.6	0.8
0	↔	↔	↔	↔	↔	↔	↑	↑	↑	↑
0.2	↔	↔	↔	↔	↔	↓	↔	↔	↔	↔
0.4	↔	↔	↔	↔	↔	↓	↔	↔	↔	↔
0.6	↔	↔	↔	↔	↔	↓	↔	↔	↔	↔
0.8	↔	↔	↔	↔	↔	↓	↔	↔	↔	↔

Table 11.9 Statistical comparison among different threshold values by stages - F5

	Stage 0					Stage 1				
	0	0.2	0.4	0.6	0.8	0	0.2	0.4	0.6	0.8
0	↔	↔	↔	↔	↔	↔	↔	↑	↑	↑
0.2	↔	↔	↔	↑	↑	↔	↔	↔	↑	↑
0.4	↔	↔	↔	↔	↑	↓	↔	↔	↔	↔
0.6	↔	↓	↔	↔	↔	↓	↓	↔	↔	↔
0.8	↔	↓	↓	↔	↔	↓	↓	↔	↔	↔

	Stage 2					Stage 3				
	0	0.2	0.4	0.6	0.8	0	0.2	0.4	0.6	0.8
0	↔	↑	↑	↑	↑	↔	↑	↑	↑	↑
0.2	↓	↔	↔	↔	↔	↓	↔	↔	↔	↔
0.4	↓	↔	↔	↔	↔	↓	↔	↔	↔	↔
0.6	↓	↔	↔	↔	↔	↓	↔	↔	↔	↔
0.8	↓	↔	↔	↔	↔	↓	↔	↔	↔	↔

they show if the row configuration is statistically better (↑), not different (↔), or worse (↓), than the corresponding column configuration. For each problem, the statistical tests show that differences among configurations depend on the considered optimisation stage. For instance, taking into account the stage number 0 of the problem F4, the configuration that uses $th = 0.8$ is better than the one which uses $th = 0$. However, in the stage number 3 the configuration with $th = 0$ performs better than the configuration with $th = 0.8$. In the case of the problem F5, $th = 0$ seems to be the most appropriate value for the whole run because there is not a better value in any of the analysed stages. For the same reasons, in the problem F11 the value $th = 0.8$ seems to be the most appropriate.

Table 11.10 Statistical comparison among different threshold values by stages - F11

	Stage 0					Stage 1				
	0	0.2	0.4	0.6	0.8	0	0.2	0.4	0.6	0.8
0	↔	↔	↔	↔	↔	↔	↔	↔	↔	↔
0.2	↔	↔	↔	↔	↔	↔	↔	↔	↔	↔
0.4	↔	↔	↔	↔	↔	↔	↔	↔	↔	↔
0.6	↔	↔	↔	↔	↔	↔	↔	↔	↔	↔
0.8	↔	↔	↔	↔	↔	↔	↔	↔	↔	↔

	Stage 2					Stage 3				
	0	0.2	0.4	0.6	0.8	0	0.2	0.4	0.6	0.8
0	↔	↔	↔	↔	↔	↔	↔	↔	↔	↓
0.2	↔	↔	↔	↔	↓	↔	↔	↔	↔	↓
0.4	↔	↔	↔	↔	↔	↔	↔	↔	↔	↓
0.6	↔	↔	↔	↔	↔	↔	↔	↔	↔	↓
0.8	↔	↑	↔	↔	↔	↑	↑	↑	↑	↔

Table 11.11 Number of evaluations to achieve a fixed quality level

	F1	F2	F3	F4	F5	F6
HH_imp	$1.45 \cdot 10^6$	$1.35 \cdot 10^6$	$1.3 \cdot 10^6$	$1.15 \cdot 10^6$	$1.4 \cdot 10^6$	$1.25 \cdot 10^6$

	F7	F8	F9	F10	F11
HH_imp	$1.5 \cdot 10^6$	$1.4 \cdot 10^6$	$1.55 \cdot 10^6$	$1.2 \cdot 10^6$	$1.55 \cdot 10^6$

11.6.3   Rising the Robustness of Multiobjectivisation

An experiment focused on analysing the benefits of mixing hyperheuristics and multiobjectivisation with parameters has also been performed. The hyperheuristic (HH_imp) has been applied to each one of the benchmark problems, considering a global adaptation level, i.e. using the parameterisation $k = \infty$. The value of β has been fixed in a way that 10% of the decisions performed by the hyperheuristic follows a uniform distribution, i.e. $\beta \cdot n_h = 0.1$. The low-level configurations have been the five configurations ($n_h = 5$) defined in the previous experiment, i.e. the NSGA-II with DCN-THR considering the following threshold ratios: 0, 0.2, 0.4, 0.6, and 0.8. The local stopping criterion has been fixed to $1 \cdot 10^4$ evaluations. The global stopping criterion has been fixed to $2.5 \cdot 10^6$ evaluations.

Table 11.11 shows the number of evaluations required by HH_imp to obtain a 50% of success ratio. It considers the same quality level than the one fixed in Table 11.7. The benefits of incorporating the hyperheuristic are clear. In fact, it is the only one model that has been able to achieve the fixed quality level in less than $1.55 \cdot 10^6$ evaluations for all the considered problems. Moreover, in two of the problems (F4 and F10) it has been the model that has required fewer evaluations to obtain the considered quality level. This means that the hyperheuristic has been able to grant

Table 11.12 Resources assignment in problems without variability in the threshold

	0	0.2	0.4	0.6	0.8
F5	56.98%	15.77%	10.82%	9.79%	6.64%
F11	15.96%	16.05%	13.82%	20.89%	33.28%

Table 11.13 Statistical comparison among hyperheuristic and different threshold values

	F1	F2	F3	F4	F5	F6	F7	F8	F9	F10	F11
$\uparrow$	4	1	2	5	4	3	4	0	4	5	4
$\leftrightarrow$	0	3	3	0	0	2	0	3	0	0	0
$\downarrow$	1	1	0	0	1	0	1	2	1	0	1

more resources to the fittest low-level strategy in the different optimisation stages. In addition, for the remaining problems, HH_imp has been always among the best-behaved approaches, and the impact over the performance when compared with the best configuration has not been large. Considering that the hyperheuristic obtains the solutions in a single run, while for the configurations with a fixed threshold value the parameter tuning must be performed, the benefits are very clear.

The hyperheuristic requires some time to realise which configuration is the fittest approach for each stage of the optimisation process. In addition, the hyperheuristic ensures that each considered low-level configuration will be executed with a probability higher or equal than $\beta \cdot n_h$. Thus, at the end of the run, every low-level configuration will have been executed several times. Consequently, in cases where a given threshold value is better than the remaining ones in every stage, the hyperheuristic has not been able to converge to high-quality values as fast as the best approach. However, it is interesting to know the resources which have been granted to the best approach for this kind of problems. Table 11.12 shows the percentage of resources granted to each low-level configuration for the problems F5 and F11. Such problems have been selected because for both cases there is a configuration that behaves properly in every analysed optimisation stage. In the case of F5, the best-behaved configuration has used $th = 0$. More than a 50% of the resources have been granted to such a configuration. In the case of F11 the best-behaved configuration has used $th = 0.8$. The hyperheuristic has granted more resources to such a configuration than to any other. However, the difference among the granted resources has not been as large as in F5. The reason is that the statistical differences among the configurations have not been so clear (Table 11.10). Therefore, the hyperheuristic has granted more resources to other configurations.

Finally, in order to better analyse the benefits of HH_imp, a comparison in terms of the achieved fitness has also been performed. Table 11.13 compares HH_imp with its corresponding five low-level configurations. For each problem it shows the number of low-level configurations that are statistically worse ($\uparrow$), not different ($\leftrightarrow$) or better ($\downarrow$) than HH_imp. In four problems none of the low-level configurations has been statistically better than HH_imp. In six problems only one low-level configuration has been statistically better than HH_imp. Finally, only in F8 two configurations have been statistically better than HH_imp.

11.6.4   Analysing the Performance of Hyperheuristics with a Large Number of Variables

The last experiment has focused on analysing the behaviour of the aforementioned approaches when a larger number of variables are used. In this case, the F5 benchmark problem has been multiobjectivised by the usage of the DCN-THR objective function. The same threshold values than in the previous experiments have been considered. The NSGA-II configurations used in the previous experiments have been executed. At the same time, the HH_imp model has also been applied, considering the NSGA-II configurations as low-level approaches. Experiments have been performed considering the following number of variables $D = 1000, 2000, .., 5000$. In every case, the global stopping criterion has been fixed to $2.5 \cdot 10^6$ evaluations. In the case of the hyperheuristic, the local stopping criterion has been fixed to $1 \cdot 10^4$ evaluations. For the analyses, the ideas of the RLD have been used. The quality level has been fixed so that all the NSGA-II configurations have been able to achieve a 50% of success ratio. Thus, the median of the fitness achieved by the worst NSGA-II configuration has been used as the quality level for each considered value of D.

Table 11.14 shows, for each one of the models, the number of evaluations required to achieve a 50% of success ratio. Data is shown for each value of D. It can be observed that, independently of the value of D, the behaviour of the HH_imp model has been adequate. In fact, it has been among the best models and it has got close to the best one, for each value of D. Also, it is worthy to mention that the most suitable threshold value depends on the considered number of variables. For instance, considering $D = 5000$, the worst-behaved configuration has applied a threshold value equal to 0. However, this value has been the best-behaved one, when a number of variables equal to 500 has been considered (Table 11.6).

Finally, it is also important to remark that when a higher number of variables has been considered, the number of evaluations performed by the HH_imp model has become similar to the number of evaluations performed by the rest of the models. The main reason is that when a higher number of variables is taken into account, a higher number of evaluations is required in order to obtain statistical differences among the models. Table 11.15 shows, for each value of D, the minimum number of evaluations for which one of the considered models has been statistically better than the rest of the models. It can be observed that, for a larger number of variables, the number of evaluations is clearly higher. By this way, in order to profit from the HH_imp model when a higher value of D is used, a higher number of evaluations must be performed.

Table 11.14 Evaluations to achieve a fixed quality level by changing the number of variables (F5)

	$D = 1000$	$D = 2000$	$D = 3000$	$D = 4000$	$D = 5000$
HH_imp	$1.15 \cdot 10^6$	$1.6 \cdot 10^6$	$1.9 \cdot 10^6$	$2.1 \cdot 10^6$	$2.15 \cdot 10^6$
$th = 0$	$1.2 \cdot 10^6$	$1.9 \cdot 10^6$	$2.35 \cdot 10^6$	$2.5 \cdot 10^6$	$2.5 \cdot 10^6$
$th = 0.2$	$9.5 \cdot 10^5$	$1.35 \cdot 10^6$	$1.75 \cdot 10^6$	$1.95 \cdot 10^6$	$2.1 \cdot 10^6$
$th = 0.4$	$1.35 \cdot 10^6$	$1.6 \cdot 10^6$	$1.8 \cdot 10^6$	$2 \cdot 10^6$	$2.1 \cdot 10^6$
$th = 0.6$	$1.55 \cdot 10^6$	$1.9 \cdot 10^6$	$2.05 \cdot 10^6$	$2.15 \cdot 10^6$	$2.15 \cdot 10^6$
$th = 0.8$	$2.5 \cdot 10^6$	$2.5 \cdot 10^6$	$2.5 \cdot 10^6$	$2.4 \cdot 10^6$	$2.4 \cdot 10^6$

Table 11.15 Number of evaluations to obtain statistical differences among configurations (F5)

	$D = 1000$	$D = 2000$	$D = 3000$	$D = 4000$	$D = 5000$
Evaluations	$2 \cdot 10^5$	$3 \cdot 10^5$	$5.5 \cdot 10^5$	$6.5 \cdot 10^5$	$1.4 \cdot 10^6$

11.7   Conclusions

Metaheuristics are a family of Approximation Techniques that have become popular to solve optimisation problems. Among them, EAs are one of the most popular techniques. They have shown great promise to calculate solutions to large and difficult optimisation problems. One of the main drawbacks of EAs is their parameter setting. There are several components and parameters which must be specified in order to completely define a particular configuration of an EA. Usually, the quality of the obtained solutions highly depends on such components and parameters. Therefore, selecting the appropriate components and parameters might be an arduous task. Another disadvantage is that in some cases, EAs may have a tendency to converge towards local optima.

In order to deal with stagnation, several mechanisms have been designed. Among them, multiobjectivisation tries to avoid stagnation by transforming a mono-objective problem into a multi-objective one. In order to deal with a multiobjectivised problem, a multi-objective technique must be applied. In this work, the NSGA-II has been used. Several ways of multiobjectivising a problem have been applied. A comparison among several mono-objective EAs and the NSGA-II has been performed in this Chapter. Comparisons have been carried out with a set of scalable benchmark problems. The experimental evaluation has demonstrated the validity of multiobjectivisation, both in terms of quality and saved resources. In most of the problems, multiobjectivised techniques have obtained higher quality solutions than the mono-objective algorithms. However, in some of them, the usage of multiobjectivisation has produced a negative effect.

More complex multiobjectivisation techniques have considered the usage of parameters. The addition of parameters has increased the quality of the obtained solutions for several kinds of problems. However, the parameter setting of such schemes can be even more time-consuming, since a larger amount of parameters must be set. In this work, a novel multiobjectivisation which requires the specification of a parameter (DCN-THR) has been proposed. It starts from the DCN alternative objective, but it incorporates the usage of a threshold ratio (th) which must be specified by the user. In this Chapter a deep analysis of the threshold ratio parameterisation has been performed. From this analysis the following conclusions have been drawn. First, the usage of multiobjectivisation with parameters has provided high-quality results faster than multiobjectivisation techniques without parameters. However, the usage of non proper values for th has slowed down the convergence to high-quality solutions in some problems. In addition, the adequate configuration depends on the problem or instance to solve, i.e. there is not a general value that fits to every problem or instance. Finally, the statistical tests have shown that the performance of different configurations depends on the optimisation process stage. This hinders the usage of optimisation schemes in which multiobjectivisation with parameters is used.

A novel model that tries to facilitate the application of multiobjectivisation with parameters, and at the same time, tries to increase its robustness, has been proposed. This new model is based on the usage of parameter control. In parameter control the aim is to design a control strategy that automatically adapts the parameter values along the run. By this way, optimisation strategies start the execution with a set of initial values, and modify those values during the run, depending on the behaviour of the approach in the different stages of the optimisation process. Hyperheuristics are a general method that can be used to deal with the parameter control. In this work, the hyperheuristic HH_imp has been used to control the parameterisation of the DCN-THR approach. Computational results have clearly demonstrated the advantages of the new model. The robustness of multiobjectivisation with parameters has been increased by the usage of the hyperheuristic HH_imp. In some problems it has been the fastest approach in obtaining a prefixed quality level. This means that the hyperheuristic has been able to grant more resources to the fittest low-level strategy in the different optimisation stages. Moreover, for the remaining problems, HH_imp has been always among the best-behaved approaches. The analysis of the resources assignment has shown that HH_imp has been able to grant more resources to the most promising configurations. Finally, the analysis of the relationship between the number of considered variables and the performance of the different proposals has revealed that even for very large values of D, the behaviour of the HH_imp model has been adequate. In addition, it has shown that the number of evaluations required to profit from the HH_imp model, when a higher value of D is used, must be larger.

Acknowledgements. This work has been supported by the EC (FEDER) and the Spanish Ministry of Science and Innovation as part of the 'Plan Nacional de I+D+i', with contract number TIN2011-25448. The work of Carlos Segura has been funded by grant FPU-AP2008-03213. The work of Eduardo Segredo has been funded by grant FPU-AP2009-0457 The work has also been funded by the HPC-EUROPA2 project (project number: 228398) with the support of the European Commission - Capacities Area - Research Infrastructures.

References

1. Abbass, H.A., Deb, K.: Searching under Multi-evolutionary Pressures. In: Fonseca, C.M., Fleming, P.J., Zitzler, E., Deb, K., Thiele, L. (eds.) EMO 2003. LNCS, vol. 2632, pp. 391–404. Springer, Heidelberg (2003)
2. Alba, E., Cervantes, A., Gómez, J.A., Isasi, P., Jaraíz, M.D., León, C., Luque, C., Luna, F., Miranda, G., Nebro, A.J., Pérez, R., Segura, C.: Metaheuristic Approaches for Optimal Broadcasting Design in Metropolitan MANETs. In: Moreno Díaz, R., Pichler, F., Quesada Arencibia, A. (eds.) EUROCAST 2007. LNCS, vol. 4739, pp. 755–763. Springer, Heidelberg (2007)
3. Alba, E., Dorronsoro, B.: Cellular Genetic Algorithms. Springer (2008)
4. Baeck, T., Fogel, D.B., Michalewicz, Z. (eds.): Advanced Algorithms and Operations (Evolutionary Computation), 1st edn. Taylor & Francis (November 2000)
5. Brockhoff, D., Friedrich, T., Hebbinghaus, N., Klein, C., Neumann, F., Zitzler, E.: Do Additional Objectives Make a Problem Harder? In: Proceedings of the 9th Annual Conference on Genetic and Evolutionary Computation, GECCO 2007, pp. 765–772. ACM, New York (2007)
6. Bui, L.T., Abbass, H.A., Branke, J.: Multiobjective optimization for dynamic environments. In: The 2005 IEEE Congress on Evolutionary Computation, vol. 3, pp. 2349–2356 (2005)
7. Burke, E.K., Kendall, G., Newall, J., Hart, E., Ross, P., Schulenburg, S.: Handbook of Meta-heuristics. Hyper-heuristics: An Emerging Direction in Modern Search Technology. Kluwer (2003)
8. Burke, E.K., Kendall, G., Soubeiga, E.: A Tabu-Search Hyperheuristic for Timetabling and Rostering. Journal of Heuristics 9(6), 451–470 (2003)
9. Burke, E.K., McCollum, B., Meisels, A., Petrovic, S., Qu, R.: A graph-based hyper-heuristic for educational timetabling problems. European Journal of Operational Research 176(1), 177–192 (2007)
10. Burke, E.K., Kendall, G., Landa Silva, J.D., O'Brien, R., Soubeiga, E.: An Ant Algorithm Hyperheuristic for the Project Presentation Scheduling Problem. In: Proceedings of the 2005 IEEE Congress on Evolutionary Computation (CEC 2005), Edinburgh, Scotland, September 2-5, vol. 3, pp. 2263–2270 (2005)
11. Caamaño, P., Prieto, A., Becerra, J.A., Bellas, F., Duro, R.J.: Real-Valued Multimodal Fitness Landscape Characterization for Evolution. In: Wong, K.W., Mendis, B.S.U., Bouzerdoum, A. (eds.) ICONIP 2010, Part I. LNCS, vol. 6443, pp. 567–574. Springer, Heidelberg (2010)
12. Chen, P.-C., Kendall, G., Vanden Berghe, G.: An Ant Based Hyper-heuristic for the Travelling Tournament Problem. In: Proceedings of IEEE Symposium of Computational Intelligence in Scheduling (CISched 2007), Honolulu, Hawaii, pp. 19–26 (April 2007)
13. Ciftcioglu, Ö., Bittermann, M.S.: Solution diversity in multi-objective optimization: A study in virtual reality. In: IEEE Congress on Evolutionary Computation, pp. 1019–1026. IEEE (2008)
14. Cowling, P., Kendall, G., Han, L.: An investigation of a hyperheuristic genetic algorithm applied to a trainer scheduling problem. In: Proceedings of the 2002 IEEE Congress on Evolutionary Computation (CEC 2002), Honolulu, Hawaii, pp. 1185–1190. IEEE Computer Society (2002)
15. Cowling, P., Kendall, G., Soubeiga, E.: A parameter-free hyperheuristic for scheduling a sales summit. In: Proceedings of 4th Metahuristics International Conference (MIC 2001), Porto, Portugal, July 16-20, pp. 127–131 (2001)

16. Cowling, P.I., Kendall, G., Soubeiga, E.: Hyperheuristics: A Robust Optimisation Method Applied to Nurse Scheduling. In: Guervós, J.J.M., Adamidis, P.A., Beyer, H.-G., Fernández-Villacañas, J.-L., Schwefel, H.-P. (eds.) PPSN 2002. LNCS, vol. 2439, pp. 851–860. Springer, Heidelberg (2002)

17. De Jong, K.: Parameter Setting in EAs: a 30 Year Perspective. In: Lobo, F.G., Lima, C.F., Michalewicz, Z. (eds.) Parameter Setting in Evolutionary Algorithms, pp. 1–18. Springer (2007)

18. Deb, K., Agrawal, R.B.: Simulated binary crossover for continuous search space. Complex Systems 9, 115–148 (1995)

19. Deb, K., Pratap, A., Agarwal, S., Meyarivan, T.: A fast and elitist multiobjective genetic algorithm: NSGA-II. IEEE Transactions on Evolutionary Computation 6, 182–197 (2002)

20. Dorigo, M., Maniezzo, V., Colorni, A.: The ant system: Optimization by a colony of cooperating agents. IEEE Transactions on Systems, Man, and Cybernetics-Part B 26(1), 29–41 (1996)

21. Dowsland, K., Soubeiga, E., Burke, E.K.: A Simulated Annealing Hyper-heuristic for Determining Shipper Sizes. European Journal of Operational Research 179(3), 759–774 (2007)

22. Eiben, A.E., Hinterding, R., Michalewicz, Z.: Parameter control in evolutionary algorithms. IEEE Transactions on Evolutionary Computation 3(2), 124–141 (1999)

23. Eiben, A.E., Smith, J.E.: Introduction to Evolutionary Computing. Natural Computing Series. Springer (October 2008)

24. Glover, F.W., Kochenberger, G.A.: Handbook of Metaheuristics. International Series in Operations Research & Management Science. Springer (January 2003)

25. Gratch, J., Chien, S.: Learning search control knowledge for the deep space network scheduling problem. Technical report, Champaign, IL, USA (1993)

26. Handl, J., Lovell, S.C., Knowles, J.D.: Multiobjectivization by Decomposition of Scalar Cost Functions. In: Rudolph, G., Jansen, T., Lucas, S., Poloni, C., Beume, N. (eds.) PPSN 2008. LNCS, vol. 5199, pp. 31–40. Springer, Heidelberg (2008)

27. Hoos, H.H., Stützle, T.: Stochastic local search: foundations and applications. The Morgan Kaufmann Series in Artificial Intelligence. Morgan Kaufmann Publishers (2005)

28. Jain, A., Fogel, D.B.: Case studies in applying fitness distributions in evolutionary algorithms. ii. comparing the improvements from crossover and gaussian mutation on simple neural networks. In: Proc. of the 2000 IEEE Symposium on Combinations of Evolutionary Computation and Neural Networks, pp. 91–97 (2000)

29. Kendall, G., Cowling, P., Soubeiga, E.: Choice function and random hyperheuristics. In: Proceedings of the 4th Asia-Pacific Conference on Simulated Evolution and Learning (SEAL 2002), Singapore, pp. 667–671 (November 2002)

30. Knowles, J.D., Watson, R.A., Corne, D.W.: Reducing Local Optima in Single-Objective Problems by Multi-objectivization. In: Zitzler, E., Deb, K., Thiele, L., Coello Coello, C.A., Corne, D.W. (eds.) EMO 2001. LNCS, vol. 1993, pp. 269–283. Springer, Heidelberg (2001)

31. León, C., Miranda, G., Segura, C.: METCO: A Parallel Plugin-Based Framework for Multi-Objective Optimization. International Journal on Artificial Intelligence Tools 18(4), 569–588 (2009)

32. Lobo, F.G., Lima, C.F., Michalewicz, Z. (eds.): Parameter Setting in Evolutionary Algorithms. SCI, vol. 54. Springer (2007)

33. Lozano, M., Molina, D., Herrera, F.: Special Issue on Scalability of Evolutionary Algorithms and Other Metaheuristics for Large-Scale Continuous Optimization Problems. In: Soft Computing - A Fusion of Foundations, Methodologies and Applications, pp. 1–3 (2010)

34. Luna, F., Estébanez, C., León, C., Chaves-González, J.M., Alba, E., Aler, R., Segura, C., Vega-Rodríguez, M.A., Nebro, A.J., Valls, J.M., Miranda, G., Gómez-Pulido, J.A.: Metaheuristics for Solving a Real-World Frequency Assignment Problem in GSM Networks. In: Genetic and Evolutionary Computation Conference, Atlanta, U.S.A, pp. 1579–1586. ACM (July 2008)

35. Mendes, S.P., Molina, G., Vega-Rodríguez, M.A., Gómez-Pulido, J.A., Sáez, Y., Miranda, G., Segura, C., Alba, E., Isasi, P., León, C., Sánchez-Pérez, J.M.: Benchmarking a Wide Spectrum of Meta-Heuristic Techniques for the Radio Network Design Problem. IEEE Transactions on Evolutionary Computation, 1133–1150 (2009)

36. Mouret, J.-B.: Novelty-Based Multiobjectivization. In: Doncieux, S., Bredèche, N., Mouret, J.-B. (eds.) New Horizons in Evolutionary Robotics. SCI, vol. 341, pp. 139–154. Springer, Heidelberg (2011)

37. Ong, Y.-S., Lim, M.-H., Zhu, N., Wong, K.-W.: Classification of Adaptive Memetic Algorithms: A Comparative Study. IEEE Transactions on Systems, Man, and Cybernetics - Part B 36(1), 141–152 (2006)

38. Segura, C., Cervantes, A., Nebro, A.J., Jaraíz-Simón, M.D., Segredo, E., García, S., Luna, F., Gómez-Pulido, J.A., Miranda, G., Luque, C., Alba, E., Vega-Rodríguez, M.Á., León, C., Galván, I.M.: Optimizing the DFCN Broadcast Protocol with a Parallel Cooperative Strategy of Multi-Objective Evolutionary Algorithms. In: Ehrgott, M., Fonseca, C.M., Gandibleux, X., Hao, J.-K., Sevaux, M. (eds.) EMO 2009. LNCS, vol. 5467, pp. 305–319. Springer, Heidelberg (2009)

39. Segura, C., Miranda, G., León, C.: Parallel hyperheuristics for the frequency assignment problem. Memetic Computing, 1–17 (2010)

40. Segura, C., Segredo, E., González, Y., León, C.: Multiobjectivisation of the Antenna Positioning Problem. In: Abraham, A., Corchado, J.M., González, S.R., De Paz Santana, J.F. (eds.) International Symposium on Distributed Computing and Artificial Intelligence. AISC, vol. 91, pp. 319–327. Springer, Heidelberg (2011)

41. Segura, C., Segredo, E., León, C.: Parallel island-based multiobjectivised memetic algorithms for a 2D packing problem. In: Proceedings of the 13th Annual Conference on Genetic and Evolutionary Computation, GECCO 2011, pp. 1611–1618. ACM, New York (2011)

42. Smit, S.K., Eiben, A.E.: Comparing parameter tuning methods for evolutionary algorithms. In: Proceedings of the Eleventh Congress on Evolutionary Computation, CEC 2009, Piscataway, NJ, USA, pp. 399–406. IEEE Press (2009)

43. Terashima-Marín, H., Ross, P.: Evolution of Constraint Satisfaction Strategies in Examination Timetabling. In: Proceedings of the Genetic and Evolutionary Computation Conference (GECCO 1999), pp. 635–642. Morgan Kaufmann (1999)

44. Toffolo, A., Benini, E.: Genetic diversity as an objective in multi-objective evolutionary algorithms. Evolutionary Computation 11, 151–167 (2003)

45. Vinkó, T., Izzo, D.: Learning the best combination of solvers in a distributed global optimization environment. In: Proceedings of Advances in Global Optimization: Methods and Applications (AGO), Mykonos, Greece, pp. 13–17 (June 2007)

46. Zielinski, K., Laur, R.: Adaptive parameter setting for a multi-objective particle swarm optimization algorithm. In: The 2007 IEEE Congress on Evolutionary Computation, pp. 3019–3026 (September 2007)

Chapter 12
A Comparative Study of Heuristic Conversion Algorithms, Genetic Programming and Return Predictability on the German Market

Esther Mohr, Günter Schmidt, and Sebastian Jansen

Abstract. This paper evaluates the predictability of the heuristic conversion algorithms *Moving Average Crossover* and *Trading Range Breakout* in the German stock market. Hypothesis testing and a bootstrap procedure are used to test for predictive ability. Results show that the algorithms considered do not have predictive ability. Further, *Genetic Programming* is used to adapt the buying and selling rules of the investigated algorithms resulting in a new algorithm. Results show that a genetic programming approach does not lead to good new algorithms. We extend former works by using the *Sortino Ratio* as a measure of risk, and by applying competitive analysis.

12.1 Introduction

Heuristic conversion algorithms are developed to achieve a preferably high empirical-case performance, and are mainly based on data from technical analysis. Unfortunately, these algorithms are comprised of predefined buying and selling rules, and can not be adapted on runtime. For this reason genetic programming approaches are used to evolve algorithms (for an introduction see [4]). Genetic programming is basically a symbolic regression which is done by the use of evolutionary algorithms. Evolutionary algorithms are search methods that take their inspiration from natural selection and survival of the fittest in the biological world [3]. The term symbolic regression represents a process in which measured data is fitted by a suitable mathematical formula [45], and is mainly used when data of an

Esther Mohr · Günter Schmidt
Saarland University, P.O. Box 151150, D-66041 Saarbrücken, Germany
e-mail: {em,gs}@itm.uni-sb.de

Sebastian Jansen
University of Hohenheim, Banking and Financial Services, D-70593 Stuttgart, Germany
e-mail: sebastian_yansen@yahoo.com

E. Tantar et al. (Eds.): EVOLVE- A Bridge between Probability, SCI 447, pp. 393–414.
springerlink.com   © Springer-Verlag Berlin Heidelberg 2013

unknown process is obtained. In the work related two well known methods which can be used for symbolic regression exist: genetic programming (*GP*) introduced by [19, 20], and grammatical evolution introduced by [34]. We limit to investigating the ability of *GP* to solve conversion problems.

Several results from the literature have shown that heuristic conversion algorithms and their related *GP* variants have predictive ability in a variety of exchange markets: returns to be expected are considered to be 'predictable' in the sense that it is possible to generate excess returns in a particular future time interval. Beginning with [9] several authors investigated the predictive ability of heuristic conversion algorithms, cf. [5, 17, 39, 27, 6, 18, 23, 40, 11, 35, 15, 12, 21, 33, 10, 8, 26, 28, 16, 24, 22, 44]. Applications of genetic programming to heuristic conversion algorithms can be found in [32, 1, 7, 37, 2, 31, 38, 14, 30]. These contributions evaluate an algorithms' predictability through backtesting. However, this approach can often not be applied in a reasonable way, as distributions are rarely known precisely.

This leads to assuming uncertainty about asset prices and analyzing an algorithms' performance considering worst-case scenarios. This approach does not demand that inputs come from some known distribution. Competitive analysis compares the performance of an online algorithm (*ON*) to that of an adversary, the optimal offline algorithm (*OPT*) considering worst-case scenarios. According to the definition of [13] also heuristic conversion algorithms are online algorithms; they work without any knowledge of future input. We suggest analyzing heuristic conversion algorithms using competitive analysis.

In case the input data processed by an online algorithm does not represent the worst-case scenario, its performance is considerably better than the competitive ratio tells (cf. [41]). For this reason we also backtest the considered algorithms.

Our aim is twofold. First, we want to evaluate the predictive ability of the algorithms *Moving Average Crossover* (*MA*) and *Trading Range Breakout* (*TRB*)[1] on the German stock market, which has not been done in previous studies. Second, we evolve *MA* and *TRB* through genetic programming, resulting in a *GP* algorithm. We 1) backtest *MA*, *TRB* and *GP* and compare their (excess) returns. For risk analysis we 2) calculate the *Sortino Ratio* (cf. [42]) and apply competitive analysis (cf. [13]) considering 3) empirical-cases as well as 4) worst-cases. So far, works on *MA*, *TRB* and *GP* do not take into account worst-case scenarios. Classical buy-and-hold (*BH*) is used as a benchmark.

The remainder of this paper is organized as follows. In the next section we give a literature overview on heuristic conversion algorithms and genetic programming. In Section 3 the problem is formulated and the algorithms considered are presented in detail. Section 4 presents the experimental design assumed, and the worst-case competitive ratio of *MA*, *TRB* and *GP* is given. Detailed experimental findings from our simulation runs are presented in Section 5. We finish with some conclusions and suggestions for future research in the last section.

---

[1] Also known as resistance and support levels.

In the following section we present work related to our study on MA, TRB and GP. A survey on the profitability of technical analysis in general can be found in [36] investigating a total of 95 modern studies.

12.2   Related Work

There are many results on experimental studies on the considered heuristic conversion algorithms. The comparison to BH or to GP is of prime interest. Either results of VMA, FMA and TRB are compared to BH, or GP results are compared to BH. Unfortunately, all these works are restricted to empirical-case results and do not take into account worst-case results from competitive analysis. In the work related, by carrying out a t-test and/or a bootstrap procedure, the question whether the backtested algorithms $ON \in \{VMA, FMA, TRB, GP\}$ have predictive ability or not is answered. It is assumed that if the null hypothesis $H_0 : \mu_{ON} \leq \mu_{BH}$ is rejected, there is a good (but not certain) chance that ON performs better than benchmark BH again in the future. The alternative hypothesis $H_1 : \mu_{ON} > \mu_{BH}$ is confirmed indirectly. In case results show that the (excess) returns generated by ON are not significant, this suggests that predictability is not economically significant.

Reference [9] suggests the algorithms VMA, FMA and TRB, and conduct experiments with a price-weighted index (Dow Jones Industrial Average (DJIA)) for the investment horizon from 1897 to 1986. The returns on buy (sell) signals on the DJIA are compared to returns from simulated comparison series generated by the following models: autoregressive (AR(1)), generalized autoregressive conditional heteroskedasticity in mean (GARCH-M), and exponential GARCH. The returns obtained from the algorithms are not likely to be generated by these three models. The results provide empirical support for utilizing the heuristic conversion algorithms as they outperform not only BH but also the AR(1), the GARCH-M, and the exponential GARCH model. [9] conclude that VMA, FMA and TRB have predictive ability.

Reference [25] presents evidence on the profitability and statistical significance of VMA, FMA and TRB in the foreign exchange market utilizing currency futures contracts for the period 1976-1990. A new testing procedure based on bootstrap methodology is implemented. Results suggest that simple technical trading rules have very often led to profits that are highly unusual. Splitting the entire sample period into three 5-year periods reveals that (on average) the profitability of some trading rules declines in the latest period. Although profits remain positive (on average) and significant in many cases.

Reference [5] tests whether VMA, FMA and TRB can predict stock price movements in Asian markets. The result is that the algorithms are 'quite successful' in the emerging markets of Malaysia, Thailand and Taiwan, but have less explanatory power in more developed markets such as Hong Kong and Japan. Transactions costs which could eliminate gains are estimated to be 1.57%.

Reference [17] tests whether the findings of [9] – that VMA, FMA and TRB have predictive ability – is replicable on the FT30 (Financial Times Ordinary) index from July 1935 to January 1994. Further, the authors test whether the algorithms generate excess returns in a costly trading environment. [17] conclude that although VMA, FMA and TRB do have predictive ability in terms of UK data, their use would not generate excess returns in the presence of costs. In general, the results presented are remarkably similar to those of [9]. Thus, one conclusion to be drawn from both studies is that VMA, FMA and TRB have predictive ability if sufficiently long series of stock indices are considered.

Reference [39] implements VMA, FMA and TRB to test whether they result in excess returns on the Hang Seng Futures Index, traded at the Hong Kong Futures Exchange. It is found that MA does not produce significant excess returns, but four out of six TRB rules result in significant positive returns.

Reference [27] compares VMA, FMA and TRB to BH by conducting experiments on the FT30 index for the time intervals 1935-1954 and 1975-1994. In addition, trading signals generated by a geometric MA are calculated. The geometric MA gave an almost identical set of buying and selling signals as the conventional (arithmetic) MA. Until 1980 all algorithms outperform BH. The results of [27] are consistent, in almost every respect, with those of [9] and [17]. But from 1980 on BH clearly dominates all other algorithms. The sample used by [9] ends in 1986; so [27] concludes that there was not the data to analyze structural shifts that might have taken place starting in 1982.

Reference [6] tests the same trading rules as [9] on dividend-adjusted DJIA data for the period 19261991. In an attempt to avoid data snooping problems the profitability and statistical significance of the returns is evaluated on 1) portfolios of the trading rules, and 2) returns of individual trading rules. Results show that it is unlikely that traders could have used VMA, FMA and TRB to earn net profits after transaction costs.

Reference [18] tests the same trading rules as in [9] in Pacific-Basin equity markets by using equilibrium asset pricing models with time-varying expected returns. Results on the Japanese, US, Canadian, Indonesian, Mexican and Taiwanese equity indices indicate that the technical rules have significant forecast power for all countries, except for the US. However, the results from the bootstrap tests indicate that some equilibrium asset pricing models (mainly, the asset pricing model under mild segmentation) are consistent with the observed trading rule returns for Japan, the US, the recent period of Canada and Taiwan.

Reference [23] discusses the evidence that simple rules used by traders have some predictive value over the future movement of foreign exchange prices. The profitability of VMA, FMA and TRB is analyzed in connection with central bank activity using intervention data from the Federal Reserve. The objective is to find out to what extent foreign exchange predictability can be confined to periods of central bank activity in the foreign exchange market. The results indicate that after removing periods in which the Federal Reserve is active, exchange rate predictability is dramatically reduced.

Reference [40] compares *VMA* and *FMA* to *BH* by investigating ten emerging equity markets in Latin America and Asia from 1982 to 1995 under transaction costs using the S&P500 and Nikkei225 indices. Results show that *VMA* and *FMA* applied to emerging markets do not have the ability to outperform *BH*.

Reference [11] investigates the applicability and validity of *VMA*, *FMA* and *TRB* in the Hang Seng index on the Hong Kong Stock Exchange for the period January 1985 to June 1997, and for two subsamples of equal lengths, partitioned from the whole sample. Statistical significance of the results can be shown when the rules are applied to data periods shorter than used in previous studies. A tendency for potentially profitable trading rules is documented.

Reference [35] tests *VMA*, *FMA* and *TRB* on the Chilean stock market using the *IPSA* (Indice de Precio Selectivo de Acciones) index from January 1987 to September 1998. The results are similar to the ones of [9], providing strong support for *VMA*, *FMA* and *TRB*.

Reference [15] tests *VMA* and *FMA* in four emerging South Asian capital markets from January 1990 to March 2000, i.e. the Bombay Stock Exchange, the Colombo Stock Exchange, the Dhaka Stock Exchange and the Karachi Stock Exchange. The findings indicate that the algorithms have predictive ability in these markets, and reject the null hypothesis that the returns to be expected are equal to those achieved from *BH*. They conclude that *VMA* and *FMA* are able to generate excess returns in South Asian markets.

Reference [12] re-examine the findings of [9] by adjusting the daily returns from trading portfolios of DJIA stocks for both dividends and the interest earned on the proceeds from short sales. Results show that previous estimates of trading profits may be biased by the inclusion of nonsynchronous prices in the closing index levels. Estimates of trading profits (based on the true closing levels of the index) are not significantly different from *BH* returns. Further, bootstrap simulation results suggest that significance levels for estimates of trading profits based on closing index levels are overstated by the inclusion of nonsynchronous prices.

Reference [21] extends the work of [9] in two ways. First by investigating the predictive ability of *VMA*, *FMA* and *TRB* on the New York Stock Exchange (NYSE) index from July 1962 to December 1996, as well as on the National Association of Security Dealers Automatic Quotations (NASDAQ) index from January 1972 to December 1996. Second by including a further *MA* algorithm, called *Moving Average with Trading Volume* (*MAV*). The results show that the suggested algorithms have potential to outperform *BH*. The results support the results of [9] showing that the suggested algorithms outperform *BH*.

Reference [33] characterizes the temporal pattern of *VMA*, *FMA* and *TRB* returns and official intervention for Australian, German, Swiss and US data to investigate whether intervention generates profits. The hypothesis that intervention generates inefficiencies from which technical rules profit is rejected. In particular, high frequency data show that abnormally high trading rule returns precede German, Swiss and US intervention. Australian intervention precedes high trading rule returns, but trading/intervention patterns make it implausible that intervention actually generates those returns. Rather, intervention responds to exchange rate trends from which trading rules have recently profited.

Reference [10] tests whether returns generated by VMA, FMA and TRB are predictable in eleven emerging stock markets in the US and Japan considering data from January 1991 to January 2004. Predictability is analyzed by means of multivariate variance ratios using bootstrap procedures. VMA, FMA and TRB are employed and compared to BH. Results show that there is some evidence of predictive power but no significance. When costs are taken into account only a few variants of the algorithms generate excess returns, and thus [10] conclude that although the algorithms show some predictive ability this is not statistically significant.

Reference [8] investigates the predictive ability and profitability of VMA, FMA and TRB for different company sizes considering different indices form January 1987 to July 2002. Results on different Financial Times Stock Exchange (FTSE) indices, namely FTSE 100, FTSE 250 and FTSE Small Cap, show that the algorithms have a progressively higher predictive ability the smaller the size of the company, but are not profitable assuming transaction costs.

Reference [26] tests the profitability of twelve variants of VMA, FMA and TRB on the New Zealand stock market. The nature and regulations suggest that the New Zealand equity market may be less efficient than overseas markets. This raises the possibility that the algorithms are profitable in New Zealand. Using a bootstrapping procedure, results show that the returns achieved in New Zealand follow a similar pattern than those in large offshore markets.

Reference [28] tests the profitability of VMA, FMA and TRB on nine Asian stock market indices from January 1988 to December 2003. The results provide strong support for VMA and FMA in China, Thailand, Taiwan, Malaysia, Singapore, Hong Kong, Korea, and Indonesia.

Reference [16] aims to characterize the stock return dynamics of four Latin American and four Asian emerging capital market economies and test the profitability of VMA and TRB. Using the Morgan Stanley Capital International (MSCI) index, the BH benchmark is outperformed in all markets before transaction costs, and in Asian markets after transaction costs.

Reference [24] tests the profitability of different algorithms by evaluating their ability to outperform BH. Different VMA, FMA, Filter rule, Bollinger Band, and TRB variants are tested on the S&P/TSX 300 index, the DJIA, the NASDAQ composite index, and the Canada/US spot exchange rate. A bootstrap procedure is used to determine the statistical significance of the results. Considering transaction costs, excess returns are generated by VMA, FMA and TRB for the S&P/TSX 300 index, the NASDAQ composite index and the Canada/US spot exchange rate. The Filter rules also earn excess returns when applied on the Canada/US spot exchange rate.

Reference [22] tests the Efficient Market Hypothesis (EMH), in seven emerging Middle-Eastern North African ($MENA$) stock markets from January 1998 to December 2004. The result of a random-walk test, and the returns of VMA, FMA and TRB are aggregated into a single efficiency index. The impact of market development, corporate governance and economic liberalization on the latter using a multinomial ordered logistic regression is to be analyzed. The results highlight heterogeneous levels of efficiency in the *MENA* stock markets. The efficiency index seems to be affected by market depth, although corporate governance factors also have explanatory power. By contrast, the impact of overall economic liberalization does not appear significant.

Reference [44] investigates the predictive power of *VMA*, *FMA* and *TRB* for the Brazilian exchange rate from 2003 to 2006. A bootstrap procedure is employed to test predictability. Furthermore, the ability of the trading algorithms to generate significant higher returns compared to *BH* is tested. Results show that the excess return generated by the algorithms is not significant, suggesting that predictability is not economically significant. Their results are consistent with those of [10].

Table 12.1 gives a year wise summary on the above 24 works related to *VMA*, *FMA* and *TRB*. A hyphen (-) in column 4 indicates that transaction costs are not considered in the respective study.

Table 12.1 Literature Summary: *VMA*, *FMA* and *TRB* compared to *BH*

Year	Reference	H_1 confirmed?	Profitable after transaction costs?
1992	[9]	yes	yes
1993	[25]	yes	-
1995	[5]	yes	yes
1996	[17]	yes	no
1996	[39]	yes	yes
1997	[27]	no	no
1998	[6]	yes	no
1999	[18]	yes	yes
1999	[23]	yes	yes
1999	[40]	no	no
2000	[11]	yes	no
2000	[35]	yes	no
2001	[15]	yes	yes
2002	[12]	no	no
2002	[21]	yes	-
2002	[33]	yes	-
2004	[10]	no	no
2005	[8]	yes	no
2005	[26]	yes	no
2006	[28]	yes	no
2007	[16]	yes	no
2007	[24]	yes	yes
2008	[22]	yes	-
2009	[44]	no	-

Several authors use genetic programming (*GP*) to adapt and evolve heuristic conversion algorithms. The comparison to *BH* is of prime interest.

Reference [32] applies the *GP* approach to the FX market and report excess returns for the evolved trading rules. [1] apply *GP* to the S&P 500 index and conclude that the market is efficient since *GP* generally fails to outperform the *BH* benchmark. This finding is supported by [31]. In contrast, [7] report *GP* excess returns over the *BH* benchmark. [37] investigates the Australian stock market and finds that excess returns from *GP* are marginal but have some forecasting power in terms of market direction. [2] report excess returns compared to *BH* in case of high market volatility. In contrast, predictive ability is found for long term market trends on a risk-adjusted basis. [38] apply a *GP* framework to the Toronto Stock Exchange focusing on single stocks, and report excess returns for nine out of fourteen stocks. [14] investigate *GP* performance for the S&P 500, the S&P 500 Auto, and the S&P 500 Bank indices and report a *GP* underperformance for the S&P 500 index and the S&P 500 Bank index. The *GP* on the S&P 500 Auto index outperformed *BH* on a risk-adjusted basis. [30] investigate *GP* performance on the NYSE index and report mixed results.

12.3   Problem Formulation

Let us consider multiple conversions, i.e. we convert an asset more than once. We assume an online conversion algorithm $ON \in \{MA, TRB, GP, BH\}$ consists of buying and selling rules, and each i-th trade ($i = 1, \ldots, p$) consists of exactly one buying and one selling transaction. *ON* obtains price quotations $q_t \in [m, M]$ ($0 < m \leq M$) at points of time $t = 1, \ldots, T$ where T is the length of the investment horizon. For each i-th trade, *ON* must first decide whether to buy at price q_t on day t or not, and second whether to sell at a later price q_{t+x} on day $t + x$ or not.[2] Open buy positions must be sold at the last possible price q_T. We fix some notation. Let

- number of trades p, with $i = 1, \ldots, p$,
- prices q_t, with $t = 1, \ldots, T$,
- data points used n, with $n \leq T$,
- moving average $MA(n)_t$ on day t, calculated for $n \in \{L, S\}$ data points, and L (S) long (short) *MA* with $S < L$,
- percentage band $b \in [0.00, \infty]$,[3]
- global minimum of prices m,
- global maximum of prices M,
- local minimum of prices $q^{min}(n)_t$ on day t calculated for n data points,
- local maximum of prices $q^{max}(n)_t$ on day t calculated for n data points,
- upper band $UB(n)_t$ and lower band $LB(n)_t$ lagging $MA(n)_t$, $q^{min}(n)_t$, or $q^{max}(n)_t$ on day t,
- fixed number of days x chosen for selling, with $1 \leq x \leq T - 1$.

---

[2] Short selling is not considered since it is prohibited in Germany since May 18, 2010.

[3] The band is used to reduce the amount of trading and therefore the trading costs.

In the following the *MA* and *TRB* algorithms of [9] as well as the *GP* approach of [38] are presented.

12.3.1   *Moving Average Crossover (MA)*

The *MA* algorithm consists of the following rules [9, p. 1736]:

> Buy if the short *MA* crosses the long *MA* from below, and sell if the short *MA* crosses the long *MA* from above.

Two variants of this algorithm, called *Variable-length Moving Average* (*VMA*) and *Fixed-length Moving Average* (*FMA*) exist. The variants differ in the way their performance is measured. From this we assume a *MA* algorithm comprised of the following steps:

Algorithm 12.1. *MA* of [9]

1: **while** $t \leq T$ **do**
2:    **if** $t \geq n$ **then**
3:      $MA(n)_t = \frac{\sum_{i=t-n+1}^{t} q_i}{n} \{n \in \{L, S\} \text{ with } S < L\}$
4:      Calculate $MA(S)_t$ and $MA(S)_{t-1}$
5:      Calculate $MA(L)_t$ and $MA(L)_{t-1}$
6:      Calculate $UB(L)_t = MA(L)_t \times (1 + b)$ and $UB(L)_{t-1} = MA(L)_{t-1} \times (1 + b)$
7:      Calculate $LB(L)_t = MA(L)_t \times (1 - b)$ and $LB(L)_{t-1} = MA(L)_{t-1} \times (1 - b)$
8:      **if** $MA(S)_t > UB(L)_t$ and $MA(S)_{t-1} \leq UB(L)_{t-1}$ **then**
9:        buy on day t
10:     **end if**
11:     **if** $MA = VMA$ **then**
12:       **if** $MA(S)_t < LB(L)_t$ and $MA(S)_{t-1} \geq LB(L)_{t-1}$ **then**
13:         sell on day t
14:       **else**
15:         **if** $t + x \leq T$ **then**
16:           sell on day $t + x\{x \geq 1 \text{ days}\}$
17:         **else**
18:           sell on day T
19:         **end if**
20:       **end if**
21:     **end if**
22:   **end if**
23:   $t = t + 1$
24: **end while**

12.3.2 *Trading Range Breakout (TRB)*

The TRB algorithm consists of the following rules [9, p. 1736]:

> Buy if the price cuts the local maximum price from below, and sell if the price cuts the local minimum price from above.

Analogously to *FMA*, a fixed time interval of x days for selling can be chosen. For *TRB* [9] only consider this case. Instead, we assume a *TRB* algorithm comprised of the following steps:

Algorithm 12.2. *TRB* of [9]

 1: **while** $t \leq T$ **do**
 2:     **if** $t > n$ **then**
 3:         Calculate $q^{min}(n)_t = \min q_i | i = t - n, \ldots, t - 1$ and $q^{min}(n)_{t-1}$
 4:         Calculate $q^{max}(n)_t = \max q_i | i = t - n, \ldots, t - 1$ and $q^{max}(n)_{t-1}$ $\{n \leq T$ days$\}$
 5:         Calculate $LB(n)_t = q^{min}(n)_t \times (1 - b)$ and $LB(n)_{t-1} = q^{min}(n)_{t-1} \times (1 - b)$
 6:         Calculate $UB(n)_t = q^{max}(n)_t \times (1 + b)$ and $UB(n)_{t-1} = q^{max}(n)_{t-1} \times (1 + b)$
 7:         **if** $q_t > UB(n)_t$ and $q_{t-1} \leq UB(n)_{t-1}$ **then**
 8:             buy on day t
 9:         **end if**
10:         **if** TRB =variable **then**
11:             **if** $q_t < LB(n)_t$ and $q_{t-1} \geq LB(n)_{t-1}$ **then**
12:                 sell on day t
13:             **else**
14:                 **if** $t + x \leq T$ **then**
15:                     sell on day $t + x \{x \geq 1$ days$\}$
16:                 **else**
17:                     sell on day T
18:                 **end if**
19:             **end if**
20:         **end if**
21:     **end if**
22:     $t = t + 1$
23: **end while**

GP is used to evolve the rules given in Alg. 12.1 and Alg. 12.2. The GP approach is presented in the following.

12.3.3   Genetic Programming (GP)

GP aims to generate and adapt (buying and selling) rules using evolutionary processes as found in nature. Similarly to [38] we consider a genetic algorithm as a randomized search procedure on a population of individuals encoded as linear bit strings. The population evolves over time by selection of the fittest, crossover, mutation, and reproduction. For the evolution of heuristic conversion algorithms [38] suggested the following approach to find rules that suit the prevailing market conditions best: *GP* starts with a set of randomly generated rules within the very first generation. This generation of rules is measured in terms of their fitness, i.e. the achieved return compared to a benchmark. The fittest rules are subject to crossover, mutation and reproduction, and build the next generation of rules. These new rules are evaluated again in terms of their fitness. The fittest rules are again subject to crossover, mutation and reproduction, and built again the basis for the next generation, etc. In theory, this approach results in better rules in each generation. The fittest rules of the final generation are denoted as the *GP* algorithm. For further details on *GP* the reader is referred to [38].

We modify the *GP* approach given in [38, p. 1035] as follows: instead of a rank-based selection process, the classical roulette wheel selection is used (details can be found in [38, Sec. 2.4]). As *MA* and *TRB* are calculated using prices q_t, moving averages $MA(n)_t$, bands (lagged prices) b, minima of prices $q^{min}(n)_t$, and maxima of prices $q^{max}(n)_t$, only these values are valid input data when creating the *GP* algorithm. [38] also allow other inputs, like volumes or technical indicators itself. Further, the function set linking the inputs is limited to the operators used by *MA* and *TRB*.

In the following the experimental design, assumptions made, and the performance measure used is presented.

12.4   Experimental Design

Our experiments are based on Dax-30 index data from 01-01-2002 to 12-31-2007. All algorithms are run on this 6-year investment horizon. For *GP* a 11-year time horizon from 01-01-1997 to 12-31-2001 is used to enable training data subsets to evolve the *GP* algorithm: the time horizon is divided into different subsets using a rolling time frame approach consisting of 1) training data subsets of length 5 years, followed by 2) out-of-sample data subsets of length 1 year. The first training subset equals the years 1997-2001 followed by the year 2002 as first out-of-sample subset. Next, the time frame is rolled one year forward resulting in a 1998-2002 training subset followed by the year as 2003 out-of-sample subset, etc. The last out-of-sample subset considered is 2007. With this approach we ensure the *GP* algorithm is run on the same 6-year out-of-sample time interval as all other algorithms. In contrast to [38], we only consider index data, rather than considering stocks offered by an individual company. The following assumptions apply for all algorithms:

1. There is an initial amount of cash meeting the demands to execute the algorithms in a profitable way.
2. All algorithms convert non-preemptive, i.e. either the whole amount available is converted in one transaction, or nothing is converted. In other words 'all or nothing' is invested in the Dax-30 index.
3. The algorithms OPT, VMA, FMA, TRB and GP carry out as many trades p as possible.
4. Similarly to [9, p. 1736 and p. 1740] we assume holding periods. One holding period ranges from a buying signal to a selling signal, further buying signals during a holding period must be ignored by all algorithms.
5. Weekends and country-specific holidays are excluded from the 6-year time interval considered, resulting in $T = 1531$ days for conversion.
6. Possible transaction prices are daily closing prices.
7. Each i-th trade (buying and selling) costs 50 basis points, i.e. 25 points equaling 0.25% for each buying (selling) transaction.
8. Interest rate on cash is assumed to be zero.

12.4.1   Algorithms Considered

In our experiments we investigate the heuristic conversion algorithms MA, TRB and GP presented in Sect. 12.3 using the following input parameters and assumptions. As a benchmark we compare to OPT and BH. The buying and selling rules are Boolean, i.e. if a buy (sell) rule evaluates as true on day t, a buy (sell) position is set up (or maintained) for day t, else the algorithm sells (buys) or stays out of the market as no short positions are allowed.

12.4.1.1   Moving Average Crossover (VMA and FMA)

VMA and FMA each buy and sell according to the MA algorithm presented in Sect. 12.3.1. VMA considers buying and selling signals. FMA only considers buying signals with $x = 10$, i.e. a fixed 10-day holding period follows a buying signal on day t. On day $t + 10$ at price q_{t+10} all index is sold. When calculating short moving average $MA(S)_t$ and long moving average $MA(L)_t$ we assume $S \in \{1,2,5\}$ days, $L \in \{50,150,200\}$ days and band $b \in \{0.00,0.01\}$. Resulting in ten MA variants (S,L,b): $(1,50,1),(1,50,0),(1,150,0),(1,150,1),(5,150,0),(5,150,1),(1,200,0)$ $(1,200,1),(2,200,0),(2,200,1)$ as suggested by [9]. Thus, overall 20 MA simulation runs were performed, 10 for VMA and 10 for FMA.

12.4.1.2   Trading Range Breakout (TRB)

TRB buys and sells according to the TRB algorithm presented in Sect. 12.3.2. In contrast to [9] instead of a fixed holding period TRB considers buying and selling signals. Local minimum prices $q^{min}(n)_t$, and local maximum prices $q^{max}(n)_t$ are calculated for $n \in \{50, 150, 200\}$ days and band $b \in \{0.00, 0.01\}$. Resulting in six TRB variants (n, b): $(50, 0), (50, 1), (150, 0), (150, 1), (200, 0), (200, 1)$.

12.4.1.3   Genetic Programming (GP)

To create and evolve the GP algorithm buying and selling rules prices q_t, moving averages $MA(n)_t$, bands (lagged prices) b, and minima and maxima of prices $q^{min}(n)_t$ and $q^{max}(n)_t$ are taken from the MA and TRB rules as possible input parameters with $n \in \{3, 5, 6, 10, 20, 100, 150, 200, 250\}$ days. The function set linking these inputs consists of the operators $[+, -, \times, /, \ln, \sqrt{}, <, >, \leq, \geq, AND, OR, XOR]$. In addition we allow constants s enabling GP to compare an input variable to a fixed value, for example '$Buy\ if\ MA(n)_t < s$'. For each training data subset the 'best' buying and selling rules are evolved by choosing inputs from the above data pool. These rules are subsequently applied to each out-of-sample data subset, and the GP algorithm is evolved in order to replace outdated rules.

12.4.1.4   Buy and Hold (BH)

BH buys the index on the first day of the considered 6-year time interval and sells it on the last day T.

12.4.1.5   Optimum (OPT)

OPT is an offline algorithm which achieves the best possible return. It is assumed that OPT knows all future prices. OPT buys (sells) at each local minimum $m(i) \geq m$ (local maximum $M(i) \leq M$) within T, i.e. exploits upward price movements $i = 1, \ldots, p$ times while buying on the first day of the run, and selling on the last day of the run.

12.4.2   Performance Measurement

The performance of the above algorithms is measured by the return achieved. Let r_t be the daily log returns for each day a heuristic conversion algorithm holds an asset; given by

$$r_t = \ln \frac{q_t}{q_{t-1}}. \tag{12.1}$$

We consider transaction costs. For each buying transaction on day $t - 1$ Eq. (12.1) modifies to

$$r_t = \ln \frac{q_t}{q_{t-1} \times 1.0025} \tag{12.2}$$

as each buying costs 25 basis points, and thus the buying price q_{t-1} is 0.25% higher. Analogously, for each selling transaction on day t Eq. (12.1) modifies to

$$r_t = \ln \frac{q_t \times 0.9975}{q_{t-1}} \tag{12.3}$$

as each selling costs 25 basis points, and thus the selling price q_t is 0.25% smaller. Daily log returns r_t are time additive. The overall return r_T after the last day T then equals

$$r_T = \sum_{t=1}^{T} r_t. \tag{12.4}$$

From Eq. (12.4) we get the achieved annualized return $R(y)$

$$R(y) = (1 + r_T)^{(1/y)} \tag{12.5}$$

where y equals the number of years within T, i.e. for the considered 6-year investment horizon $y = 6$: then $R(y)$ tells us which empirical-case return we could achieve within one year. Further, excess returns are calculated by subtracting the achieved $R(y)$ of BH from the achieved $R(y)$ of algorithms OPT and $ON \in \{VMA, FMA, TRB, GP\}$.

For risk analysis we apply competitive analysis (cf. [13]) considering empirical-cases as well as worst-cases, and we calculate the *Sortino Ratio* (cf. [42]).

Competitive analysis assumes each input is represented as a finite input sequence I with $t = 1, \ldots, T$ elements, and a feasible output is also represented as a finite sequence with T elements. ON computes online if for each $t = 1, \ldots, T - 1$, it computes an output for t before the input for $t + 1$ is given. OPT computes offline if it computes a feasible output given the entire input sequence I in advance. An online algorithm ON is c-competitive if for any I [13, Eq. (1), p. 104]

$$ON(I) \geq \frac{1}{c} \times OPT(I) \tag{12.6}$$

$$c \geq \frac{OPT(I)}{ON(I)}$$

where $ON \in \{VMA, FMA, TRB, GP, BH\}$.

Any c-competitive algorithm is guaranteed a value of at least the fraction $\frac{1}{c}$ of the optimal offline value $OPT(I)$, no matter how unfortunate or uncertain the future will be. We consider converting assets as a maximization problem, i.e. $c \geq 1$. The smaller c the more effective is ON.

Let c_{wc} be the worst-case competitive ratio, and let c_{ec} be the empirical-case competitive ratio. In order to derive c_{wc} we assume each i-th trade ON achieves a worst-case return of $r(i) = \frac{m}{M}$, resulting in a worst-case competitive ratio of $c_{wc}(i) = \left(\frac{M}{m}\right)^2$ (cf. Eq. (12.6)). Further, we assume ON is confronted with the worst possible sequence of prices p times and upper and lower bounds of prices, m and M, are constants. Assuming an identical number of $i = 1, \ldots, p$ trades for OPT and ON the overall worst-case competitive ratio c_{wc} of an algorithm ON then equals

$$c_{wc} = \left(\frac{M}{m}\right)^{2p}.$$

(12.7)

The proof of c_{wc} given in Eq. (12.7) can be found in [29].

To compute c_{wc} for the considered 6-year investment horizon the overall high equals $8105.69 = M$ and the overall low equals $2202.93 = m$. From Eq. (12.7) we get $c_{wc} = (4.0068)^{2p}$, i.e. an an annualized basis $c_{wc} = (1.2603)^{2p}$ for all ON. For GP we assume the same c_{wc} as for MA and TRB because GP is comprised of the elements of the MA and TRB rules.

The empirical-case competitive ratio c_{ec} is calculated in the same manner as c_{wc}. But instead of the worst-case return of $r(i) = \frac{m}{M}$, the $R(y)$ (cf. Eq. (12.4)) achieved by ON and OPT is used to calculate c_{ec}. Note that $c_{ec} \leq c_{wc}$.

The *Sortino Ratio* is a variation of the *Sharpe Ratio* but only considers the downside deviation σ_d that occurs when returns fall below the minimum acceptable rate of return, denoted by r_{MAR} (cf. [43]). We set r_{MAR} to zero as we want to know whether the risk incorporated is rewarded with respect to never losing money. The *Sortino Ratio SR* is defined as follows

$$SR = \frac{\bar{r}_T - r_{MAR}}{\sigma_d}$$

(12.8)

$$= \frac{\bar{r}_T}{\sigma_d}$$

with average daily return (Eq. (12.4))

$$\bar{r}_T = \frac{r_T}{T},$$

(12.9)

and downside deviation

$$\sigma_d = \sqrt{\frac{1}{T} \sum_{t=1}^{T} (\min(r_t, 0))^2}$$

(12.10)

as $r_{MAR} = 0$. In the following experimental results are presented.

12.5   Results

We carried out simulation runs in order to find out how the following measures compare:

1. The empirical-case performance, measured by the annualized return $R(y)$,
2. the empirical-case competitive ratio c_{ec},
3. the worst-case risk, measured by the competitive ratios c_{wc},
4. the empirical-case risk, measured by the *Sortino Ratio SR*.

Tables 12.2 to 12.5 present the computational results of the overall 28 simulation runs. Each algorithm ON results in a different number of trades p. Clearly, ON cannot beat OPT. Table 12.2 presents the results for OPT, BH and GP. The results for VMA, FMA and TRB are given in Tables 12.3 to 12.5. Table 12.3 presents the results of the different VMA variants of Alg. 12.1. Table 12.4 presents the results of the different FMA variants of Alg. 12.1. Table 12.5 presents the results of the different TRB variants of Alg. 12.2.

Table 12.2 Computational results for OPT, BH and GP

2002-2007	$R(y)$	p	c_{ec}	c_{wc}	SR
OPT	1.4363	394	-	-	-
BH	1.0622	1	1.3514	1.59	0.0111
GP	1.0392	14	1.3821	650.25	0.0141

Table 12.3 Computational results for VMA

2002-2007	$R(y)$	p	c_{ec}	c_{wc}	SR
$VMA(1,50,0)$	1.0660	39	1.3474	68617522.13	0.3664
$VMA(1,50,1)$	1.0620	19	1.3524	6572.80	0.0621
$VMA(1,150,0)$	1.0755	22	1.3354	26336.20	0.1370
$VMA(1,150,1)$	1.0893	9	1.3186	64.33	0.3819
$VMA(5,150,0)$	1.0821	12	1.3273	257.76	0.2912
$VMA(5,150,1)$	1.0945	6	1.3123	16.05	1.1418
$VMA(1,200,0)$	1.0985	11	1.3075	162.28	0.5658
$VMA(1,200,1)$	1.0910	7	1.3165	25.50	0.4588
$VMA(2,200,0)$	1.0944	9	1.3124	64.33	0.4879
$VMA(2,200,1)$	1.0935	6	1.3135	16.05	0.6256

Table 12.4 Computational results for FMA

2002-2007	$R(y)$	p	c_{ec}	c_{wc}	SR
$FMA(1,50,0)$	1.0027	26	1.4324	167606.18	0.0022
$FMA(1,50,1)$	0.9564	32	1.5018	2690886.69	-0.0224
$FMA(1,150,0)$	1.0305	15	1.3938	1032.79	0.0785
$FMA(1,150,1)$	1.0128	19	1.4182	6572.80	0.0273
$FMA(5,150,0)$	0.9965	11	1.4414	162.28	-0.0153
$FMA(5,150,1)$	1.0047	15	1.4296	1032.79	0.0102
$FMA(1,200,0)$	0.9994	8	1.4371	40.50	-0.0024
$FMA(1,200,1)$	1.0032	11	1.4317	162.28	0.0131
$FMA(2,200,0)$	0.9991	8	1.4376	40.50	-0.0043
$FMA(2,200,1)$	1.0147	8	1.4154	40.50	0.1067

Table 12.5 Computational results for TRB

2002-2007	$R(y)$	p	c_{ec}	c_{wc}	SR
$TRB(50,0)$	1.0528	9	1.3642	64.33	-0,0003
$TRB(50,1)$	0.9558	25	1.5027	105525.10	-0.0147
$TRB(150,0)$	1.0966	2	1.3097	2.52	0.0009
$TRB(150,1)$	0.9738	18	1.4748	4138.24	-0.0248
$TRB(200,0)$	1.1090	1	1.2951	1.59	0.0011
$TRB(200,1)$	0.9784	17	1.4679	2605.45	-0,0267

Question 1: How do the annualized returns $R(y)$ of the algorithms ON compare? To answer **Question 1** column 1 and 2 in Tables 12.2 to 12.5 are of main interest. The overall best annualized return is achieved by $TRB(200,0)$ (10.9%), while conducting only one trade resulting in the lowest possible transaction costs. But only 3 of 6 TRB variants generate positive returns, namely those without band ($b = 0.00$), and 2 of these 3 variants outperform BH. All TRB variants lagged by a band ($b = 0.01$) generate losses, and are outperformed by BH and GP. $TRB(50,1)$ generates the overall worst result (-4.42%). The second best result is achieved by $VMA(1,200,0)$ (9.85%), but 11 trades are conducted resulting in high transaction costs. In general, all 10 VMA variants generate positive returns from minimum 6.20% to maximum 9.85%. These relatively high returns are 'bought' by a large number of trades p, e.g. $VMA(1,50,0)$ achieves a return of 6.6% while conducting 39 trades. Overall between 6 and 39 trades were performed. All VMA variants without band, and 2 of 3 VMA variants with band ($b = 0.01$) outperform BH. Only $VMA(1,50,1)$ is outperformed by BH. All VMA variants outperform GP. The results of FMA worse than the results of the VMA, BH, GP and TRB without band. We conclude that defining holding periods of fixed length does not lead to good results. 4 of 10 FMA variants generate losses, and the 6 positive returns only range from minimum 0.27% to maximum 3.05%. The number of trades ranges between 8 and 32.

GP is outperformed by BH. This might be due to the different length of the training data subset (5 years) and the out-of-sample data subset (1 year). GP has developed trading rules fitting a 5-year time interval. We conclude that training data subsets and out-of-sample data subsets should have an equal length. In order to clarify the above findings in addition we computed the excess returns by subtracting the annualized BH result (6.28%) from the $R(y)$ achieved by each other algorithm, as suggested in [38]. Fig. 12.1 shows the excess returns of all algorithms. The results for VMA provide evidence that the greater the number of days for calculating $MA_t(L)$, the higher the achieved return of VMA. For FMA the results are biased. We conclude that further experiments should be carried out. The results for TRB provide evidence that a band of 1% corrupts the results. BH outperforms exactly 16 of the 28 variants of the algorithms including GP, i.e. 16 excess returns are negative.

To test for significance we performed a t-test, the null hypothesis to be rejected is that BH outperforms ON in terms of the return to be expected μ, i.e. $H_0 := \mu_{ON} \leq$

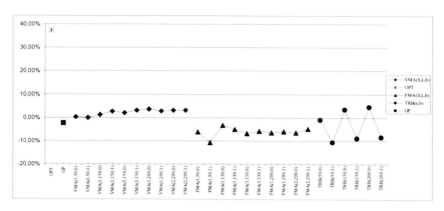

Fig. 12.1 Annualized excess returns versus BH on the time interval 2002 to 2007

μ_{BH} and $ON \in \{VMA, FMA, TRB, GP\}$ at a significance level of 0.05 (5%). All p-values are greater than 0.05, i.e. H_0 can not be rejected. Thus, we employed the moving block bootstrap procedure to test for predictability (cf. [44]). We generated 1000 time series (bootstrap samples) of prices and evaluated the performance of all MA and TRB variants on each sample. Compared to BH, for all variants and samples the above null hypothesis H_0 can not be rejected. Thus, we assume $ON \in \{VMA, FMA, TRB\}$ will not perform better than BH in the future. Further, the return to be expected μ_{ON} equals the average return of ON over the 1000 samples. We can not provide support for VMA, FMA and TRB, and conclude that these heuristic conversion algorithms do not have predictive ability in the German market. Our results are consistent with the results of [40, 10, 8, 44] as VMA, FMA and TRB are not profitable under transaction costs when compared to BH (cf. Sect. 12.2). Further, when comparing GP to BH our results are consistent with the results of [1, 31, 14] (cf. Sect. 12.2) as GP is outperformed by BH.

Question 2: How do the empirical-case competitive ratios c_{ec} of the algorithms ON compare?

To answer **Question 2** column 1 and 4 in Tables 12.2 to 12.5 are of main interest. When calculating c_{ec} the achieved $R(y)$ of ON is compared to OPT, where $ON \in \{VMA, FMA, TRB, GP, BH\}$. For example GP achieves a $c_{ec} = OPT/GP = 1.4363/1.0392 = 1.3821$ (cf. Eq. (12.6)). The best result is achieved by $TRB(200,0)$, and the worst result by $TRB(50,1)$. Clearly, the answers to **Question 1** regarding the performance comparison of the algorithms are also true for **Question 2** as the numerator OPT when calculating c_{ec} is constant.

Question 3: How do the worst-case competitive ratios c_{wc} which could have been possible from the experimental data compare?

To answer **Question 3** column 5 in Tables 12.2 to 12.5 is of main interest. Answering this question we calculated c_{wc} by Eq. (12.7). As ON is confronted with the worst possible sequence of prices p times, the ratio c_{wc} grows exponential with p. Growth is shown in Fig. 12.2, with $f(x) = x^{2p}$ and $x = 1.2603$. The greater the number of trades p, the worse (greater) c_{wc} gets as it grows exponential with p.

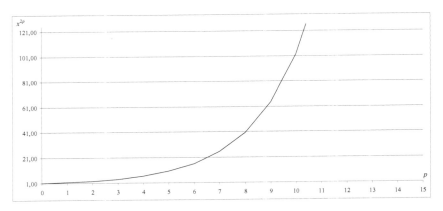

Fig. 12.2 Exponential growth of the worst-case competitive ratio c_{wc}

Question 4: How do the *Sortino Ratios SR* compare?

To answer **Question 4** column 6 in Tables 12.2 to 12.5 is of main interest. A $SR \leq 0$ indicates that the risk incorporated is not rewarded in terms of a higher return achieved: the return achieved is close to r_{MAR}. A $SR > 0$ indicates that the risk incorporated is rewarded: the greater the SR the lower the risk of large losses occurring. We calculated the SR of $ON \in \{VMA, FMA, TRB, GP, BH\}$ by Eq. (12.8). 4 of 6 TRB variants achieve a $SR < 0$, and the remaining 2 are close to zero ($r_{MAR} = 0$). Although $TRB(200, 0)$ achieves the overall best annualized return its SR is 0.0011. All VMA variants generate a $SR > 0$, from minimum 0.0621 to maximum 1.1418. $VMA(5, 150, 0)$ generates the overall best SR (1.1418). 4 of 10 FMA variants achieve a $SR < 0$, and the remaining 6 are close to zero. Summing up, for FMA and TRB the risk incorporated is not rewarded in terms of a higher return achieved. We conclude that only for VMA the incorporated risk is rewarded by never losing money as $r_{MAR} = 0$.

12.6   Conclusions

In order to answer the four questions raised in this paper simulation runs with different heuristic conversion algorithms were performed. For the algorithms FMA and GP we can not provide support in the German market. The results for GP show that it is not beneficial to automatically generate trading rules as GP is outperformed by BH. Further, GP algorithms can hardly be interpreted economically. We found support for VMA and TRB. For practical use we conclude that a VMA algorithm should be calculated using a greater number of days for the long moving average, this leads to a higher return and less trades. Similarly, a TRB algorithm should be computed without band using a greater number of days for local minima and maxima prices. Considering the *Sortino Ratio*, we only found support for using

VMA, as all *SR* values are greater zero. The incorporated risk is rewarded by avoiding losses. From 2002 to 2007 the Dax-30 (as well as other indices) was in a strong trending mode. Thus, in order to generalize our results, further experiments on a non-trending market, including statistical and worst-case analysis are necessary. The question whether the considered algorithms generate significant positive excess returns in the long run is to be answered. Besides a more extensive data set, more realistic entry and exit signals with some rudimentary money management strategy are further fields of future research.

Acknowledgements. The authors thank two anonymous reviewers and the editors for helpful comments.

References

1. Allen, F., Karjalainen, R.: Using genetic algorithms to find technical trading rules. Journal of Financial Economics 51(2), 245–271 (1999)
2. Ammann, M., Zenkner, C.: Tactical asset allocation mit genetischen algorithmen. Schweizerische Zeitschrift für Volkswirtschaft und Statistik 139(1), 1–40 (2003)
3. Bäck, T.: Evolutionary Algorithms in Theory and Practice: Evolution Strategies, Evolutionary Programming, Genetic Algorithms. Oxford University Press (1996)
4. Banzhaf, W., Nordin, P., Keller, R.E., Francone, F.D.: Genetic Programming: An Introduction. Morgan Kaufmann, San Francisco (1998)
5. Bessembinder, H., Chan, K.: The profitability of technical trading rules in the Asian stock markets. Pacific-Basin Finance Journal 3(2-3), 257–284 (1995)
6. Bessembinder, H., Chan, K.: Market efficiency and the returns to technical analysis. Financial Management 27(2), 5–17 (1998)
7. Bhattacharyya, S., Mehta, K.: Evolutionary induction of trading models. In: Chen, S.-H. (ed.) Evolutionary Computation in Economics and Finance. STUDFUZZ, pp. 311–331. Physica-Verlag (2002)
8. Bokhari, J., Cai, C., Hudson, R., Keasey, K.: The predictive ability and profitability of technical trading rules: Does company size matter? Economics Letters 86(1), 21–27 (2005)
9. Brock, W.A., Lakonishok, J., LeBaron, B.: Simple technical trading rules and the stochastic properties of stock returns. Journal of Finance 47(5), 1731–1764 (1992)
10. Chang, E.J., Lima, E.J.A., Tabak, B.M.: Testing for predictability in emerging equity markets. Emerging Markets Review 5(3), 295–316 (2004)
11. Coutts, J.A., Cheung, K.: Trading rules and stock returns: some preliminary short run evidence from the hang seng 1985-1997. Applied Financial Economics 10(6), 579–586 (2000)
12. Day, T.E., Wang, P.: Dividends, nonsynchronous prices, and the returns from trading the dow jones industrial average. Journal of Empirical Finance 9(4), 431–454 (2002)
13. El-Yaniv, R., Fiat, A., Karp, R.M., Turpin, G.: Optimal search and one-way trading algorithm. Algorithmica 30(1), 101–139 (2001)
14. Fyfe, C., Marney, J.P., Tarbert, H.: Risk adjusted returns from technical trading: A genetic programming approach. Applied Financial Economics 15, 1073–1077 (2005)

15. Gunasekarage, A., Power, D.M.: The profitability of moving average trading rules in South Asian stock markets. Emerging Markets Review 2(1), 17–33 (2001)
16. Hatgioannides, J., Mesomeris, S.: On the returns generating process and the profitability of trading rules in emerging capital markets. Journal of International Money and Finance 26(6), 948–973 (2007)
17. Hudson, R., Dempsey, M., Keasey, K.: A note on the weak form efficiency of capital markets: The application of simple technical trading rules to UK stock prices - 1935 to 1994. Journal of Banking and Finance 20(6), 1121–1132 (1996)
18. Ito, A.: Profits on technical trading rules and time-varying expected returns: evidence from pacific-basin equity markets. Pacific-Basin Finance Journal 7(3-4), 283–330 (1999)
19. Koza, J.R.: Genetic Programming II: Automatic Discovery of Reusable Programs. MIT Press (1994)
20. Koza, J.R., Bennet, F.H., Andre, D., Keane, M.: Genetic Programming III: Darwinian Invention and Problem Solving. Morgan Kaufmann (1999)
21. Kwon, K.-Y., Kish, R.J.: A comparative study of technical trading strategies and return predictability: An extension of Brock, Lakonishok, and LeBaron (1992) using NYSE and NASDAQ indices. Quarterly Review of Economics and Finance 42(3), 611–631 (2002)
22. Lagoarde-Segot, T., Lucey, B.M.: Efficiency in emerging markets - evidence from the MENA region. Journal of International Financial Markets, Institutions and Money 18(1), 94–105 (2008)
23. LeBaron, B.: Technical trading rule profitability and foreign exchange intervention. Journal of International Economics 49(1), 125–143 (1999)
24. Lento, C., Gradojevic, N.: The profitability of technical trading rules: A combined signal approach. Journal of Applied Business Research 23(1), 13–28 (2007)
25. Levich, R.M., Thomas, L.R.: The significance of technical trading-rule profits in the foreign exchange market: A bootstrap approach. Journal of International Money and Finance 12(5), 451–474 (1993)
26. Marshall, B.R., Cahan, R.H.: Is technical analysis profitable on a stock market which has characteristics that suggest it be inefficient? Research in International Business and Finance 19(3), 384–398 (2005)
27. Mills, T.C.: Technical analysis and the London Stock Exchange: Testing trading rules using the FT30. International Journal of Finance & Economics 2(4), 319–331 (1997)
28. Ming-Ming, L., Siok-Hwa, L.: The profitability of the simple moving averages and trading range breakout in the Asian stock markets. Journal of Asian Economics 17(1), 144–170 (2006)
29. Mohr, E.: Online Algorithms for Conversion Problems. PhD thesis, Saarland University and Max Planck Institute for Informatics (July 2011)
30. Navet, N., Chen, S.-H.: On predictability and profitability. would gp induced trading rules be sensitive to the observed entropy of time series? In: Brabazon, T., O'Neill, M. (eds.) Natural Computing in Computational Finance, pp. 197–210. Springer, Berlin (2008)
31. Neely, C.: Risk-adjusted, ex ante, optimal technical trading rules in equity markets. International Review of Economics & Finance 12(1), 69–87 (2003)
32. Neely, C., Weller, P., Dittmar, R.: Is technical analysis in the foreign exchange market profitable? a genetic programming approach. Journal of Financial and Quantitative Analysis 32(4), 405–426 (1997)
33. Neely, C.J.: The temporal pattern of trading rule returns and exchange rate intervention: intervention does not generate technical trading profits. Journal of International Economics 58(1), 211–232 (2002)

34. O'Neill, M., Ryan, C.: Grammatical Evolution: Evolutionary Automatic Programming in an Arbitrary Language. Kluwer Academic Publishers (2003)
35. Parisi, F., Vasquez, A.: Simple technical trading rules of stock returns: Evidence from 1987 to 1998 in Chile. Emerging Markets Review 1(2), 152–164 (2000)
36. Park, C.-H., Irwin, S.H.: What do we know about the profitability of technical analysis. Journal of Economic Surveys 21(4), 786–826 (2007)
37. Pereira, R.: Forecasting ability but no profitability: An empirical evaluation of genetic algorithm-optimised technical trading rules. In: Chen, S.-H. (ed.) Evolutionary Computation in Economics and Finance. STUDFUZZ, pp. 287–310. Physica-Verlag (2002)
38. Potvin, J.-Y., Sorianoa, P., Vallée, M.: Generating trading rules on the stock markets with genetic programming. Computers & Operations Research 31(7), 1033–1047 (2004)
39. Raj, M., Thurston, D.: Effectiveness of simple technical trading rules in the hong kong futures markets. Applied Economics Letters 3(1), 33–36 (1996)
40. Ratner, M., Leal, R.P.C.: Tests of technical trading strategies in the emerging equity markets of Latin America and Asia. Journal of Banking and Finance 23(12), 1887–1905 (1999)
41. Schmidt, G., Mohr, E., Kersch, M.: Experimental analysis of an online trading algorithm. Electronic Notes in Discrete Mathematics 36, 519–526 (2010)
42. Sortino, F.A., Price, L.: Performance measurement in a downside risk framework. Journal of Investing 3(3), 59–64 (1994)
43. Sortino, F.A., van der Meer, R.: Downside risk. The Journal of Portfolio Management 17(4), 27–31 (1991)
44. Tabak, B.M., Lima, E.J.A.: Market efficiency of Brazilian exchange rate: Evidence from variance ratio statistics and technical trading rules. European Journal of Operational Research 194(3), 814–820 (2009)
45. Zelinka, I., Oplatkova, Z., Nolle, L.: Analytic programming - symbolic regression by means of arbitrary evolutionary algorithms. Journal of Simulation 6(9), 44–56 (2005)

Printed by Publishers' Graphics LLC
MO20120920-274-245